UCF
109.00

Adaptive Control of Nonsmooth Dynamic Systems

Springer
*London
Berlin
Heidelberg
New York
Barcelona
Hong Kong
Milan
Paris
Singapore
Tokyo*

Gang Tao and Frank L. Lewis (Eds)

Adaptive Control of Nonsmooth Dynamic Systems

With 102 Figures

Springer

Gang Tao, PhD
Department of Electrical and Computer Engineering, University of Virginia, Charlottesville, VA 22903, USA

Frank L. Lewis, PhD
Automation and Robotics Research Institute, University of Texas at Arlington, Fort Worth, TX 76118, USA

ISBN 1-85233-384-7 Springer-Verlag London Berlin Heidelberg

British Library Cataloguing in Publication Data
Adaptive control of nonsmooth dynamic systems
  1.Adaptive control systems
  I.Tao, Gang II.Lewis, Frank L. (Frank Leroy), 1949-
  629.8'36
  ISBN 1852333847

Library of Congress Cataloging-in-Publication Data
Adaptive control of nonsmooth dynamic systems / Gang Tao and Frank L. Lewis (eds.).
  p. cm.
  Includes bibliographical references.
  ISBN 1-85233-384-7 (alk. paper)
  1. Adaptive control systems. 2. Dynamics. I. Tao, Gang. II. Lewis Frank L.
  TJ217.A319 2001
  629.8'36--dc21   2001020311

Apart from any fair dealing for the purposes of research or private study, or criticism or review, as permitted under the Copyright, Designs and Patents Act 1988, this publication may only be reproduced, stored or transmitted, in any form or by any means, with the prior permission in writing of the publishers, or in the case of reprographic reproduction in accordance with the terms of licences issued by the Copyright Licensing Agency. Enquiries concerning reproduction outside those terms should be sent to the publishers.

© Springer-Verlag London Limited 2001
Printed in Great Britain

The use of registered names, trademarks etc. in this publication does not imply, even in the absence of a specific statement, that such names are exempt from the relevant laws and regulations and therefore free for general use.

The publisher makes no representation, express or implied, with regard to the accuracy of the information contained in this book and cannot accept any legal responsibility or liability for any errors or omissions that may be made.

Typesetting: Camera ready by editors
Printed and bound by the Athenæum Press Ltd., Gateshead, Tyne & Wear
69/3830-543210 Printed on acid-free paper SPIN 10783155

Preface

Nonsmooth nonlinearities such as backlash, dead-zone, component failure, friction, hysteresis, saturation and time delays are common in industrial control systems. Such nonlinearities are usually poorly known and may vary with time, and they often limit system performance. Control of systems with nonsmooth nonlinearities is an important area of control systems research. A desirable control design approach for such systems should be able to accommodate system uncertainties. Adaptive methods for the control of systems with unknown nonsmooth nonlinearities are particularly attractive in many applications because adaptive control designs are able to provide adaptation mechanisms to adjust controller parameters in the presence of parametric, structural and environmental uncertainties. Most adaptive or nonlinear control techniques reported in the literature are for linear systems or for some classes of systems with smooth nonlinearities, but not for nonsmooth nonlinearities. The need for effective control methods to deal with nonsmooth nonlinear systems has motivated growing research activities in adaptive control of systems with such common practical nonsmooth nonlinearities. Recently, there have been many encouraging new results on adaptive control problems with backlash, dead-zone, failures, friction, hysteresis, saturation, and time delays. This book, entitled *Adaptive Control of Nonsmooth Dynamic Systems*, is aimed at reflecting the state of the art in designing, analyzing and implementing adaptive control methods which are able to accommodate uncertain nonsmooth nonlinearities in industrial control systems.

Backlash, dead-zone, component failure, friction, hysteresis, saturation, and time delays are the most common nonsmooth nonlinearities in industrial control systems. Backlash, a dynamic (with memory) characteristic, exists in mechanical couplings such as gear trains, and always limits the accuracy of servo-mechanisms. Dead-zone is a static input-output relationship which for a range of input values gives no output; it also limits system performance. Dead-zone characteristics are often present in amplifiers, motors, hydraulic valves and even in biomedical actuation systems. Failures of different types in actuators, sensors and other components of a control system can cause major system performance deterioration. Friction exists wherever there is motion or tendency for motion between two physical components. Friction can cause a steady-state error or a limit cycle near the reference position and stick-slip phenomenon at low speed in the conventional linear control of positioning systems.

Hysteresis, another dynamic characteristic, exists in electromagnetic and piezoelectric actuators which are used for micromotion control and high-accuracy positioning. Saturation is always a potential problem for actuators of control systems—all actuators do saturate at some level. Actuator saturation affects the transient performance and even leads to system instability. Time delays are also important factors to deal with in order to improve control system performance such as for teleoperations and in real-time computer control systems.

Although backlash, dead-zone, failure, friction, hysteresis, saturation, and time delay characteristics are different, they are all nonsmooth in nature. Therefore, most existing adaptive control methods are not applicable. Unfortunately these nonlinearities can severely limit the performance of feedback systems if not compensated properly. Moreover, adaptive control of dynamic systems with each of these nonsmooth characteristics is a control problem that needs a systematic treatment. It makes the control problem even more challenging when there are more than one nonlinear characteristic present in the control system.

In this book it will be shown how nonsmooth nonlinear industrial characteristics can be adaptively compensated and how desired system performance is achieved in the presence of such nonlinearities. The book has 16 chapters on issues including system modeling, control design, analysis of stability and robustness, simulation and implementation:

Chapter One: *New Models and Identification Methods for Backlash and Gear Play*, by M. Nordin, P. Bodin and P.-O. Gutman

Chapter Two: *Adaptive Dead Zone Inverses for Possibly Nonlinear Control Systems*, by E.-W. Bai

Chapter Three: *Deadzone Compensation in Motion Control Systems Using Augmented Multilayer Neural Networks*, by R. R. Selmic and F. L. Lewis

Chapter Four: *On-line Fault Detection, Diagnosis, Isolation and Accommodation of Dynamical Systems with Actuator Failures*, by M. A. Demetriou and M. M. Polycarpou

Chapter Five: *Adaptive Control of Systems with Actuator Failures*, by G. Tao and S. M. Joshi

Chapter Six: *Multi-mode System Identification*, by E. I. Verriest

Chapter Seven: *On Feedback Control of Processes with "Hard" Nonlinearities*, by B. Friedland

Chapter Eight: *Adaptive Friction Compensation for Servo Mechanisms*, by J. Wang, S. S. Ge and T. H. Lee

Chapter Nine: *Relaxed Controls and a Class of Active Material Actuator Models*, by A. Kurdila

Chapter Ten: *Robust Adaptive Control of Nonlinear Systems with Dynamic Backlash-like Hysteresis*, by C.-Y. Su, M. Oya and X.-K. Chen

Chapter Eleven: *Adaptive Control of a Class of Time-delay Systems in the Presence of Saturation*, by A. M. Annaswamy, S. Evesque, S.-I. Niculescu and A. P. Dowling

Chapter Twelve: *Adaptive Control for Systems with Input Constraints: A Survey*, by J.-W. John Cheng and Y.-M. Wang

Chapter Thirteen: *Robust Adaptive Control of Input Rate Constrained Discrete Time Systems*, by G. Feng

Chapter Fourteen: *Adaptive Control of Linear Systems with Poles in the Closed Left Half Plane with Constrained Inputs*, by D. A. Suarez-Cerda and R. Lozano

Chapter Fifteen: *Adaptive Control with Input Saturation Constraints*, by C.-S. Zhang

Chapter Sixteen: *Adaptive Control of Linear Systems with Unknown Time Delay*, by C.-Y. Wen, Y.-C. Soh and Y. Zhang

The authors of the chapters in this book are experts in their areas of interest and their chapters present new solutions to important issues in adaptive control of industrial systems with nonsmooth nonlinearities such as backlash, dead-zone, failure, friction, hysteresis, saturation, and time delay. These solutions result from recent work in these areas and are believed to be attractive to people from both academia and industry. Adaptive control of nonsmooth dynamical systems is theoretically challenging and practically important. This book is the first book on adaptive control of such systems, addressing all these nonsmooth nonlinear characteristics: backlash, dead-zone, failure, friction, hysteresis, saturation and time delays. Such a book is also aimed at motivating more research activities in the important field of adaptive control of nonsmooth nonlinear industrial systems.

Recent advances in adaptive control of nonsmooth dynamic systems have shown that those practical nonsmooth nonlinear characteristics such as backlash, dead-zone, component failure, friction, hysteresis, saturation and time delays can be adaptively compensated when their parameters are uncertain, as is common in real-life control systems. Rigorous designs have been given for selecting desirable controller structures to meet the control objectives and for deriving suitable algorithms to tune the controller parameters for control of systems with uncertainties in dynamics and nonsmooth nonlinearities. There have been increasing interest and activities in these areas of research, as evidenced by recent conference invited sessions and journal special issues on related topics. It is clear that this is a promising direction of research and there have been many encouraging results. Given the practical importance and theoretical significance of such research, it is time to summarize, unify, and develop advanced techniques for adaptive control of nonsmooth dynamic systems.

Since this book is about some important and new areas of adaptive control research, its contents are intended for people from both academia and industry, including professors, researchers, graduate students who will use this book for research and advanced study, and engineers who are concerned with the fast and precision control of motion systems with imperfections (such as backlash, dead-zone, component failure, friction, hysteresis, saturation and time delays) in mechanical connections, hydraulic servovalves, piezoelectric translators, and electric servomotors, and biomedical actuators systems. The book can be useful for people from aeronautical, biomedical, civil, chemical, electrical, industrial, mechanical and systems engineering, who are working on aircraft flight control, automobile control, high performance robots, materials growth process control, precision motor control, radar and weapons system pointing platforms, VLSI assembly. The adaptive system theory developed in this book is also of interest to people who work on communication systems, signal processing, real-time computer system modeling and control, biosystem modeling and control.

The first editor would like to gratefully acknowledge the partial support from National Science Foundation under grant ECS-9619363 and National Aeronautics and Space Administration under grant NCC-1342 to this project. He would also like to thank his graduate student Xidong Tang for his editorial assistance on this project. The second editor acknowledges the vital support of the Army Research Office under grant DAAD19-99-1-0137.

Gang Tao
Charlottesville, Virginia

Frank L. Lewis
Fort Worth, Texas

Contents

Contributors xvii

1. **New Models and Identification Methods for Backlash and Gear Play**
   Mattias Nordin, Per Bodin and Per-Olof Gutman 1
   1. Introduction 2
   2. Backlash Models 2
      2.1 Physical Model for the Shaft Torque T 2
      2.2 Phase Plane Backlash Model 6
      2.3 The Dead Zone Model 8
      2.4 Model Simulation Comparisons 9
   3. Model for Backlash with Rubber 11
      3.1 Shaft Model for Backlash with Rubber 12
      3.2 Simulations and Measurements: Backlash with Rubber .. 14
   4. Describing Function Analysis 16
      4.1 DIDF Calculations for the Phase Plane Model 18
      4.2 Phase Plane vs. Dead Zone Model 19
   5. A Backlash Gap Estimation Method 21
      5.1 Preliminaries and Assumptions 22
      5.2 Describing Function Analysis for Identification 23
      5.3 The Estimation Method 25
      5.4 Simulation Experiments 26
   6. Conclusions 28
      References 29

2. **Adaptive Dead Zone Inverses for Possibly Nonlinear Control Systems**
   Er-Wei Bai 31
   1. Introduction 31

	2.	Problem Formulation	33
	3.	Adaptive Dead Zone Inverse	34
	4.	Adaptive Dead Zone Inverse for Linear Control Systems	42
		4.1 With Known $G(s)$	42
		4.2 With Unknown $G(s)$	45
	5.	Extension to Nonlinear Control Systems	45
	6.	Concluding Remark	47
		References	47

3. Deadzone Compensation in Motion Control Systems Using Augmented Multilayer Neural Networks

Rastko R. Šelmic and Frank L. Lewis 49

	1.	Introduction	50
	2.	Background	51
		2.1 Dynamics of Mechanical Motion Tracking Systems	51
	3.	Neural Networks For Deadzone Compensation	52
		3.1 Background on Neural Networks	53
		3.2 NN Approximation of Continuous Functions	54
		3.3 NN Approximation of Jump Functions	55
		3.4 Structure of Augmented Multilayer NN	59
		3.5 Simulation of the Augmented Multilayer NN	60
	4.	Compensation of Deadzone Nonlinearity	61
		4.1 Deadzone Nonlinearity	61
		4.2 NN Deadzone Precompensator	62
	5.	NN Controller With Deadzone Compensation	66
		5.1 Tracking Controller with NN Deadzone Compensation	66
	6.	Simulation of NN Deadzone Compensator	71
	7.	Conclusions	77
	8.	Appendix	78
		References	81

4. On-line Fault Detection, Diagnosis, Isolation and Accommodation of Dynamical Systems with Actuator Failures

Michael A. Demetriou and Marios M. Polycarpou 85

1.	Introduction	85
2.	Problem Formulation	87
3.	Adaptive Diagnostic Observers	89
	3.1 Properties and Analysis of Detection/Diagnostic Observer	91
4.	Fault Accommodation	95
5.	Special Case: Vector Second-order Systems	96
6.	Examples and Numerical Results	102
7.	Conclusions	108
	References	108

5. Adaptive Control of Systems with Actuator Failures

Gang Tao and Suresh M. Joshi 111

1.	Introduction	111
2.	Problem Statement	113
3.	Plant-Model State Matching Control	115
	3.1 Basic Plant-Model Matching Conditions	116
	3.2 Actuator Failure Compensation	117
4.	Adaptive State Tracking Control	119
5.	Compensation Designs for Parametrizable Time-varying Failures	123
	5.1 Plant-Model Matching Control Design	124
	5.2 Adaptive Control Design	126
6.	Compensation Designs for Unparametrizable Time-varying Failures	128
	6.1 Stabilizing Control	128
	6.2 Adaptive Control Design	131
7.	Plant-Model Output Matching Control	133
	7.1 State Feedback Designs	133
	7.2 Output Feedback Design	142
8.	Adaptive Output Tracking Control	145
	8.1 State Feedback Designs	145
	8.2 Output Feedback Design	151
9.	Concluding Remarks	153
	References	155

6. Multi-mode System Identification

Erik I. Verriest 157

1. Introduction 157
2. Subspace Identification Degradation 160
   - 2.1 Total Least Squares Identification 160
   - 2.2 Degradation in Bi-modal Systems 162
3. Multi-mode ARMA Identification........................... 163
   - 3.1 Determination of the Mode Parameters 164
   - 3.2 Resolution of the Mode Parameters 167
4. Multi-mode State Space Identification...................... 172
   - 4.1 Mode Transitions and Transition Modes............. 172
   - 4.2 Identifiability Canonical Form...................... 174
   - 4.3 Performance 177
5. Markov Chain and Hybrid Modeling 177
   - 5.1 Hidden Markov Model Identification 179
   - 5.2 Hybrid System Modeling 183
6. Applications 184
   - 6.1 Adaptive Control of Nonlinear and Nonsmooth Systems 184
   - 6.2 Fault Detection and Isolation 185
7. Examples... 186
8. Conclusions... 187
   - References 189

7. On Feedback Control of Processes with 'Hard' Nonlinearities

Bernard Friedland 191

1. Introduction 191
2. The State-dependent Algebraic Riccati Equation Method 192
3. Modeling Hard Nonlinearities 193
4. Controlling Systems with State Variable Constraints 195
   - 4.1 Illustrative Examples 195
5. Friction and Backlash Control............................. 199
   - 5.1 Illustrative Example 199
   - 5.2 Backlash Control 205
6. Conclusions... 208
   - References 209

8. **Adaptive Friction Compensation for Servo Mechanisms**

   J. Wang, S. S. Ge and T. H. Lee 211

   1. Introduction 211
   2. Friction Models 215
      2.1 Static models 215
      2.2 Dynamic models 220
   3. Adaptive Friction Compensation 222
      3.1 Problem Statement 222
      3.2 Static Friction Model Based Adaptive Controller Design 222
      3.3 LuGre Friction Model Based Adaptive Controller Design 225
   4. Simulation Studies 234
      4.1 Static Model Based Adaptive Control 234
      4.2 Dynamic LuGre Friction Model Based Adaptive Control 235
   5. Conclusion 246
      References 246

9. **Relaxed Controls and a Class of Active Material Actuator Models**

   Andrew Kurdila 249

   1. Introduction 249
   2. The Governing Dynamical Equations 252
      2.1 Infinite Dimensional Structural Systems 253
      2.2 The Control Influence Operator 255
      2.3 Existence and Uniqueness of Solutions 258
   3. Convergent Approximations 261
      3.1 Finite Dimensional Approximation 261
      3.2 Approximation of Hysteresis Operator 264
      References 271

10. **Robust Adaptive Control of Nonlinear Systems with Dynamic Backlash-like Hysteresis**

    Chun-Yi Su, Masahiro Oya and Xinkai Chen 273

    1. Introduction 273
    2. Problem Statement 274

- 3. Backlash-like Hysteresis Model and Its Properties 275
- 4. Lyapunov-based Control Structure 276
- 5. Function Approximation Using Gaussian Functions 278
- 6. Adaptive Controller Design 280
- 7. Simulation Studies 283
- 8. Conclusion 286
  - Appendix 286
  - References 288

11. Adaptive Control of a Class of Time-delay Systems in the Presence of Saturation

A. M. Annaswamy, S. Evesque, S.-I. Niculescu and A. P. Dowling 289

- 1 Introduction 289
- 2 Statement of the Problem 290
- 3 The Controller in the Presence of Saturation 291
  - 3.1 A First-order Plant 292
  - 3.2 Output Feedback with $n^* = 1$ 293
  - 3.3 Output Feedback with $n^* = 2$ 296
  - 3.4 A Low Order Adaptive Controller 297
- 4 The Adaptive Controller in the Presence of a Delay 298
  - 4.1 The Controller without Saturation Constraints 298
  - 4.2 The Controller in the Presence of Saturation and Time-delay 302
- 5 Simulation Results 305
- 6 Summary 308
  - References 310

12. Adaptive Control for Systems with Input Constraints – A Survey

J.-W. John Cheng and Yi-Ming Wang 311

- 1. Introduction 311
- 2. History of the Constrained Control Theory 312
  - 2.1 Existence of Stabilizing Magnitude Constrained Control 312
  - 2.2 Designs of Effective Magnitude Constrained Control .. 313

3.	Adaptive Constrained Control 316
	3.1	Adaptive Control Designs with Input Magnitude Constraint 317
	3.2	Adaptive Control Design with Input Magnitude and Rate Constraints 323
4.	Conclusions 328
	References 329

13. Robust Adaptive Control of Input Rate Constrained Discrete Time Systems

Gang Feng 333

1.	Introduction 333
2.	Rate Constrained Pole Placement Adaptive Control 334
	2.1 Plant Model 334
	2.2 Pole Placement Control 335
	2.3 Rate Constrained Pole Placement Adaptive Control .. 336
3.	Stability Analysis 341
4.	A Simulation Example 344
5.	Conclusions 347
	References 347

14. Adaptive Control of Linear Systems with Poles in the Closed LHP with Constrained Inputs

Dionisio A. Suarez-Cerda and Rogelio Lozano 349

1.	Introduction 349
2.	Control of Systems with Constrained Inputs 351
	2.1 Plants with no Poles at the Origin 351
	2.2 Double Integrator 352
	2.3 Plant with an Integrator and a Stable Pole 353
3.	Adaptive Control of Second-order Plants with Constrained Input 354
	3.1 Plant Parameter Identification 354
	3.2 Adaptive State Observation 355
4.	Simulations Results 356

15. Adaptive Control with Input Saturation Constraints

Cishen Zhang 361

1. Introduction 361
2. Amplitude Constrained Direct Adaptive Control 362
3. Stability of Amplitude Constrained Adaptive Control Systems 366
4. Amplitude Constrained Adaptive Control of Type-1 Plants .. 370
5. Rate Constrained Direct Adaptive Control 373
6. Stability of Rate Constrained Direct Adaptive Control Systems 374
7. Rate Constrained Adaptive Control of Stable Plants 379
8. Conclusion 380
   References 380

16. Adaptive Control of Linear Systems with Unknown Time Delay

Changyun Wen, Yeng Chai Soh and Ying Zhang 383

1. Introduction 383
2. System Models 384
3. Adaptive Control Scheme 388
   - 3.1 Parameter Estimator 389
   - 3.2 Control Law Synthesis 391
4. Stability Analysis....................................... 392
5. Conclusion 400
   References 400

Index 403

5. Conclusions 358
   References 359

Contributors

A. M. Annaswamy
Adaptive Control Laboratory
Department of Mechanical Engineering
Massachusetts Institute of Technology
Cambridge, MA 02139, USA

Er-Wei Bai
Department of Electrical and Computer Engineering
The University of Iowa
Iowa City, IA 52242, USA

Per Bodin
Space Systems Division
Swedish Space Corporation
S-171 04 Solna, Sweden

Xinkai Chen
Department of Information Science
Tokyo Denki University
Hatoyama-machi, hiki-gun
Saitama 350-0394, Japan

J.-W. John Cheng
Department of Mechanical Engineering
National Chung Cheng University
Chia Yi, 621, Taiwan, R.O.C.

Michael A. Demetriou
Department of Mechanical Engineering
Worcester Polytechnic Institute
Worcester, MA 01609-2280, USA

A. P. Dowling
Department of Engineering
University of Cambridge
Cambridge, UK

S. Evesque
Department of Engineering
University of Cambridge
Cambridge, UK

Gang Feng
Department of Manufacturing Engineering and Engineering Management
City University of Hong Kong
Tat Chee Ave., Kowloon
Hong Kong

Bernard Friedland
Department of Electrical and Computer Engineering
New Jersey Institute of Technology
Newark, NJ 07102, USA

S. S. Ge
Department of Electrical and Computer Engineering
National University of Singapore
117576, Singapore

Per-Olof Gutman
Agricultural Engineering
Technion – Israel Institute of Technology
Haifa 32000, Israel

Suresh M. Joshi
Mail Stop 161
NASA Langley Research Center
Hampton, VA 23681, USA

Andrew Kurdila
Department of Aerospace Engineering, Mechanics and Engineering Science
University of Florida
Gainesville, FL 32611-6250, USA

T. H. Lee
Department of Electrical and Computer Engineering
National University of Singapore
117576, Singapore

Frank L. Lewis
Automation and Robotics Research Institute
The university of Texas at Arlington
Fort Worth, Texas, USA

Rogelio Lozano
Heudiasyc UMR CNRS 6599
Universit de Technologie de Compigne
BP 20529
60205 Compigne cedex, France

S.-I. Niculescu
HEUDIASYC
University of Compiegne
Compiegne, France

Mattias Nordin
Metals and Mining Division
ABB Automation Systems AB
S-721 67 Västeras, Sweden

Masahiro Oya
Department of Control Engineering
Kyushu Institute of Technology
1-1 Sensui, Tobata, Kitakyushu, 804-8550, Japan

Marios M. Polycarpou
Department of Electrical and Computer Engineering and Computer Science
University of Cincinnati
Cincinnati, OH 45221-0030, USA

Rastko R. Šelmic
Signalogic, Inc.
9617 Wendell
Dallas, Texas 75243, USA

Yeng Chai Soh
School of Electrical and Electronic Engineering
Nanyang Technological University
Nanyang Ave. 639798, Singapore

Chun-Yi Su
Department of Mechanical Engineering
Concordia University
Montreal, Quebec, H3G 1M8, Canada

Dionisio A. Suarez-Cerda
Instituto de Investigaciones Electricas
Av. Reforma No. 113, Col. Palmira
62490 Temixco, Morelos, Mexico

Gang Tao
Department of Electrical and Computer Engineering
University of Virginia
Charlottesville, VA 22903, USA

Erik I. Verriest
School of Electrical and Computer Engineering
Georgia Institute of Technology
Atlanta, GA 30332-0250, USA

J. Wang
Department of Electrical and Computer Engineering
National University of Singapore
Singapore 117576

Yi-Ming Wang
Department of Mechanical Engineering
National Chung Cheng University
Chia Yi, 621
Taiwan, R.O.C.

Changyun Wen
School of Electrical and Electronic Engineering
Nanyang Technological University
Nanyang Ave. 639798, Singapore

Cishen Zhang
Department of Electrical and Electronic Engineering
The University of Melbourne
VIC 3010, Australia

Ying Zhang
Division of Automation Technology
Gintic Institute of Manufacturing Technology
638075, Singapore

New Models and Identification Methods for Backlash and Gear Play

Mattias Nordin[1], Per Bodin[2], and Per-Olof Gutman[3]

[1] Metals and Mining Division, ABB Automation Systems AB, S-721 67 Västerås, Sweden mattias.c.nordin@se.abb.com
[2] Space Systems Division, Swedish Space Corporation, S-171 04 Solna, Sweden per.bodin@ssc.se
[3] Faculty of Agricultural Engineering, Technion — Israel Institute of Technology, Haifa 32000, Israel peo@tx.technion.ac.il

Abstract

Backlash is the most important nonlinearity that limits the performance of speed control in industrial drives, and is an important impediment to position control as well.

A simple model based on phase plane analysis is derived for an elastic shaft with internal damping connected to a backlash element. The model is extended to include the case of a flexible coupling partly filled with rubber. Measurements are compared with simulations and found to fit very well.

With internal damping, the classical dead zone model predicts non-physical behavior of the dynamics, a feature that is corrected in the new model.

In most industrial systems the backlash appears in closed loop, without direct access to measurements of either the input or the output of the backlash. Typically there is access only to motor speed measurements. In order to design an efficient controller for mechanical drive systems including backlash, it is of benefit to have a good estimate of the backlash gap.

A method to identify the backlash gap in speed controlled elastic two-mass systems with backlash is presented. The idea is to inject the same sinewave signal as the motor torque into the mechanical plant, and into the simulation model. By adapting the backlash gap of the model until the measured and the simulated motor speed amplitudes coincide, an estimate of the backlash gap size is obtained. Describing function analysis is used to choose the frequency and amplitude of the input motor torque. The method is fast, has high accuracy, and requires only motor speed measurements.

1 Introduction

In industrial drives, elements like gear boxes and flexible couplings introduce backlash. For instance a commonly used flexible coupling gives a backlash of about $10°$. This gap is then partly filled with rubber.

Traditionally backlash connected with an elastic shaft has been modeled as a dead zone [23,13,8,10], which gives the output shaft torque as a function of the displacement between the input and output shafts, $\theta_d = \theta_1 - \theta_2$. Here the classical dead zone model is shown to be wrong when the shaft has non-neglible internal damping: the transmitted torque is erroneously computed, even with the wrong sign. It is not possible to model the shaft damping as an equivalent motor damping and/or load damping, or when the damping is included in series with the dead zone [3,7,4–6].

Another traditional backlash model is the hysteresis model [12,20–22] where output position of the backlash is given as a function of the time history of the input position. When the output position is influenced by disturbances the causality is however lost. In our model, as well as the dead zone model, all disturbances enter at the input of the model.

This section first presents a new, exact backlash model that takes the internal damping into account. After simplification the torque is found to be a function of the displacement θ_d, and its time derivative. The model is extended to the case of a flexible coupling, i.e. when a part of the backlash gap is filled with rubber. The models are also analyzed using the describing function method.

In the last part of the chapter a new method is presented to estimate the unknown backlash gap of a system operating in closed loop.

It should be pointed out that the shaft moment of inertia is assumed to be neglible in comparision with the motor and load moments of inertia. This is a valid assumption in most industrial speed control systems, and in many robotics and servo control applications.

2 Backlash Models

In this section the behavior of an inertiafree elastic shaft with backlash is analyzed. Based on phase plane analysis, an approximate shaft model is derived. Comparisons between the new model, the exact solution and the classical dead zone model are presented, both with analysis and simulations.

2.1 Physical Model for the Shaft Torque T

Consider an inertiafree shaft consisting of a backlash gap 2α [rad] and a spring with elasticity k [Nm/rad] and viscous damping c [Nm/rad/s], as in Figure 1. Note that with an inertiafree shaft the torque T on the left side is equal to the torque T on the right side in the figure. Denote time [s] by t; when convenient the explicit time dependence is suppressed. Let $\theta_1(t)$ [rad] be the angle of the motor, $\theta_3(t)$ [rad] the angle of the driving axis at the backlash, and $\theta_2(t)$ [rad] the angle of the driven member. The goal is to acquire the torque

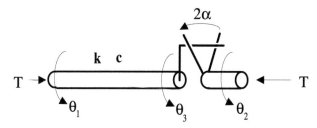

Fig. 1. The physical system: a shaft connected to a backlash.

T [Nm] as a function of the *displacement*, defined as $\theta_d(t) = \theta_1(t) - \theta_2(t)$ and its time derivative $\dot{\theta}_d(t) = \dot{\theta}_1(t) - \dot{\theta}_2(t)$ i.e. without using the state $\theta_3(t)$. Define the *shaft twist* $\theta_s(t) = \theta_1(t) - \theta_3(t)$ and the *backlash angle* $\theta_b(t) = \theta_3(t) - \theta_2(t)$, defined symmetrically within the backlash gap such that $|\theta_b| \leq \alpha$. An examination of the model in Figure 1 gives an exact expression for T, using $\theta_s = \theta_d - \theta_b$

$$T(t) = k\theta_s + c\dot{\theta}_s = k(\theta_d - \theta_b) + c(\dot{\theta}_d - \dot{\theta}_b) \tag{1}$$

The impact when the backlash gap is closed is assumed to be inelastic. There can be contact between the driving and the driven member on either side of the backlash. Define the case with $\theta_b = \alpha$ and $\dot{\theta}_b = 0$ as *right contact* and the case with $\theta_b = -\alpha$ and $\dot{\theta}_b = 0$ as *left contact*. We say that there is *contact* if there is either *right contact* or *left contact*.

It holds that $T > 0$ implies *right contact*, because otherwise no positive torque could be transmitted. Together with (1) this gives

$$T > 0 \Rightarrow T = k(\theta_d - \alpha) + c\dot{\theta}_d > 0 \tag{2}$$

Similarly $T < 0$ implies *left contact*. Together with (1) this gives

$$T < 0 \Rightarrow T = k(\theta_d + \alpha) + c\dot{\theta}_d < 0 \tag{3}$$

Logical negation of (2) and (3) yields

$$\begin{cases} k\theta_d + c\dot{\theta}_d \leq k\alpha & \Rightarrow T \leq 0 \\ k\theta_d + c\dot{\theta}_d \geq -k\alpha & \Rightarrow T \geq 0 \end{cases} \tag{4}$$

respectively. Equations (1)–(4) give

$$\begin{cases} T = 0 \text{ or } T = k(\theta_d - \alpha) + c\dot{\theta}_d & \theta_d + (c/k)\dot{\theta}_d > \alpha \\ T = 0 & |\theta_d + (c/k)\dot{\theta}_d| \leq \alpha \\ T = 0 \text{ or } T = k(\theta_d + \alpha) + c\dot{\theta}_d & \theta_d + (c/k)\dot{\theta}_d < -\alpha \end{cases} \tag{5}$$

which can be expressed in terms of the phase plane $(\theta_d, \dot{\theta}_d)$, see Figure 2a. In the non-shaded areas of Figure 2a the $T = 0$ cases of (5) are valid only

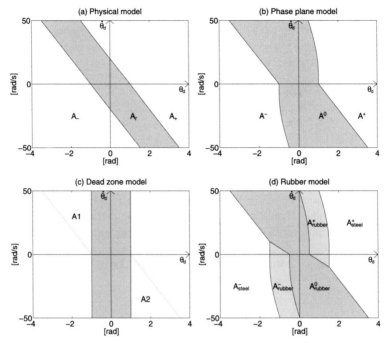

Fig. 2. Phase plane plots of T for $\alpha = 1$ [rad], $k/c = 0.1$ s^{-1} and $\beta = 0.5$ [rad]. (a): the exact model (5), (b): the phase plane model (23), (c): the dead zone model (26) and (d): the rubber model (40).

if there is no contact, because otherwise $\theta_b = \pm\alpha$ and $\dot{\theta}_d = 0$, which would cause $T \neq 0$, see (1),(2) and (3). This means that, if it were known whether or not there is contact, T could be computed exactly from θ_d and $\dot{\theta}_d$. This problem can then be solved by first analyzing when contact is lost and then when contact is achieved.

Define the following areas of the phase plane in Figure 2a.

$$A_+ = \{(\theta_d, \dot{\theta}_d) : k\theta_d + c\dot{\theta}_d \geq k\alpha\} \tag{6}$$

$$A_r = \{(\theta_d, \dot{\theta}_d) : |k\theta_d + c\dot{\theta}_d| < k\alpha\} \tag{7}$$

$$A_- = \{(\theta_d, \dot{\theta}_d) : k\theta_d + c\dot{\theta}_d \leq -k\alpha\} \tag{8}$$

A_r is the interior of the shaded area in Figure 2a.

Lemma 1. *There can be persistent (during a nonzero interval) right contact only in A_+ and persistent left contact only in A_-.*

Proof by contradiction: Assume *right contact* outside A_+. From (1) and (6) it follows that
$$T = \underbrace{k\theta_d + c\dot{\theta}_d}_{<k\alpha} - k\alpha < 0 \tag{9}$$
Hence a persistent negative torque with right contact implies a pull force which is physically impossible. Likewise the case of *left contact* outside A_- is impossible. □

Lemma 2. *If the system state $(\theta_d, \dot{\theta}_d)$ at the initial time $t = t_0$ lies in A_+ with $\theta_b(t_0) = \alpha$ (right contact), then $\theta_b(t_1) = \alpha$ for all times $t_1 > t_0$ such that $(\theta_d(t), \dot{\theta}_d(t)) \in A_+$ for all $t \in [t_0, t_1]$. If $(\theta_d(t_0), \dot{\theta}_d(t_0)) \in A_-$ with $\theta_b(t_0) = -\alpha$ (left contact), then $\theta_b(t_1) = -\alpha$ for all times $t_1 > t_0$ such that $(\theta_d(t), \dot{\theta}_d(t)) \in A_-$ for all $t \in [t_0, t_1]$.*

Proof: *Right contact* in the interior of A_+ together with (1) implies $T > 0$, see (2). As long as there is a positive torque, *right contact* cannot be lost for obvious reasons. On the border of A_+, i.e. with $k\theta_d + c\dot{\theta}_d = k\alpha$, it holds from (5) that $T = 0$ and together with equation (1) this yields
$$\dot{\theta}_b(t) = k(\alpha - \theta_b(t))/c \tag{10}$$
The solution of (10) is
$$\theta_b(t) = \alpha + (\theta_b(t_0) - \alpha)e^{(-k(t-t_0)/c)} \tag{11}$$
Starting with *right contact*, i.e. $\theta_b = \alpha$, equation (11) becomes a constant $\theta_b = \alpha$, i.e. the *right contact* is preserved. Symmetric proof for *left contact* and trajectory in A_-. □

The following theorem can now be proved:

Theorem 1 (Release Condition). *Assume that $\theta_b(t_0) = \alpha$ or $\theta_b(t_0) = -\alpha$ (contact at time t_0). Contact is lost at the first time $t_1 > t_0$ such that the trajectory $(\theta_d(t), \dot{\theta}_d(t))$ reaches the release set A_r.*

Proof: This follows from the fact that right respectively left contact cannot be lost in A_+ respectively A_-, and that in A_r, which lies in between, contact is not possible. □

When contact is lost we know that $T = 0$. Equation (1) then gives
$$\dot{\theta}_d - \dot{\theta}_b = -k(\theta_d - \theta_b)/c \tag{12}$$
whose solution is
$$\theta_b(t) - \theta_d(t) = (\theta_b(t_0) - \theta_d(t_0))e^{-k(t-t_0)/c} \tag{13}$$
Since θ_d is given, (13) gives θ_b. Furthermore it holds that *contact* is achieved again when $|\theta_b| = \alpha$. An exact solution of the trajectory of θ_b is now given

by solving for θ_b as a one state model. Using equation (12), $|\theta_b| \leq \alpha$ and the release condition we get

$$\dot{\theta}_b = \begin{cases} \max(0, \dot{\theta}_d + \frac{k}{c}(\theta_d - \theta_b)) & \theta_b = -\alpha \ (T \leq 0) \\ \dot{\theta}_d + \frac{k}{c}(\theta_d - \theta_b) & |\theta_b| < \alpha \ \text{eq. (12)} \\ \min(0, \dot{\theta}_d + \frac{k}{c}(\theta_d - \theta_b)) & \theta_b = \alpha \ (T \geq 0) \end{cases} \quad (14)$$

This state equation can be interpreted is a limited integrator with the time derivate $\dot{\theta}_d + \frac{k}{c}(\theta_d - \theta_b))$ and limit α. With θ_b and $\dot{\theta}_b$ known from (14), and $\theta_d(t), \dot{\theta}_d(t)$ given, the torque T is found by (1). We now have a nonlinear dynamical system, not a function, that gives the torque T with given θ_d and $\dot{\theta}_d$.

2.2 Phase Plane Backlash Model

After an approximation the previous phase plane analysis can be used to derive a new model. This model, called the *phase plane model*, will be a function of θ_d and $\dot{\theta}_d$.

Define t_i^r as the ith time the transmitted torque becomes zero, i.e. $T(t_i^r) = 0$ and $T(t_i^r - \epsilon) \neq 0$ with ϵ an infinitesimal time. Define also t_i^c as the ith time the transmitted torque becomes nonzero, i.e. $T(t_i^c) = 0$ and $T(t_i^c + \epsilon) \neq 0$. Let them be defined so that $t_i^r < t_i^c$, and assume without loss of generality that $t_i^r = 0$.

Contact is achieved again when $\theta_b(t_i^c) = \theta_d(t_i^c) - \theta_s(t_i^c) = \pm\alpha$. Equation (13) gives $\theta_b(t_i^c) - \theta_d(t_i^c) = (\theta_d(0) - \theta_b(0))e^{-kt_i^c/c}$. If kt_i^c/c is sufficiently large and $|\theta_d(0) - \theta_b(0)| = |\theta_s(0)|$ is sufficiently small, then $\theta_b(t_i^c) - \theta_d(t_i^c) = -\theta_s(t_i^c) \approx 0$, which implies that the shaft has reached steady state. Then contact is achieved for $\theta_d(t_i^c) = \theta_b(t_i^c) = \pm\alpha$.

For large values of $|\theta_s(0)|$ the shaft is not in steady state at t_i^c, so $\theta_s(t_i^c)$ needs to be calculated. Integration gives

$$\dot{\theta}_d(t) = \dot{\theta}_d(0) + \int_0^t \ddot{\theta}_d(s)\,ds \quad \forall t \in [0, t_i^c] \quad (15)$$

Since $T = 0$ holds here, $\ddot{\theta}_d(t)$ is typically an acceleration produced by motor and load torques, and cannot be *a priori* known. Assume therefore that $\ddot{\theta}_d(t) = 0 \ \forall t \in [0, t_i^c]$, which gives

$$\theta_d(t) = \theta_d(0) + \dot{\theta}_d(0)t \quad \forall t \in [0, t_i^c] \quad (16)$$

For large values of $|\dot{\theta}_d(0)|$ the integrand of (15) is small relative to the constant term, since the integrand can be assumed bounded and t_i^c will be small.

Consider now the upper half plane of the phase plane $(\theta_d, \dot{\theta}_d)$, starting at $t_i^r = 0$ on the border of A_r. From Theorem 1 and the definition of A_r it follows that $\dot{\theta}_d(0) = -k(\theta_d(0) - \alpha)/c > 0$. All motion is now in the positive θ_d direction (15),(16). Contact will be achieved again when

$$\theta_b(t_i^c) = \alpha = \theta_d(t_i^c) - \theta_s(0)e^{-kt_i^c/c} \qquad (17)$$

From (16) it follows that

$$t_i^c = (\theta_d(t_i^c) + \alpha)/\dot{\theta}_d(0) + c/k \qquad (18)$$

Now (17) and (18) gives

$$f(\theta_d(t_i^c) + \alpha, \dot{\theta}_d(0)) = (\theta_d + \alpha) + (c\dot{\theta}_d/k)e^{-k(\theta_d+\alpha)/(c\dot{\theta}_d)-1} = 2\alpha \qquad (19)$$

which corresponds to $\theta_b(t_i^c) - \theta_b(0) = 2\alpha$. For a given value of $\dot{\theta}_d(0)$ there exists a unique solution $\theta^*(\dot{\theta}_d(0)) \in [\alpha - c\dot{\theta}_d(0)/k, \alpha]$ (which is outside A_r), that satisfies $f(\theta^*(\dot{\theta}_d(0)) + \alpha, \dot{\theta}_d(0)) = 2\alpha$.

The backlash will not be in contact, with $T = 0$, as long as $\theta_d < \theta^*(\dot{\theta}_d)$, but when $\theta_d = \theta^*(\dot{\theta}_d)$ is reached, the backlash gap is closed, and right contact with $\theta_b(t_i^c) = \alpha$ is achieved. Note that for small values of $c\dot{\theta}_d(0))/k$ the exponential part of (19) will be small, and $\theta^* \approx \alpha$. This means that although the assumption that the integral part of (15) is not neglible holds, the approximation that contact is achieved again for $\theta_d = \theta^*(\dot{\theta}_d)$ is still valid.

Right contact will then be preserved and the linear model $T = k(\theta_d - \alpha) + c\dot{\theta}_d$, given by (5) will hold. T will remain positive until the release condition of theorem 1 is fulfilled. Note that this can only happen for $\dot{\theta}_d < 0$, i.e. in the lower half plane, because of the directional constraints in the phase plane $(\theta_d, \dot{\theta}_d)$. By noting that $\theta_d < \theta^*$ is equivalent to $f(\theta_d + \alpha, \dot{\theta}_d) < 2\alpha$ for $k(\theta_d - \alpha) + c\dot{\theta}_d \geq 0$, it holds that there is right contact for $(\theta_d, \dot{\theta}_d) \in A^+$ defined by,

$$A^+ = \left\{ (\theta_d, \dot{\theta}_d) : \begin{cases} f(\theta_d + \alpha, \dot{\theta}_d) \geq 2\alpha & \dot{\theta}_d > 0 \\ k(\theta_d - \alpha) + c\dot{\theta}_d \geq 0 & \forall \dot{\theta}_d \end{cases} \right\} \qquad (20)$$

see Figure 2b. Symmetry conditions holds for left contact with

$$A^- = \left\{ (\theta_d, \dot{\theta}_d) : \begin{cases} f(\theta_d - \alpha, \dot{\theta}_d) \leq -2\alpha & \dot{\theta}_d < 0 \\ k(\theta_d + \alpha) + c\dot{\theta}_d \leq 0 & \forall \dot{\theta}_d \end{cases} \right\} \qquad (21)$$

Finally we can can conclude that there is no contact in A^0 defined by

$$A_0 = \{(\theta_d, \dot{\theta}_d)\} \setminus (A^+ \cup A^-) \qquad (22)$$

Remember that the difficult problem was to know if there was contact or not, which we now have solved approximately. The torque can now be given as a function of θ_d and $\dot{\theta}_d$. The phase plane model is summarized in Figure 2b and in equation (23).

$$T_{\text{phase}} = \begin{cases} k(\theta_d - \alpha) + c\dot{\theta}_d & (\theta_d, \dot{\theta}_d) \in A^+ \\ 0 & (\theta_d, \dot{\theta}_d) \in A^0 \\ k(\theta_d + \alpha) + c\dot{\theta}_d & (\theta_d, \dot{\theta}_d) \in A^- \end{cases} \qquad (23)$$

Note that in (23) T_{phase} is independent of θ_b. Note also that T_{phase} is discontinuous in θ_d for $\theta_d = \theta^*$, which describes what happens when the backlash achieves contact with a nonzero speed difference.

Remark 1. Note that we have made two approximations for the shaft:
   (i) the shaft is inertia free, section 2.1, and
   (ii) the approximation of $\theta_d(t_i^c)$, given by (20) and (21).

Note, however, that when the shaft is not in backlash contact ($T = 0$), and only connected to the driving member on one side, approximation (i) might not be valid and (1) and (13) should be at least of second order.

In the derivation of the release condition we used approximation (i) only to prove that contact could not be achieved in A_r and that contact was lost as soon as we reached A_r. It seems that this is still a good approximation, i.e. that contact ceases when $(\theta_d, \dot\theta_d)$ leaves the area in the phase plane that gives a nonzero torque. For the exact solution (14) it is critical that (1) or (i) holds also when there is no backlash contact, because this is used to predict when contact is achieved again, and (ii) is not necessarily a worse approximation.

These observations imply that the solutions (14) and (23) are both somewhat inexact with respect to real shafts with inertia.

2.3 The Dead Zone Model

An often proposed model, found in almost any basic control course, for analysis and simulations of an elastic shaft with backlash is the classical dead zone model [23,13,12,10,8], where the torque is given by

$$T = k D_\alpha(\theta_d) \qquad (24)$$

where

$$D_\alpha(x) = \begin{cases} x - \alpha & x > \alpha \\ 0 & |x| < \alpha \\ x + \alpha & x < -\alpha \end{cases} \qquad (25)$$

The model of the shaft is here a pure spring, i.e. it is inertia free and has no internal damping. The shaft is assumed to be in steady state when it is free, i.e. when there is no backlash contact. If internal damping is introduced in the shaft a commonly used modification of the dead zone model, see Brandenburg in [3,7,4–6] is presented in Figure 2c and in equation (26). This model will from now on, in this text, be referred to as the *dead zone model*.

$$T_{\text{dead zone}} = \begin{cases} k(\theta_d - \alpha) + c\dot\theta_d & \theta_d > \alpha \\ k(\theta_d + \alpha) + c\dot\theta_d & \theta_d < -\alpha \\ 0 & |\theta_d| < \alpha \end{cases} \qquad (26)$$

This model, just as the phase plane model, gives the torque $T_{\text{dead zone}}$ as a function of θ_d and $\dot\theta_d$. Compare the phase plane analysis of the physical model (23) with the phase plane analysis of the phase plane model and the dead zone model (26), see Figure 2. It is seen that $T_{\text{dead zone}} \neq 0$ in area A1 and A2 of Figure 2c, but this violates (5) and Lemma 1! Furthermore this is a very non-physical model, because $T_{\text{dead zone}}$ has the wrong sign in areas A1 and A2, where we know from analysis of the physical model that $T = 0$, i.e. the dead zone model gives a pull force at the side of the backlash where only a push force is possible, see Figure 2c. A modification of the dead zone model that does not violate Lemma 1 is given by

$$T = kD_\alpha(\theta_d + c\dot{\theta}_d/k) \tag{27}$$

An interpretation of (27) is that there is always right contact for $(\theta_d, \dot{\theta}_d) \in A_+$ with $T = k(\theta_d - \alpha) + c\dot{\theta}_d$ and always left contact for $(\theta_d, \dot{\theta}_d) \in A_-$ with $T = k(\theta_d) + \alpha) + c\dot{\theta}_d$. Compare equations (6)–(8), (23) and Figures 2a and 2b.

Remark 2. Note that if $k/c \to \infty$ the solution of (13) immediately goes to $\theta_d = \theta_b$ so now approximation (ii) in Remark 1 made for the phase plane model is exact, namely $\theta_s(t_i^c) = 0$. Furthermore the slope of the left respectively right border of A1 respectively A2 goes to $-\infty$, and hence the areas A1 and A2 to zero, so the phase plane model converges to the classical dead zone model (24), to which the dead zone models (26) and (27) also converge.

Hence, the phase plane model and the dead zone models coincide and are an exact description of the torque behaviour, when the shaft is modeled as a pure spring without damping.

2.4 Model Simulation Comparisons

Let us compare simulations of the true model (1),(14) the phase plane model (23), the dead zone model (26) and the revised dead zone model (27).

In Figure 3 a simulation of the two-mass system with an elastic shaft with internal damping and backlash can be seen. Remember from previous analysis that the release condition is exact for the phase plane model (23) and the revised dead zone model (27) but bad for the dead zone model (26), which can clearly be seen in Figure 3: the shaft torque predicted by the dead zone model sometimes is negative although only one side of the backlash is touched. Note that the phase plane model torque T_{phase} and the exact torque T almost overlap in the figure.

The ratio $k/c = 100$ gives a half time of the exponential term in (13) of about $\ln 2/(k/c) \approx 0.0069$s. For this simulation the shortest time between contacts is about 0.05s, and the exponential term would then be diminished by a factor $e^{-0.05k/c} = e^{-5} \approx 0.0067$. This means that the prediction of when contact is achieved is very good for both (23) and (26), but it is bad for (27), see Figure 3. Simulations of this kind made the need obvious for a better model than the dead zone model.

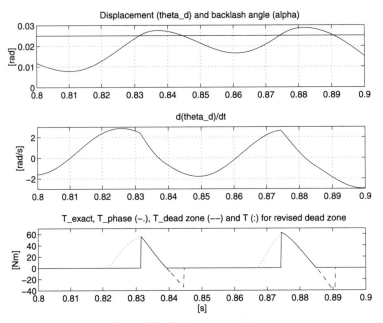

Fig. 3. Simulation comparisons.

The simulation shown in Figure 3 used the following mathematical model:

$$J_m \ddot{\theta}_m = T_m - T(\theta_d, \dot{\theta}_d) - T_{\text{preload}} + a_{\text{dist}} \sin(40\pi t)$$
$$J_l \ddot{\theta}_l = T_l + T(\theta_d, \dot{\theta}_d) + T_{\text{preload}}$$
$$\theta_d = \theta_m - \theta_l$$

The exact torque $T(\theta_d, \dot{\theta}_d)$ is used for the simulation, and the other torques, namely $T_{\text{phase}}(\theta_d, \dot{\theta}_d)$ (23), $T_{\text{deadzone}}(\theta_d, \dot{\theta}_d)$ (26) and T (27) are calculated from the θ_d and $\dot{\theta}_d$ given by this simulation. The following parameter values, most of them approximations of a laboratory system at ABB Industrial Systems AB, were used:

$J_m = 0.4$ [kgm^2]   motor inertia
$J_l = 5.6$ [kgm^2]   load inertia
$k = 5895$ [Nm/rad]   shaft elasticity
$c = k\, 0.01 s^{-1} = 58.95$ [Nm/rad/s]   inner damping of the shaft
$\alpha = 0.0025$ [rad]   backlash angle
$a_{\text{dist}} = 19$ [Nm]   disturbance amplitude
$T_{\text{preload}} = 19$ [Nm]   preload torque

One method to compare the different models is to measure the the integral of the transmitted torque over one cycle, and compare with the exact model. For the simulation presented in Figure 3 the following results were obtained

Backlash model	Integrated torque [10^{-3}Nms]	error [%]
T from eq. (5)	1.5327	-
T_{phase} from eq. (23)	1.5331	+0.29
T from eq. (26)	0.9486	-38
T from eq. (27)	2.997	+95

We can see that the phase plane model fits well to the exact model, and that the errors of both dead zone models are large. This means that even though the revised dead zone model (27) is exact with respect to the release condition, it predicts badly when contact is achieved, and is really no better than the dead zone model (26).

3 Model for Backlash with Rubber

In an industrial standard flexible coupling there is rubber, in principle as in Figure 4. In this work the specific coupling is manufactured by Benzlers [9], but similar couplings are very common in many applications throughout the world. Therefore a model for backlash with rubber is needed. In this

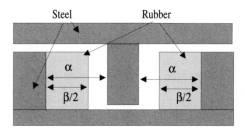

Fig. 4. Backlash with rubber.

section we derive a simple model for an inertiafree elastic shaft with internal damping, connected to a backlash partly filled with rubber. We are not aiming for an exact and complicated model, but rather a workable approximation. The new model is derived as an extension of the phase plane model (23). In section 3.2 we simulate a laboratory setup with the new model, and compare with measurements. Striking similarities are found.

3.1 Shaft Model for Backlash with Rubber

Extensive measurements were performed on a laboratory setup (described in section 3.2). After considering the differences between these measurements and those expected from a setup with a rubberfree coupling we were led to introduce the following features into the new model for backlash with rubber:

1. The motor speed measurements seemed to be differentiable for the experiments. Since the motor speed is proportional to the integral of the difference of motor and shaft torques (42), the shaft torque $T_{\text{rubber}}(\theta_d, \dot{\theta}_d)$ should be modeled as a continuous, not necessarily differentiable, function.
2. The width of the rubber, β [rad], should enter the model.
3. The rubber is soft when not compressed, but gets harder as it becomes compressed. For large twists of the shaft, the model should behave as if it were just a steel shaft. Hence we should have $dT_{\text{rubber}}(\theta_s, \dot{\theta}_s)/d\theta_s = k$ and $dT_{\text{rubber}}(\theta_s, \dot{\theta}_s)/d\dot{\theta}_s = c$, when the rubber is compressed, i.e. for large $|\theta_s|$ or equivalently for large $|\theta_d|$.

To achieve features 1–3 from above we extend the phase plane model (23) to include areas of 'rubber effect' in the phase plane $(\theta_d, \dot{\theta}_d)$, as follows:

First assume that right contact is achieved for $\theta_d = \theta^*(\dot{\theta}_d) - \beta/2$, i.e. as predicted by the phase plane model (23), modified by the rubber thickness, see Figure 4 and equation (19). Then introduce a band of thickness β, (double the rubber thickness), for $\theta^*(\dot{\theta}_d) - \beta/2 < \theta_d < \theta^*(\dot{\theta}_d) + \beta/2$. In this band let the damping and elasticity k and c vary linearly with θ_d from 0 to the parameter values of the steel shaft, which reflects feature 3, i.e. that the rubber gets harder as it is compressed. The choice to let the band of compression be double the rubber thickness is *ad hoc*, but as will be seen later, fits well with measurements.

When right contact is lost at high speed, i.e for $|\dot{\theta}_d| > c\beta/(2k)$, assume that there is no time for the rubber to decompress, i.e. no rubber effect. For $|\dot{\theta}_d| < c\beta/(2k)$ let the damping, and elasticity constant vary linearly from zero to the parameter values of the steel shaft for $\theta_d \in (\alpha - \beta/2 - 2c\dot{\theta}_d/k, \alpha + \beta/2)$, see Figure 2d. The idea is that when contact is lost at low speeds, there is time for the rubber to decompress somewhat, and for very low speeds it decompresses totally. Let us now formalize these assumptions of rubber behavior into equations:

Note first that $|\theta_d - \theta^*(\dot{\theta}_d)| < \beta/2 \iff f(\theta_d + \alpha - \beta/2, \dot{\theta}_d) < 2\alpha < f(\theta_d + \alpha + \beta/2, \dot{\theta}_d)$, see the derivation of equation (19). Define now the area of right contact with rubber effect as

$$A^+_{\text{rubber}} = \left\{ (\theta_d, \dot{\theta}_d) : \begin{array}{c} f(\theta_d + \alpha - \beta/2, \dot{\theta}_d) < 2\alpha < f(\theta_d + \alpha + \beta/2, \dot{\theta}_d) \quad \theta_d > 0 \\ \alpha - \beta/2 - 2c\dot{\theta}_d/k < \theta_d < \alpha + \beta/2 \quad \forall \dot{\theta}_d \end{array} \right\}$$
(28)

see Figure 2, and the remaining area of right contact

$$A^+_{\text{steel}} = A^+ \setminus A^+_{\text{rubber}} \qquad (29)$$

as the area of right contact where the shaft behaves as a steel shaft (23). Symmetrically we have

$$A^-_{\text{rubber}} =$$

$$= \left\{ (\theta_d, \dot\theta_d) : \begin{cases} f(\theta_d - \alpha + \beta/2, \dot\theta_d) < -2\alpha < f(\theta_d - \alpha - \beta/2, \dot\theta_d) & \theta_d < 0 \\ -\alpha - \beta/2 < \theta_d < -\alpha + \beta/2 - 2c\dot\theta_d/k & \forall \dot\theta_d \end{cases} \right\} \tag{30}$$

$$A^-_{\text{steel}} = A^- \setminus A^-_{\text{rubber}} \tag{31}$$

The area of no contact is now given by

$$A^0_{\text{rubber}} = \{(\theta_d, \dot\theta_d)\} \setminus (A^-_{\text{rubber}} \cup A^-_{\text{steel}} \cup A^+_{\text{rubber}} \cup A^+_{\text{steel}}) \tag{32}$$

The areas A^0_{rubber}, A^-_{rubber}, A^-_{steel}, A^+_{rubber} and A^+_{steel} are depicted in Figure 2d. In areas $A^+_{\text{steel}} \subset A^+$ and $A^-_{\text{steel}} \subset A^-$ the shaft torque should be given by (23), i.e.

$$T^+_{\text{steel}} = k(\theta_d - \alpha) + c\dot\theta_d \tag{33}$$

$$T^-_{\text{steel}} = k(\theta_d + \alpha) + c\dot\theta_d \tag{34}$$

and in A^0_{rubber} T should be zero. We have now accomplished feature 2, i.e. included the rubber thickness into the model. Let the stationary elasticity (i.e. for $\dot\theta_s = 0$) of the shaft torque $T_{\text{rubber}}(\theta_s, \dot\theta_s)$ satisfy

$$dT_{\text{rubber}}(\theta_s, 0)/d\theta_s = \begin{cases} k|\theta_s|/\beta & |\theta_s| < \beta \\ k & |\theta_s| \geq \beta \end{cases} \tag{35}$$

By integrating (35), and remembering the continuity feature 1, we obtain the following models, valid in A^+_{rubber} and A^-_{rubber} respectively.

$$T^+_{\text{rubber}} =$$
$$= \begin{cases} \left(\frac{k(\theta_d - \theta^*(\dot\theta_d) + \beta/2)}{2} + k(\theta^*(\dot\theta_d) - \alpha) + c\dot\theta_d \right) \frac{\theta_d - \theta^*(\dot\theta_d) + \beta/2}{\beta} & \dot\theta_d > 0 \\ \left(\frac{k(\theta_d - \alpha + \beta/2)}{2} + c\dot\theta_d \right) \frac{\theta_d - \alpha + \beta/2 + 2c\dot\theta_d/k}{\beta/2 + 2c\dot\theta_d/k} & \dot\theta_d \leq 0 \end{cases} \tag{36}$$

$$T^-_{\text{rubber}} =$$
$$= \begin{cases} \left(\frac{k(-\theta_d + \theta^*(\dot\theta_d) + \beta/2)}{2} + k(-\theta^*(\dot\theta_d) - \alpha) + c\dot\theta_d \right) \frac{\theta_d - \theta^*(\dot\theta_d) - \beta/2}{\beta} & \dot\theta_d > 0 \\ \left(\frac{k(-\theta_d - \alpha + \beta/2)}{2} + c\dot\theta_d \right) \frac{\theta_d + \alpha - \beta/2 + 2c\dot\theta_d/k}{\beta/2 - 2c\dot\theta_d/k} & \dot\theta_d \leq 0 \end{cases} \tag{37}$$

For $\dot\theta_d \to 0$ or for $c \to 0$ equations (36) and (37) reduce to

$$T^+_{\text{rubber}}(\theta_d, 0) = (k(\theta_d - \alpha + \beta/2)/2)(\theta_d - \alpha + \beta/2)/\beta \tag{38}$$

$$T^-_{\text{rubber}}(\theta_d, 0) = (k(-\theta_d - \alpha + \beta/2)/2)(\theta_d + \alpha - \beta/2)/\beta \tag{39}$$

which satisfy (35). The extended phase plane model for backlash with rubber, T_{rubber} is now given by

$$T_{\text{rubber}}(\theta_d, \dot\theta_d) = \begin{cases} T^+_{\text{steel}} & (\theta_d, \dot\theta_d) \in A^+_{\text{steel}} \\ T^+_{\text{rubber}} & (\theta_d, \dot\theta_d) \in A^+_{\text{rubber}} \\ 0 & (\theta_d, \dot\theta_d) \in A^0_{\text{rubber}} \\ T^-_{\text{rubber}} & (\theta_d, \dot\theta_d) \in A^-_{\text{rubber}} \\ T^-_{\text{steel}} & (\theta_d, \dot\theta_d) \in A^-_{\text{steel}} \end{cases} \tag{40}$$

Note that $T_{\text{rubber}}(\theta_d, \dot{\theta}_d)$ is continuous in θ_d and $\dot{\theta}_d$. $T_{\text{rubber}}(\theta_d, \dot{\theta}_d)$ is graphically expressed in figure 2d. Note also that if $\beta \to 0$, then the rubber model (40) converges to the phase plane model (23).

For the case where $c = 0$, the torque is independant of $\dot{\theta}_d$. A simplified model for $c = 0$ is then given by (33),(34),(38),(39) to be

$$T(\theta_d) = T_{\text{rubber}}(\theta_d, 0) =$$

$$= \begin{cases} k(\theta_d - \alpha) & \theta_d \geq \alpha + \beta/2 \\ (k(\theta_d - \alpha + \beta/2)/2)(\theta_d - \alpha + \beta/2)/\beta & \alpha - \beta/2 < \theta_d < \alpha + \beta/2 \\ 0 & |\theta_d| \leq \alpha - \beta/2 \\ (k(-\theta_d - \alpha + \beta/2)/2)(\theta_d + \alpha - \beta/2)/\beta & \alpha - \beta/2 < \theta_d < \alpha + \beta/2 \\ k(\theta_d + \alpha) & \theta_d \leq -\alpha - \beta/2 \end{cases}$$
(41)

This is a 'smoothed' dead zone model, where $dT(\theta_d)/d\theta_s$ is continuous.

3.2 Simulations and Measurements: Backlash with Rubber

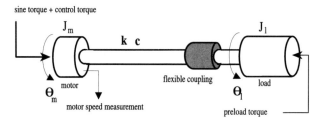

Fig. 5. Laboratory setup.

In order to test the model derived in section 3, let us compare simulations and measurements of a speed controlled two-mass system connected with a shaft with flexible coupling including a rubber part. The measurements were performed on a 60 kW laboratory setup at ABB Industrial Systems, modeled and described in (42) and Figure 5:

$$\begin{cases} J_m \ddot{\theta}_m = T_{\text{control}} + a_{\text{dist}} T_{\text{base}} \sin(2\pi f_{\text{dist}} t) - T_{\text{rubber}}(\theta_d, \dot{\theta}_d) \\ J_l \ddot{\theta}_l = m T_{\text{base}} + T_{\text{rubber}}(\theta_d, \dot{\theta}_d) \\ \theta_d = \theta_m - \theta_l \end{cases}$$
(42)

with the parameters

$J_m = 0.4 \ [\text{kgm}^2]$ — motor inertia
$J_l = 5.6 \ [\text{kgm}^2]$ — load inertia
$k = 5895 \ [\text{Nm/rad}]$ — shaft elasticity
$c = 3 \ [\text{Nm/rad/s}]$ — inner damping of shaft
$\alpha = 0.05 \ [\text{rad}]$ — backlash gap
$\beta = 0.05 \ [\text{rad}]$ — rubbber thickness
$T_{\text{base}} = 380 \ [\text{Nm}]$ — nominal motor torque
$\omega_{\text{ref}} = 1438 \ [\text{rpm}] = 150.6 \ [\text{rad/s}]$ — speed reference
$a_{\text{dist}} = 0.05$ — disturbance amplitude factor
$f_{\text{dist}} = 6$ or $9.5 \ [\text{Hz}]$ — disturbance frequency
$m = -0.06$ — preload gain

The control input is the motor torque T_{control} and the measured output is the motor speed $\dot{\theta}_m$. Some of the parameter values above are given in the mechanical data sheets of the laboratory system. Others were adjusted due to the results of identification experiments which were performed using high preload torques causing the backlash to remain closed, and hence the system could be considered as essentially linear. The backlash gap α and rubber thickness β were measured with a ruler. The original purpose of the measurements was to find an approximation of the frequency response of the system, rather than validating a simulation model. Therefore the measurements consisted of 49 swept sine experiments for different working conditions, parameterized by different constant preload gains m and different amplitudes a_{dist} of the injected sinewave. The control system was a PI-controller with very low bandwidth, which in practice will give a constant control $T_{\text{control}} = -mT_{\text{base}}$. The motor speed θ_m and load speed θ_l will then, in steady state, vary around the speed reference ω_{ref}.

To test the simulation model we chose a number of time domain measurements and then compared these measurements to simulations of the same system including the new model for a flexible coupling. In Figures 6 and 7 simulations and measurements can be compared. In Figure 7 it can be seen that the shaft bounces between *right contact* and *left contact*. Here the limit cycle is reasonably close to a sinewave, and the energy content of higher harmonics is small. In Figure 6 a case with persistent *right contact* is shown. Here both higher and lower harmonics than the base frequency are large. Note that here the load displacement satisfies $(\theta_d, \dot{\theta}_d) \in A^+_{\text{rubber}}$ for this case, see (40). This means that the measured behavior reflects the fact that there is rubber in the coupling.

It seems that the new model approximates well the behavior of this two-mass system. Numerous other simulations and measurements for different loads and frequencies fit very well to measurements. Some of these are presented in [15], including modes with persisting right contact, where the measured behavior depends only on the rubber thickness.

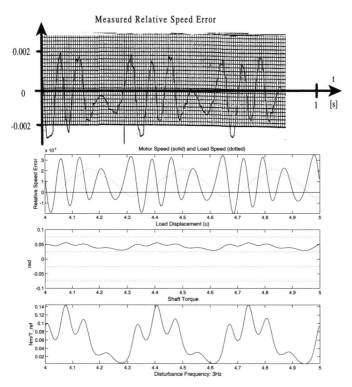

Fig. 6. Measurement and simulation of system (42): 3.0 Hz. The speeds are relative to ω_{base}.

4 Describing Function Analysis

A method commonly used to analyze certain nonlinearities is the describing function method [12]. As pointed out in [3,7,4–6], a constant *output* torque, here denoted T_0, best describes the working point for a shaft with backlash. Using Brandenburg's methodology, first calculate the *dual-input describing function* **DIDF**, where the input is a sinewave with a constant offset B given as

$$\theta_d(t) = B + A\sin(\omega t) \tag{43}$$

and the output is approximated by a constant offset $N_B B$ and the first harmonic (see [12] for a theoretical background):

$$T(t) = T(\theta_d, \dot{\theta}_d) \approx N_B B + A N_p \sin(\omega t) + A N_q \cos(\omega t) \tag{44}$$

from which the **DIDF**s $N_B(A, B, \omega)$ and $N_A(A, B, \omega)$ are calculated with

$$N_A(A, B, \omega) = N_p(A, B, \omega) + i N_q(A, B, \omega) \tag{45}$$

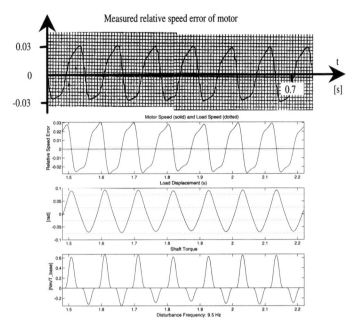

Fig. 7. Measurement and simulation of system (42): 9.5 Hz. The speeds are relative to ω_{base}.

Then consider the equation

$$T_0 = BN_B(A, B, \omega) \qquad (46)$$

with B as the independent variable. If there exists a unique solution $B^*(A, T_0, \omega)$ for (46), the system around a working point T_0 can now be described by $B^*(A, T_0, \omega)$ and $N_A(A, T_0, \omega)$ with

$$N_A(A, T_0, \omega) = N_A(A, B^*(A, T_0, \omega), \omega) \qquad (47)$$

in a **DIDF** setting.

In this section these calculations are performed for the phase plane model, and comparisions with the dead zone model are made. Since the model for backlash with rubber is quite complicated an analytic expression is not calculated for this **DIDF**. It is however straightforward to express them as integrals suitable for numerical evaluation.

4.1 DIDF Calculations for the Phase Plane Model

Note that (43) gives $\dot{\theta}_d = A\omega \cos(\omega t)$ and that this input is an ellipse in the $(\theta_d, \dot{\theta}_d)$-plane, see Figure 8. With $\theta = \omega t$ and T_{phase} from equation (23) the

DIDFs are now given by

$$N_B = \frac{1}{2\pi B} \int_0^{2\pi} T_{\text{phase}}(B + A\sin\theta, A\omega\cos\theta) d\theta \tag{48}$$

$$N_p = \frac{1}{A} \int_0^{2\pi} T_{\text{phase}}(B + A\sin\theta, A\omega\cos\theta) \sin\theta d\theta \tag{49}$$

$$N_q = \frac{1}{A} \int_0^{2\pi} T_{\text{phase}}(B + A\sin\theta, A\omega\cos\theta) \cos\theta \, d\theta \tag{50}$$

Consider the trajectory in Figure 8. In the nonshaded areas we know from

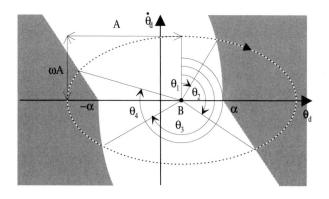

Fig. 8. A phase plane trajectory for the **DIDF** computation.

previous sections that $T_{\text{phase}} = 0$ holds. In the right shaded area, along the phase plane curve it holds that

$$T^+(\theta) = T_{\text{phase}}(B + A\sin\theta, A\omega\cos\theta) = k(B - \alpha) + kA\sin\theta + A\omega c\cos\theta \tag{51}$$

and similarly in the left shaded area

$$T^-(\theta) = T_{\text{phase}}(B + A\sin\theta, A\omega\cos\theta) = k(B + \alpha) + kA\sin\theta + A\omega c\cos\theta \tag{52}$$

Now equations (48), (50),(51) and (52) give

$$N_B = \frac{1}{2\pi B} \int_{\theta_1}^{\theta_2} T^+(\theta) d\theta + \frac{1}{2\pi B} \int_{\theta_3}^{\theta_4} T^-(\theta) d\theta \tag{53}$$

$$N_p = \frac{1}{A} \int_{\theta_1}^{\theta_2} T^+(\theta) \sin\theta d\theta + \frac{1}{A} \int_{\theta_3}^{\theta_4} T^-(\theta) \sin\theta d\theta \tag{54}$$

$$N_q = \frac{1}{A} \int_{\theta_1}^{\theta_2} T^+(\theta) \cos\theta d\theta + \frac{1}{A} \int_{\theta_3}^{\theta_4} T^-(\theta) \cos\theta d\theta \qquad (55)$$

Geometry gives $\theta_1 - \theta_4$ in Figure 8. N_B and N_A are now given by inserting $\theta_1 - \theta_4$ in (53)–(55). It is straightforward to see that $N_B(A, B, \omega)B$ is zero for $B = 0$ and monotonically increasing as a function of B. Hence there exist a unique solution of equation (46) for the phase plane model. The full expressions for N_A and N_B are given in [15].

4.2 Phase Plane vs. Dead Zone Model

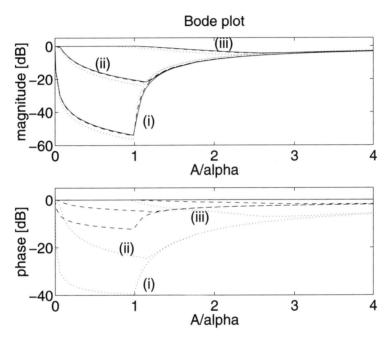

Fig. 9. Bode plot of $N_A(A, T_0, \omega)/(k + j\omega c)$. 0 Hz represented by solid, 5 Hz by dashed and 15 Hz by dotted curve. (i): $T_0 = 0.001k\alpha$ (ii): $T_0 = 0.05k\alpha$ (iii): $T_0 = k\alpha$. Note that the 0 Hz case is equivalent to the dead zone case.

The describing function method can be used to predict limit cycles by plotting them in Nyquist or Nichols charts together with the frequency function of the linear system surrounding the nonlinearity. We will now compare the phase plane model and the dead zone model by plotting their **DIDF**s.

Most control engineers are used to the describing function of the classical dead zone model (24). Note that dead zone models (26) and (27) are equivalent to the classical dead zone model (24) in series with a linear block $k + cs$, and have the same describing functions. Therefore we plot $N_A(A, T_0, \omega)/(k+$

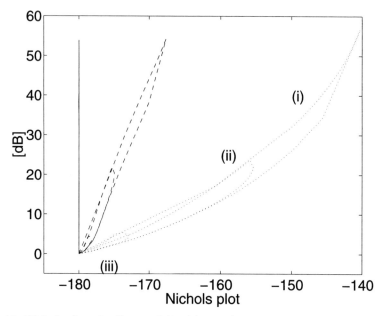

Fig. 10. Nichols plot of $-(k+j\omega c)/N_A(A,T_0,\omega)$. 0 Hz represented by solid, 5 Hz by dashed and 15 Hz by dotted curve. (i): $T_0 = 0.001k\alpha$ (ii): $T_0 = 0.05k\alpha$ (iii): $T_0 = k\alpha$. Note that the 0 Hz case is equivalent to the dead zone case.

$j\omega c$) versus the dual input describing function of the classical dead zone model in a Bode plot and their negative inverses in a Nichols chart. Note that they coincide for $\omega = 0$, and that the classical dead zone model (24) is frequency independent. In Figures 9–10 plots of $N_A(A, T_0, \omega)/(k + j\omega c)$ are presented. The parameter values used are $k = 5895$ [Nm/rad], $\alpha = 0.05$ [rad], $c = 60$ [Nm/rad/s]. T_0 had the three values $0.001k\alpha$, $0.05k\alpha$ and $k\alpha$ and ω had the three values 0, 5 and 15 [Hz]. Note that the gain of the **DIDF**s are approximately the same for the phase plane model and the dead zone model. On the other hand there is a quite substantial extra phase shift *not* predicted by the dead zone model. This phase shift might give previously unpredicted limit cycles. Note also that there is a remaining phase shift even for very large amplitudes, although it vanishes as $A \to \infty$. This means that, for increasing amplitudes, a system with backlash converges much more slowly to a linear system without backlash than the dead zone model predicts.

5 A Backlash Gap Estimation Method

Consider a simple elastic two-mass system with backlash, see Figure 11. The system can be divided into a nonlinear part consisting of a dead zone, and a linear part containing the motor, shaft and load dynamics. Note that the

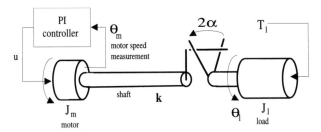

Fig. 11. Elastic two-mass system with backlash.

shaft damping here is assumed to be zero, and hence the dead zone model is correct, see previous sections. However, the methodology described in this section will work also for the other, more general, backlash models.

There exist model-based design methods for such systems, see [17,3–7,2], that require knowledge of the size of the backlash gap.

For identification of the linear part there are both parametric and non-parametric methods such as RLS or direct frequency response measurements. The linear dynamics are not complicated and the measurements can be performed with a preload under controlled conditions so that the backlash is not active. The gap size, however, is in most cases an uncertain parameter.

The case of backlash in series with a linear system is treated in [22], where adaptive control algorithms are presented together with rigorous proofs for stability. In many practical applications there is no access to the input or the output of the dead zone.

For the case considered in this chapter, where only the motor speed is measured, a naïve approach would be to turn the motor while the system is at rest and measure the angle between the backlash contacts. Engineers having tried this in practice know that the accuracy is low because it is difficult to know when contact is achieved due to friction and stiction.

With the parameters of the linear system known with reasonable accuracy, we develop an on-line identification method for the backlash gap. The idea is to add a smart choice of sine input to the motor torque. By doing this in a way guaranteeing that the speeds of the motor and load always are positive, stiction effects do not appear. Friction is then well modeled by dampings and constant torques. This can easily be achieved on speed controlled systems. It is then possible to adapt the gap size estimate until the gains of a model plant and the physical plant coincide. The adaptation idea is inspired by the MRAS method [1]. The input amplitude is adapted to a value yielding a plant gain sensitive to the changes in the backlash gap size. Describing function analysis is used to choose suitable gain for the plant, frequency of the input and also to indicate convergence of the adaptations.

5.1 Preliminaries and Assumptions

The plant class on which we estimate the backlash gap α consists of elastic two-mass systems, with motor moment of inertia J_m [kgm^2], load moment of inertia J_l [kgm^2], shaft elasticity k_s [Nm/rad], motor damping d_m [Nms/rad] and load damping d_l [Nms/rad]. The plant is schematically shown in Figure 11 and the corresponding block diagram shown in Figure 12.

The system dynamics of the plant are

$$J_m \dot{\omega}_m = -d_m \omega_m + u - T_s$$
$$J_l \dot{\omega}_l = -d_l \omega_l + T_s - T_l$$
$$\dot{\theta} = \omega_m - \omega_l$$
$$T_s = k_s D_\alpha(\theta) \tag{56}$$

where u [Nm] is the control (motor) torque, T_s [Nm] is the transmitted torque

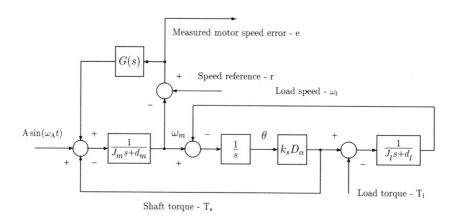

Fig. 12. Block diagram of the PI-controlled two-mass system.

in the shaft and T_l [Nm] is a constant load torque. ω_m [rad/s] and ω_l [rad/s] is the motor and load speed respectively, and θ [rad] is the displacement of the motor with respect to the load. The backlash is modeled as a dead zone, defined by

$$D_\alpha(\theta) = \begin{cases} \theta - \alpha & \theta > \alpha \\ 0 & |\theta| \leq \alpha \\ \theta + \alpha & \theta < -\alpha \end{cases} \tag{57}$$

see [23,10,8,11,13]. The system is speed controlled with a PI-controller, $G(s)$, so that

$$u = G(s)(r - \omega_m) + A\sin(\omega_A t)$$
$$= (k_p + k_i/s)(r - \omega_m) + A\sin(\omega_A t) \tag{58}$$

where r is a positive speed reference and $A\sin(\omega_A t)$ is an injected sine signal. If r is sufficiently large relative to A, no changes in sign of speeds will appear. The friction effects are then accurately modeled by the dampings d_m and d_l and a constant torque T_l. A low gain PI-controller should be used, in order to stabilize the system, without affecting the high frequency behavior.

5.2 Describing Function Analysis for Identification

As shown above in section 4, the describing function method is a powerful approximative method for analyzing nonsmooth nonlinearities see e.g. [12], including backlash. In this section we will use describing function analysis to indicate that the plant gain depends monotonically on the input amplitude, A, and the backlash gap α for a proper choice of input frequency ω_A. Using monotonicity, a gradient adaptation scheme can be applied to estimate α.

Assume for simplicity that $r = 0$ and $T_l = 0$. A more detailed analysis of the case $r \neq 0$ and $T_l \neq 0$ can be found in [4,14]. Let

$$N(\theta) = \begin{cases} 0 & |\theta| < \alpha \\ 1 - \frac{2}{\pi}\left[\arcsin(\frac{\alpha}{|\theta|}) + \frac{\alpha}{|\theta|}\sqrt{1 - (\frac{\alpha}{|\theta|})^2}\right] & |\theta| \geq \alpha \end{cases} \quad (59)$$

be the describing function of the dead zone (57). Note that $N(\theta)$ varies monotonically from 0 to 1. Approximate all signals with pure sinewaves of frequency ω_A, i.e. their first harmonic. Let

$$L_1(s) = \frac{1}{J_m s + d_m + G(s)}$$
$$L_2(s) = \frac{1}{J_l s + d_l}$$
$$G(s) = k_p + \frac{k_i}{s}. \quad (60)$$

Then $T_s = N(\theta)\theta$, which inserted into (56) gives

$$|\theta(i\omega_A)| = \left|\frac{L_1(i\omega_A)A}{i\omega_A + k_s(L_1(i\omega_A) + L_2(i\omega_A))}\right| \quad (61)$$

$$P(i\omega_A) = \frac{L_1(i\omega_A)(i\omega_A + L_2(i\omega_A)k_s N(\theta))}{i\omega_A + k_s N(\theta)(L_1(i\omega_A) + L_2(i\omega_A))} \quad (62)$$

where $P(i\omega_A)$ is the "transfer function" from the injected sinewave motor torque to to the motor speed ω_m.

If a preload $T_l > 0$ is introduced, the describing function deviates from N given by (59), see [4,14]. However, for reasonably large input amplitudes, this deviation is small and can be neglected.

$|P(i\omega)|$ explicitly depends on N but not on A (61). Thus proving monotonicity of $|P(i\omega_A)|$ with respect to A is equivalent to showing that $|P(i\omega)|$ depends monotonically on N, and that N depends monotonically on A.

It is rather straightforward to show that $|P(i\omega)|$ depends monotonically on N above a frequency close to the open loop resonance

$$\omega_r = \sqrt{k_s(\frac{1}{J_m} + \frac{1}{J_l})}$$

by checking the sign of the derivative.

N as a function of A can be obtained by solving (61) and (59), however not in explicit form. It is tedious to prove that N grows monotonically with A for a specific frequency. However, all systems of this structure have the same principal frequency response properties. Therefore, we investigate monotonicity numerically on a test example. With parameter values from (5.4) in section 5.3 it is straightforward to repeat this for any specific design example in order to ensure monotonicity. Consider the frequency range above the resonance ω_r. In the upper plot of Figure 13 we see that $|P(i\omega)|$ depends monotonically on N; the curves do not intersect. In the lower plot we see that N depends monotonically on A for $\omega_A > \omega_r$. Thus, we have indicated that $|P(i\omega)|$ depends monotonically on A. Note that changing α is equivalent to scaling A and hence $|P(i\omega)|$ also depends monotonically on α for $\omega_A > \omega_r$.

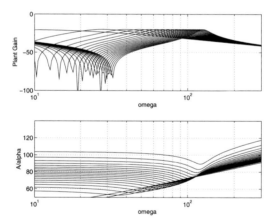

Fig. 13. Top: Magnitude plots of $|P(i\omega)|$ for $N \in [0, 1]$. The $N = 0$ case is the plot without resonance peak and corresponds to driving a free motor. Bottom: $A(i\omega)$ corresponding to the same N-values as in the upper plots.

From earlier frequency response measurements we have noted that multiple steady state solutions exist between the plant's zero and its resonance, depending on the direction of the frequency sweep. Above this frequency only one steady state solution has been observed. An interpretation of this is that $|P(i\omega)|$ is not monotonic for $\omega_A < \omega_r$, see Figure 13.

Note that for frequencies above, but rather close to the resonance frequency ω_r, $|P(i\omega)|$ is both sensitive and monotone with respect to changes in A. This is then a choice of frequency, suitable for adaptation of the input gain and identification of α.

5.3 The Estimation Method

Based on the DF-analysis of the previous section we propose the following method to estimate the backlash gap α. The experimental setup is shown in Figure 14 and the method is summarized in the following three steps:

1. Use Bode plots of the system, such as in Figure 13 to pick a frequency ω_A, where $|P(i\omega)|$ is monotone in A. Then choose a gain P_0 such that $P_{\min} < P_0 < P_{\max}$ where $P_{\min}$ and $P_{\max}$ are given by $|P(j\omega_A)|$ for $N = 1$ and $N = 0$ respectively. A reasonable choice is $P_0 = P(j\omega_A)|_{N=0.4}$.
2. Use Fourier Integrals [12] to measure the motor speed amplitude a. For a given input amplitude A, the gain $|P(i\omega_A)| = a/A$ can be calculated. This gives an input–output relation for the plant. Then adapt the amplitude A until $|P(i\omega_A)| = P_0$.
3. Remember that scaling α is equivalent to scaling A in the input–output relation that gives $|P|$. Construct a model plant $P_m(i\omega)$ with all parameters the same as in P except for the unknown backlash gap. Adapt the backlash gap estimate $\hat{\alpha}$, so that the output amplitude a_m of the plant model equals the measured output amplitude a. This can be done on-line during the adaptation of $|P(i\omega_A)|$, or off-line afterwards. Since $|P(i\omega_A)|$ depends monotonically on α this gives an exact estimation of α, given that all the other parameters are known and that $T_l = 0$. A straightforward modification that deals with the $T_l \neq 0$ case is presented in section 5.4.

Let A be the input amplitude, a the output amplitude of the real system and a_m the output amplitude of the model system. Denoting the amplitudes at time $2\pi k N_c/\omega_A$ by A^k, a^k, a_m^k, and similarly the α-estimate $\hat{\alpha}^k$, the algorithm can be formalized as:

$$A^{k+1} = A^k + \gamma_P(P_0 - a/A) \qquad (63)$$

$$\hat{\alpha}^{k+1} = \hat{\alpha}^k - \gamma_\alpha(a^k - a_m^k) \qquad (64)$$

The accuracy and speed of the Fourier Integrals are affected by the number of cycles N_c each estimation is integrated over. A high N_c gives low noise sensitivity but takes a longer time. We have used $N_c = 10$ throughout the paper.

Two tuning parameters, besides the PI-controller parameters k_p and k_i, are needed. First γ_P, that is the gain of the feedback from the gain error $P_0 - |P(j\omega_A)|$, and secondly γ_α, which is the feedback gain for the α-estimation.

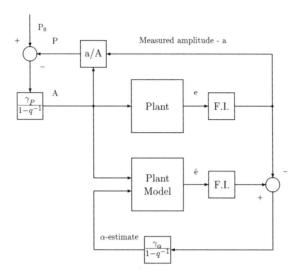

Fig. 14. Schematic block diagram of the estimation method.

5.4 Simulation Experiments

The following parameter values, taken from a laboratory system at ABB Industrial Systems AB, Sweden, used in simulations and numerical examples are

$J_m = 0.4$ [kgm^2]   motor inertia
$J_l = 5.6$ [kgm^2]   load inertia
$k = 5895$ [Nm/rad]   shaft elasticity
$d_m = 0$ [Nm/rad/s]   motor damping
$d_l = 0.056$ [Nm/rad/s]   load damping
$\alpha = 0.0025$ [rad]   backlash gap
$r = 150.6$ [rad/s]   speed reference

Note that $d_m = 0$ is usually a good approximation, since motors are constructed to have low friction.

Let us try a numerical experiment with $T_l = k_s \alpha$, for which the shaft twist in steady state is equal to the gap α. The other parameter settings were $k_p = 10$, $k_i = 1$, $\omega_A = 150$, $\gamma_P = 150\, \omega_A J_m$ and $\gamma_\alpha = 0.0008$. Figure 15 shows a simulation of the identification scheme. A steady state error of 3.3% remains in the estimate of α. This is due to the fact that we have used $T_l = 0$ in the plant model. A remedy for this is to estimate the mean control torque from the PI-controller with a Fourier Integral. The injected sinewave does not affect this estimate. Assuming neglible friction on the motor side, this mean torque is equivalent to a constant load torque, due to the integrator in the PI-controller. This estimate can now be introduced in the plant model,

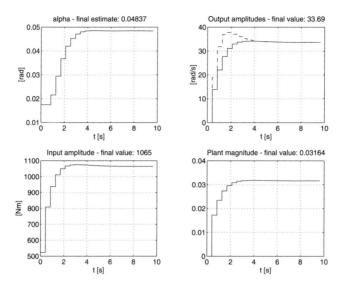

Fig. 15. Simulations without load torque compensation.

in order to eliminate the error in the α-estimation. In Figure 16, this modification is utilized, and now the estimation error vanishes. Note that the whole estimation takes only 10 seconds.

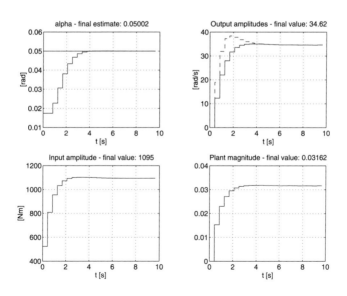

Fig. 16. Simulations with load torque compensation.

What if the linear parameters are uncertain? A straightforward way to test the sensitivity of the α-estimate, is to repeat the experiment for different parameter values. Note that it is not necessary to redo the measurement on the real system, but rather use the steady state values of A and a, given by the first estimation and repeat the estimation

$$\hat{\alpha}^{k+1} = \hat{\alpha}^k - \gamma_\alpha (a - a_m^k) \tag{65}$$

for the different parameter values we are interested in. Let us perform this sensitivity test where k_s vary with 1% and J_l with 10%. For these parameter variations, the α-estimates vary from 0.0483 to 0.0511 around the true value 0.05. The maximum deviation is 3.4 %. Note that it is often easy to estimate the resonance frequency ω_r, which motivates the small uncertainty in k_s.

6 Conclusions

In order to design an efficient control system, by any method, for a mechanical system including nonsmooth nonlinearities such as backlash, it seems necessary to have a correct model. Hitherto the dead zone model [23,13,10,8] has been used for designs reported in the literature, such as classical design using the describing function [12], robust linear design using the describing function [18], linear design with a load observer [4], a switching robust design with a load sensor [2], or adaptive control using an inverse model [20–22]. In this paper it is shown that the dead zone model is wrong, even giving the wrong sign of the driving torque, for the case of an elastic shaft with internal damping. Designs based on the hysteresis model are reported in [12,20–22]. The hysteresis model is, however, not suitable when there are disturbances acting on the load side, due to loss of causality.

A new, exact model for an inertiafree elastic shaft with internal damping connected to a backlash is presented in this paper. The model includes a state for the shaft twist. A simplified state-less model is derived from the exact model, giving the shaft torque as a function of the displacement between the driving and driven members, θ_d and its time derivative $\dot{\theta}_d$.

The model is also extended to the 'flexible coupling' case when the backlash gap is partly filled with rubber.

After some easily satisfied simplifying assumptions, the new model is compared by analysis and by simulations with the exact torque solution and with the dead zone model.

It is shown that (i) the new model gives a negligable torque error in comparision with the exact torque solution. In [15] a longer version of this paper is presented. There, describing function methods are used to analyze the phase plane model. Comparisons with the dead zone model show that , the new model exhibits a significant phase lag that might indicate the possibility of previously unpredicted limit cycles.

For the flexible coupling case, simulations of the new model were compared with measurements of a 60 kW laboratory system. The results were strikingly similar, see [15] for further examples.

The new model seems to represent the physical backlash sufficiently well for control purposes, and should replace the prevalent but erroneous dead zone model for analysis and design. In contrast to the dead zone model, the new model cannot be replaced by an equivalent uncertain linear gain plus a bounded disturbance [18,19,2]. Hence the above mentioned robust and adaptive techniques probably have to be modified. An attempt in this direction is reported in [16].

We believe that the new model will serve well as the basis for development of robust and adaptive control algorithms for mechanical systems with backlash.

Moreover, in order to design an efficient control system for a mechanical drive system including backlash [17,4,2] it seems necessary to have a good estimate of the backlash gap.

For the case of backlash in series with a linear system there are carefully investigated adaptive control algorithms [22], with rigorous proofs for stability.

In many mechanical systems, however, the backlash appears in closed loop, without direct access to the input or output of the dead zone that models the backlash, and often there is access only to motor speed measurements.

We have developed a method to identify the backlash gap α for the case of an elastic two-mass system with motor speed measurements. The method is analyzed with the help of the describing function method, but the actual estimation scheme does not use it, in order to avoid the inherent approximations of this method. The method works well on simulation experiments.

An additional feature of the method is that it can use existing frequency response measurements to estimate the backlash gap.

References

1. K.J. Åström and B. Wittenmark. *Adaptive Control*. Addison-Wesley, Reading, Massachusetts, 1989.
2. R. Boneh and O. Yaniv. Reduction of limit cycle amplitude in the presence of backlash. *J. of Dynamic Systems, Measurement and Control*, 121(2): 278–284, 1999.
3. G. Brandenburg. Stability of a speed controlled elastic two-mass system with backlash and Coulomb friction and optimization by a load observer. In *Symposium: Modelling and Simulation for Control of Lumped and Distributed Parameter Systems*, pages 107–113, Lille, 1986. IMACS-IFACS.
4. G. Brandenburg, H. Hertle, and K. Zeiselmair. Dynamic influence and partial compensation of coulomb friction in a position- and speed-controlled elastic two-mass systems. In *IFAC World Congress, Munich*, 1987.

5. G. Brandenburg and U. Schäfer. Influence and partial compensation of backlash for a position controlled elastic two-mass system. In *Proc. European Conf. on Power Electronics and Applications*, pages 1041–1047, Grenoble, Sept. 1987. EPE.
6. G. Brandenburg and U. Schäfer. Influence and adaptive compensation simultaneously acting backlash and Coulomb friction in elastic two-mass systems of robots and machine tools. In *Proc. ICCON '89*, pages WA-4-5, Jerusalem, 1989. IEEE.
7. G. Brandenburg, H. Unger, and A. Wagenpfeil. Stability problems of a speed controlled drive in an elastic system with backlash and corrective measures by a load observer. In *Proc. International Conference on Electrical Machines*, pages 523–527, München, Sept. 1986. Technische Universität München.
8. H. Chestnut and R. W. Mayer. *Servomechanisms and Regulating System Design*, pages 251–257,301–321. John Wiley & sons, 1955.
9. *Flexible Coupling No 2245 5268-7 with Rubber Element No 2245 5195-4*. Benzlers AB, Box 922, S-251 09 Helsingborg, Sweden.
10. R. L. Cosgriff. *Nonlinear Control Systems*, pages 121–123,193–194. McGraw-Hill, Reading, Massachusetts, 1958.
11. E. A. Freeman. The effect of speed dependent friction and backlash on the stability of automatic control systems. *Transaction of the American IEE*, 7:680–690, 1959.
12. A. Gelb and W.E. Vander Velde. *Multiple-Input Describing Functions and Nonlinear System Design*. McGraw-Hill, 1968.
13. J. H. Liversidge. *Backlash and Resilience within Closed Loop of Automatic Control Systems*. Academic Press, 1952.
14. M. Nordin. Uncertain systems with backlash: Analysis, identification and synthesis. Licenciate thesis, trita/mat-95-os3, Royal Institute of Technology, 1995.
15. M. Nordin, J. Galić, and P.O. Gutman. New models for backlash and gear play. *Int. J. of Adaptive Control and Signal Processing*, 1:9–63, 1997.
16. M. Nordin and P.-O. Gutman. A robust linear design of an uncertain two-mass system with backlash. In *Proc. of 1st IFAC Workshop in Automotive Control*, pages 183–188, Ascona, Schweiz, 1995. IFAC.
17. S. Oldak, C. Baril, and P. O. Gutman. Quantative design of a class of nonlinear systems with parameter uncertainty. *Int. J. of Robust and Nonlinear Control*, 4:101–117, 1994.
18. S. Oldak and P. O. Gutman. Self oscillating adaptive design of system with dry friction and significant parametric uncertainty. *Int. J. of Adaptive Control and Signal Processing*, 9:239–253, 1995.
19. J. Peschon. *Disciplines and Techniques of Systems Control*. Blanisdell, New York, 1965.
20. G. Tao and P. Kokotovic. Adaptive control of systems with backlash. *Automatica*, 29(2):323–335, 1993.
21. G. Tao and P. V. Kokotovic. Adaptive control of systems with unknown output backlash. In *Proc. of 32nd CDC*, pages 2635–2640, San Antonio, Texas, USA, 1993. IEEE.
22. G. Tao and P. V. Kokotovic. *Adaptive Control of Systems with Actuator and Sensor Nonlinearities*. John Wiley & sons, 1996.
23. A. Tustin. Effects of backlash and of speed dependent friction on the stability of closed-cycle control systems. *J. IEE*, 94(IIA):143–151, 1947.

Adaptive Dead Zone Inverses for Possibly Nonlinear Control Systems

Er-Wei Bai

Department of Electrical and Computer Engineering
The University of Iowa
Iowa City, IA 52242,
er-wei-bai@uiowa.edu

Abstract

In this chapter, an adaptive dead zone inverse is proposed for control of systems containing an unknown dead zone. Assuming that the output of the dead zone is measurable, it is shown in the chapter that the effect of the unknown dead zone on the closed loop system can be eliminated asymptotically with no assumptions on the persistent excitation of the input signal. The results do not rely on the linearity property of the system and therefore can be easily applied to some nonlinear systems.

1 Introduction

Some physical components have nonsmooth nonlinear characteristics referred to as the "hard" nonlinearities. The dead zone nonlinearity is a typical one for which the output is zero until the magnitude of the input exceeds a certain value, a phenomenon called dead zone nonlinearity [8]. The dead zone phenomena occur in various components of control systems including sensors, amplifiers and actuators, especially in valve-controlled pneumatic actuators, in hydraulic components and in electric servo motors. Dead zones have a number of possible effects on control systems and the most common effect is to decrease the control accuracy. They may also lead to limit cycles or system instability.

In this chapter, we study a novel approach to deal with an unknown dead zone, called the adaptive dead zone inverse (ADI). The idea is to cancel completely the effect of the dead zone so that linear system analysis and design can be applied. This approach was pioneered by Recker, Kokotovic, Rhode and Winkelman [5] and by Tao and Kokotovic [6,7]. In [5], an unknown dead zone is divided into several regions and each region is represented by a set of unknown parameters. Then, a Lyapunov-based adaptive dead zone inverse with switching logic is proposed. Under the assumption that each region can be identified on-line accurately, perfect cancellation of the dead zone can be achieved. If this assumption is violated, errors occur. This approach was further explored by Tao and Kokotovic in [6] in the setting of the model reference adaptive controls to include controls of unknown systems with an unknown dead zone. Boundness of error has been achieved. Moreover, simulations indicate that the tracking performance is greatly improved by using an adaptive dead zone inverse, though an analytical proof of the convergence still remains open.

This chapter is a continuation of the above research. In the chapter, asymptotical adaptive cancellation of an unknown dead zone is achieved. The results obtained in this chapter are stronger than those in the literature but at the price of assuming that both the input and output of the dead zone are available for measurement, a condition much stronger than that assumed in the literature [5,6]. Therefore, the results presented here are applicable only to systems in which the output of a dead zone is measurable. It should be pointed out that there does exist a large class of systems with dead zones whose outputs are available for measurement. For instance, in a pulsatile pressure stimulus apparatus [3], in order to investigate nerve activity under some physiologic conditions, the control input $u(t)$ to the blood vessel needs to be changed accordingly. The input $u(t)$ is the output of a flexible tube connected to a pressure bottle with an inlet and an outlet valve as controls. The characteristics from control valves to the output $u(t)$ are well modeled by a non-symmetric dead zone [3] and moreover, $u(t)$, the output of the dead zone is readily measurable by a pressure transducer. We would also like to point out that relaxing the measurability assumption on the dead zone does not trivialize the problem. Note that the idea of an adaptive dead zone inverse is built on the certainty equivalence principle. The unknown parameters of a dead zone are estimated on-line and a dead zone inverse is constructed as if these estimates were the true values. If these estimates are not the true values, an error would occur. Usually, to guarantee the convergence of the estimates, some kind of persistent excitation (PE) assumptions are required. This greatly limits the application of the adaptive dead zone inverse to the case where the command signal is not PE, a very common situation in practice. The key question is, without assuming the convergence of parameter estimates, can the performance of the adaptive dead zone inverse be guaranteed? In this chapter, a perfect adaptive dead zone inverse is obtained independently of the convergence of the parameter estimates. The only assumption required is the availability of either the upper bounds or the lower bounds on the unknown parameters of the dead zone.

Finally, we would like to point out that the results obtained in the chapter are based on our previous works for the linear systems [1]. The results are extended in this chapter to possibly nonlinear systems. The key result is that if the closed loop system in the absence of a dead zone is robust with respect to L_2 noises, then the closed loop system is stable in the presence of an unknown dead zone and its adaptive inverse. Note that many nonlinear systems have this property. The outline of the chapter is as follows: Section 2 formulates the problem. An adaptive dead zone inverse is proposed in Section 3 along with its convergence analysis. Its applications to linear and nonlinear systems are discussed in Section 4 and Section 5, respectively. Some concluding remarks are given in Section 6.

2 Problem Formulation

Let us consider Figure 1, where G represents a possibly unknown nonlinear system, $u(t)$ and $y(t)$ are the input and output of G respectively. $u(t)$ is also the output of the dead zone D with input $v(t)$.

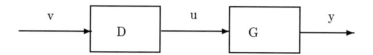

Fig. 1. System with a dead-zone.

The dead zone D is graphically described in Figure 2a for some unknown constants $0 < b_l$, b_r, m_r, $m_l < \infty$.

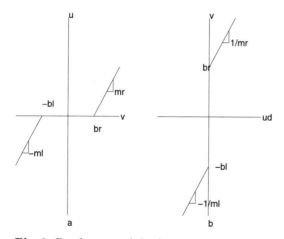

Fig. 2. Dead-zone and dead-zone inverse.

Mathematically, the dead zone can be described by

$$u(t) = \begin{cases} m_r v(t) - m_r b_r, & if\ v(t) > b_r\ or\ u(t) > 0 \\ 0, & if\ -b_l \leq v(t) \leq b_r\ or\ u(t) = 0 \\ m_l v(t) + m_l b_l, & if\ v(t) < -b_l\ or\ u(t) < 0. \end{cases} \quad (1)$$

Since $u(t)$ is assumed to be measurable, we define the vector

$$\phi(t) = \begin{cases} (v(t),\ sgn\ v(t),\ 0,\ 0)^T, & if\ u(t) > 0\ or\ v > b_r \\ (0,\ 0,\ 0,\ 0)^T, & if\ u(t) = 0\ or\ -b_l \le v \le b_r \\ (0,\ 0,\ v(t),\ sgn\ v(t))^T, & if\ u(t) < 0\ or\ v < -b_l \end{cases} \quad (2)$$

where sgn is the sign function, i.e., $sgn(x) = 1\ if\ x > 0$, $sgn(x) = -1\ if\ x < 0$ and $sgn(x) = 0\ if\ x = 0$. Define the true parameter vector

$$\theta^T = (m_r,\ -m_r b_r,\ m_l,\ -m_l b_l)$$

Then, the output $u(t)$ of the dead zone can be rewritten in a compact form as

$$u(t) = \theta^T \phi(t) \quad (3)$$

The objective of the dead zone inverse is to cancel the dead zone so that $u(t) \equiv u_d(t)$ for any $u_d(t)$ which is the desired input to the system G. Now let us suppose we want to apply a desired control signal $u_d(t)$ to the system G. If m_r, m_l, b_r, b_l are known, by constructing $v(t)$ as shown in Figure 2b, we can cancel the dead zone so that the output of the dead zone $u(t) \equiv u_d(t)$. Mathematically, the relation between $v(t)$ and $u_d(t)$, which specifies the dead zone inverse, is defined as follows:

$$v(t) = \begin{cases} \frac{u_d + m_r b_r}{m_r} = \frac{u_d}{m_r} + b_r, & if\ u_d > 0 \\ \frac{u_d - m_l b_l}{m_l} = \frac{u_d}{m_l} - b_l, & if\ u_d < 0 \\ 0, & if\ u_d = 0 \end{cases} \quad (4)$$

It can be easily verified that with this choice of $v(t)$, $u(t) \equiv u_d(t)$. The problem is that the parameters m_r, m_l, b_r, and b_l are unknown. To this end, we use the adaptive dead zone inverse based on the estimates to produce $v(t)$,

$$v(t) = \begin{cases} \frac{u_d + \widehat{m_r b_r}}{\widehat{m_r}} = \frac{u_d}{\widehat{m_r}} + \widehat{b_r}, & if\ u_d(t) > 0 \\ \frac{u_d - \widehat{m_l b_l}}{\widehat{m_l}} = \frac{u_d}{\widehat{m_l}} - \widehat{b_l}, & if\ u_d(t) < 0 \\ 0, & if\ u_d(t) = 0 \end{cases} \quad (5)$$

where $\widehat{m_r}$, $\widehat{m_l}$, $\widehat{m_r b_r}$ and $\widehat{m_l b_l}$ are the estimates of m_r, m_l, $m_r b_r$ and $m_l b_l$ respectively. And $\widehat{b_r} = \frac{\widehat{m_r b_r}}{\widehat{m_r}}$ and $\widehat{b_l} = \frac{\widehat{m_l b_l}}{\widehat{m_l}}$ are the estimates of b_r and b_l. What remains is that we need to find a parameter update law for the estimates $\widehat{m_r}$, $\widehat{m_l}$, $\widehat{m_r b_r}$ and $\widehat{m_l b_l}$ so that all signals are bounded and the error between the desired control signal $u_d(t)$ and the actual output of the dead zone $u(t)$

$$\Delta u(t) = u_d(t) - u(t) \quad (6)$$

converges to zero.

3 Adaptive Dead Zone Inverse

In this section, we will propose two parameter update laws and show their convergence. Let

$$\widehat{\theta}(t) = (\widehat{m_r}, -\widehat{m_r b_r}, \widehat{m_l}, -\widehat{m_l b_l})^T \tag{7}$$

be the estimates of $\theta = (m_r, -m_r b_r, m_l, -m_l b_l)^T$ and $\Delta u(t) = u_d(t) - u(t)$. Then we have the following theorem.

Theorem 3.1
Assume that $|u_d(t)| \le \beta e^{\alpha t}$ for some $\alpha, \beta > 0$. Consider the adaptive dead zone inverse (5) and the parameter update law

$$\dot{\widehat{\theta}}(t) = -\phi(t) \Delta u(t) \tag{8}$$

where $\phi(t)$ is defined in equation (2). Suppose the initial condition of $\widehat{\theta}(0)$ satisfies

$$\widehat{m_r}(0) \ge m_r, \ \widehat{m_l}(0) \ge m_l, \ \widehat{b}_r(0) = \frac{\widehat{m_r b_r}(0)}{\widehat{m_r}(0)} \ge b_r \tag{9}$$

$$\widehat{b}_l(0) = \frac{\widehat{m_l b_l}(0)}{\widehat{m_l}(0)} \ge b_l$$

Then, all the parameter estimates $\widehat{m_r}(t), \widehat{m_r b_r}(t), \widehat{m_l}(t), \widehat{m_l b_l}(t)$ are bounded. Moreover, $\Delta u(t) \in L_2$ and $|\Delta u(t)| = |u_d(t) - u(t)| \le k_1 + k_2 |u_d(t)|$ for some $k_1, k_2 \ge 0$.

Remark 3.1 In (8), $\phi(t)$ is used. In real implementations of (8), however, $\phi(t)$ can be replaced by $\phi_1(t)$(will be defined in (11)) that only depends on $u_d(t)$, an easily obtainable signal, because $\phi(t) \equiv \phi_1(t)$ under the condition of the theorem, see the proof of the theorem.

Remark 3.2 The assumption on the initial conditions are easily satisfied. Let $\bar{m}_r, \bar{m}_l, \bar{b}_r$, and $\bar{b}_l$ be the upper bounds on m_r, m_l, b_r and b_l respectively. Then, any initial conditions

$$\widehat{m_r}(0) \ge \bar{m}_r, \ \widehat{m_l}(0) \ge \bar{m}_l, \ \frac{\widehat{m_r b_r}(0)}{\widehat{m_r}(0)} \ge \bar{b}_r, \ \frac{\widehat{m_l b_l}(0)}{\widehat{m_l}(0)} \ge \bar{b}_l$$

satisfy the assumption. In essence, this assumption is equivalent to the availability of the upper bounds on the unknown parameters.

Proof of Theorem 3.1 First we show that the asserted results are obtained under the conditions $\widehat{m_r}(t) \ge m_r > 0, \widehat{m_l}(t) \ge m_l > 0, \widehat{b}_r(t) \ge b_r > 0$ and

$\widehat{b}_l(t) \geq b_l > 0$ for all $t \geq 0$. Then we show that the above conditions are guaranteed to be met if the initial estimates are chosen to be as in (9).

Suppose $\widehat{m}_r(t) \geq m_r > 0$, $\widehat{m}_l(t) \geq m_l > 0$, $\widehat{b}_r(t) \geq b_r > 0$ and $\widehat{b}_l(t) \geq b_l > 0$ for all $t \geq 0$. From equation (5), $u_d(t)$ may be expressed in terms of $v(t)$ and $sgn\ v(t)$, i.e.,

$$u_d(t) = \begin{cases} \widehat{m}_r(t)v(t) - \widehat{m_r b_r}(t) sgn\ v(t), & if\ u_d(t) > 0 \\ 0, & if\ u_d(t) = 0 \\ \widehat{m}_l(t)v(t) - \widehat{m_l b_l}(t) sgn\ v(t), & if\ u_d(t) < 0 \end{cases} \quad (10)$$

If we define the vector $\phi_1(t)$ as

$$\phi_1(t) = \begin{cases} (v(t),\ sgn\ v(t),\ 0,\ 0)^T, & if\ u_d(t) > 0 \\ (0,\ 0,\ 0,\ 0)^T, & if\ u_d(t) = 0 \\ (0,\ 0,\ v(t),\ sgn\ v(t))^T, & if\ u_d(t) < 0 \end{cases} \quad (11)$$

$u_d(t)$ can be written in a compact form as

$$u_d(t) = \widehat{\theta}^T(t) \phi_1(t)$$

where $\widehat{\theta}^T(t) = (\widehat{m}_r(t), -\widehat{m_r b_r}(t), \widehat{m}_l(t), -\widehat{m_l b_l}(t))$. Under the condition that $\widehat{m}_r(t) \geq m_r > 0$, $\widehat{m}_l(t) \geq m_l > 0$, $\widehat{b}_r(t) \geq b_r > 0$ and $\widehat{b}_l(t) \geq b_l > 0$, $sgn(u_d(t)) \equiv sgn(u(t))$, hence $\phi_1(t) \equiv \phi(t)$. With equation (3), we have

$$\Delta u(t) = u_d(t) - u(t) = \phi_1^T \widehat{\theta} - \phi^T \theta = \phi^T \widetilde{\theta}$$

This implies

$$\dot{\widetilde{\theta}}(t) = -\phi \phi^T \widetilde{\theta}(t) \quad (12)$$

where $\widetilde{\theta}(t) = \widehat{\theta}(t) - \theta$ denotes the parameter estimate error. It is easily verified that

$$\frac{d}{dt}(\frac{1}{2}\widetilde{\theta}^T \widetilde{\theta}) = -(\widetilde{\theta}^T \phi)^2 = -\Delta u^2$$

Thus, the parameter error $\widetilde{\theta}(t)$ is non-increasing. Together with the fact that $\widehat{m}_r(t) \geq m_r$ and $\widehat{m}_l(t) \geq m_l$, it follows that $|\Delta u| \leq k_1 + k_2|u_d|$. Moreover, $\int_0^\infty \Delta u^2 dt < \infty$ implies that $\Delta u(t) \in L_2$.

Now, we show that if $\widehat{\theta}(0)$ satisfies (9), then, $\widehat{m}_r(t) \geq m_r$, $\widehat{m}_l(t) \geq m_l$, $\widehat{b}_r(t) \geq b_r$ and $\widehat{b}_l(t) \geq b_l$ for all $t \geq 0$. To see that, let

$$\theta^T = (m_r,\ -m_r b_r,\ m_l,\ -m_l b_l) = (\theta_1, \theta_2, \theta_3, \theta_4),\ \widehat{\theta}^T(t) = (\widehat{\theta}_1, \widehat{\theta}_2, \widehat{\theta}_3, \widehat{\theta}_4)$$

Suppose the initial estimates are as given in (9). As long as $\widehat{m}_r(t) \geq m_r$, $\widehat{m}_l(t) \geq m_l$, $\widehat{b}_r(t) \geq b_r$ and $\widehat{b}_l(t) \geq b_l$, the update law (8) is identical to (12) and can be rewritten as

$$\dot{\widetilde{\theta}}(t) = -\begin{pmatrix} v^2 & |v| & 0 & 0 \\ |v| & 1 & 0 & 0 \\ 0 & 0 & 0 & 0 \\ 0 & 0 & 0 & 0 \end{pmatrix} \widetilde{\theta}(t), \quad if\ u(t) > 0$$

$$\dot{\hat{\theta}}(t) = -\begin{pmatrix} 0 & 0 & 0 & 0 \\ 0 & 0 & 0 & 0 \\ 0 & 0 & v^2 & |v| \\ 0 & 0 & |v| & 1 \end{pmatrix} \tilde{\theta}(t), \qquad if \ u(t) < 0$$

and

$$\dot{\hat{\theta}}(t) = 0, \qquad if \ u(t) = 0$$

Equivalently,

$$\begin{pmatrix} \dot{\hat{\theta}}_1(t) \\ \dot{\hat{\theta}}_2(t) \end{pmatrix} = -\begin{pmatrix} v^2 & |v| \\ |v| & 1 \end{pmatrix} \begin{pmatrix} \tilde{\theta}_1(t) \\ \tilde{\theta}_2(t) \end{pmatrix} \qquad and$$

$$\begin{pmatrix} \dot{\hat{\theta}}_3(t) \\ \dot{\hat{\theta}}_4(t) \end{pmatrix} = 0, \qquad if \ u(t) > 0 \tag{13}$$

$$\begin{pmatrix} \dot{\hat{\theta}}_1(t) \\ \dot{\hat{\theta}}_2(t) \end{pmatrix} = 0 \qquad and$$

$$\begin{pmatrix} \dot{\hat{\theta}}_3(t) \\ \dot{\hat{\theta}}_4(t) \end{pmatrix} = -\begin{pmatrix} v^2 & |v| \\ |v| & 1 \end{pmatrix} \begin{pmatrix} \tilde{\theta}_3(t) \\ \tilde{\theta}_4(t) \end{pmatrix}, \qquad if \ u(t) < 0 \tag{14}$$

And finally

$$\dot{\hat{\theta}}(t) = 0, \qquad if \ u(t) = 0 \tag{15}$$

Clearly, the parameter estimates $\hat{\theta}_1$ and $\hat{\theta}_2$ are decoupled from that of $\hat{\theta}_3$ and $\hat{\theta}_4$. So are $\hat{\theta}_3$ and $\hat{\theta}_4$. Therefore, we will show only that $\widehat{m_r}(t) \geq m_r$ and $\hat{b}_r(t) \geq b_r$ for all $t \geq 0$ if $\widehat{m_r}(0) \geq m_r$ and $\hat{b}_r(0) \geq b_r$. The verification for $\widehat{m_l}(t) \geq m_l$ and $\hat{b}_l(t) \geq b_l$ is symmetric. The proof is shown by contradictions. First, notice that at a finite time t, $v(t)$ is finite because $|u_d(t)| \leq \beta e^{\alpha t}$. So, if $\widehat{m_r}(t) > 0$, $\widehat{m_r b_r}(t)$ and $\widehat{m_r}(t)$ are finite at time t and therefore, solutions of $\widehat{mb_r}(t)$ and $\widehat{m_r}(t)$ are continuous at time t. Now, for the sake of contradictions, suppose $\widehat{m_r}(0) \geq m_r$ and $\hat{b}_r(0) \geq b_r$ but $\widehat{m_r}(t) < m_r$ and/or $\hat{b}_r(t) < b_r$ for some $t \geq 0$. By continuity argument, $\widehat{m_r}(t)$ and/or $\hat{b}_r(t)$ have to cross m_r and/or b_r at some time t_1. More precisely, there exists some time $t_1 \geq 0$ and some sufficiently small positive constant $\delta \geq 0$ such that one of the following scenarios occurs,

1. $\hat{b}_r(t_1) = b_r$, $\widehat{m_r}(t_1) = m_r$ and

   $\hat{b}_r(t_1 + \Delta t) < b_r$ or $\widehat{m_r}(t_1 + \Delta t) < m_r$ or both for all $0 < \Delta t < \delta$

2. $\hat{b}_r(t_1) > b_r$, $\widehat{m_r}(t_1) = m_r$ and

   $\hat{b}_r(t_1 + \Delta t) > b_r$, $\widehat{m_r}(t_1 + \Delta t) < m_r$ for all $0 < \Delta t < \delta$

3. $\widehat{b}_r(t_1) = b_r$, $\widehat{m_r}(t_1) > m_r$ and

$$\widehat{b}_r(t_1 + \Delta t) < b_r, \quad \widehat{m_r}(t_1 + \Delta t) > m_r \quad \text{for all } 0 < \Delta t < \delta$$

We will show contradictions in each case.

Case 1) If $\widehat{b}_r(t_1) = b_r$ and $\widehat{m_r}(t_1) = m_r$, then $\widetilde{\theta}_1(t_1) = \widehat{\theta}_1(t_1) - \theta_1 = 0$ and $\widetilde{\theta}_2(t_1) = -m_r \widehat{b}_r(t_1) + m_r b_r = 0$. Note that $\widetilde{\theta}_1(t_1) = 0$ and $\widetilde{\theta}_2(t_1) = 0$ is an equilibrium point of the parameter update law (8) or (12). That means $\widehat{b}_r(t) \equiv b_r$ and $\widehat{m_r}(t) \equiv m_r$ for all $t \geq t_1$. This contradicts the hypothesis.

Case 2) Let $\widehat{b}_r(t_1) > b_r$ and $\widehat{m_r}(t_1) = m_r$. If $u(t_1) > 0$, then $\dot{\widehat{m_r}}(t_1) > 0$, a contradiction. For $u(t_1) \leq 0$, there are two possibilities. $u(t) \leq 0$, $t \in [t_1, t_1+\delta]$ and $u(t)$ is not always less than or equal to zero for all $t \in [t_1, t_1+\delta]$. If $u(t) \leq 0$, $t \in [t_1, t_1 + \delta]$, then from (14)

$$\begin{pmatrix} \dot{\widehat{\theta}}_1(t) \\ \dot{\widehat{\theta}}_2(t) \end{pmatrix} \equiv 0, \quad t \in [t_1, t_1 + \delta]$$

and this implies $\widehat{m_r}(t) = m_r(t_1)$ and $\widehat{b}_r(t) = b_r(t_1)$, $t \in [t_1, t_1 + \delta]$. A contradiction occurs. If $u(t)$ is not always less than or equal to zero for all $t \in [t_1, t_1 + \delta]$, say, $u(t) \leq 0$, $t \in [t_1, \tau) \subset [t_1, t_1 + \delta]$ and $u(\tau) > 0$, then, $v(\tau) \neq 0$ and $\widehat{m_r}(t_1) = \widehat{m_r}(\tau) = m_r$, $\widehat{b}_r(t_1) = \widehat{b}_r(\tau) > b_r$. Note at time τ,

$$\begin{pmatrix} \dot{\widehat{\theta}}_1(\tau) \\ \dot{\widehat{\theta}}_2(\tau) \end{pmatrix} = -\begin{pmatrix} v^2 & |v| \\ |v| & 1 \end{pmatrix} \begin{pmatrix} \widehat{m_r}(\tau) - m_r \\ \widehat{\theta}_2(\tau) + m_r b_r \end{pmatrix}$$

$$= -\begin{pmatrix} v^2 & |v| \\ |v| & 1 \end{pmatrix} \begin{pmatrix} 0 \\ -\widehat{b}_r(\tau) + b_r \end{pmatrix} m_r$$

It follows that

$$\dot{\widehat{m_r}}(\tau) = \dot{\widehat{\theta}}_1(\tau) = |v|(\widehat{b}_r(\tau) - b_r)m_r > 0$$

This contradicts the hypothesis that $\widehat{m_r}(t) < m_r$, $t \in [t_1, t_1 + \delta]$.

Case 3) If $\widehat{b}_r(t_1) = b_r$ and $\widehat{m_r}(t_1) > m_r$, there are again two possibilities. $u(t) \leq 0$, $t \in [t_1, t_1 + \delta]$ and $u(t)$ is not always less than or equal to zero for all $t \in [t_1, t_1 + \delta]$. If $u(t) \leq 0$ for $t \in [t_1, t_1 + \delta]$, we have from (14)

$$\begin{pmatrix} \dot{\widehat{\theta}}_1(t) \\ \dot{\widehat{\theta}}_2(t) \end{pmatrix} \equiv 0, \quad t \in [t_1, t_1 + \delta]$$

and this implies $\widehat{m_r}(t) = \widehat{m_r}(t_1) > m_r$ and $\widehat{b}_r(t) = \widehat{b}_r(t_1) = b_r$. A contradiction occurs. If $u(t)$ is not always less that or equal to zero for all $t \in [t_1, t_1+\delta]$, say, $u(t) \leq 0$, $t \in [t_1, \tau) \subset [t_1, t_1 + \delta]$ and $u(\tau) > 0$, then, at time τ,

$$\begin{pmatrix} \dot{\widehat{\theta}}_1(\tau) \\ \dot{\widehat{\theta}}_2(\tau) \end{pmatrix} = -\begin{pmatrix} v^2 & |v| \\ |v| & 1 \end{pmatrix} \begin{pmatrix} \widehat{m_r}(\tau) - m_r \\ \widehat{\theta}_2(\tau) + m_r b_r \end{pmatrix}$$

$$= -\begin{pmatrix} v^2 & |v| \\ |v| & 1 \end{pmatrix} \begin{pmatrix} 1 \\ -b_r \end{pmatrix} (\widehat{m_r}(\tau) - m_r)$$

And this implies

$$\dot{\hat{b}}_r(\tau) = \frac{d}{dt}\left(-\frac{\widehat{\theta}_2(\tau)}{\widehat{\theta}_1(\tau)}\right) = \frac{-\dot{\widehat{\theta}}_2(\tau)\widehat{\theta}_1(\tau) + \dot{\widehat{\theta}}_1(\tau)\widehat{\theta}_2(\tau)}{\widehat{\theta}_1^{\,2}(\tau)}$$

$$= \frac{(|v| - b_r)\widehat{m_r}(\tau) + |v|(|v| - b_r)\widehat{m_r}(\tau)\widehat{b}_r(\tau)}{\widehat{m_r}^{\,2}(\tau)}(\widehat{m_r}(\tau) - m_r)$$

$u(\tau) > 0$ implies $|v| - b_r > 0$ and $\widehat{m_r}(\tau) = \widehat{m_r}(t_1) > m_r$. It follows that

$$\dot{\hat{b}}_r(\tau) > 0$$

This again contradicts the hypothesis. Combining all three cases together, we have $\widehat{m_r}(t) \geq m_r$, $\widehat{b}_r(t) \geq b_r$ for all $t \geq 0$ if $\widehat{m_r}(0) \geq m_r$ and $\widehat{b}_r(0) \geq b_r$. Similarly we have $\widehat{m_l}(t) \geq m_l$, $\widehat{b}_l(t) \geq b_l$ for all $t \geq 0$ if $\widehat{m_l}(0) \geq m_l$, and $\widehat{b}_l(0) \geq b_l$. This completes the proof.

Theorem 3.1 requires some upper bounds on the unknown parameters. The following results assume some lower bounds on these parameters. Let us first define two half spaces in R^4 and a projection operator. Suppose $\underline{m_r}$, $\underline{m_l}$, $\underline{b_r}$, $\underline{b_l}$ are some lower bounds on m_r, m_l, b_r, b_l, i.e., $0 < \underline{m_r} \leq m_r$, $0 < \underline{m_l} \leq m_l$, $0 < \underline{b_r} \leq b_r$, $0 < \underline{b_l} \leq b_l$. Define

$$Q_\alpha = \left\{ \begin{pmatrix} x_1 \\ x_2 \\ x_3 \\ x_4 \end{pmatrix} \in R^4;\; x_1 \geq \alpha \underline{m_r},\; x_2 \leq -\alpha \underline{m_r}\underline{b_r}, \right. \quad (16)$$

$$x_3 \geq \alpha \underline{m_l},\; x_4 \leq -\alpha \underline{m_l}\underline{b_l} \Big\}$$

for some $0 < \alpha < 1$ and

$$Q_1 = \left\{ \begin{pmatrix} x_1 \\ x_2 \\ x_3 \\ x_4 \end{pmatrix} \in R^4;\; x_1 \geq \underline{m_r},\; x_2 \leq -\underline{m_r}\underline{b_r},\; x_3 \geq \underline{m_l},\; x_4 \right. \quad (17)$$

$$\leq -\underline{m_l}\underline{b_l} \Big\}$$

Clearly,

$$\widehat{\theta}(t) = \begin{pmatrix} \widehat{m_r}(t) \\ -\widehat{m_r b_r}(t) \\ \widehat{m_l}(t) \\ -\widehat{m_l b_l}(t) \end{pmatrix} \in Q_\alpha$$

if and only if $\widehat{m_r}(t) \geq \alpha \underline{m_r} > 0$, $\widehat{m_l}(t) \geq \alpha \underline{m_l} > 0$, $\widehat{m_r b_r}(t) \geq \alpha \underline{m_r}\underline{b_r} > 0$, and $\widehat{m_l b_l}(t) \geq \alpha \underline{m_l}\underline{b_l} > 0$. Further, let us define a projection operator $Proj_{Q_1} \widehat{\theta}(t)$

on $\widehat{\theta}(t)$ as follows if $\widehat{\theta}(t)$ is inside Q_α, then $Proj_{Q_1}\widehat{\theta}(t) = \widehat{\theta}(t)$. And if $\widehat{\theta}(t)$ is outside Q_α, then $Proj_{Q_1}\widehat{\theta}(t)$ is the projection of $\widehat{\theta}(t)$ on Q_1. Mathematically,

$$Proj_{Q_1}\widehat{\theta}(t) = \begin{cases} \widehat{\theta}(t) & if\ \widehat{\theta}(t) \in Q_\alpha \\ \begin{pmatrix} max(\widehat{m_r}(t), \underline{m_r}) \\ min(-\widehat{m_rb_r}(t), -\underline{m_rb_r}) \\ max(\widehat{m_l}(t), \underline{m_l}) \\ min(-\widehat{m_lb_l}(t), -\underline{m_lb_l}) \end{pmatrix} & if\ \widehat{\theta}(t) \notin Q_\alpha \end{cases} \quad (18)$$

We see that $\widehat{\theta}(t) = Proj_{Q_1}\widehat{\theta}(t)$ guarantees all the estimates

$$\widehat{m_r}(t),\ \widehat{m_l}(t),\ \widehat{m_rb_r}(t),\ \widehat{m_lb_l}(t) > 0\ for\ all\ t$$

Theorem 3.2
Assume that $|u_d| \leq \beta e^{\alpha t}$ for some $\alpha,\ \beta > 0$. Also assume that some lower bounds $0 < \underline{m_r},\ \underline{m_l},\ \underline{b_r},\ \underline{b_l}$ on the unknowns $m_r,\ m_l,\ b_r,\ b_l$ are available. Consider the adaptive dead-zone inverse (5) with the following parameter update law.

$$\dot{\widehat{\theta}}(t) = -\phi_1(t)\Delta u(t),\quad \widehat{\theta}(0) \in Q_\alpha \qquad (19)$$

and

$$\widehat{\theta}(t) = Proj_{Q_1}\widehat{\theta}(t) \qquad (20)$$

where $\phi_1(t)$ is defined in (11). Then the parameter estimation error is bounded and $\Delta u \in L_2$.

Proof of Theorem 3.2 By projection, it follows that (1) $\widehat{m_r}(t),\ \widehat{m_l}(t),\ \widehat{m_rb_r}(t),$ and $\widehat{m_lb_l}(t)$ are guaranteed to be larger than 0. (2) $sgn\ u_d(t) = sgn\ v(t)$ and (3) $\Delta u(t)$ can be written as

$$\Delta u(t) = \begin{cases} \phi_1^T(t)\tilde{\theta}(t) & if\ v \geq b_r\ or\ v \leq -b_l \\ \phi_1^T(t)\widehat{\theta}(t) & if\ -b_l < v < b_r \end{cases} \qquad (21)$$

Thus,

$$\dot{\tilde{\theta}} = \dot{\widehat{\theta}} = \begin{cases} -\phi_1\phi_1^T\tilde{\theta} & if\ v \geq b_r\ or\ v \leq -b_l \\ -\phi_1\phi_1^T\widehat{\theta} & if\ -b_l < v < b_r \end{cases} \qquad (22)$$

Let $V = \frac{1}{2}\tilde{\theta}^T\tilde{\theta}$ then,

$$\dot{V} = \begin{cases} -(\phi_1^T\tilde{\theta})^2 = -\Delta u^2 & if\ v \geq b_r\ or\ v \leq -b_l \\ -(\Delta u)\phi_1^T\tilde{\theta} & if\ -b_l < v < b_r \end{cases} \qquad (23)$$

We now show that $-(\Delta u)\phi_1^T\tilde{\theta} \leq -(\Delta u)^2\ if\ -b_l < v < b_r$. To see that, we divide $-b_l < v < b_r$ into three cases, $u_d > 0,\ u_d < 0,$ and $u_d = 0$.

Case 1) $u_d > 0$. Note that $sgn\ u_d = sgn\ v$, $u = 0$, $u_d = \Delta u$, and $v > 0$. This implies $0 < v < b_r$.

$$\phi_1^T \tilde{\theta} = \widehat{m_r v} - \widehat{m_r b_r} + m_r b_r - m_r v = \Delta u + m_r(b_r - v) > \Delta u > 0$$

Case 2) $u_d < 0$. Note that in this case, $-b_l < v < 0$. Thus

$$\phi_1^T \tilde{\theta} = \widehat{m_l v} + \widehat{m_l b_l} - m_l b_l - m_l v = \Delta u - m_l(b_l + v) < \Delta u < 0$$

Case 3) $u_d = 0$. This implies $v(t) = 0$, and $\Delta u \phi_1^T \tilde{\theta} = 0 = \Delta u$.

In all three cases, we have $-(\Delta u)\phi_1^T \tilde{\theta} \leq -(\Delta u)^2$. Hence $\dot{V} \leq -\Delta u^2$. This shows that the parameter estimates error is bounded and $\Delta u \in L_2$.

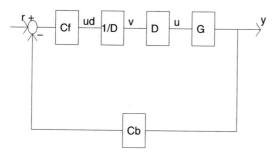

Fig. 3. Control system.

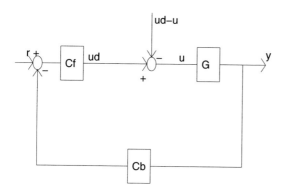

Fig. 4. Equivalent system.

Remark 3.3 The key for the update law (19) and (20) is that all parameter estimates are kept positive by projection which assures $sgn\ u_d(t) = sgn\ v(t)$.

In fact, projection can only happen a finite number of times for $0 < \alpha < 1$ simply because the initial error is bounded and each projection would reduce parameter estimation error by some fixed amount.

Remark 3.4 The projection (18) makes discontinuous changes in $\widehat{\theta}$ whenever $\widehat{\theta} \notin Q_\alpha$. If continuous projection is desirable, we can adopt the following projection scheme. First, let $\widehat{\theta}(0) \in Q_1$. Now, at any time t, if $\widehat{\theta}(t)$ is on the boundary of Q_1 and is to leave Q_1, then we assign the corresponding component of $\dot{\widehat{\theta}}(t)$ to zero. Since the parameter error does not increase, it is easily verified that the same results hold.

Remark 3.5 Theorems 3.1 and 3.2 show only that $\Delta u(t)$ is in L_2. $\Delta u(t)$ may not converge to zero as $t \to \infty$. We would like to make two comments here.

- $\Delta u(t)$ in L_2 is enough for the control of strictly proper systems. Notice that any bounded L_2 input to a stable and strictly proper system produces only an asymptotically decaying signal in the output. Therefore, $\Delta u(t)$ in L_2 is sufficient as far as the control of strictly proper systems is concerned. (For details, see section 4.)
- With some additional assumptions on $u_d(t)$, namely (1) $u_d(t)$ is piecewise continuous and the length of each piece does not go to zero and (2) $\dot{u}_d(t)$ is bounded, it can be shown that $\Delta u(t) \to 0$ as $t \to \infty$. The proof is straightforward. Therefore, if the control of non-strictly proper systems is concerned, the additional assumptions on $u_d(t)$ as mentioned before are needed.

Remark 3.6 The importance of Theorems 3.1 and 3.2 lies in that only the exponential boundedness assumption on $u_d(t)$ is required. Therefore, the adaptive dead zone inverse can be combined with almost any controller that is designed for the system G without considering the dead zone.

Remark 3.7 In this section, robustness issue was not discussed. If noise is present, the schemes proposed in this section can be robustified by many standard methods. This issue will be studied in the next section.

4 Adaptive Dead Zone Inverse for Linear Control Systems

4.1 With Known $G(s)$

Now, we will apply the ADI to linear control systems where $G(s)$ is a transfer function. First, we consider the case where the transfer function $G(s)$ is known. Since $G(s)$ is known, let $C_f(s)$ and $C_b(s)$ be the controllers designed for $G(s)$ such that the closed loop system is stable and certain performance requirements are satisfied, where $C_f(s)$ stands for the controller in the forward loop and $C_b(s)$ stands for the controller in the backward loop. In some cases, $C_f(s)$ or $C_b(s)$ can be chosen to be 1. For instance, $C_b(s) = 1$ in the unit feedback structure. Now, because the actual system is not $G(s)$ but $DG(s)$ with some unknown dead zone D as shown in Figure 1, direct application of the controller $C_f(s)$ and $C_b(s)$ to the system $DG(s)$ may not achieve the desired performance and may in fact lead to instability. To this end, we use the ADI in the control system as in Figure 3. From the results of the last section, Figure 3 is equivalent to Figure 4 with some $|\Delta u(t)| \leq k_1 + k_2|u_d(t)|$.

Thus, the output $y(t)$ can be written as

$$Y(s) = \frac{C_f(s)G(s)}{1 + C_f(s)C_b(s)G(s)} R(s) - \frac{G(s)}{1 + C_f(s)C_b(s)G(s)} \Delta U(s)$$

$$= Y^*(s) - \Delta Y(s)$$

The first term $Y^*(s)$ is the desired output in the absence of the dead zone. The second term $\Delta Y(s)$ is the total effect of the dead zone with the adaptive dead zone inverse.

Theorem 4.1

Let $G(s)$ be a strictly proper transfer function and let $C_f(s)$ and $C_b(s)$ be the controllers designed for $G(s)$ in the absence of the dead zone such that the closed loop system is stable. Now, consider the control system including the unknown dead zone D, the controllers $C_f(s)$ and $C_b(s)$ and the adaptive dead zone inverse D^{-1} proposed in either Theorem 3.1 or Theorem 3.2 as shown in Figure 3 or equivalently in Figure 4. Suppose $r(t)$ is bounded, then $\Delta y(t)$ is bounded and converges to zero as $t \to \infty$.

Proof. Consider Figure 4. Let $\Delta y(t)$ denote the error signal at the output due to $\Delta u = u_d - u$. By the assumption that the closed loop system is stable in the absence of D and D^{-1}, $\Delta y(t)$ is also in L_2. Moreover, the closed loop transfer function from Δu to Δy is strictly proper together with that $\Delta u \in L_2$, we have that $\Delta \dot{y}(t)$ is bound by the Schwarz inequality. Therefore, $\Delta y(t) \to 0$ as $t \to \infty$.

Remark 4.1 So far, it is assumed that $y(t)$ and $u(t)$ can be measured exactly. In reality, of course, noise always exists. In the presence of output noise, it can be seen obviously that it would only produce a small deviation at the output because the update on the inverse dead zone is decoupled from the output measurement and the closed system is exponentially stable. Not so obvious is when the measurement on $u(t)$ has some error that contributes to the adaptation of parameter estimate. The proposed scheme, however, can be easily made robust by using many standard methods as in [2] and [4], e.g., normalizing the signals, dead zone projection, use of persistent excitation signal etc. We only discuss one method, dead zone projection here to illustrate the idea. Let $w(t)$ be the measurement noise bounded by some constant $\epsilon > 0$, i.e., $|w(t)| < \epsilon$. Then, only $\Delta u_m \equiv \Delta u + w$ is available. Instead of (19), we use the following update law.

$$\dot{\hat{\theta}} = \begin{cases} -\phi_1 \Delta u_m & if\ |\Delta u_m| \geq 2\epsilon \\ 0 & otherwise \end{cases} \tag{24}$$

and

$$\hat{\theta}(0) \in Q_\alpha, \quad \hat{\theta}(t) = Proj_{Q_1}\hat{\theta}(t)$$

Theorem 4.2

Assume the noise $w(t)$ is bounded by some known positive constant ϵ. If we apply the update law (24), then

$$\Delta u(t) = e_{L_2}(t) + e_b(t) \tag{25}$$

where $e_{L_2}(t) \in L_2$ and $|e_b(t)| < 3\epsilon$ for all t.

Proof of Theorem 4.2

$\Delta u_m(t)$ can be written as

$$\Delta u_m(t) \equiv \Delta u(t) + w(t) = \begin{cases} \phi_1^T(t)\tilde{\theta}(t) + w(t) & if\ v \geq b_r\ or\ v \leq -b_l \\ \phi_1^T(t)\hat{\theta}(t) + w(t) & if\ -b_l < v < b_r \end{cases} \tag{26}$$

Thus, if $|\Delta u_m| \geq 2\epsilon$,

$$\dot{\tilde{\theta}} = \dot{\hat{\theta}} = \begin{cases} -\phi_1\phi_1^T\tilde{\theta} - \phi_1 w & if\ v \geq b_r\ or\ v \leq -b_l \\ -\phi_1\phi_1^T\hat{\theta} - \phi_1 w & if\ -b_l < v < b_r \end{cases} \tag{27}$$

and

$$\dot{V} = \begin{cases} -(\phi_1^T\tilde{\theta})^2 - \phi_1^T\tilde{\theta}w = -\Delta u^2 - w\Delta u & if\ v \geq b_r\ or\ v \leq -b_l \\ -(\Delta u)\phi_1^T\tilde{\theta} - \phi_1^T\tilde{\theta}w & if\ -b_l < v < b_r \end{cases} \tag{28}$$

where $V = \frac{1}{2}\tilde{\theta}^T\tilde{\theta}$ as before.
Now, note the following two facts.
(i) If $|\Delta u| \geq 3\epsilon$, then $|\Delta u_m| \geq 2\epsilon$.
(ii) Whenever parameter updating occurs $|\Delta u| > \epsilon$, then from (28), we see that the parameter error never increases.
Let us define

$$e(t) = \begin{cases} \Delta u(t) - 3\epsilon & if\ \Delta u(t) \geq 3\epsilon > 0 \\ 0 & if\ -3\epsilon < \Delta u < 3\epsilon \\ \Delta u(t) + 3\epsilon & if\ \Delta u(t) \leq -3\epsilon < 0 \end{cases} \tag{29}$$

Clearly,

$$|\Delta u(t)| \leq |e(t)| + 3\epsilon \quad for\ all\ t \tag{30}$$

In order to show the result, it suffices to show that $e(t) \in L_2$. We complete this by observing all the cases.

Case 1) $v \geq b_r$ or $v \leq -b_l$, and $\Delta u \geq 3\epsilon$
$$\dot{V} = -\Delta u^2 - \Delta u w = -(e+3\epsilon)^2 - (e+3\epsilon)w = -(e+3\epsilon)(e+3\epsilon+w) < -e^2$$

Case 2) $v \geq b_r$ or $v \leq -b_l$, and $\Delta u \leq -3\epsilon$
$$\dot{V} = -\Delta u^2 - \Delta u w = -(e-3\epsilon)^2 - (e-3\epsilon)w = -(e-3\epsilon)(e-3\epsilon+w) < -e^2$$

Case 3) $-b_l < v < b_r$, and $\Delta u \geq 3\epsilon$
$$\dot{V} = -(\Delta u)\phi_1^T \tilde{\theta} - \phi_1^T \tilde{\theta} w = -\phi_1^T \tilde{\theta}(e+3\epsilon+w) < -e^2$$

Note that $\phi_1^T \tilde{\theta} > \Delta u > 0$ as shown in the proof of Theorem 3.2.

Case 4) $-b_l < v < b_r$, and $\Delta u \leq -3\epsilon$
$$\dot{V} = -(\Delta u)\phi_1^T \tilde{\theta} - \phi_1^T \tilde{\theta} w = -\phi_1^T \tilde{\theta}(e-3\epsilon+w) < -e^2$$

Here, we used $\phi_1^T \tilde{\theta} < \Delta u < 0$ as shown in the proof of Theorem 3.2.
Since the parameter error never increases (even if $-3\epsilon < \Delta u < 3\epsilon$), it follows that $e(t) \in L_2$. Then, the conclusion follows.

From this theorem, we see that the measurement noise $w(t)$ on $u(t)$ contributes only a small bounded signal $e_b(t)$ that in turn produces a small output error. In other words, the update law (24) is robust with respect to the measurement noise.

4.2 With Unknown $G(s)$

If $G(s)$ is unknown, an adaptive controller can be used. The structure is the same as in Figure 3. Notice that C_f and C_b are now adaptive controllers that are in general, time varying and nonlinear. By the same arguments for the case of known $G(s)$, Figure 3 is equivalent to Figure 4 with some L_2 disturbance Δu. Thus, the total effect of the dead zone with the ADI is again an L_2 noise in the adaptive system. Obviously, if the adaptive controllers C_f and C_b designed for the unknown system $G(s)$ are robust with respect to L_2 noise, they will work in the presence of Δu. In particular, the effect of dead zone disappears asymptotically. A large class of adaptive control schemes is robust with respect to L_2 noise, for examples, schemes with σ-modification, dead zone modification, projection, and normalization.

5 Extension to Nonlinear Control Systems

We have presented two ADI schemes for linear control systems. By a closer look, however, we will find that the linearity assumption is not necessary. Note that the convergence results of Theorems 3.1 and 3.2 hold whether the control system is linear or not. What is necessary is that the closed loop linear or nonlinear system in the absence of dead zone is robust with respect to L_2 noises. Therefore, the proposed ADI schemes can easily be applied to many nonlinear control systems containing a dead zone with measurable outputs. To this end, consider a nonlinear system

$$\dot{x} = f(x, u, \theta_1), \ y = h(x, \theta_2) \tag{31}$$

with the controller

$$\begin{pmatrix} \dot{\hat{\theta}}_1 \\ \dot{\hat{\theta}}_2 \end{pmatrix} = p(u, y, r, \hat{\theta}_1, \hat{\theta}_2), \ u = g(y, r, \hat{\theta}_1, \hat{\theta}_2) \tag{32}$$

where r is a bounded input and f, h, p and g are nonlinear functions of proper dimension. It is assumed that the closed loop system is globally asymptotically stable. Moreover, it is assumed that the closed loop system is robust with respect to L_2 noise on the input u, i.e., for any L_2 noise Δu on the input u, all the signals are bounded and the deviation Δy in the output due to Δu goes to zero asymptotically. Then, exactly as in the linear cases, we have

Theorem 5.1
Let the nonlinear system in (31) be with an unknown dead zone

$$\dot{x} = f(x, u, \theta_1), \ y = h(x, \theta_2), \ u = D(v)$$

Consider the control law described in (32) together with an adaptive dead zone inverse D^{-1} proposed in either Theorem 3.1 or Theorem 3.2,

$$v = D^{-1}(u_d), \ u_d = g(y, r, \hat{\theta}_1, \hat{\theta}_2), \ \begin{pmatrix} \dot{\hat{\theta}}_1 \\ \dot{\hat{\theta}}_2 \end{pmatrix} = p(u, y, r, \hat{\theta}_1, \hat{\theta}_2)$$

Then, all the signals in the closed loop are bounded and the output error Δy due to the dead zone and adaptive dead zone inverse is bounded and converges to zero as $t \to \infty$.

To illustrate the idea, consider a scalar nonlinear system with an unknown dead zone

$$\dot{y} = y^2 + u$$

where u is the output of the dead zone $D(v)$ with the input v as shown in Figure 2a. The goal is to design a controller such that the output y follows the reference output y_m asymptotically, where the reference output is generated by

$$y_m = \frac{1}{s+1} r$$

for some bounded input signal r. In the absence of dead zone, $v = u$ and the control design can be done easily by choosing

$$u = v = -y^2 - y + r$$

This choice of v certainly does not work in the presence of dead zone. However, by the result of Theorem 5.1, we do not have to modify the design. Instead, we may simply add an ADI and this results in a closed loop system

$$\dot{y} = y^2 - y^2 - y + r + \Delta u = -y + r + \Delta u$$

for some $\Delta u \in L_2$.

Define the tracking error $\tilde{y} = y - y_m$. It follows that

$$\dot{\tilde{y}} = -\tilde{y} + \Delta u$$

Clearly, the error $\tilde{y}$ is bounded and moreover

$$|\tilde{y}(t)| \leq |e^{-t/2}\tilde{y}(t/2)| + \int_{t/2}^{t} |e^{-(t-\tau)} \Delta u| d\tau$$

$$\leq |e^{-t/2}\tilde{y}(t/2)| + [\int_{t/2}^{t} |e^{-(t-\tau)}|^2 d\tau]^{1/2} [\int_{t/2}^{t} |\Delta u|^2 d\tau]^{1/2}$$

$$\to 0 \quad as \ t \to \infty$$

for any bounded r. This result is actually a directly consequence of Theorem 5.1.

6 Concluding Remark

In the chapter, we did not discuss the robustness issue with model uncertainty in detail because our intention was to present to readers a simple and clear picture of the method of the adaptive dead zone inverse. In fact, many standard fixes for robustifying adaptive systems can be easily incorporated into our schemes. We also believe that the idea presented in the chapter can be used for other hard nonlinearities, including backlash and hysteresis.

References

1. H. Cho and E.W. Bai, Convergence results for an adaptive dead zone inverse, *Int. J. Adaptive Control and Signal Processing*, 451-466, 1998
2. P.A. Ioannou and K.S. Tsakalis, A robust direct adaptive controller, *IEEE Trans. on Automatic Control*, 1033-1043, 1986
3. J. Lu, Investigation of baroreceptor function using real time model based adaptive control and estimation, Ph.D. Thesis, University of Iowa, 1992
4. K.S. Narendra and A. M. Annaswamy, *Stable Adaptive Systems*, Prentice-Hall, 1989
5. D.A. Recker, P.V. Kokotovic, D. Rhode and J. Winkelman, Adaptive nonlinear control of systems containing a dead zone, *Proc. of CDC*, 2111-2115, 1991
6. G. Tao and P. V. Kokotovic, Adaptive control of plants with unknown dead zones, *Proc. of ACC*, 1992
7. G. Tao and P. V. Kokotovic, Adaptive control of systems with backlash, Tech. Report CCEC-910108, UCSB, 1991
8. J.G. Truxal, *Control Engineers' Handbook*, McGraw-Hill, New York, 1958

Deadzone Compensation in Motion Control Systems Using Augmented Multilayer Neural Networks

Rastko R. Šelmic[1] and Frank L. Lewis[2]

[1] Signalogic, Inc., Dallas, Texas, USA, Email: **rselmic@ieee.org**
[2] Automation and Robotics Research Institute, The University of Texas at Arlington, Fort Worth, Texas, USA, Email: **flewis@controls.uta.edu**

Abstract

A novel neural network (NN) structure is given for approximation of piecewise continuous functions of the sort that appear in friction, deadzone, backlash and other motion control actuator nonlinearities. The novel NN consists of neurons having standard sigmoid activation functions, plus some additional neurons having a special class of non-smooth activation functions termed 'jump approximation basis functions'. This augmented NN with additional neurons having 'jump approximation' activation functions can approximate any piecewise continuous function with discontinuities at a finite number of known points.

A compensation scheme is presented for general nonlinear actuator deadzones of unknown width. The compensator uses two NN, one to estimate the unknown deadzone and another to provide adaptive compensation in the feedforward path. The compensator is an augmented multilayer NN for approximating piecewise continuous functions like the deadzone inverse. Rigorous proofs of closed-loop stability for the deadzone compensator are provided, and yield tuning algorithms for the weights of the two NN. The technique provides a general procedure for using NN to determine the preinverse of an unknown right-invertible function.

1 Introduction

A general class of industrial motion control systems has the structure of a dynamical system, usually of the Lagrangian form, preceded by some *nonlinearities* in the actuator (deadzone, backlash, saturation), and actuator dynamics [9]. This includes xy-positioning tables [28], robot manipulators [22], overhead crane mechanisms, and more. The problems are particularly exacerbated when the required accuracy is high, as in micropositioning devices. Due to the nonanalytic nature of the actuator nonlinearities and the fact that their exact nonlinear functions are unknown, such systems present a challenge for the control design engineer. Proportional-derivative (PD) controllers have been observed to result in limit cycles if the actuators have deadzones. Standard techniques for overcoming deadzone are variable structure control [54], dithering [8], etc. Rigorous results for motion tracking of such systems are notably sparse, though ad hoc techniques relying on simulations for verification of effectiveness are prolific. A neural net scheme for deadzone compensation appears in [21], but no proof of performance is offered. Stability proofs and design of deadzone compensator for industrial positioning systems using a fuzzy logic controller is given in [26]. Fuzzy logic deadzone compensator is given in [14, 16, 17, 18, 19, 26]. The experimental evolutionary programming is used to obtain the fuzzy rules in [14].

Recently, in seminal work several rigorously derived adaptive schemes have been given [52]. Backlash compensation is considered in [46, 48, 50] and hystereses in [49]. Compensation for nonsymmetric deadzones is considered in [47, 51] for linear systems, in [36] for nonlinear systems in Brunovsky form with known nonlinear functions, and for unknown nonlinear canonical form systems in [53] where a backstepping approach is used. All of the known approaches in deadzone compensation assume that the deadzone function can be parametrized using a few parameters such as deadzone width, slope, etc. Standard adaptive techniques require that assumption.

NN have been used extensively in feedback control systems. Most applications are ad hoc with no demonstrations of stability. The stability proofs that do exist rely almost invariably on the universal approximation property for NN [3, 24, 25, 27, 34, 37, 38, 39]. However, in most real industrial control systems there are nonsmooth functions (piecewise continuous) for which NN approximation results in the literature are sparse. Examples include deadzone, friction, backlash, and so on. Though there do exist some results for piecewise continuous functions, it is found that attempts to approximate jump functions using smooth activation functions require many NN nodes and many training iterations, and still do not yield very good results.

Here a general model of the deadzone is assumed. Other actuator dynamics are neglected in this analysis. It is not required to be symmetric, and the function outside the dead-band may not be a linear function. Moreover, the proposed method can be applied for compensation of any *invertible, bounded, unknown, nonlinear function*. The generality of the method and its applicability to a broad range of nonlinear functions, make this approach a useful tool for compensation of deadzone and other static invertible nonlinearities.

The deadzone compensator consists of two NN, one used as an estimator of the nonlinear deadzone function, and the other used for the compensation itself. The NN used for deadzone compensation is a *modified multilayer perceptron* [41], capable of approximating the piecewise continuous functions of the sort that appear in deadzone, backlash, friction, and other motion control actuator nonlinearities. It is found that to approximate such functions suitably, it is necessary to augment the standard NN that uses smooth activation functions with extra nodes containing a certain jump function approximation basis set of nonsmooth activation functions.

2 Background

Let S be a compact simply connected set of $\Re^n$. With map $f: S \to \Re^m$, define $C(S)$ as the space such that f is continuous. The space of functions whose rth derivative is continuous is denoted by $C^r(S)$, and the space of smooth functions is $C^\infty(S)$.

By $\| \ \|$ is denoted any suitable vector norm. When it is required to be specific we denote the p-norm by $\| \ \|_p$. The supremum norm of $f(x)$, over S, is defined as [2]

$$\sup_{x \in S} \|f(x)\|, \ f: S \to \Re^m. \tag{2.1}$$

Given $A=[a_{ij}]$, $B \in \Re^{m \times n}$ the Frobenius norm is defined by

$$\|A\|_F^2 = tr(A^T A) = \sum_{i,j} a_{ij}^2, \tag{2.2}$$

with $tr(\ )$ the trace. The associated inner product is $<A,B>_F = tr(A^T B)$. The Frobenius norm is compatible with the two-norm so that $\|Ax\|_2 \leq \|A\|_F \|x\|_2$.

When $x(t) \in \Re^n$ is a function of time we use the standard L_p norms. It is said that $x(t)$ is bounded if its L_∞ norm is bounded. Matrix $A(t) \in \Re^{m \times n}$ is bounded if its induced matrix ∞-norm is bounded.

Consider the nonlinear system

$$\dot{x} = g(x,u,t), \ y = h(x,t) \tag{2.3}$$

with state $x(t) \in \Re^n$. The equilibrium point x_e is said to be uniformly ultimately bounded (UUB) if there exists a compact set $S \subset \Re^n$, so that for all $x_0 \in S$ there exists an $\varepsilon > 0$, and a number $T(\varepsilon, x_0)$ such that $\|x(t) - x_e\| \leq \varepsilon$ for all $t \geq t_0 + T$. That is, after a transition period T, the state $x(t)$ remains within the ball of radius ε around x_e.

2.1 Dynamics of Mechanical Motion Tracking Systems

The dynamics of mechanical systems with no vibratory modes can be written [22] as

$$M(\dot{q})\ddot{q} + V_m(q,\dot{q})\dot{q} + G(q) + F(\dot{q}) + \tau_d = \tau, \tag{2.4}$$

where $q(t) \in \Re^n$ is a vector describing position and orientation, $M(q)$ is the inertia matrix, $V_m(q,\dot{q})$ is the coriolis/centripetal matrix, $F(q,\dot{q})$ are the friction terms, $G(q)$ is the gravity vector, $\tau_d(t) \in \Re^n$ represents disturbances.

The dynamics (2.4) satisfy some important physical properties as a consequence of the fact that they are a Lagrangian system. These properties are important in control system design and are as follows:

Property 1. $M(q)$ is a positive definite symmetric matrix bounded by $m_1 I \leq M(q) \leq m_2 I$, where m_1, m_2 are known positive constants.

Property 2. The norm of the matrix $V_m(q,\dot{q})$ is bounded by $v_m(q)\|\dot{q}\|$ with $v_m(q)$ known.

Property 3. The matrix $\dot{M}-2V_m$ is skew-symmetric. This is equivalent to the fact that the internal forces do no work.

Property 4. The unknown disturbance satisfies $\|\tau_d\| \leq \tau_m$, with $\tau_m(t)$ a known positive constant.

To design a motion controller that causes the mechanical system to track a prescribed trajectory $q_d(t)$, define the *tracking error* by

$$e(t) = q_d(t) - q(t) , \qquad (2.5)$$

and the *filtered tracking error* by

$$r = \dot{e} + \Lambda e , \qquad (2.6)$$

where $\Lambda = \Lambda^T > 0$ is a design parameter matrix. Common usage is to select Λ diagonal with large positive entries. Then, (2.6) is a stable system so that $e(t)$ is bounded as long as the controller guarantees that the filtered error $r(t)$ is bounded. In fact,

$$\|e\| \leq \frac{\|r\|}{\sigma_{\min}(\Lambda)}, \|\dot{e}\| \leq \|r\| , \qquad (2.7)$$

with $\sigma_{\min}(\Lambda)$ the minimum singular value of Λ.

Differentiating (2.6) and invoking (2.4) it is seen that the robot dynamics are expressed in terms of the filtered error as

$$M\dot{r} = -V_m r - \tau + f(x) + \tau_d \qquad (2.8)$$

where the nonlinear robot function is

$$f(x) = M(\dot{q})(\ddot{q}_d + \Lambda\dot{e}) + V_m(q,\dot{q})(\dot{q}_d + \Lambda e) + G(q) + F(\dot{q}) . \qquad (2.9)$$

Vector x contains all the time signals needed to compute $f(x)$, and may be defined for instance as $x \equiv \begin{bmatrix} e^T & \dot{e}^T & q_d^T & \dot{q}_d^T & \ddot{q}_d^T \end{bmatrix}^T$. It is noted that the function $f(x)$ contains all the potentially unknown functions, except for M, V_m appearing in (2.8); these latter terms cancel out in the stability proof.

3 Neural Networks For Deadzone Compensation

NN have been used extensively in feedback control systems. Most applications are ad hoc with no demonstrations of stability. The stability proofs that do exist rely almost invariably on the universal approximation property for NN [3, 24, 25, 27, 34, 37, 38, 39]. However, in most real industrial control systems there are nonsmooth functions with jumps for which approximation results in the literature are sparse. Examples include friction, deadzone, backlash, and so on. Though there do exist some results for piecewise continuous functions, it is found that attempts to approximate jump functions using smooth activation functions require many NN nodes and many training iterations, and still do not yield very good results. In this section we review some background in NN and present a new result on NN approximation of piecewise continuous functions. It is found that to approximate such functions suitably, it is necessary to augment the standard NN that uses smooth activation functions with extra nodes containing a certain *jump function approximation basis set* of nonsmooth activation functions.

3.1 Background on Neural Networks

The two-layer NN in Fig. 3.1 consists of two layers of tunable weights. The hidden layer has L neurons, and the output layer has m neurons. The multilayer NN is a nonlinear mapping from input space $\Re^n$ into output space $\Re^m$.

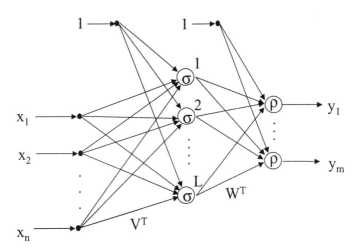

Fig. 3.1. Two-layer NN.

If x_j is the jth input to the network, and v_{jk} the weight between the jth input and kth hidden node, then the output y_i of the output layer is given by

$$y_k = \rho\left(\sum_{k=1}^{L} w_{ki}\sigma\left(\sum_{j=1}^{n} v_{ji}x_j + v_{0i}\right) + w_{0i}\right) \; ; \quad k=1, 2,..., m, \qquad (3.1)$$

where $\sigma(\cdot)$ is a hidden layer activation function, v_{jk} are the first-layer weights, v_{0k} are the bias weights for the first layer, $\rho(\cdot)$ is the output layer activation function, w_{ij} are the second-layer weights, and w_{0i} are the bias weights for the second layer. There are many different ways to choose the activation functions $\sigma(\cdot)$, including sigmoid, hyperbolic tangent, etc. We choose $\sigma(\cdot)$ to be sigmoid function. The output layer activation function $\rho(\cdot)$ is usually chosen to be a linear function. Define the weight matrices for the first and second layers as $V=[v_{ji}]$, $j=1, 2, ..., n$; $i=1, 2, ..., L$ and $W=[w_{ik}]$ $i=0, 1, 2, ..., L$; $k=1, 2, ..., m$. Define vectors $x=[x_1\ x_2\ ...\ x_n]$, $y=[y_1\ y_2\ ...\ y_m]$, one can write (3.1) in matrix form as

$$y = W^T \sigma(V^T x + v_0) , \qquad (3.2)$$

where the vector of the thresholds v_0 is defined as $v_0=[v_{01}\ v_{02}\ ...\ v_{0L}]^T$. In order to include the thresholds in the matrix W, the vector activation function is defined as $\sigma(w) = [1\ \sigma(w_1)\ \sigma(w_2)\ ...\ \sigma(w_L)]^T$, where $w \in \Re^L$. Tuning of the weights W then includes tuning of the thresholds too.

3.2 NN Approximation of Continuous Functions

Many well-known results say that any sufficiently smooth function can be approximated arbitrarily closely on a compact set using a two-layer NN with appropriate weights [7, 10, 13]. Given any function $f \in C(S)$, with S compact subset of $\Re^n$, and any $\varepsilon > 0$, there exists a sum $G(x)$ of the form

$$G(x) = \sum_{k=0}^{L} \alpha_k \sigma(m_k^T x + n_k) , \qquad (3.3)$$

for some $m_k \in \Re^n$, $n_k \in \Re$, $\alpha_k \in \Re$, such that

$$|G(x) - f(x)| < \varepsilon \qquad (3.4)$$

for all $x \in S$. Function $\sigma(\cdot)$ could be any continuous sigmoidal function [7]. This result shows that any continuous function can be approximated arbitrarily well using a linear combination of sigmoidal functions. This is known as the NN **universal approximation property**.

Combining (3.2), (3.3), and (3.4), one has

$$f(x) = W^T \sigma(V^T x + v_0) + \varepsilon(x) , \qquad (3.5)$$

where the $\varepsilon(x)$ is the NN approximation error. The reconstruction error is bounded on a compact set S by $\|\varepsilon(x)\| < \varepsilon_N$. Moreover, for any ε_N one can find a NN such that $\|\varepsilon(x)\| < \varepsilon_N$ for all $x \in S$.

The NN given by (3.5) has two layers of tunable weights. If the first-layer of NN weights V and v_0 are selected by some *a priori* method, then the NN has only one tunable layer of weights W. Without ambiguity, by one-layer NN we mean one tunable layer NN. One can define the fixed function $\phi(x) \equiv \sigma(V^T x)$, and then one layer NN has the form of

$$y = W^T \phi(x) . \qquad (3.6)$$

This type of NN is often called a *functional link neural network* (*FLNN*), [38]. The FLNN architecture appears to be identical to that of the conventional two-layer NN, except for the fundamental difference that only weights of the output layer or second layer are adjusted. The consequence is that FLNN is a linear in the parameters W, and therefore much easier for mathematical analysis. On the other side one should carefully choose the first-layer weights, depending on the application that FLNN is designed for. That usually involves some preprocessing or some *a priori* knowledge about the function that has to be approximated. If the first-layer weights are selected randomly, the approximation result holds [12] for such NN, with approximation error convergence to zero of order $O(C/\sqrt{L})$, where L is the number of the hidden layer nodes (basis functions), and C is independent of L. The approximating weights W are ideal target weights, and it is assumed that they are bounded such that $\|W\|_F \leq W_M$.

3.3 NN Approximation of Jump Functions

There are few results for approximation of nonsmooth functions [32, 33, 42]. Here we presented our main results for approximation of piecewise continuous functions or functions with jumps. It is found that to approximate such functions suitably, it is necessary to augment the set of functions $\sigma(\cdot)$ used for approximation. In addition to continuous sigmoidal functions, one requires a set of discontinuous basis functions. We propose two suitable sets- the 'polynomial jump approximation functions', and the 'sigmoidal jump approximation functions'.

Theorem 1 (Approximation of piecewise continuous functions)
Let there be given any bounded function $f: S \to \Re$ which is continuous and analytic on a compact set $S \subset \Re$, except at $x=c$ where the function f has a finite jump and is continuous from the right. Then, given any $\varepsilon > 0$, there exists a sum F of the form

$$F(x) = g(x) + \sum_{k=0}^{N} a_k f_k(x-c), \qquad (3.7)$$

such that

$$|F(x) - f(x)| < \varepsilon \qquad (3.8)$$

for every x in S, where g is a function in $C^\infty(S)$, and the **polynomial jump approximation basis functions** f_k are defined as

$$f_k(x) = \begin{cases} 0, & \text{for } x < 0 \\ \dfrac{x^k}{k!}, & \text{for } x \geq 0. \end{cases} \qquad (3.9)$$

Proof. See Appendix.

Remark. We use the terminology 'basis set' for (3.9) somewhat loosely. Note that $f_k(x-c)$ is a basis set for approximation of functions with discontinuities at a specified value $x=c$.

Corollary 1
If the function f is continuous from the left, then f_k is defined as

$$f_k(x) = \begin{cases} 0, & \text{for } x \leq 0 \\ \dfrac{x^k}{k!}, & \text{for } x > 0. \end{cases} \tag{3.10}$$

◇

It is desired now to replace the polynomial jump approximation basis functions f_k by another set of basis functions that are bounded, although both activation functions could be used. To accomplish this, some technical lemmas are needed.

Lemma 1

$$\sum_{i=0}^{m} \binom{m}{i}(-1)^i(-i)^k = 0 \tag{3.11}$$

for every $m > k$, where $\binom{m}{i}$ is defined as usual $\binom{m}{i} = \dfrac{m!}{i!(m-i)!}$.

Proof. See Appendix.

Lemma 2
Any linear combination of polynomial jump approximation functions f_k, $\sum_{k=0}^{N} a_k f_k(x)$, can be represented as

$$\sum_{k=0}^{N} a_k f_k(x) = z(x) + \sum_{k=0}^{N} b_k \varphi_k(x), \tag{3.12}$$

where the φ_k (**sigmoid jump approximation functions**) are defined as

$$\varphi_k(x) = \begin{cases} 0, & \text{for } x < 0 \\ (1 - e^{-x})^k, & \text{for } x \geq 0, \end{cases} \tag{3.13}$$

and where $z(x)$ is a function which belongs to $C^N(S)$. Coefficients b_k are given by

$$b_k = \dfrac{a_k - \sum_{m=0}^{k-1} b_m \sum_{i=0}^{m} \binom{m}{i}(-1)^i(-i)^k}{\sum_{i=0}^{k} \binom{k}{i}(-1)^i(-i)^k} \tag{3.14}$$

for $k=1, 2, \ldots, N$, and with $b_0 = a_0$.

Proof. See Appendix.

The next main result provides a set of bounded jump approximation basis functions that is extremely useful for feedback control purposes.

Theorem 2 (Approximation using sigmoid basis functions)
Let there be given any bounded function $f: S \to \Re$ which is continuous and analytic on a compact set S, except at $x=c$ where function f has a finite jump and is continuous from the right. Then, given any $\varepsilon > 0$, there exists a sum F of the form

$$F(x) = g(x) + \sum_{k=0}^{N} a_k \varphi_k(x-c), \qquad (3.15)$$

such that

$$|F(x) - f(x)| < \varepsilon \qquad (3.16)$$

for every x in S, where g is a function in $C^N(S)$, and the **sigmoid jump approximation basis functions** φ_k are defined as

$$\varphi_k(x) = \begin{cases} 0, & \text{for } x < 0 \\ (1-e^{-x})^k, & \text{for } x \geq 0 \end{cases}. \qquad (3.17)$$

Proof. See Appendix.

Theorem 2 says that any bounded function with a finite jump can be approximated arbitrarily well by a sum of a function with continuous first N derivatives, and a linear combination of jump basis functions φ_k. Functions φ_k for $k=0, 1, 2, 3$ are shown in Fig. 3.2.

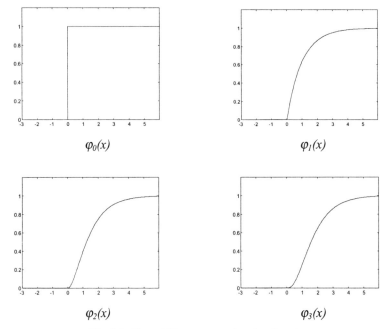

Fig. 3.2. Sigmoid jump approximation functions.

Using known results for continuous function approximation, one can formulate the following result for approximation of functions with a finite jump.

Theorem 3 (General approximation result)
Let there be given a bounded function $f: S \to \Re$ which is continuous and analytic on a compact set S, except at $x=c$ where function f has a finite jump and is continuous from the right. Given any $\varepsilon > 0$, there exists a sum F of the form

$$F(x) = \sum_{k=0}^{L} \alpha_k \sigma(m_k x + n_k) + \sum_{k=0}^{N} a_k \varphi_k (x-c), \qquad (3.18)$$

such that

$$|F(x) - f(x)| < \varepsilon \qquad (3.19)$$

for every x in S, where $\sigma(x)$ is a sigmoid function, e.g., (3.3), and the sigmoidal jump approximation functions φ_k are defined as (3.17).

Proof. Using Theorem 2 and applying existing results for continuous function approximation to function $g \in C^N(S)$ one obtains expression (3.18).
◇

Similarly as in the Theorem 2 it can be shown that instead of the sigmoidal jump approximation functions, one can use either

$$\varphi_k(x) = \begin{cases} 0, & \text{for } x < 0 \\ \left(\dfrac{1-e^{-x}}{1+e^{-x}}\right)^k, & \text{for } x \geq 0, \end{cases} \quad (3.20)$$

or jump basis functions based on the hyperbolic tangent

$$\varphi_k(x) = \begin{cases} 0, & \text{for } x < 0 \\ \left(\dfrac{e^x - e^{-x}}{e^x + e^{-x}}\right)^k, & \text{for } x \geq 0. \end{cases} \quad (3.21)$$

3.4 Structure of Augmented Multilayer NN

Here we present the structure and training algorithms for an augmented multilayer NN that is capable of approximating functions with jumps, provided that the points of discontinuity are known. Since the points of discontinuity are known in many nonlinear characteristics in industrial motion systems, this augmented NN is a useful tool for compensation of parasitic effects and actuator nonlinearities in industrial control systems. There are many hard nonlinearities, for which models are discontinuous with the points of discontinuity known. Examples include deadzone inverse that has a finite jump at zero, backlash that is discontinuous with respect to the velocity, etc., see [52].

The augmented multilayer NN shown in Fig. 3.3 is capable of approximating any piecewise continuous function $f: S \subset \Re \to \Re$, provided that the points of discontinuity are known.

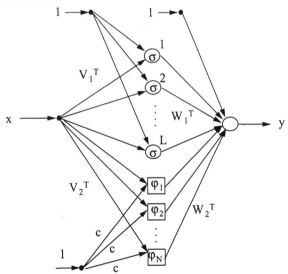

Fig. 3.3. Augmented multilayer NN.

Comparing with the standard NN given in Fig. 3.1, one can see that the augmented NN has two sets of hidden activation functions, σ and φ; two sets of weights for the first layer, V_1^T and V_2^T; and two set of weights for the second layer, W_1^T and W_2^T. With this structure of NN, one has

$$y = W_1^T \sigma(V_1^T x) + W_2^T \varphi(V_2^T x) \ . \tag{3.22}$$

If the hidden layer activation functions standard sigmoid functions, and one takes the jump basis functions φ_i in (3.17), then using Theorem 3 for approximation of jump functions one can see that the augmented NN is capable of approximating any continuous function with a finite jump, provided that the jump location is known.

The standard NN tuning algorithms must be modified since the structure of the NN is changed. The set of second-layer weights, W_1^T and W_2^T are trained as in usual multilayer NN. Both of them are treated the same way, including threshold weights.

3.5 Simulation of the Augmented Multilayer NN

Here we present some simulation results that compare the approximation capabilities of the standard NN in Fig. 3.1, and our augmented NN in Fig. 3.3. Both NN have $L=20$ hidden nodes, both are trained for the same number of iterations, and the augmented NN has two additional jump approximation nodes. Two samples of jump functions are chosen. Both NN are first trained to approximate the discontinuous function defined as $y=sin(x)$ for $x<-1$, else $y=1+sin(x)$. The results are shown in Fig. 3.4 (a) and (b). Next, the function defined as $y=x$ for $x<0$, else $y=0.5x+1$, was approximated. The results are shown in Fig. 3.5 (a) and (b).

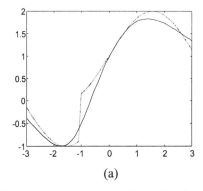
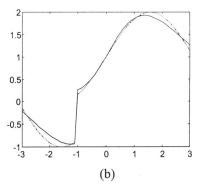

(a) (b)

Fig. 3.4. Approximation of jump functions using standard NN (a), and using augmented NN (b); NN function (full), desired function (dash).

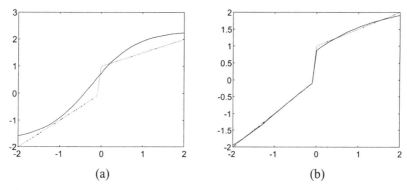

Fig. 3.5. Approximation of jump functions using standard NN (a), and using augmented NN (b); NN function (full), desired function (dash).

4 Compensation of Deadzone Nonlinearity

In this section a NN precompensator for a general model of the deadzone is given. It is not required to be symmetric, and the function outside the dead-band may not be a linear function. The proposed method can be applied for compensation of any *pre-invertible, bounded, unknown, nonlinear function*.

4.1 Deadzone Nonlinearity

Fig. 4.1 shows a nonsymmetric deadzone nonlinearity $D(u)$ where u and τ are scalars. In general, u and τ are vectors. It is assumed that the deadzone has a nonlinear form, which is more general than in [46, 47, 49–51, 53].

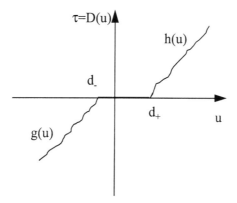

Fig. 4.1. Nonsymmetric deadzone nonlinearity.

A mathematical model for the deadzone characteristic of Fig. 4.1 is given by

$$\tau = D(u) = \begin{cases} g(u) < 0, & u \leq d_- \\ 0, & -d_- < u < d_+ \\ h(u) > 0, & u \geq d_+ \end{cases}. \qquad (4.1)$$

Functions $h(u)$, $g(u)$ are smooth, nonlinear functions, so this describes a very general class of $D(u)$. All of $h(u)$, $g(u)$, d_+, and d_- are unknown, so that compensation is difficult.

Assumption 1. The functions $h(u)$ and $g(u)$ are smooth and invertible continuous functions.

◇

Most application schemes cover only the case of symmetric deadzones, where $d_+ = d_-$, or the case where functions $h(u)$ and $g(u)$ are linear. These assumptions are required because of the limitations of standard adaptive control techniques. In our approach, the general deadzone model is analyzed, and Assumption 1 requires only that the deadzone nonlinearities be invertible.

4.2 NN Deadzone Precompensator

To offset the deleterious effects of deadzone, one may place a precompensator as illustrated in Fig. 4.3. There, the desired function of the precompensator is to cause the composite throughput from w to τ to be unity. In order to accomplish this, it is necessary to generate the inverse of the deadzone nonlinearity [36, 51]. By assumption, the function (4.1) is invertible, therefore there exist $D^{-1}(w)$, such that

$$D(D^{-1}(w)) = w. \qquad (4.2)$$

The function $D^{-1}(w)$ is shown in Fig. 4.2. This type of pre-compensation is also called pre-load.

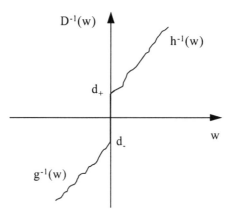

Fig. 4.2. Deadzone inverse.

The mathematical model for the function shown in Fig. 4.2 is given by

$$D^{-1}(w) = \begin{cases} g^{-1}(w), & w < 0 \\ 0, & w = 0 \\ h^{-1}(w), & w > 0 \end{cases} \quad (4.3)$$

The deadzone inverse $D^{-1}(w)$ can be expressed in equivalent form as

$$D^{-1}(w) = w + w_{NN}(w), \quad (4.4)$$

where the *modified deadzone inverse* w_{NN} is given by

$$w_{NN}(w) = \begin{cases} g^{-1}(w) - w, & w < 0 \\ 0, & w = 0 \\ h^{-1}(w) - w, & w > 0 \end{cases} \quad (4.5)$$

Equation (4.4) is a direct feedforward term plus a correction term. Function $w_{NN}(w)$ is discontinuous at zero, as is $u(w)$.

Based on the NN approximation property, one can approximate the deadzone function by

$$\tau = D(u) = W^T \sigma(V^T u + v_0) + \varepsilon(u). \quad (4.6)$$

Using the modified NN with activation functions shown in section 3, one can design a NN for the approximation of the modified inverse function given in (4.5) by

$$w_{NN}(w) = W_i^T \sigma(V_i^T w + v_{0i}) + \varepsilon_i(u). \quad (4.7)$$

In these equations $\varepsilon(u)$, $\varepsilon_i(w)$ are the NN reconstruction error and W, W_i, are ideal target weights. The reconstruction error is bounded by $\|\varepsilon\| < \varepsilon_N(x)$, $\|\varepsilon_i\| < \varepsilon_{Ni}(x)$, where x is equal to u and w respectively. We consider here x restricted to a compact set, and in that case these bounds are constant, i.e., $\varepsilon_N(x) = \varepsilon_N$, $\varepsilon_{Ni}(x) = \varepsilon_{Ni}$. The case where these bounds are not constant and not restricted to the compact set is treated in [23], and it that case an additional saturation term has to be added to the robustifying signal. The first-layer weights V, V_i, v_0, v_{0i} in both (4.6), and (4.7) are fixed, and if they are properly chosen, the approximation property of the NN is still valid [12].

Note we use two NN. The structure of the NN deadzone precompensator and deadzone estimator are shown in Fig. 4.3. The first NN is used as a deadzone estimator, while the second is used as a deadzone compensator. Note that only the output of the NN II directly affects the input u, while the NN I is a higher level 'performance evaluator', and is used for tuning the NN II.

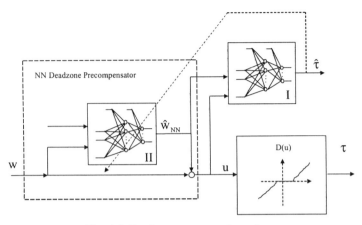

Fig. 4.3. Deadzone compensation scheme.

Define $\hat{W}$, $\hat{W}_i$ as estimates of the ideal NN weights, which are given by the NN tuning algorithms. Define the weight estimation errors as

$$\tilde{W} = W - \hat{W}, \quad \tilde{W}_i = W_i - \hat{W}_i, \tag{4.8}$$

and the approximations of the nonlinear deadzone and modified deadzone inverse functions as

$$\hat{\tau} = \hat{D}(u) = \hat{W}^T \sigma(V^T u + v_0), \tag{4.9}$$

$$\hat{w}_{NN}(w) = \hat{W}_i^T \sigma(V_i^T w + v_{0i}). \tag{4.10}$$

Note that expressions (4.9) and (4.10) represent respectively a NN approximation of the deadzone function (4.1) and of the modified deadzone inverse (4.5). Signal $\hat{w}_{NN}(w)$ is used for deadzone compensation, and $\hat{\tau}$ represents the estimated value of signal τ. Note that

$$u = w + \hat{w}_{NN}(w). \tag{4.11}$$

Assumption 2 (Bounded ideal NN weights). The ideal weights W, W_i are bounded such that $\|W\|_F \leq W_M$, $\|W_i\|_F \leq W_{iM}$, with W_M, and W_{iM} known bounds.

◇

The next result shows the effectiveness of the proposed NN structure, by providing an expression for the *composite throughput error* of the compensator plus deadzone. It shows that, as the estimates $\hat{W}$, $\hat{W}_i$ approach the actual neural network parameters W, W_i, the NN precompensator effectively provides a *preinverse* for the deadzone nonlinearity.

Theorem 4 (Throughput error using NN deadzone compensation)
Given the NN deadzone compensator (4.10), (4.11), and the NN observer (4.9), the throughput of the compensator plus the deadzone is given by

$$\tau = w - \hat{W}^T \sigma'(V^T u + v_0) V^T \tilde{W}_i^T \sigma_i (V_i^T w + v_{0i}) + \tilde{W}^T \sigma'(V^T u + v_0) V^T \hat{w}_{NN} + d(t) \quad (4.12)$$

where the modeling mismatch term $d(t)$ is given by

$$\begin{aligned}
d(t) = &-\tilde{W}^T \sigma'(V^T u + v_0) V^T W_i^T \sigma_i (V_i^T w + v_{0i}) \\
&- W^T \sigma'\left[V^T (w + \hat{W}_i^T \sigma_i (V_i^T w + v_{0i})) + v_0\right] V^T \varepsilon_i(w) \\
&- W^T R_1(\tilde{W}_i, w) - \varepsilon(w + w_{NN}) + \varepsilon(u),
\end{aligned} \quad (4.13)$$

and $R_1(\tilde{W}_i, w)$ is the remainder of the first Taylor polynomial.

Proof. See [43].

The form of (4.12) is crucial in Section 5 in deriving the NN tuning laws that guarantee closed-loop stability. The first term has known factors multiplying $\tilde{W}_i$, the second term has known factors multiplying $\tilde{W}$, and a suitable bound can be found for $d(t)$. The form of (4.12) is similar to the form in [47], but instead of parametrizing the direct and the inverse functions with the same parameters, we use different function approximators for the direct and inverse functions together with crucial fact (4.2), which is actually a connection between them. The expression (4.2) couples the information inherent in NN I and NN II. Intuitively, compensating the unknown effect (NN II) depends on what one observes (NN I), and vice versa; observation of the unknown effect (NN I) depends on how one modifies the system (NN II). This will later, in Section 5, be very clearly seen in the tuning laws derived for $\tilde{W}_i$, and $\tilde{W}$, where the differential equations for tuning NN I and NN II are mutually coupled.

In general, this proposed NN compensation scheme could be used for inverting **any** continuous invertible function. Therefore it is a powerful result for compensation of general actuator nonlinearities in motion control systems.

The next result gives us the upper bound of the norm of $d(t)$. It is an important result used in the stability proof.

Lemma 3
The norm of the modeling mismatching term $d(t)$ in (4.12) is bounded on a compact set by

$$\|d(t)\| \le a_1 \|\tilde{W}\|_F + a_2 \|\tilde{W}_i\|_F^2 + a_3 \|\tilde{W}_i\|_F + a_5, \quad (4.14)$$

where a_1, a_2, a_3, a_5 are computable constants.

Proof. See [43].

5 NN Controller With Deadzone Compensation

Deadzone compensation is considered in [36] for nonlinear systems in Brunovsky form with known nonlinear functions, and for linear systems in [47, 51]. In this section we show how to provide NN deadzone compensation for deadzones in mechanical systems, including robotic systems, with inexactly known nonlinearities. The deadzone precompensator given by [26, 28] used fuzzy logic. Static deadzone is assumed without any additional dynamics. Actuator dynamics that follow deadzone nonlinearity could be included as part of the unknown mechanical system. Any dynamics that precede the deadzone nonlinearity should be separately considered and compensator would include dynamic inversion [15].

In this section it is shown how to tune or learn the weights of the NN in (4.9), (4.10) on-line so that the tracking error is guaranteed small and all internal states are bounded. It is assumed, of course, that the actuator output $\tau(t)$ is not measurable.

5.1 Tracking Controller with NN Deadzone Compensation

If $f(x)$ in (2.9) is unknown, it can be estimated using adaptive control techniques [6, 22, 44, 45] or the neural network controller in [24].

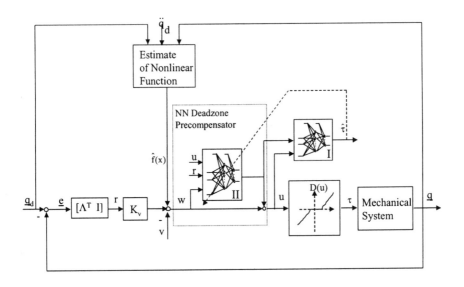

Fig. 5.1. Tracking controller with NN deadzone compensation.

A robust compensation scheme for unknown terms in $f(x)$ is provided by selecting the tracking controller

$$w = \hat{f}(x) + K_V r - v , \qquad (5.1)$$

with $\hat{f}(x)$ an estimate for the nonlinear terms $f(x)$ and $v(t)$ a robustifying term to be selected for the disturbance rejection. The estimate $\hat{f}(x)$ is fixed and will not be adapted, as is common in robust control techniques [5, 22]. The feedback gain matrix $K_V>0$ is often selected diagonal. Deadzone compensation is provided using

$$u = w + \hat{w}_{NN} = w + \hat{W}_i^T \sigma_i (V_i^T w + v_{0i}) . \qquad (5.2)$$

The multi-loop control structure implied by this scheme is shown in Fig. 5.1, where $\underline{q} = [q^T \, \dot{q}^T]^T$, $\underline{q}_d = [q_d^T \, \dot{q}_d^T]^T$, $\underline{e} = [e^T \, \dot{e}^T]^T$. The controller has a proportional-derivative (PD) tracking loop with gains $K_V r = K_V \dot{e} + K_V \Lambda e$, where the deadzone effect is ameliorated by the NN feedforward compensator. The estimate $\hat{f}(x)$ is computed by an inner nonlinear control loop.

In order to design a NN system such that the tracking error $r(t)$ is bounded and all internal states are stable, one must examine the error dynamics. Substituting (5.1) and (4.12) into (2.8) yields the *closed-loop error dynamics*

$$\begin{aligned}M\dot{r} = &-V_m r - K_V r + \hat{W}^T \sigma'(V^T u + v_0) V^T \tilde{W}_i^T \sigma_i (V_i^T w + v_{0i}) \\ &- \tilde{W}^T \sigma'(V^T u + v_0) V^T \hat{w}_{NN} - d(t) + \tilde{f} + \tau_d + v,\end{aligned} \qquad (5.3)$$

where the nonlinear function estimation error is given by $\tilde{f}(x) = f(x) - \hat{f}(x)$.

Assumption 3 (Bounded estimation error) The nonlinear function $f(x)$ is assumed to be unknown, but a fixed estimate $\hat{f}(x)$ is assumed known such that the functional estimation error $\tilde{f}(x)$ satisfies

$$\|\tilde{f}\| \le f_M(x) , \qquad (5.4)$$

for some known bounding function $f_M(x)$.

◇

This is not unreasonable [5, 22, 40] as in practical systems the bound $f_M(x)$ can be computed knowing the upper bound on payload masses, frictional effects, and so on.

Assumption 4 (Bounded Reference Trajectory). The desired trajectory is bounded so that

$$\begin{Vmatrix} q_d(t) \\ \dot{q}_d(t) \\ \ddot{q}_d(t) \end{Vmatrix} \le q_B , \qquad (5.5)$$

with q_B a known scalar bound.

◇

The next theorem provides algorithms for tuning the NN weights for the deadzone precompensator with guaranteed closed-loop stability.

Theorem 5 (Tuning of NN adaptive deadzone compensator)
Given the system in (2.8) and Assumptions 1–3, select the tracking control law (5.1), plus the deadzone compensator (5.2). Choose the robustifying signal as

$$v(t) = -(f_M(x) + \tau_M)\frac{r}{\|r\|}, \qquad (5.6)$$

where the $f_M(x)$ and τ_M are bounds on functional estimation error and disturbance respectively. Let the estimated NN weights be provided by the NN tuning algorithm

$$\dot{\hat{W}} = -S\sigma'(V^T u + v_0)V^T \hat{W}^T \sigma_i(V_i^T w + v_{0i})r^T - k_1 S\|r\|\hat{W}, \qquad (5.7)$$

$$\dot{\hat{W}}_i = T\sigma_i(V_i^T w + v_{0i})r^T \hat{W}^T \sigma'(V^T u + v_0)V^T - k_1 T\|r\|\hat{W}_i - k_2 T\|r\|\|\hat{W}_i\|_F \hat{W}_i, \qquad (5.8)$$

with any constant matrices $S=S^T>0$, $T=T^T>0$, and k_1, $k_2>0$ small scalar design parameters. Then the filtered tracking error $r(t)$ and NN weight estimates $\hat{W}$, $\hat{W}_i$ are UUB, with bounds given by (5.28), (5.29), (5.30). Moreover, the tracking error may be kept as small as desired by increasing the gains K_v.

Proof. Select the Lyapunov function candidate

$$V = \frac{1}{2}r^T Mr + \frac{1}{2}tr[\tilde{W}^T S^{-1}\tilde{W}] + \frac{1}{2}tr[\tilde{W}_i^T T^{-1}\tilde{W}_i]. \qquad (5.9)$$

Differentiating V yields

$$\dot{V} = r^T M\dot{r} + \frac{1}{2}r^T \dot{M}r + tr[\tilde{W}^T S^{-1}\dot{\tilde{W}}] + tr[\tilde{W}_i^T T^{-1}\dot{\tilde{W}}_i], \qquad (5.10)$$

and using (5.3) yields

$$\dot{V} = -r^T V_m r - r^T K_V r + r^T \hat{W}^T \sigma'(V^T u + v_0)V^T \tilde{W}_i^T \sigma_i(V_i^T w + v_{0i})$$
$$- r^T \tilde{W}^T \sigma'(V^T u + v_0)V^T \hat{w}_{NN} - r^T d(t) + r^T \tilde{f} + r^T \tau_d + r^T v \qquad (5.11)$$
$$+ \frac{1}{2}r^T \dot{M}r + tr[\tilde{W}^T S^{-1}\dot{\tilde{W}}] + tr[\tilde{W}_i^T T^{-1}\dot{\tilde{W}}_i].$$

Applying *Property 3* and the tuning rules one has

$$\dot{V} = -r^T K_V r + tr[\tilde{W}^T(S^{-1}\dot{\tilde{W}}^T - \sigma'(V^T u + v_0)V^T \hat{w}_{NN} r^T)]$$
$$+ tr[\tilde{W}_i^T(T^{-1}\dot{\tilde{W}}_i^T + \sigma_i(V_i^T w + v_{0i})r^T \hat{W}^T \sigma'(V^T u + v_0)V^T)] \qquad (5.12)$$
$$+ r^T(v - d(t) + \tilde{f} + \tau_d),$$

$$\dot{V} = -r^T K_V r + k_1 \|r\| tr[\tilde{W}^T(W - \tilde{W})]$$
$$+ \|r\| tr[\tilde{W}_i^T k_1(W_i - \tilde{W}_i) + \tilde{W}_i^T k_2\|\hat{W}_i\|_F(W_i - \tilde{W}_i)] - r^T d(t) + r^T(v + \tilde{f} + \tau_d). \qquad (5.13)$$

Using the inequality
$$tr[\tilde{X}^T(X-\tilde{X})] \le \|\tilde{X}\|_F \|X\|_F - \|\tilde{X}\|_F^2 , \tag{5.14}$$
and Lemma 3, the expression (5.13) can be modified as

$$\dot{V} \le -K_{V\min}\|r\|^2 + k_1\|r\|\|\tilde{W}\|_F\left(W_M - \|\tilde{W}\|_F\right)$$
$$+ k_1\|r\|\|\tilde{W}_i\|_F\left(W_{iM} - \|\tilde{W}_i\|_F\right) + k_2\|r\|\|\tilde{W}_i\|_F\|W_i - \tilde{W}_i\|_F\left(W_{iM} - \|\tilde{W}_i\|_F\right) \tag{5.15}$$
$$+ \|r\|\left(a_1\|\tilde{W}\|_F + a_2\|\tilde{W}_i\|_F^2 + a_3\|\tilde{W}_i\|_F + a_5\right) - \|r\|(f_M + \tau_M) + \|r\|\|\tilde{f} + \tau_d\|,$$

$$\le -K_{V\min}\|r\|^2 + k_1\|r\|\|\tilde{W}\|_F\left(W_M - \|\tilde{W}\|_F\right) + k_1\|r\|\|\tilde{W}_i\|_F\left(W_{iM} - \|\tilde{W}_i\|_F\right)$$
$$+ k_2\|r\|\|\tilde{W}_i\|_F\left(\|W_i\|_F + \|\tilde{W}_i\|_F\right)W_{iM} - k_2\|r\|\|\tilde{W}_i\|_F^2\left(\|\tilde{W}_i\|_F - \|W_i\|_F\right) \tag{5.16}$$
$$+ \|r\|\left(a_1\|\tilde{W}\|_F + a_2\|\tilde{W}_i\|_F^2 + a_3\|\tilde{W}_i\|_F + a_5\right) - \|r\|(f_M + \tau_M) + \|r\|\|\tilde{f} + \tau_d\|,$$

$$\le -K_{V\min}\|r\|^2 + k_1\|r\|\|\tilde{W}\|_F\left(W_M - \|\tilde{W}\|_F\right) + k_1\|r\|\|\tilde{W}_i\|_F\left(W_{iM} - \|\tilde{W}_i\|_F\right)$$
$$+ k_2\|r\|\|\tilde{W}_i\|_F W_{iM}^2 + 2k_2\|r\|\|\tilde{W}_i\|_F^2 W_{iM} - k_2\|r\|\|\tilde{W}_i\|_F^3 \tag{5.17}$$
$$+ \|r\|\left(a_1\|\tilde{W}\|_F + a_2\|\tilde{W}_i\|_F^2 + a_3\|\tilde{W}_i\|_F + a_5\right) - \|r\|(f_M + \tau_M) + \|r\|\|\tilde{f} + \tau_d\|,$$

$$\dot{V} \le -\|r\|\left\{K_{V\min}\|r\| - k_1\|\tilde{W}\|_F\left(W_M - \|\tilde{W}\|_F\right) - k_1\|\tilde{W}_i\|_F\left(W_{iM} - \|\tilde{W}_i\|_F\right)\right.$$
$$- k_2\|\tilde{W}_i\|_F W_{iM}^2 - 2k_2\|\tilde{W}_i\|_F^2 W_{iM} + k_2\|\tilde{W}_i\|_F^3 \tag{5.18}$$
$$\left. - a_1\|\tilde{W}\|_F - a_2\|\tilde{W}_i\|_F^2 - a_3\|\tilde{W}_i\|_F - a_5 + f_M + \tau_M - \|\tilde{f} + \tau_d\|\right\},$$

$$\dot{V} \le -\|r\|\left\{K_{V\min}\|r\| + k_1\|\tilde{W}\|_F^2 - (k_1 W_M + a_1)\|\tilde{W}\|_F\right.$$
$$+ k_2\|\tilde{W}_i\|_F^3 + (k_1 - 2k_2 W_{iM} - a_2)\|\tilde{W}_i\|_F^2 - (k_1 W_{iM} + k_2 W_{iM}^2 + a_3)\|\tilde{W}_i\|_F \tag{5.19}$$
$$\left. - a_5 + f_M + \tau_M - \|\tilde{f} + \tau_d\|\right\},$$

$$\dot{V} \le -\|r\|\left\{K_{V\min}\|r\| + k_1\left[\|\tilde{W}\|_F - \frac{1}{2}\left(W_M + \frac{a_1}{k_1}\right)\right]^2 - k_1\frac{1}{4}\left(W_M + \frac{a_1}{k_1}\right)^2\right.$$
$$\left. + g(\|\tilde{W}_i\|_F) - a_5 + f_M + \tau_M - \|\tilde{f} + \tau_d\|\right\}, \tag{5.20}$$

where the function $g(x)$ is defined as
$$g(x) = k_2 x^3 + (k_1 - 2k_2 W_{iM} - a_2)x^2 - (k_1 W_{iM} + k_2 W_{iM}^2 + a_3)x . \tag{5.21}$$
Let the constant C be defined by

$$C = \inf\{g(x), x > 0\}. \quad (5.22)$$

Defining
$$h(x) = g(x) - C, \quad (5.23)$$
one has $h(x) \geq 0$ for every $x>0$. Then, (5.20) is equivalent to

$$\dot{V} \leq -\|r\| \left\{ K_{V\min} \|r\| + k_1 \left[\|\tilde{W}\|_F - \frac{1}{2}\left(W_M + \frac{a_1}{k_1}\right) \right]^2 - k_1 \frac{1}{4}\left(W_M + \frac{a_1}{k_1}\right)^2 \right.$$
$$\left. + h\left(\|\tilde{W}_i\|_F\right) + C - a_5 - \|\tilde{f} + \tau_d\| + f_M + \tau_M \right\}. \quad (5.24)$$

Therefore, $\dot{V}$ is guaranteed negative as long as

$$\|r\| \geq \frac{\frac{1}{4}k_1\left(W_M + \frac{a_1}{k_1}\right)^2 - C + a_5}{K_{V\min}}, \quad (5.25)$$

or

$$k_1\left[\|\tilde{W}\|_F - \frac{1}{2}\left(W_M + \frac{a_1}{k_1}\right)\right]^2 \geq \frac{1}{4}k_1\left(W_M + \frac{a_1}{k_1}\right)^2 - C + a_5, \quad (5.26)$$

or

$$h\left(\|\tilde{W}_i\|_F\right) \geq \frac{1}{4}k_1\left(W_M + \frac{a_1}{k_1}\right)^2 - C + a_5. \quad (5.27)$$

The last three inequalities are equivalent to

$$\|r\| \geq \frac{\frac{1}{4}k_1\left(W_M + \frac{a_1}{k_1}\right)^2 - C + a_5}{K_{V\min}}, \quad (5.28)$$

$$\|\tilde{W}\|_F \geq \sqrt{\frac{1}{4}\left(W_M + \frac{a_1}{k_1}\right)^2 + \frac{a_5 - C}{k_1}} + \frac{1}{2}\left(W_M + \frac{a_1}{k_1}\right), \quad (5.29)$$

$$\|\tilde{W}_i\|_F \geq \max\left\{h^{-1}\left(\frac{1}{4}k_1\left(W_M + \frac{a_1}{k_1}\right)^2 - C + a_5\right)\right\}. \quad (5.30)$$

◇

The mutual dependence between NN I and NN II results in *coupled tuning law equations*. This mathematical result followed from the fact that the information stored in NN I and NN II are dependent on each other. In fact, the proposed NN compensator with two NNs can be viewed as a NN compensator of *second order*.

The first terms of (5.7), (5.8), are modified versions of the standard backpropagation algorithm. The k_1 terms correspond to the e-modification [29], to guarantee bounded parameter estimates. Note that the term corresponding to k_2 in

(5.8) is a *second-order e-modification*, which is an efficient way to compensate for the second-order modeling mismatching term $a_2 \|\tilde{W}_i\|_F^2$ in (4.14).

The right-hand side of (5.28) can be taken as a practical bound on the tracking error in the sense that $r(t)$ will never stray far above it. Note that the stability radius may be decreased any amount by increasing the PD gain K_V. It is noted that PD control without deadzone compensation requires much higher gain in order to achieve a similar performance- that is, eliminating the NN feedforward compensator will result in degraded performance. Moreover, it is difficult to guarantee the stability of such a highly nonlinear system using only PD. Using the NN deadzone compensation, stability of the system is proven, and the tracking error can be kept arbitrarily small by increasing the gain K_v. The NN weight errors are fundamentally bounded in terms of W_M, W_{iM}. The tuning parameters k_1, k_2 offer a design tradeoff between the relative eventual magnitudes of $\|r\|$ and $\|\tilde{W}\|_F$, $\|\tilde{W}_i\|_F$.

NN weights and at the same time deadzone inverse do not necessary converge to their true values. Still, tracking error stays bounded, thus achieving effective deadzone compensation.

The weights V, V_i are set to random values; they are not tuned. It is shown in [22] that for such NN, termed random variable functional link (RVFL) NN, the approximation property holds. The weights W are initialized to random values, W_i are initialized at zero. Then the PD loop in Fig. 5.1 holds the system stable until the NN begins to learn.

Proposed deadzone compensator does not require linearity in the unknown deadzone parameters (LIP). The standard techniques for deadzone compensation usually require such assumption. The LIP requirement is a severe restriction for practical systems. We use here the NN approximation property, augmented for piecewise continuous functions, which holds over a compact set. The nonlinear nonsymmetrical deadzone model is linear with respect to the nonlinear NN activation functions, which is a fundamental difference from the LIP condition.

A compensator scheme consists of a NN estimator and a NN compensator. The NN compensator is placed in the feedforward loop. The NN estimator serves as a performance evaluator for the NN compensator, or *adaptive critic* for the NN in the feedforward loop. The technique provides a general procedure for using NN to determine the preinverse of an unknown right-invertible function in the feedforward loop.

6 Simulation of NN Deadzone Compensator

To illustrate the performance of the NN deadzone compensator, the two-link robot arm (Fig. 6.1) is used. The model of the system shown in Fig. 6.1 is given in [22]. The system parameters are chosen as $\ell_1=1$, $\ell_2=1$, $m_1=1.8$, $m_2=1.3$. The manipulator acts in the horizontal plane. In order to examine the effects of the deadzone, gravity is not included in the system model.

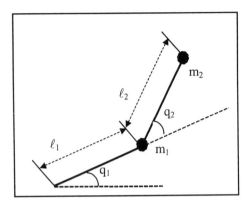

Fig. 6.1. Two-link robot arm.

The NN I has $L=20$ hidden-layer nodes with sigmoidal activation functions. The first-layer weights V are selected randomly [12]. They are uniformly randomly distributed between -1 and $+1$. These weights represent the stiffness of the sigmoidal activation function. The threshold weights for the first layer v_0 are uniformly randomly distributed between -35 and $+35$. The threshold weights represent the bias in activation functions positions. Therefore they should cover the range of the deadzone. Since the deadzone width is not known, it is recommended that this range is large enough so it covers the deadzone width.

The tuning law requires that the second-layer weights W and W_i cannot both be initialized at zero, because it is clear from (5.7) and (5.8) that in that case NN weights would stay at zero forever. Therefore, the second-layer weights for the NN I W are uniformly randomly initialized between -50 and 50. Note that this weight initialization will not affect system stability since the weights W_i are initialized at zero, and therefore there is initially no input to the system except for the PD loop.

The NN II is augmented for approximation of discontinuous functions, and has $L=20$ hidden-layer nodes with sigmoidal activation functions, and four additional nodes with jump function basis. The first-layer weights V are uniformly randomly distributed between -1 and $+1$ as in NN I, and the thresholds weights v_0 - uniformly randomly distributed between -35 and $+35$. The second-layer weights W_i are initialized at zero. We simulated the two-link robot arm with deadzones in both links.

The size of the NNs should be selected to satisfy the system performance, but not too large. The right way to do it is to select smaller number of the hidden-layer neurons, and then increase the number in successive simulations until we get satisfactory performance of the system behavior.

To focus on the deadzone compensation, we selected the disturbance as $\tau_d(t)=0$ and $\hat{f} = f$ so the robust term $v(t)=0$. We simulated several cases.

i) Nonsymmetrical deadzone with different slopes, step input

The step signal is applied at the input. The NN weight tuning parameters are chosen as $S=\text{diag}\{240, 240\}$, $T=\text{diag}\{500, 500\}$, $k_1=0.001$, $k_2=0.0001$. The controller parameters are chosen as $\Lambda=\text{diag}\{7, 7\}$, $K_v=\{15, 15\}$.

The deadzone is assumed to have linear functions with the same slopes outside the deadband, i.e., $d_+ = 25$, $d_- = -20$, $h(u) = 1.5(u - d_+)$, $g(u) = 0.8(u + d_-)$. Deadzone nonlinearity is the same for both joints. The position errors to unit step inputs for the first and second joints are shown in Fig. 6.2 and Fig. 6.3 using only PD without deadzone compensation (using equation (5.2) with $\hat{w}_{NN} = 0$) and using PD plus NN compensator. Control signal $u(t)$ is shown in Fig. 6.4 and Fig. 6.5.

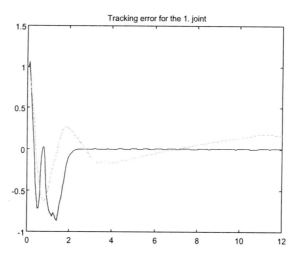

Fig. 6.2. Position error for the first joint: without deadzone compensation (dash), and with NN deadzone compensator (full).

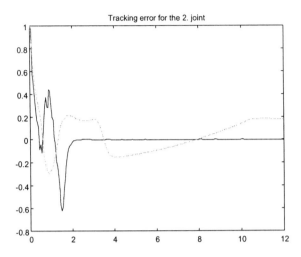

Fig. 6.3. Position error for the second joint: without deadzone compensation (dash), and with NN deadzone compensator (full).

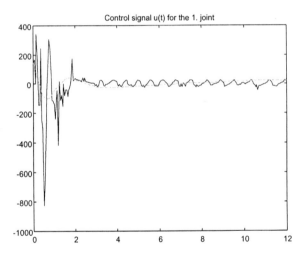

Fig. 6.4. Control signal $u(t)$ for the first joint: without deadzone compensation (dash), and with NN deadzone compensator (full).

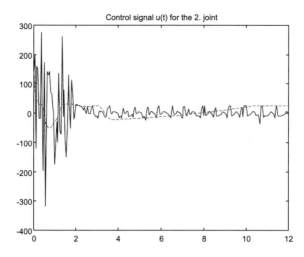

Fig. 6.5. Control signal $u(t)$ for the second joint: without deadzone compensation (dash), and with NN deadzone compensator (full).

ii) Nonsymmetrical deadzone with different slopes, sinusoidal input

The same compensator is simulated when the sinusoidal input is applied. The structure of the compensator and controller is the same as in the previous case. The tracking errors for the first and second joints are shown in Fig. 6.6 and Fig. 6.7 using only PD without NN deadzone compensation, and using PD plus NN compensator. Control signal $u(t)$ is shown in Fig. 6.8 and Fig. 6.9.

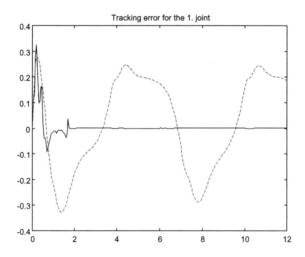

Fig. 6.6. Tracking error for the first joint: without deadzone compensation (dash), and with NN deadzone compensator (full).

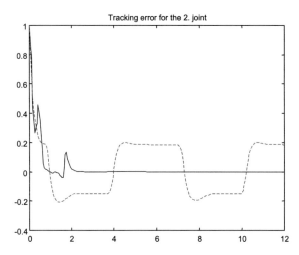

Fig. 6.7. Tracking error for the second joint: without deadzone compensation (dash), and with NN deadzone compensator (full).

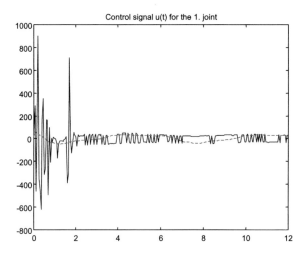

Fig. 6.8. Control signal $u(t)$ for the first joint: without deadzone compensation (dash), and with NN deadzone compensator (full).

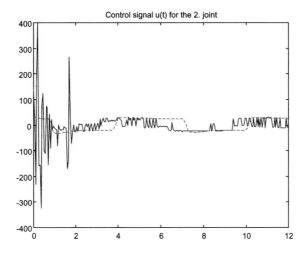

Fig. 6.9. Control signal $u(t)$ for the second joint: without deadzone compensation (dash), and with NN deadzone compensator (full).

One can see that after the transient period of 1.5 seconds, NNs adapt their weights in order to decrease the filter tracking error. It is interesting to note that the control $u(t)$ has a superimposed high-frequency component that is very similar to that injected using dithering techniques [11]. In fact, the signal from NN is injected at the same position in the control loop that dithering signals are injected. Therefore, one could consider the adaptive NN deadzone compensator as an *adaptive dithering technique*.

From this simulation it is clear that the proposed NN deadzone compensator is an efficient way to compensate for deadzone nonlinearities of all kinds, without any restrictive assumptions on the deadzone model itself.

7 Conclusions

A new technique for inverting an unknown right-invertible function is presented. It is applied to actuator deadzone compensation, and does not require any restrictive assumptions on the deadzone nonlinearity (e.g., linearity outside the dead-band). A compensator scheme consisting of a NN estimator and a NN compensator is developed. The NN controller does not require preliminary off-line training. Rigorous stability proofs are given using Lyapunov theory. Simulation results show that the proposed compensation scheme is efficient for both symmetrical and unsymmetrical deadzone nonlinearities.

8 Appendix

Proof of Theorem 1. Let *f* be a smooth bounded function on *S*, except at *x=c*, where the function has a finite discontinuity. Let *g* be an analytic function in $C^\infty(S)$, such that $f(x)=g(x)$ for all $x<c$ as shown in Fig. 8.1.

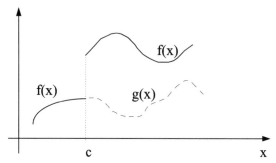

Fig. 8.1. Functions *f* and *g*.

Then for $a \geq c$ and $x \geq c$, expand *f* and *g* into Taylor series as

$$f(x) = f(a) + \frac{f'(a)}{1!}(x-a) + \frac{f''(a)}{2!}(x-a)^2 + \ldots + \frac{f^{(N)}(a)}{n!}(x-a)^N + R_f^N(x,a) \quad (8.1)$$

$$g(x) = g(a) + \frac{g'(a)}{1!}(x-a) + \frac{g''(a)}{2!}(x-a)^2 + \ldots + \frac{g^{(N)}(a)}{n!}(x-a)^N + R_g^N(x,a). \quad (8.2)$$

Combining these two equations yields

$$f(x) = g(x) + f(a) - g(a) + \frac{f'(a) - g'(a)}{1!}(x-a) + \frac{f''(a) - g''(a)}{2!}(x-a)^2 + \ldots$$
$$\ldots + \frac{f^{(N)}(a) - g^{(N)}(a)}{N!}(x-a)^N + R_f^N(x,a) - R_g^N(x,a). \quad (8.3)$$

Letting $a \to c$, and knowing that the first *N* derivatives of *g* are continuous and that *f* is continuous from the right, results in

$$f(x) = g(x) + f(c) - g(c) + \frac{f'(c) - g'(c)}{1!}(x-c) + \frac{f''(c) - g''(c)}{2!}(x-c)^2 + \ldots$$
$$\ldots + \frac{f^{(N)}(c) - g^{(N)}(c)}{N!}(x-c)^N + R_f^N(x,c) - R_g^N(x,c), \quad (8.4)$$

for $x \geq c$. Therefore one has

$$f(x) = g(x) + a_0 f_0(x-c) + a_1 f_1(x-c) + \ldots + a_N f_N(x-c)$$
$$+ R_f^N(x,c) - R_g^N(x,c). \quad (8.5)$$

Since $R_f^N(x,c)$ and $R_g^N(x,c)$ go to zero as *N* approaches infinity, the proof is complete.

Proof of Lemma 1. We will prove this expression using regressive mathematical induction.
a) For $k=1$ (3.11) is simplified into

$$\sum_{i=0}^{m} \binom{m}{i}(-1)^{i}(-i) = m\sum_{j=0}^{m-1}\binom{m-1}{j}(-1)^{j} \qquad (8.6)$$

$$= m(1-1)^{m-1} = 0,$$

where $j=i-1$. Therefore for $k=1$ expression (3.11) is true.
b) Assume expression (3.11) is true for some $k=p$, $k=p-1$, $k=p-2$,..., $k=1$. One needs to show it is true for $k=p+1$.
For $k=p+1$,

$$\sum_{i=0}^{m}\binom{m}{i}(-1)^{i}(-i)^{p+1} = \sum_{i=0}^{m}\frac{m(m-1)...(m-i+1)}{i!}(-1)^{i}(-i)^{p}(-i)$$

$$= -\sum_{i=0}^{m}\frac{m(m-1)...(m-i+1)}{(i-1)!}(-1)^{i}(-i)^{p}. \qquad (8.7)$$

Putting $j=i-1$ and knowing that the term for $i=0$ is equal to zero, using the binomial formula one can transform (8.7) into

$$-m\sum_{j=0}^{m-1}\frac{(m-1)...(m-j)}{j!}(-1)^{j-1}(-j+1)^{p}$$

$$= m\sum_{j=0}^{m-1}\binom{m-1}{j}(-1)^{j}\sum_{k=0}^{p}\binom{p}{k}(-j)^{k} \qquad (8.8)$$

$$= m\sum_{k=0}^{p}\binom{p}{k}\left[\sum_{j=0}^{m-1}\binom{m-1}{j}(-1)^{j}(-j)^{k}\right] = 0.$$

Expression (8.8) is equal to zero for every $m-1>k$, $k=0, 1,..., p$ because of the induction assumption. Therefore, (3.11) is true for $k=p+1$.

Proof of Lemma 2. It is enough to prove that there exist coefficients b_k, such that first N derivatives of the expression

$$\sum_{k=0}^{N}a_{k}f_{k}(x) - \sum_{k=0}^{N}b_{k}\varphi_{k}(x) \qquad (8.9)$$

are continuous.
For $x<0$, expression (8.9) has the constant value 0. Therefore one must show that there exist coefficients b_i, such that for $x>0$, the first N derivatives of the expression (8.9) are equal to zero.
For $x>0$ one has

$$\sum_{k=0}^{N}a_{k}f_{k}(x) - \sum_{k=0}^{N}b_{k}\varphi_{k}(x) = \sum_{k=0}^{N}a_{k}\frac{x^{k}}{k!} - \sum_{k=0}^{N}b_{k}(1-e^{-x})^{k}. \qquad (8.10)$$

Using the binomial formula one can expand expression (8.10) into

$$\sum_{k=0}^{N} a_k \frac{x^k}{k!} - \sum_{k=0}^{N} b_k \left[\sum_{i=0}^{k} \binom{k}{i} (-e^{-x})^i \right]$$

$$= \sum_{k=0}^{N} a_k \frac{x^k}{k!} - \sum_{k=0}^{N} b_k \left[\sum_{i=0}^{k} \binom{k}{i} (-1)^i (e^{-ix}) \right]$$

$$= \sum_{k=0}^{N} a_k \frac{x^k}{k!} - \sum_{k=0}^{N} \sum_{i=0}^{k} b_k \binom{k}{i} (-1)^i \sum_{j=0}^{\infty} (-i)^j \frac{x^j}{j!}$$

$$= \sum_{k=0}^{N} a_k \frac{x^k}{k!} - \sum_{k=0}^{N} \sum_{i=0}^{k} \sum_{j=0}^{\infty} b_k \binom{k}{i} (-1)^i (-i)^j \frac{x^j}{j!}$$

$$= \sum_{k=0}^{N} a_k \frac{x^k}{k!} - \sum_{j=0}^{\infty} \sum_{k=0}^{N} \sum_{i=0}^{k} b_k \binom{k}{i} (-1)^i (-i)^j \frac{x^j}{j!}$$

$$= \sum_{k=0}^{N} a_k \frac{x^k}{k!} - \sum_{k=0}^{\infty} \sum_{m=0}^{N} \sum_{i=0}^{m} b_m \binom{m}{i} (-1)^i (-i)^k \frac{x^k}{k!}. \tag{8.11}$$

In order that first N derivatives of expression (8.9) be zero, the coefficients in (8.11) of x^k for $k=0, 1,..., N$ should be zero. Therefore, one requires

$$a_k - \sum_{m=0}^{N} \sum_{i=0}^{m} b_m \binom{m}{i} (-1)^i (-i)^k = 0, \tag{8.12}$$

$$a_k - \sum_{m=0}^{N} b_m \sum_{i=0}^{m} \binom{m}{i} (-1)^i (-i)^k = 0. \tag{8.13}$$

Using Lemma 1, one obtains

$$a_k - \sum_{m=0}^{k} b_m \sum_{i=0}^{m} \binom{m}{i} (-1)^i (-i)^k = 0 \tag{8.14}$$

for $k=1, 2,..., N$, or

$$a_k - b_k \sum_{i=0}^{k} \binom{k}{i} (-1)^i (-i)^k - \sum_{m=0}^{k-1} b_m \sum_{i=0}^{m} \binom{m}{i} (-1)^i (-i)^k = 0, \tag{8.15}$$

for $k=1,2,..., N$. Therefore, one obtains the recurrent relation (3.14).

Proof of Theorem 2. According to Theorem 1 there exists a sum F of the form

$$F(x) = g(x) + \sum_{k=0}^{N} a_k f_k(x-c), \tag{8.16}$$

such that

$$|F(x) - f(x)| < \varepsilon. \tag{8.17}$$

Using Lemma 2

$$\sum_{k=0}^{N} a_k f_k(x) = z(x) + \sum_{k=0}^{N} b_k \varphi_k(x), \tag{8.18}$$

one obtains the result. ◇

Acknowledgment
This research is supported by the ARO Grant DAAD19-99-1-0137.

References

1. R. Barron, Universal approximation bounds for superpositions of a sigmoidal function, *IEEE Trans. Inf. Theory*, 39, no. 3, 930-945, 1993.
2. R. G. Bartle, *The Elements of Real Analysis*, Wiley, New York, 1964.
3. F.-C. Chen and H. K. Khalil, Adaptive control of nonlinear systems using neural networks, *Int. J. Control*, 55, no. 6, 1299-1317, 1992.
4. S. Commuri and F. L. Lewis, CMAC neural networks for control of nonlinear dynamical systems: structure, stability and passivity, *Proc. IEEE Int. Symp. Intell. Control*, Monterey, 1995, 123-129.
5. M. J. Corless and G. Leitmann, Continuous state feedback guaranteeing uniform ultimate boundedness for uncertain dynamic systems, *IEEE Trans. Automat. Control*, 26, no. 5, 850-861, 1982.
6. J. J. Craig, *Adaptive Control of Robot Manipulators*, Addison-Wesley, Reading, MA, 1988.
7. G. Cybenko, Approximation by superpositions of a sigmoidal function, *Math. Control Signals, Syst.*, 2, no. 4, 303-314, 1989.
8. C. A. Desoer and S. M. Shahruz, Stability of dithered nonlinear systems with backlash or hysteresis, *Int. J. Control*, 43, no. 4, 1045-1060, 1986.
9. B. Friedland, *Advanced Control System design*, Prentice-Hall, New Jersey, 1996.
10. K. Funahashi, "On the approximate realization of continuous mappings by neural networks," *Neural Networks*, vol. 2, pp. 183-192, 1989.
11. J. W. Gilbart and G. C. Winston, Adaptive compensation for an optical tracking telescope, *Automatica*, 10, 125-131, 1974.
12. B. Igelnik and Y. H. Pao, Stochastic choice of basis functions in adaptive function approximation and the functional-link net, *IEEE Trans. Neural Networks*, 6, no. 6, 1320-1329, 1995.
13. K. Hornik, M. Stinchombe, and H. White, Multilayer feedforward networks are universal approximators, *Neural Networks*, 2, 359-366, 1989.
14. J.-Y. Jeon, J.-H. Kim, and K. Koh, Experimental evolutionary programming-based high-precision control, *IEEE Control Syst. Mag.*, 17, no. 2, 66-74, 1997.
15. B. S. Kim and A. J. Calise, Nonlinear flight control using neural networks, *Journal of Guidance, Control, and Dynamics*, 20, no. 1, 1997.
16. J. -H. Kim, H. -K. Chae, J. -Y. Jeon, S.-W. Lee, Identification and control of systems with friction using accelerated evolutionary programming, *IEEE Control Syst. Mag.*, 16, no. 4, 38-47, 1996.
17. J. -H. Kim, K. -C. Kim, and E. K. P. Chong, Fuzzy precompensated PD controllers, *IEEE Trans. Control Syst. Technol.*, 2, no. 4, 406-411, 1994.

18. J.-H. Kim, J.-H. Park, S.-W. Lee, and E. K. P. Chong, Fuzzy precompensation of PD controllers for systems with deadzones, *J. Int. Fuzzy Syst.*, 1, 125-133, 1993.
19. J.-H. Kim, J.-H. Park, S.-W. Lee, and E. K. P. Chong, A two-layered fuzzy logic controller for systems with deadzones, *IEEE Trans. Industrial Electron.*, 41, no. 2, 155-162, 1994.
20. B. Kosko, *Neural Networks and Fuzzy Systems*, Prentice Hall, New Jersey, 1992.
21. S.-W. Lee and J.-H. Kim, Control of systems with deadzones using neural-network based learning control, *Proc. IEEE Int. Conf. Neural Networks*, 1994, 2535-2538.
22. F. L. Lewis, C. T. Abdallah, and D. M. Dawson, *Control of Robot Manipulators*, Macmillan, New York, 1993.
23. F. L. Lewis, S. Jagannathan, and A. Yesildirek, *Neural Network Control of Robot Manipulators and Nonlinear Systems*, Taylor and Francis, Philadelphia, PA, 1999.
24. F. L. Lewis, A. Yesildirek, and K. Liu, Multilayer neural-net robot controller with guaranteed tracking performance, *IEEE Trans. Neural Networks*, 7, no. 2, 1-11, 1996.
25. F. L. Lewis, K. Liu, and A. Yesilidrek, Neural net robot controller with guaranteed tracking performance, *IEEE Trans. Neural Networks*, 6, no. 3, 703-715, 1995.
26. F. L. Lewis, K. Liu, R. R. Selmic, and Li-Xin Wang, Adaptive fuzzy logic compensation of actuator deadzones, *J. Robot. Syst.*, 14, no. 6, 501-511, 1997.
27. F. L. Lewis, Neural network control of robot manipulators, *IEEE Expert special track on "Intelligent Control"* Eds. K. Passino and Ü. Özgüner, 64-75, 1996.
28. W. Li and X. Cheng, Adaptive high-precision control of positioning tables--theory and experiment, *IEEE Trans. Control Syst. Technol.*, 2, no. 3, 265-270, 1994.
29. K. S. Narendra and A. M. Annaswamy, A new adaptive law for robust adaptation without persistent excitation, *IEEE Trans. Automat. Control*, 32, no. 2, 134-145, 1987
30. K. S. Narendra, Adaptive Control Using Neural Networks, *Neural Networks for Control*, 115-142. Eds. W. T. Miller, R. S. Sutton, P. J. Werbos, MIT Press, Cambridge, MA, 1991.
31. K. S. Narendra and K. Parthasarathy, Identification and control of dynamical systems using neural networks, *IEEE Trans. Neural Networks*, 1, 4-27, 1990.
32. J. Park and I. W. Sandberg, Criteria for the approximation of nonlinear systems, *IEEE Trans. Circuits Syst.*, 39, no. 8, 673-676, 1992.
33. J. Park and I. W. Sandberg, Nonlinear approximations using elliptic basis function networks, *Circuits, Systems, and Signal Processing*, 13, no. 1, 99-113, 1993.
34. M. M. Polycarpou, Stable adaptive neural control scheme for nonlinear systems, *IEEE Trans. Automat. Control*, 41, no. 3, 447-451, 1996.

35. M. M. Polycarpou and P. A. Ioannou, Modeling, identification and stable adaptive control of continuous-time nonlinear dynamical systems using neural networks, *Proc. Amer. Control Conf.*, 1992, 1, 36-40.
36. D. A. Recker, P. V. Kokotovic, D. Rhode, and J. Winkelman, Adaptive nonlinear control of systems containing a dead-zone, *Proc. IEEE Conf. Decis. Control*, 1991, 2111-2115.
37. G. A. Rovithakis and M. A. Christodoulou, Adaptive control of unknown plants using dynamical neural networks, *IEEE Trans. Systems, Man, and Cybernetics*, 24, no. 3, 400-412, 1994.
38. N. Sadegh, A perceptron network for functional identification and control of nonlinear systems, *IEEE Trans. Neural Networks*, 4, no. 6, 982-988, 1993.
39. R. M. Sanner and J. -J. E. Slotine, Stable adaptive control and recursive identification using radial gaussian networks, *Proc. IEEE Conf. Decis. and Control*, Brighton, 1991.
40. J. R. Schilling, *Fundamentals of Robotics Analysis and Control*, Prentice Hall, New Jersey.
41. R. R. Selmic and F. L. Lewis, Neural network approximation of piecewise continuous functions: application to friction compensation, *Proc. IEEE Int. Symp. Intell. Control*, Istanbul, July 1997.
42. R. R. Selmic and F. L. Lewis, Neural network approximation of piecewise continuous functions: application to friction compensation, *Soft Computing and Intelligent Systems: Theory and Applications*, Eds. N. K. Sinha and M. M. Gupta, Academic Press, 1999.
43. R. R. Selmic and F. L. Lewis, Deadzone compensation in motion control systems using neural networks, *IEEE Trans. Automat. Control*, 45, no. 4, 602-613, 2000.
44. J.-J. E. Slotine and W. Li, *Applied Nonlinear Control*, Prentice-Hall, New Jersey, 1991.
45. J.-J. E. Slotine and W. Li, Adaptive manipulator control: a case study, *IEEE Trans. Automat. Control*, 33, no. 11, 995-1003, 1988.
46. G. Tao, Adaptive backlash compensation for robot control, *Proc. IFAC World Congress*, San Francisco, 1996, pp. 307-312.
47. G. Tao and P. V. Kokotovic, Adaptive control of plants with unknown dead-zones, *IEEE Trans. Automat. Control*, 39, no. 1, 59-68, 1994.
48. G. Tao and P. V. Kokotovic, Adaptive control of systems with unknown output backlash, *IEEE Trans. Automat. Control*, 40, no. 2, 326-330, 1995.
49. G. Tao and P. V. Kokotovic, Adaptive control of plants with unknown hystereses, *IEEE Trans. Automat. Control*, 40, no. 2, 200-212, 1995.
50. G. Tao and P. V. Kokotovic, Continuous-time adaptive control of systems with unknown backlash, *IEEE Trans. Automat. Control*, 40, no. 6, 1083-1087, 1995.
51. G. Tao and P. V. Kokotovic, Discrete-time adaptive control of systems with unknown deadzones, *Int. J. Control*, 61, no. 1, 1-17, 1995.
52. G. Tao and P. V. Kokotovic, *Adaptive Control of Systems With Actuator and Sensor Nonlinearities*, John Wiley & Sons, New York, 1996.

53. M. Tian and G. Tao, Adaptive control of a class of nonlinear systems with unknown deadzones, *Proc. IFAC World Congress*, San Francisco, 1996, pp. 209-214.
54. V. I. Utkin, Sliding modes and their application in variable structure systems, Mir, Moscow, 55-63, 1978.
55. L. -X. Wang, *Adaptive Fuzzy Systems and Control: Design and Stability Analysis*, Prentice-Hall, New Jersey, 1994.

On-line Fault Detection, Diagnosis, Isolation and Accommodation of Dynamical Systems with Actuator Failures

Michael A. Demetriou[1] and Marios M. Polycarpou[2]

[1] Department of Mechanical Engineering, Worcester Polytechnic Institute, Worcester, MA 01609-2280, USA
[2] Department of Electrical & Computer Engineering and Computer Science, University of Cincinnati, Cincinnati, OH 45221-0030, USA

Abstract

A general framework for the on-line detection, diagnosis, isolation and accommodation for a class of actuator faults in dynamical systems is developed. This framework, which addresses both abrupt and incipient faults, utilizes a nonlinear adaptive detection observer along with an on-line approximation scheme for failure assessment. The actuator failure is modeled as an additive perturbation of the actuator signal (actuator gain) that may be described by a nonlinear function of both the input and output signals. Robust adaptive schemes are introduced to account for modeling errors that affect the diagnosis process and may cause false alarms. Numerical studies using a neural network-based on-line approximator are presented to demonstrate the applicability of the proposed automated diagnostic scheme.

1 Introduction

Recent advances in materials science (especially in sensor and actuator technology) and stricter demands on controller design and performance have led to the development of successful controllers that employ advanced state-of-the art sensors and actuators (e.g. piezoelectric/piezoceramic, shape memory alloys, magnetostrictives/electrostrictives, [1, 2]). The combination of these advanced sensors, actuators and microprocessors, which comprise intelligent material systems, have many advantages over traditional servomechanisms. Under normal operating conditions (idealized conditions) these smart sensors and actuators have negligible effect on the passive system dynamics of the host structure (plant). Their rapid respond time with minimal destabilizing effect makes them ideal for control in several aerospace, medical, and manufacturing applications. In extreme environments though, e.g. in the presence of large thermal gradients or large pressures, they can exhibit a nonlinear behavior whereby model nonlinearities, spillover, hysteresis and phase shifts occur. Due to the high reliability demand, an automated self-diagnostic procedure that can monitor and detect these unanticipated changes (failures) is warranted. Unfortunately, for many controllers that employ smart materials this is still the topic of current research. Therefore a fault diagnostic architecture is needed to monitor these smart materials when entering the nonlinear regime and, by proper reconfiguration, account for any additional dynamics due to sensor and actuator failures.

In this paper we present a unified fault diagnosis methodology for both incipient and abrupt [3] actuator faults in nonlinear dynamical systems characterized by linear nominal models (i.e., the nonlinearities are treated as modeling uncertainties). The actuator faults are modeled as changes in the actuator gain with magnitude given *not* by a constant gain *but* by a nonlinear function of the measurable input and output signals having a time-varying failure profile [4]. An adaptive observer is proposed in order to monitor the plant for unanticipated actuator failures and an on-line approximation scheme is used to assess the nature of the failure. The design of the diagnostic observers falls in the category of *model-based analytical redundancy* approach. In summary, according to this approach, quantitative nominal models of the physical plant together with sensory measurements are used to provide estimates of measured and unmeasured variables. Deviations between the estimated and measured signals provide a *residual vector* which is used to make decisions regarding failure detection and isolation. The survey papers and books by Frank [5], Gertler [6, 7], Isermann [8], Patton [9, 10] and Willsky [11] provide detailed overviews of various model-based fault detection and diagnosis algorithms.

Modifications to the observer are incorporated to improve the robustness of the fault diagnosis scheme and reduce the possibility of false alarms that may occur due to modeling uncertainties. These modifications are in the spirit of the ones utilized in [12, 14] but applied in the current context of actuator failures with actuator gains depending on both the input and output signals. Unlike the detection and diagnostic observers in [15, 13], we present a *single* adaptive observer scheme that can serve both as a *diagnostic* and a *detection* observer. Hence, the need to first implement the detection observer in order to detect the failure and then employ another observer (diagnostic observer) for failure diagnosis is eliminated.

In this research effort we also extend our previous work on actuator failures [16] to a class of vector second-order mechanical systems. In [17] and [18] we considered a distributed parameter system (infinite dimensional system) with a simple model for actuator failures and proposed a detection and accommodation scheme based on the infinite dimensional system. Possible problems of existence of solutions and well-posedness of the infinite dimensional plant and its state estimator were bypassed and concentrated on the stability and convergence properties of those schemes. The examples presented there were given in terms of finite dimensional approximations of the actual infinite dimensional systems. We found it necessary to expand on these earlier efforts by addressing first the problem from a finite dimensional point of view; i.e. consider finite dimensional systems and propose finite dimensional estimators in order to detect possible actuator failures and present a complete theoretical framework. Due to their specific input/output nature, these collocated mechanical systems do not require the input/output transfer function to be *strictly positive real* (SPR), [19, 20, 21]. This (SPR) transfer function property amounts to having the system's input and output matrices coupled via the solution to a certain algebraic matrix Riccati equation [22] that involves the system's dynamics. This coupling between the input and output matrices is bypassed at the expense of finding a *parameter-dependent* Lyapunov function required for the stability and convergence properties of the proposed detection and diagnostic scheme.

In the next section we present the class of dynamical systems under consideration and propose a model for the actuator failures. In Section 3 we present the adaptive detection and diagnostic observer and the relevant stability and robustness results. Results on fault accommodation via an appropriate control reconfiguration are summarized in Section 4. The special case of second-order mechanical systems that do not require an SPR condition is treated in Section 5. Examples with the relevant numerical studies are presented in Section 6. Concluding remarks follow in Section 7.

2 Problem Formulation

The class of dynamical systems considered here consists of a linear (nominal) system with nonlinear modeling uncertainties and an additive actuator gain which is a nonlinear function of the input and output signals. Specifically, the system under consideration is described by the following MIMO square system

$$\dot{x} = Ax + \xi(x, u, t) + Bu,$$
$$y = Cx,$$
(2.1)

where
$x(t) \in \mathbb{R}^n$ is the state vector of the plant,
$u(t) \in \mathbb{R}^p$ $(p \geq 1)$ is the actuator input vector,
$y(t) \in \mathbb{R}^p$ is the measured plant output vector,
and the vector field $\xi : \mathbb{R}^n \times \mathbb{R}^p \times \mathbb{R}^+ \to \mathbb{R}^n$ represents any modeling errors or uncertainties that may arise due to the difference between the nominal and actual plant dynamics (under healthy operating conditions).

The above system describes a plant prior to any unanticipated actuator failures. A schematic of the system with actuator failures is depicted in Figure 2.1. The input term before and after the failure is given by

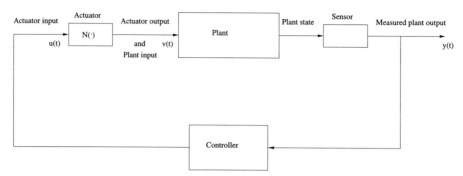

Fig. 2.1. Schematic of a system with actuator failures.

$$\begin{aligned}\text{before:} \quad & Bv(t) = Bu(t) \\ \text{during:} \quad & Bv(t) = B\left[u(t) + \Delta(t-T)\left(F_0(y,u) - u\right)\right] \\ \text{after:} \quad & Bv(t) = BF_0(y,u),\end{aligned} \quad (2.2)$$

where $v(t)$ denotes the actuator-output/plant-input. The function $F_0 : \mathbb{R}^p \times \mathbb{R}^p \mapsto \mathbb{R}^p$ is a nonlinear function of the measurable output and input signals, $y(t)$ and $u(t)$, respectively, that represents the input/output characteristics of the actuator failures. The term $\Delta(t-T) \in \mathbb{R}^{p \times p}$ denotes the *time profile* of the failures, given by

$$\Delta(\tau) = \begin{cases} 0 & \text{if } \tau < 0, \\ I - e^{-\Lambda \tau} & \text{if } \tau \geq 0, \end{cases}$$

and represents the rate at which the failures at the p actuators evolve, where Λ is a positive definite diagonal matrix. Both *abrupt* and *incipient* (slowly developing) failure profiles are considered and thus

$$\begin{aligned}\Delta_{ii}(t-T) &= H(t-T)\left(1 - e^{-\lambda_i(t-T)}\right) & \text{for incipient failures} \\ \Delta_{ii}(t-T) &= H(t-T) & \text{for abrupt failures}\end{aligned}, \quad (2.3)$$

where $\lambda_i > 0$, $i = 1, \ldots, p$ is the rate at which each actuator failure evolves, and $H(t-T)$ denotes the Heaviside function. Specifically, it is the entries of the diagonal matrix Λ that describe the rate at which the failure in each actuator channel evolves. If the diagonal elements of Λ are very large then $\Delta(\tau)$ approaches a diagonal matrix step function, thereby modeling abrupt failures. The justification for such a choice of the failure model stems from the fact that the matrix B provides information on the location of the actuators while $u(t)$ (the actuator input) provides information on the actuator dynamics. When the actuator is healthy, one has that the actuator-output/plant-input $v(t)$ is given by $v = u$, and when a failure occurs, $v = F_0(y, u)$.

Note that for $t \leq T$ we have $Bu(t)$ as the input term and for $t \gg T$ it becomes $BF_0(y(t), u(t))$. Thus the failure changes the actuator dynamics from the *linear* ($v = Iu$) regime to possibly a *highly nonlinear* ($v = F_0(y, u)$) one.

For analysis purposes the plant (2.1), (2.2) is re-written as

$$\dot{x} = Ax + \xi(x, u, t) + B[u + \Delta(t - T)F(y, u)], \qquad (2.4)$$
$$y = Cx,$$

where $F(y(t), u(t)) = F_0(y(t), u(t)) - u(t)$. Throughout this paper we will consider (2.4) as the equation that describes the plant dynamics both *before* and *after* the actuator failures.

We make the assumption that we have complete knowledge of the nominal plant in the absence of faults (healthy system)

$$\dot{x} = Ax + Bu, \qquad (2.5)$$
$$y = Cx,$$

and that the modeling uncertainty term satisfies the inequality

$$|\xi(x(t), u(t), t)| \leq \xi_0, \quad \forall (x, u) \in \mathcal{X} \times \mathcal{U}, \ \forall t \in R^+,$$

where $\mathcal{X} \subset \mathbb{R}^n, \mathcal{U} \subset \mathbb{R}^p$ are compact sets. The upper bound ξ_0 of the modeling errors is assumed to be known. Furthermore, we assume for the detection and diagnosis analysis, that the state vector x and input u remain bounded prior ($t < T$) and after ($t \geq T$) the fault occurrence.

Assumption 1. *There exist compact sets $\mathcal{X} \subset \mathbb{R}^n$ and $\mathcal{U} \subset \mathbb{R}^p$ such that $x(t) \in \mathcal{X}$ and $u(t) \in \mathcal{U}$ for all $t \geq 0$.*

Assumption 2. *There exists an upper bound for the unknown dynamics given by*

$$|\xi(x, u, t)| \leq \xi_0, \quad \forall (x, u) \in \mathcal{X} \times \mathcal{U}.$$

In the section below, we present the adaptive detection observer along with the on-line approximator for actuator failure diagnosis.

3 Adaptive Diagnostic Observers

We consider the following adaptive detection/diagnostic observer

$$\dot{\hat{x}} = A\hat{x} + Bu + B\widehat{F}(y, u; \hat{\theta}) + L\varepsilon, \qquad (3.1)$$
$$\hat{y} = C\hat{x},$$

where

$\widehat{x}(t) \in \mathbb{R}^n$ denotes the *estimated state vector*,
$\varepsilon(t) = y(t) - \widehat{y}(t)$ is the *output error*,
$\widehat{F}(\cdot, \cdot; \cdot) : \mathbb{R}^p \times \mathbb{R}^p \times \mathbb{R}^q \to \mathbb{R}^p$ is the *on-line approximation model* (OLA),
$\widehat{\theta} \in \mathbb{R}^q$ is a vector of *adjustable parameters*,
L is a gain matrix (an appropriately chosen Kalman observer gain) such that $A - LC$ is a Hurwitz matrix (i.e., it is implicitly assumed that the nominal model (2.5) has an observable pair (C, A)).

The on-line approximator is a generic function approximator having adjustable parameters. A systematic procedure for constructing nonlinear estimation algorithms and deriving a stable learning scheme using Lyapunov theory is developed via the use of this on-line approximator, see for example [4] for more details. Next, we proceed to the design of the adaptive law for the adjustable parameter vector $\widehat{\theta}(t)$.

The goal here is to derive a parameter adaptive law for $\widehat{\theta}$ based on Lyapunov redesign methods [23] so that the on-line approximator $\widehat{F}(y, u; \widehat{\theta})$ estimates the failure term $\Delta(t - T)F(y, u)$. When this is achieved, the same on-line approximator may be used for fault diagnosis and, if so desired, to accommodate the actuator failure by the appropriate control reconfiguration.

We make the assumption that the estimation filter has the following property:

Assumption 3. *We assume that the filter matrix* $C\left(sI - (A - LC)\right)^{-1} B$ *used in (3.1), is strictly positive real (SPR), [24]. This allows for the use of the Lur'e equations*

$$A_o^T \Pi + \Pi A_o = -Q, \qquad \Pi B = C^T, \qquad (3.2)$$

where Π, Q are positive definite matrices and $A_o = A - LC$.

Using Lyapunov redesign methods, we obtain the following adaptation rule for parameter adjustment

$$\dot{\widehat{\theta}}(t) = \mathcal{P}\left\{\Gamma Z^T \varepsilon\right\}, \qquad (3.3)$$

with $0 < \Gamma = \Gamma^T \in \mathbb{R}^{q \times q}$ denoting the learning rate, $\mathcal{P}$ denoting the projection operator and the $p \times q$ regressor matrix function $Z(y, u; \widehat{\theta})$ given by

$$Z = Z(y, u; \widehat{\theta}) = \left[\frac{\partial \widehat{F}(y, u; \widehat{\theta})}{\partial \widehat{\theta}}\right]. \qquad (3.4)$$

The projection operator, which constrains the parameter $\widehat{\theta}$ to some selected compact, convex region $\mathcal{M}_\theta$ of the parameter space $\mathbb{R}^q$, is utilized in order to prevent parameter drift, [24].

In addition to the above assumption, an identifiability assumption is made in order to ensure that the failure can be observed. This in essence requires that when the failure term $F_0 \neq 0$, it implies $\varepsilon \neq 0$ or equivalently that when $\varepsilon = 0$ it implies that $F_0 = 0$ whenever $\xi = 0$.

3.1 Properties and Analysis of Detection/Diagnostic Observer

In the time interval $t \in [0, T)$, i.e. prior to the failure occurrence, we have the following state and parameter error equations from (2.4), (3.1), and (3.3)

$$\dot{e}(t) = A_o e(t) + \xi(x, u, t) - B\widehat{F}(y, u; \widehat{\theta}),$$
$$\dot{\widehat{\theta}}(t) = \mathcal{P}\left\{\Gamma Z^T \varepsilon\right\}, \tag{3.5}$$

where $e = x - \widehat{x}$ denotes the state error, and the initial condition $\widehat{\theta}(0) = \widehat{\theta}_0$ is chosen so that $\widehat{F}(y, u; \widehat{\theta}_0) = 0$ for all (y, u). When no modeling errors (i.e. $\xi(x(t), u(t), t) \equiv 0$) are present, the system is in equilibrium as this was similarly established in [16]. In this case, by simply monitoring either the output error $\varepsilon(t)$ or the on-line approximator (OLA) $\widehat{F}(y, u; \widehat{\theta})$, a fault is declared when either of the two signals becomes nonzero.

In practical situations, however, the plant equation includes a nonzero $\xi(x(t), u(t), t)$ in (2.1). In the absence of further information on this term, the observer will give a false alarm due to the nonzero values of the output error. A way to remedy the above is to first activate the adaptation whenever the output error exceeds a certain threshold in order to avoid false alarms; see also [15] for a similar modification used in the context of actuator fault detection. Indeed, in the presence of modeling errors it can be seen that for $t < T$ (prior to failure) and for $\widehat{F} = 0$ we have that

$$\varepsilon(t) = C \int_0^t e^{A_o(t-\tau)} \xi(x(\tau), u(\tau), \tau)\, d\tau \neq 0.$$

Using the fact that A_o is a Hurwitz matrix and the upper bound for $\|\xi(x, u, t)\|$, given by ξ_0, we have that

$$|\varepsilon(t)| \leq N \int_0^t e^{-\alpha(t-\tau)} \xi_0\, d\tau \leq \frac{N\xi_0}{\alpha},$$

where
$$\|Ce^{A_o t}\| \leq \|C\|\|e^{A_o t}\| \leq \|C\|\mu e^{-\alpha t} = Ne^{-\alpha t}.$$

The above bound can also be calculated via $\max_{\omega \geq 0} \|C(j\omega I - A_o)^{-1}\|\xi_0$ as was presented in [13]. Since ε and $\widehat{\theta}$ are nonzero for $t < T$, a modification to the adaptation (3.6) needs to be implemented in the case that the output error is below a certain bound. When the output error is below a certain threshold (i.e. $|\varepsilon| < \frac{N\xi_0}{\alpha}$), no adaptation should take place and when it exceeds it, the on-line approximator $\widehat{F}(y,u;\widehat{\theta})$ will attempt to estimate $\Delta(t-T)F(y,u)$. The modification, known as *dead zone* [24], is given by

$$\dot{\widehat{\theta}} = \mathcal{P}\left\{\Gamma Z^T D[\varepsilon]\right\}, \qquad D[\varepsilon] = \begin{cases} 0 & \text{if } |\varepsilon| < g_0 \\ \varepsilon & \text{otherwise} \end{cases}, \qquad g_0 = \frac{N\xi_0}{\alpha}, \qquad (3.6)$$

where g_0 is the measure of the modeling error. We then have that the output error ε remains within the dead zone $\forall t \leq T$ and that the output of the on-line approximator is zero; i.e. we have that the set

$$\left\{(\varepsilon,\widehat{\theta}) : |\varepsilon| < \frac{N\xi_0}{\alpha}, \ \widehat{\theta} = \widehat{\theta}_0\right\}$$

is positively invariant, [23]. In summary, the dead zone (3.6) guarantees that $\widehat{F}(y,u;\widehat{\theta}) = 0$ for $t \in [0,T)$ as long as $|\varepsilon| \leq g_0$. It should be noted that the idea of using a threshold to achieve robustness with respect to modeling uncertainties in order to avoid false alarms has been used extensively in the fault diagnosis literature [9, 12, 13].

After the failure occurs, i.e. for $t \geq T$, the above system deviates from its equilibrium and thus $e(t) \neq 0$ and $\varepsilon(t) \neq 0$; hence the instance where ε attains a nonzero value (or, in the case of a nonzero modeling term ξ, when $|\varepsilon|$ attains a value greater than g_0), indicates the failure occurrence. Alternatively, when the on-line approximator $\widehat{F}$ becomes nonzero, then a failure occurred, since $\widehat{F} \equiv 0$ as long as $|\varepsilon| < g_0$, as this is guaranteed by the dead zone (3.6). Once a failure is declared, the dead zone is deactivated. Combining (2.4) and (3.1) we arrive at the state error equation (with $\xi \neq 0$)

$$\dot{e} = A_o e + \xi(x,u,t) + B\left[\Delta F(y,u) - \widehat{F}(y,u;\widehat{\theta})\right]. \qquad (3.7)$$

We rewrite the third term of the right-hand side of (3.7) as follows

$$B\left[\Delta(t-T)F(y,u) - \widehat{F}(y,u;\widehat{\theta})\right] = B\left[\Delta(t-T)\widehat{F}(y,u;\theta^*) - \widehat{F}(y,u;\widehat{\theta})\right]$$
$$+ B\nu(t)$$

where $\nu(t)$ is the *minimum functional approximation error* given by

$$\nu(t) = \Delta(t-T)\left[F(y,u) - \widehat{F}(y,u;\theta^*)\right]. \qquad (3.8)$$

The artificial quantity θ^* used above is an optimal parameter, in the sense that it is chosen as the value of admissible $\widehat{\theta}$ that minimizes the L_2-norm distance between the actuator failure $F(y,u)$ and its online approximator $\widehat{F}(y,u;\widehat{\theta})$ over all $(y,u) \in \mathcal{Y} \times \mathcal{U}$ subject to the constraint $\widehat{\theta} \in \mathcal{M}_{\widehat{\theta}}$, [4]. The compact set $\mathcal{Y}$ above is simply the image of $\mathcal{X}$ under C, $\mathcal{Y} = C(\mathcal{X})$.

Using the mean value theorem [25], the above error equation (3.7) can be rewritten as

$$\dot{e} = A_o e + \xi(x,u,t) - B\Phi \widehat{F}(y,u;\theta^*) - BZ(y,u)\tilde{\theta} + B\omega,$$

where $\tilde{\theta} = \widehat{\theta} - \theta^*$, $\Phi = I - \Delta(t-T)$ and $\omega = \nu - \mathcal{D}(y,u;\widehat{\theta})$ with

$$\mathcal{D}(y,u;\widehat{\theta}) = \widehat{F}(y,u;\widehat{\theta}) - \widehat{F}(y,u;\theta^*) - \frac{\partial \widehat{F}^T(y,u;\widehat{\theta})}{\partial \widehat{\theta}}(\widehat{\theta} - \theta^*).$$

The above term can be regarded to represent the higher-order terms of the Taylor series expansion of $\widehat{F}(y,u;\widehat{\theta})$ with respect to $\widehat{\theta}$. Notice that for $t \geq T$, we have that

$$\dot{\Phi}(t) = -\Lambda \Phi(t), \quad \Phi(T) = I.$$

We use the following Lyapunov function candidate

$$V = \frac{1}{2} e^T \Pi e + \frac{1}{2} \tilde{\theta}^T \Gamma^{-1} \tilde{\theta} + \frac{\lambda}{2} \text{tr}\left(\Phi^T \Lambda^{-1} \Phi\right), \quad (3.9)$$

where the constant $\lambda > 0$ will be defined below. Using the above error equation we have that

$$\dot{V} \leq -\frac{\lambda_{min}(Q)}{4}|e|^2 - \frac{\lambda}{2}|\Phi|^2 + k(|\omega|^2 + |\xi|^2). \quad (3.10)$$

In the inequality above we used $\Pi B = C^T$ from Assumption 3, the identity $b^T a = tr(ab^T)$, the completion of squares and the fact that the contribution due to the projection term is negative (see [4, 24] for similar results used according to standard adaptive control techniques). Furthermore, the smoothness assumption of $\widehat{F}$ and the uniform boundedness of y and u were used (Assumption 1) to derive the bound $c_1 = \|C\| \sup_{t \geq T} \{\widehat{F}(y,u;\theta^*)\}$. The constant λ can now be defined as

$$\lambda = \frac{3c_1^2}{\lambda_{min}(Q)} \quad \text{and} \quad k = \frac{3}{\lambda_{min}(Q)} \max\left(\|C\|^2, \|\Pi\|^2\right).$$

When

$$\frac{\lambda_{min}(Q)}{4}|e|^2 + \frac{\lambda}{2}|\Phi|^2 > k\left(|\omega|^2 + |\xi(x,u,t)|^2\right)$$

we have that $\dot{V} \leq 0$. Since the projection operator guarantees that $\widehat{\theta}$ is uniformly bounded, we then have that $V, e, \tilde{\theta}$ are uniformly bounded. When (3.10) is integrated over the interval $[T, T+\tau]$ we have that

$$V(T+\tau) + \int_T^{T+\tau} \left(\kappa_1 |e(t)|^2 + \kappa_2 |\Phi(t)|^2\right) dt \leq V(T)$$

$$+ k \int_T^{T+\tau} \left(|\omega(t)|^2 + |\xi(x(t),u(t),t)|^2\right) dt,$$

where

$$\kappa_1 = \frac{\lambda_{min}(Q)}{4} \quad \text{and} \quad \kappa_2 = \frac{\lambda}{2}.$$

By rearranging terms and performing some calculations, we can show that

$$\int_T^{T+\tau} |e(t)|^2\, dt \leq \mu_1 + \mu_2 \int_T^{T+\tau} |\tilde{\omega}(t)|^2\, dt, \tag{3.11}$$

where

$$\mu_1 = \sup_{t \geq 0} \left(\frac{V(T) - V(T+\tau)}{\kappa_1} \right),$$

$$\mu_2 = \frac{k}{\kappa_1}, \quad |\tilde{\omega}(t)|^2 = |\omega(t)|^2 + |\xi(x(t), u(t), t)|^2.$$

The above implies that the extended L_2-norm of the state error $e(t)$ over any finite time interval is at most of the same order as the extended L_2-norm of $\tilde{\omega}(t)$, i.e. e is $|\tilde{\omega}|^2$-small in the mean: $e \in \mathcal{S}(|\tilde{\omega}|^2)$, [24]. In the event that we have:

(i) no modeling uncertainties present ($\xi = 0$),
(ii) linearly parameterized approximators giving $\mathcal{D} \equiv 0$, and
(iii) exact possible matching between the approximator and the fault term (i.e. $\nu(t) \equiv 0$) leading to $\tilde{\omega} \equiv 0$,

an application of Barbălat's lemma [24], yields convergence of the state error to zero, $\lim_{t \to \infty} e(t) = 0$. The above L_2 type of bounds for the state error simplify in the case of zero modeling errors ($\xi(x, u, t) = 0$). The difference in this case is that (3.11) is now given by

$$\int_T^{T+\tau} |e(t)|^2\, dt \leq \mu_1 + \tilde{\mu}_2 \int_T^{T+\tau} |\omega(t)|^2\, dt,$$

with

$$\tilde{\mu}_2 = \frac{1}{\kappa_1} \frac{2\|C\|^2}{\lambda_{min}(Q)},$$

and the value of λ in (3.9) is changed

$$\text{from} \quad \frac{3c_1^2}{\lambda_{min}(Q)} \quad \text{to} \quad \frac{4c_1^2}{\lambda_{min}(Q)}.$$

The above diagnostic observer can be used to detect the time T of the failure occurrence by monitoring the output of the on-line approximator $\hat{F}$. When the on-line approximator starts adapting, it gives an indication of a failure. Convergence of the state error to zero cannot be inferred unless it is assumed that $\nu = \mathcal{D} = \xi = 0$. In this case, parameter convergence can be achieved by imposing the *persistence of excitation condition*, [24] and hence leading to unbiased fault diagnosis.

4 Fault Accommodation

In this section we employ the above diagnostic observer for the control reconfiguration required for failure accommodation. We first design a robust controller for the plant (2.1) taking into account the uncertainties $\xi(x, u, t)$, and then modify it in order to account for the actuator failure.

Towards this end, let's denote by $u_0(t)$ the robust controller for the plant prior to failure. Then the proposed accommodation for (2.4) takes the implicit form:
$$\text{choose } u(t) \text{ such that}: \quad \left(I(\cdot) + \widehat{F}(y, \cdot; \widehat{\theta})\right)[u] = u_0. \tag{4.1}$$

In a first inspection, it appears that this would cancel the actuator failure in the state estimator equation (3.1) giving
$$\dot{\widehat{x}} = A\widehat{x} + Bu_0 + L\varepsilon, \quad \widehat{y} = C\widehat{x}.$$

The above is valid only in the case where the implicit control (4.1) is well-defined, i.e. when the inversion
$$\left(I(\cdot) + \widehat{F}(y, \cdot; \widehat{\theta})\right)[u] = u_0$$
is feasible. Even in the special case of $F(y, u) = Q(y)u$ with $I + Q(y)$ being invertible, it must be ensured in the adaptation law (3.6) that
$$\left\| I + \widehat{Q}(y; \widehat{\theta}) \right\| \geq c$$
is enforced in order to avoid large inputs, where $\|\cdot\|$ is an appropriate norm and c is a small positive constant. To ensure that the above is satisfied, we modify the control law (4.1) which now becomes
$$u = \begin{cases} \left(I + \widehat{Q}(y; \widehat{\theta})\right)^{-1} u_0 & \text{if } \left\| I + \widehat{Q}(y; \widehat{\theta}) \right\| \geq c, \\ \dfrac{1}{c} u_0 & \text{if } \left\| I + \widehat{Q}(y; \widehat{\theta}) \right\| < c. \end{cases}$$

It is worth noting that the above modification to the control law (4.1) is implemented for both before and after the failure detection since $\widehat{Q} = 0$ before the failure which in turn yields $u = u_0$. Hence this does not require a different diagnostic observer for fault accommodation.

Failure accommodation for the general case of $F(y, u)$ (instead of $Q(y)u$) imposes additional mathematical requirements in order to guarantee that $\left(I + \widehat{F}(y, \cdot; \widehat{\theta})\right)[u] = u_0$ can be inverted. This is given as follows: We denote the function in (4.1) by
$$G(y, u, u_0, \widehat{\theta}) = Iu + \widehat{F}(y, u; \widehat{\theta}) - Iu_0 = 0,$$
and use the *implicit function theorem* in [25, 26] to solve u in terms of u_0 in a neighborhood of any point (u^*, u_0^*) at which $G(y, u^*, u_0^*, \widehat{\theta}) = 0$ and $\nabla_u G \neq 0$. It is assumed that the gradient of $G(y, u^*, u_0^*, \widehat{\theta})$ exists.

Remark 4.1. In the special case where the fault is simply a constant gain matrix of the form

$$\dot{x} = Ax + \xi(x, u) + B\left[I + \Delta(t - T)F\right]u$$

with F a diagonal $p \times p$ constant matrix, then the detection observer takes the form

$$\dot{\hat{x}} = A\hat{x} + B\left[I + \widehat{F}(t)\right]u + L\epsilon.$$

The diagonal entries of $\widehat{F}(t)$ are updated as follows

$$\dot{\hat{f}}_i(t) = \mathcal{P}\left\{D[\epsilon(t)]u_i(t)\right\}, \qquad i = 1, 2, \ldots, p,$$

and the control re-configuration then becomes

$$u = \begin{cases} \left(I + \widehat{F}(t)\right)^{-1} u_0 & \text{if } \left\|I + \widehat{F}(t)\right\| \geq c, \\ \dfrac{1}{c} u_0 & \text{if } \left\|I + \widehat{F}(t)\right\| < c. \end{cases}$$

5 Special Case: Vector Second-order Systems

Many structural/mechanical systems are generally written in terms of the vector second-order system

$$M\ddot{x}(t) + D\dot{x}(t) + Kx(t) = Bu(t), \tag{5.1}$$
$$x(0) = x_0, \quad \dot{x}(0) = x_1,$$

having output

$$y(t) = C\begin{bmatrix} x(t) \\ \dot{x}(t) \end{bmatrix}, \tag{5.2}$$

where the system matrices M, D, K are termed as the *mass, damping* and *stiffness* matrices, respectively [27, 29]. The matrices B and C are the *input* (or control influence, [30, 31]) and *output* matrices. It is assumed that the matrices $M, D, K \in \mathbb{R}^{n \times n}$ and that the system has p inputs and $2p$ outputs with $1 \leq p < n$; hence $B \in \mathbb{R}^{n \times p}$ and $C \in \mathbb{R}^{2p \times 2n}$. A special case of structural systems is the one with symmetric or structural output [30, 31], namely a system that has collocated sensors and actuators. Here we consider collocated systems with $C = \text{diag}(B^T, B^T)$. Thus we have that both displacement and velocity (collocated) outputs are available and are given by

$$y(t) = \begin{bmatrix} B^T & 0_{p \times n} \\ 0_{p \times n} & B^T \end{bmatrix} \begin{bmatrix} x(t) \\ \dot{x}(t) \end{bmatrix} = \begin{bmatrix} B^T x(t) \\ B^T \dot{x}(t) \end{bmatrix} = \begin{bmatrix} y_d(t) \\ y_v(t) \end{bmatrix}.$$

The plant with the actuator fault incorporated into its dynamics is then given by

$$M\ddot{x}(t) + D\dot{x}(t) + Kx(t) = B\left[u(t) + \Delta(t-T)F(y(t), u(t))\right], \quad (5.3)$$

where the $p \times 1$ term $\Delta(t-T)F(y,u)$ denotes the actuator fault. The diagonal $p \times p$ matrix $\Delta(t-T)$ is the time profile of the fault and the $p \times 1$ vector function $F(\cdot, \cdot) : \mathbb{R}^{2p} \times \mathbb{R}^p \to \mathbb{R}^p$ is assumed to be an unknown nonlinear function of the measurable output y and input u. In the same manner, we assume that each actuator failure has a distinct time profile and hence a diagonal matrix $\Delta(t-T)$ is assumed. Before we propose the state estimator for mechanical systems we make the following assumptions.

Assumption 4.

(A1) The matrices $(\mathcal{A}, \mathcal{B})$ in the first-order formulation of (5.1), (5.2) below

$$\frac{d}{dt}\begin{bmatrix} x(t) \\ \dot{x}(t) \end{bmatrix} = \begin{bmatrix} 0_{n \times n} & I_{n \times n} \\ -M^{-1}K & -M^{-1}D \end{bmatrix} \begin{bmatrix} x(t) \\ \dot{x}(t) \end{bmatrix} + \begin{bmatrix} 0_{n \times p} \\ M^{-1}B \end{bmatrix} u(t)$$

$$= \mathcal{A}\xi(t) + \mathcal{B}u(t)$$

$$y(t) = \begin{bmatrix} B^T & 0_{p \times n} \\ 0_{p \times n} & B^T \end{bmatrix} \begin{bmatrix} x(t) \\ \dot{x}(t) \end{bmatrix} = \mathcal{C}\xi(t) \quad (5.4)$$

form a controllable pair.

(A2) The matrices $(\mathcal{A}, \mathcal{C})$ in (5.4) above, form an observable pair.

(A3) The function $\Big(I(\cdot) + \Delta(t-T)F(y, \cdot)\Big)[u] \neq 0$ for all $t \geq T$, i.e. the actuator failure does not cause loss of controllability.

The following vector second-order state estimator is proposed for failure detection

$$M\ddot{\widehat{x}}(t) + \widetilde{D}\dot{\widehat{x}}(t) + \widetilde{K}\widehat{x}(t) = Bu(t) + B\widehat{F}(y(t), u(t); \widehat{\theta}(t)) + G_d^T y_d(t) + G_v^T y_v(t),$$

$$\widehat{x}(0) = x_0, \quad \dot{\widehat{x}}(0) = x_1,$$

$$\widehat{y}(t) = \begin{bmatrix} B^T \widehat{x}(t) \\ B^T \dot{\widehat{x}}(t) \end{bmatrix} = \begin{bmatrix} \widehat{y}_d(t) \\ \widehat{y}_v(t) \end{bmatrix}, \quad (5.5)$$

where

$\widehat{x}(t)$ is the state displacement estimate,
$\dot{\widehat{x}}(t)$ is the state velocity estimate,
$\widehat{y}(t)$ is the observer output, and
$\widehat{F}(y, u; \widehat{\theta})$ is the *on-line approximation model* or fault estimate.

The q-dimensional vector $\widehat{\theta}$ is a vector of *adjustable parameters*. The 'observer' damping and stiffness matrices are given by $\widetilde{D} = (D + G_v B^T)$ and $\widetilde{K} = (K + G_d B^T)$, respectively. The gains G_d and G_v are observer gains that are chosen so that the resulting Luenburger-type observer error, in the absence of any faults, will converge to zero. It should be noted that the above state estimator is a non-standard one, albeit natural for vector second-order systems [32], since in the design of estimators for mechanical systems, one usually writes the second-order system (5.1) as a first-order system (5.4) and *then* designs an observer for the equivalent first-order system. The reason that we prefer the above observer (5.5) is that it does not require a *positive realness* condition [19, 24, 33] imposed on its first-order form given by (5.4). This positive real condition was imposed in the general case of dynamical systems presented earlier in order to guarantee the stability of the detection scheme. Collocation along with the ability of the observer damping $\widetilde{D}$ and stiffness $\widetilde{K}$ matrices to shift the observer state error modes further to the left eliminates the need to enforce positive realness condition on (5.4).

The resulting state error equation, with $e(t) = x(t) - \widehat{x}(t)$ and $\epsilon_d(t) = y_d(t) - \widehat{y}_d(t)$, $\epsilon_v(t) = y_v(t) - \widehat{y}_v(t)$, is given by

$$M\ddot{e}(t) + \widetilde{D}\dot{e}(t) + \widetilde{K}e(t) = B\left(\Delta(t-T)F(y,u) - \widehat{F}(y,u;\widehat{\theta})\right),$$

$$\epsilon(t) = y(t) - \widehat{y}(t) = \begin{bmatrix} \epsilon_d(t) \\ \epsilon_v(t) \end{bmatrix}, \quad e(0) = \dot{e}(0) = 0.$$

(5.6)

The following adaptation rule, derived from Lyapunov redesign methods [23], is used for the parameter update

$$\dot{\widehat{\theta}}(t) = -\mathcal{P}\left\{\Gamma Z^T B^T D[\gamma\dot{e} + e]\right\} = -\mathcal{P}\left\{\Gamma Z^T D[\gamma\epsilon_v + \epsilon_d]\right\}$$
$$\widehat{\theta}(0) = \widehat{\theta}_0,$$

(5.7)

where the gain $\gamma > 0$ is a 'tuning' parameter to be defined below, the $q \times q$ symmetric positive definite matrix Γ denotes the *adaptation learning rate* [24, 33] and the $p \times q$ *regressor matrix function* $Z(y,u)$ is given by

$$Z(y,u) = \left[\frac{\partial \widehat{F}(y,u;\widehat{\theta})}{\partial \widehat{\theta}}\right].$$

(5.8)

The initial conditions $\widehat{\theta}_0$ are chosen so that $\widehat{F}(y,u;\widehat{\theta}_0) = 0$ for all y and u.

The system (5.6) prior to failure (i.e. for $t < T$) is in equilibrium and thus $e(t) = 0$ with $\widehat{\theta}(t) = \widehat{\theta}_0$ for $t \in [0, T)$. The above deviates from its equilibrium at $t \geq T$ and hence $e(t) \neq 0$ and $\widehat{\theta}(t) \neq \widehat{\theta}_0$ for $t \in [T, \infty)$. By monitoring either the position or velocity outputs, $\epsilon_d(t)$ and $\epsilon_v(t)$, respectively, or the on-line approximator $\widehat{F}(y,u;\widehat{\theta})$, an actuator fault is declared whenever any of the above signals becomes nonzero.

In order to analyze the stability of the proposed detection scheme after the failure occurrence (i.e. after $t \geq T$) using Lyapunov stability theory, we rewrite the right-hand side of the error equation (5.6) as

$$B\left(\Delta(t-T)F(y,u) - \widehat{F}(y,u;\widehat{\theta})\right) = B\left(\Delta(t-T)(F(y,u) - \widehat{F}(y,u;\theta^*))\right.$$
$$\left. + \Delta(t-T)\widehat{F}(y,u;\theta^*) - \widehat{F}(y,u;\widehat{\theta})\right)$$
$$= B\nu(t) + B\left(\Delta(t-T)\widehat{F}(y,u;\theta^*) - \widehat{F}(y,u;\widehat{\theta})\right),$$

where now the approximation error $\nu(t)$ is given by

$$\nu(t) = \Delta(t-T)\left(F(y,u) - \widehat{F}(y,u;\theta^*)\right). \tag{5.9}$$

Using the above and the mean value theorem, the state error equation (5.6) can now be rewritten as

$$M\ddot{e}(t) + \widetilde{D}\dot{e}(t) + \widetilde{K}e(t) = -B\Phi(t)\widehat{F}(y,u;\theta^*) - BZ(y,u)\widetilde{\theta}(t) + B\omega(t), \tag{5.10}$$

where

$$\widetilde{\theta}(t) = \widehat{\theta}(t) - \theta^*, \qquad \Phi(t) = I - \Delta(t),$$

$$\omega(t) = \nu(t) - \left[\widehat{F}(y,u;\widehat{\theta}) - \widehat{F}(y,u;\theta^*) - \frac{\partial \widehat{F}(y,u;\widehat{\theta})}{\partial \widehat{\theta}}(\widehat{\theta} - \theta^*)\right].$$

Once again, the last term represents the higher-order terms of the Taylor series expansion of $\widehat{F}(y,u;\widehat{\theta})$ with center θ^*. The matrix $\Phi(t)$ once again satisfies the following

$$\frac{d}{dt}\Phi(t) = -\Lambda\Phi(t), \qquad \Phi(T) = I, \qquad t \geq T. \tag{5.11}$$

We use the following *parameter-dependent* Lyapunov function

$$\begin{aligned}V(t) &= \frac{\gamma}{2}\left[e^T(t)\widetilde{K}e(t) + \dot{e}^T(t)M\dot{e}(t)\right] + \dot{e}^T(t)Me(t) \\ &\quad + \frac{1}{2}e^T(t)\widetilde{D}e(t) + \frac{1}{2}\mathrm{tr}\left(\widetilde{\theta}(t)\Gamma^{-1}\widetilde{\theta}^T(t)\right) \\ &\quad + \frac{\kappa}{2}\mathrm{tr}\left(\Phi^T(t)\Lambda^{-1}\Phi(t)\right)\end{aligned} \tag{5.12}$$

where the constant $\gamma > 0$ is chosen so as to guarantee that V is positive-definite, decrescent and radially unbounded [24, 34], and the constant $\kappa > 0$ will be defined below. Specifically, γ is chosen as

$$\gamma > \max\left\{\frac{\lambda_{max}(M)}{\lambda_{min}(M)}, \frac{\lambda_{max}(M) - \lambda_{min}(\widetilde{D})}{\lambda_{min}(\widetilde{K})}, \frac{2\lambda_{max}(M)}{\lambda_{min}(\widetilde{D})}\right\}, \tag{5.13}$$

where $\lambda_{min}(M)$ and $\lambda_{max}(M)$ denote the smallest and largest eigenvalues of a matrix M, respectively. The above Lyapunov function consists of the standard energy function (kinetic: $\frac{1}{2}\dot{x}^T M\dot{x}$ and potential: $\frac{1}{2}x^T\widetilde{K}x$) multiplied by the tuning parameter γ, plus additional terms that would allow for exponential convergence of the state error $(e, \dot{e})$ to zero in the absence of faults. When the derivative of (5.12) is evaluated at the solutions to (5.10), it yields

$$\begin{aligned}
\dot{V}(t) &= -\gamma \dot{e}^T(t) \widetilde{D} \dot{e}(t) + \dot{e}^T(t) M \dot{e}(t) - e^T(t) \widetilde{K} e(t) - \kappa \operatorname{tr}\left(\Phi^T(t) \Phi(t)\right) \\
&\quad - (\gamma \epsilon_v(t) + \epsilon_d(t))^T \Phi(t) \widehat{F}(y; \theta^*) u(t) + (\gamma \epsilon_v(t) + \epsilon_d(t))^T w(t) \\
&\leq -\left(\frac{\gamma}{2} \lambda_{min}(\widetilde{D}) - \lambda_{max}(M)\right) |\dot{e}(t)|^2 - \frac{1}{2} \lambda_{min}(\widetilde{K}) |e(t)|^2 \\
&\quad - \left[\kappa - \left(\frac{\gamma c_1^2}{\lambda_{min}(\widetilde{D})} + \frac{c_1^2}{\lambda_{min}(\widetilde{K})}\right)\right] \|\Phi(t)\|^2 \\
&\quad + \left(\frac{\gamma \|B\|^2}{\lambda_{min}(\widetilde{D})} + \frac{\|B\|^2}{\lambda_{min}(\widetilde{K})}\right) |w(t)|^2 \\
&\leq -\alpha_v |\dot{e}(t)|^2 - \alpha_d |e(t)|^2 - \frac{\kappa}{2} \|\Phi(t)\|^2 + \mu |w(t)|^2,
\end{aligned}$$
(5.14)

where

$$\alpha_v = \left(\frac{\gamma}{2} \lambda_{min}(\widetilde{D}) - \lambda_{max}(M)\right), \qquad \alpha_d = \frac{1}{2} \lambda_{min}(\widetilde{K}),$$

$$\kappa = 2 c_1^2 \left(\frac{\gamma}{\lambda_{min}(\widetilde{D})} + \frac{1}{\lambda_{min}(\widetilde{K})}\right), \qquad \mu = \|B\|^2 \left(\frac{\gamma}{\lambda_{min}(\widetilde{D})} + \frac{1}{\lambda_{min}(\widetilde{K})}\right),$$

and the constant c_1 is given by $c_1 = \|B\| \sup_{t \geq T} \left\{\widehat{F}(y, u; \theta^*)\right\}$. When

$$\alpha_v |\dot{e}(t)|^2 + \alpha_d |e(t)|^2 + \frac{\kappa}{2} \|\Phi(t)\|^2 > \mu |w(t)|^2,$$

we have that $\dot{V} \leq 0$. Then we have that $V, e, \widehat{\theta}$ are uniformly bounded. Integrating the last equation we obtain

$$V(T+\tau) + \int_T^{T+\tau} \left(\alpha_v |\dot{e}(t)|^2 + \alpha_d |e(t)|^2 + \frac{\kappa}{2} \|\Phi(t)\|^2\right) dt$$
$$\leq V(T) + \mu \int_T^{T+\tau} |w(t)|^2 dt$$

Similar to the general case, we have that the extended L_2-norms of the velocity and position state errors are at most of the same order as the extended L_2-norm of w. In the case of linearly parameterized approximators (i.e. $\widehat{F}(y, u; \widehat{\theta}) = Z(y, u) \widehat{\theta}$) and exact matching of the approximators and the failure term (i.e. $\nu \equiv 0$), we can show asymptotic convergence of the state position and velocity errors to zero.

Remark 5.1. As was similarly done in the general case of dynamical systems that require the strictly positive realness condition (Assumption 3), an extension to the above can also be implemented that takes into account additional terms in (5.1) due to modeling errors. Due to these modeling errors, the output errors ϵ_d and ϵ_v attain nonzero values prior to the unanticipated failure time $(t < T)$. This would give false alarms while monitoring either the on-line approximator or the output errors for fault detection. In this case, a dead-zone modification to the adaptation law (5.7), can be used to avoid false alarms due to these modeling errors. Briefly, it forces the adaptation (5.7) to remain at zero unless a certain threshold δ is exceeded, i.e. it guarantees that the set $\{(e, \dot{e}, \widehat{F}) : |(e, \dot{e})| < \delta, \widehat{F} = 0\}$ is positively invariant, [35, 34].

Remark 5.2. A way to improve the convergence properties of the above adaptation rule for the on-line approximator, is to use techniques from adaptive control and estimation, [24, 33]. One such modification to use is the so-called σ modification in (5.7), giving

$$\dot{\widehat{\theta}}(t) = -\Gamma Z^T(\gamma \epsilon_v + \epsilon_d) - \sigma \widehat{\theta}.$$

When either the fixed or the switching σ modification is used, they guarantee exponential convergence of the Lyapunov function (and hence $(e, \dot{e}, \widetilde{\theta}, \varPhi)$) to a certain residual set described by the size of the modeling errors and the bounds on σ. Continuing, if the parameters are known to belong to a certain set then, as was similarly done in [3, 4] for the detection of abrupt state failure, a parameter projection modification [24] can be used that would constrain the parameters $\widehat{\theta}$ to some selected compact, convex region of the parameter space. This projection method would prevent parameter drift of the adjustable parameters $\widehat{\theta}$, a phenomenon that often occurs with standard adaptive laws in the presence of modeling uncertainty and approximation error [24].

Remark 5.3. Regarding the controllability of the pair $(\mathcal{A}, \mathcal{B})$ in Assumption (A1) above, one can take advantage of the algebraic structure of the vector second-order systems and express the condition in terms of the matrices K and B, as explained in section 5.2.1.4 in [31]. Similar results hold for the observability of the pair $(\mathcal{C}, \mathcal{A})$ in Assumption (A2), see section 5.4.4 in [31].

In order to accommodate the actuator failure and avoid possible lack of performance or even instability, we propose a reconfiguration of the control law. By denoting the control law under ideal conditions (i.e. when $F(y, u) \equiv 0$) with $u_0(t)$, we propose a modification that resembles the certainty equivalence principle. We use the control law by assuming that the nature of failures is known and then replace it by the on-line approximator. A modification to the controller for (5.3) takes the form

$$u(t) = \Big(I(\cdot) + F(y,\cdot)\Big)^{-1}[u_0(t)] \qquad \text{when nature of failure is known}$$

$$u(t) = \Big(I(\cdot) + \widehat{F}(y,\cdot;\widehat{\theta})\Big)^{-1}[u_0(t)] \qquad \text{when nature of failure is unknown.}$$

This changes (5.3) to

$$\begin{aligned} M\ddot{x}(t) + D\dot{x} + Kx(t) &= B\Big(I(\cdot) + F(y,\cdot)\Big)[u(t)] \\ &= B\Big(I(\cdot) + F(y,\cdot)\Big)\Big(I(\cdot) + F(y,\cdot)\Big)^{-1}[u_0(t)] \\ &= Bu_0(t) \end{aligned}$$

when the failure is known. The above hinges on the premise that the matrix function $I(\cdot) + F(y,\cdot)$ is always invertible and that $F(y,\cdot)$ is known. Since the failure is not known, it is replaced by its on-line approximator. While it may be the case that $I + Q(y)$ is invertible (Assumption ($A3$)) for the special case $F(y,u) = Q(y)u$, there is no guarantee from the adaptation law (5.7) that the term $I + \widehat{Q}(y;\widehat{\theta})$ would be invertible. This can lead to unbounded control signals which in turn can destabilize the plant. Similar to the general case, a way to ensure boundedness of the input signal when the matrix $I + \widehat{Q}(y;\widehat{\theta})$ becomes singular (or near singular) is to use the following saturation-type control modification

$$u(t) = \begin{cases} \Big(I + \widehat{Q}(y;\widehat{\theta})\Big)^{-1} u_0(t) & \text{if } \|I + \widehat{Q}(y;\widehat{\theta})\| > c > 0 \\ \dfrac{1}{c} u_0(t) & \text{if } \|I + \widehat{Q}(y;\widehat{\theta})\| \le c. \end{cases} \qquad (5.15)$$

6 Examples and Numerical Results

As a first example, we consider the second-order system with state matrices

$$A = \begin{bmatrix} -1.0 & 0.4 \\ 0.2 & -2.0 \end{bmatrix}, \quad B = \begin{bmatrix} 1.0 \\ 0.01 \end{bmatrix}, \quad C = \begin{bmatrix} 2.01 & 1.04 \end{bmatrix},$$

with observer gain $L = \begin{bmatrix} 0.1 & 0.2 \end{bmatrix}^T$ and Lur'e matrices (cf. equation (3.2)) given by

$$\Pi = \begin{bmatrix} 2 & 1 \\ 1 & 4 \end{bmatrix}, \quad Q = \begin{bmatrix} 5.208 & 3.625 \\ 3.625 & 17.072 \end{bmatrix}.$$

The nonlinear actuator fault function is

$$\Delta(t-T)F(y) = (1 - e^{-\lambda(t-1)})\frac{10y}{1+y^2},$$

with a failure commencing at $t = 1$ seconds. We treat the case of incipient failures with a fault rate of $\lambda = 1.0$. The modeling uncertainty is given by

$$\xi(x,u,t) = \frac{0.1}{1+\|x(t)\|}\begin{bmatrix} x_1 \\ -x_1 x_2 \end{bmatrix},$$

which, using (3.6) with $N/\alpha = 0.4113$ and $\xi_0 = 0.05$, yields a threshold of $g_0 = 0.211$.

We use a radial basis function network as the on-line approximator model and thus we have

$$\widehat{F}(y;\widehat{\theta}) = \sum_{i=1}^{l} \widehat{\theta}_i(t)\exp\left(-|y-c_i|^2/\sigma^2\right) = Z^T(y)\widehat{\theta}(t)$$

with $l = 11$. The centroids are set uniformly over the interval $[-5, 5]$ and $\sigma = 5/[(l-1)\sqrt{\log(2)}]$. The adaptive gain, Γ in (3.3), is chosen as $\Gamma = 50$.

In a set of three simulations, we studied the case of no failures (healthy system), then actuator failures without any accommodation and lastly the case in which fault accommodation is utilized.

The case of no failures is presented in Figure 6.1 where we present the output error, in Figure 6.1(a), and the plant and observer outputs in Figure 6.1(b). In Figure 6.1(c), we present the evolution of the output error $\varepsilon(t)$ vs time, for the case of no accommodation, and we observe that it attains a nonzero value at the failure time $T = 1$ seconds and converges to zero afterwards. Both the plant output $y(t)$ and the observer output $\widehat{y}(t)$ (dashed) are depicted in Figure 6.1(d) where we observe that the estimator output tracks the plant output everywhere except around the failure time. The same plots are duplicated in Figure 6.1(e),(f) for the case of failure accommodation. It is observed that when the accommodation is activated, the output error gives a smaller norm compared to the case of no accommodation (compare Figures 6.1(c) and 6.1(e)). Closer examination of the above figures reveals that when failure accommodation is incorporated, the plant output converges to the steady state value of 0.5 (Figure 6.1(f)) which is the same steady state value under no failures (Figure 6.1(b)). Inspection of Figure 6.1(d) reveals that the steady state value of the plant output is roughly 2.25, well above that of 0.5 for the ideal case. The evolution of the fault term $F(y)$ and its on-line approximator $\widehat{F}(y;\widehat{\theta})$ (dashed) are presented in Figure 6.2(a) where the diagnosis of the failure can be observed. Figure 6.2(b) depicts the evolution of the normalized L_2 norm of the diagnosis error, given by

$$\left\|F(y)-\widehat{F}(y;\widehat{\theta})\right\|_{2N} = \frac{\sqrt{\int_0^{2.2} \left|F(y)-\widehat{F}(y;\widehat{\theta})\right|^2 dy}}{\sqrt{\int_0^{2.2} |F(y)|^2 dy}}.$$

From Figure 6.2(b), we conclude that the OLA $\widehat{F}(y;\widehat{\theta})$ is able to learn the fault function $F(y)$ within the domain in which y varies. This domain in

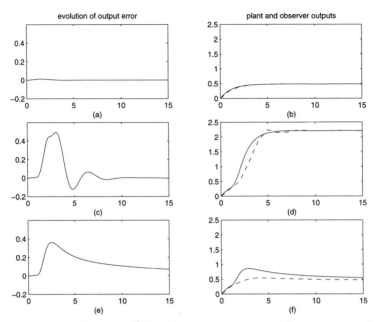

Fig. 6.1. Evolution of ε and y, $\widehat{y}$: healthy system, system without accommodation, and system with accommodation.

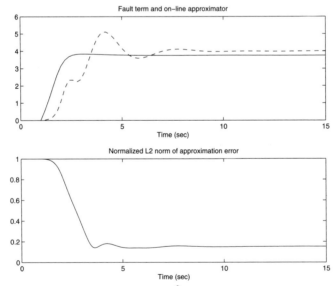

Fig. 6.2. Evolution of (a) f (solid) and $\widehat{f}$ (dashed), (b) normalized L_2 norm of approximation error.

which y varies was found to be $[0, 2.216]$ for the case of no accommodation and hence the interval $[0, 2.2]$ was used in the above norm evaluation.

As a second example, we consider a plant whose dynamics describe the transverse vibration of a uniform beam fixed at both ends with one (force) actuator located at $L/4$ as seen in Figure 6.3.

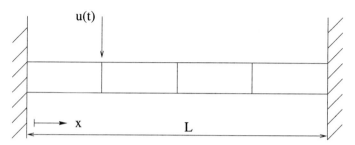

Fig. 6.3. Uniform beam fixed at both ends.

We use the same structural data as in [30]. When four beam elements are used to discretize the corresponding partial differential equation, we arrive at the following matrices

$$M = \begin{bmatrix} 312 & 0 & 54 & -13 & 0 & 0 \\ 0 & 8 & 13 & -3 & 0 & 0 \\ 54 & 13 & 312 & 0 & 54 & -13 \\ -13 & -3 & 0 & 8 & 13 & -3 \\ 0 & 0 & 54 & 13 & 312 & 0 \\ 0 & 0 & -13 & -3 & 0 & 8 \end{bmatrix},$$

$$K = \begin{bmatrix} 24 & 0 & -12 & 6 & 0 & 0 \\ 0 & 8 & -6 & 2 & 0 & 0 \\ -12 & -6 & 24 & 0 & -12 & 6 \\ 6 & 2 & 0 & 8 & -6 & 2 \\ 0 & 0 & -12 & -6 & 24 & 0 \\ 0 & 0 & 6 & 2 & 0 & 8 \end{bmatrix},$$

with damping given as a combination of air and Kelvin-Voigt viscoelastic damping, $D = \alpha M + \beta K$ with $\alpha = \beta = 0.001$. The input vector is given by

$$B = \begin{bmatrix} 0 & 1 & 0 & 0 & 0 & 0 \end{bmatrix}^T.$$

The failure term and its time profile are given by

$$F(y) = 5 \left(2 + \cos(y_d) + \frac{y_d y_v \sin(y_v)}{1 + y_d^2 + y_v^2} \right), \qquad \Delta(t-T) = H(t-2)\left(1 - e^{-(\frac{t-2}{2})}\right).$$

As the observer gains G_v, G_d in (5.5) we have chosen $G_d = -B$ and $G_v = -50B$ and as the input to the system we used

$$u(t) = -K \begin{bmatrix} \widehat{x}(t) \\ \dot{\widehat{x}}(t) \end{bmatrix} + 10\sin(\pi t/10),$$

where the feedback gain is derived from solution of an appropriate Riccati equation for (5.4).

In Figure 6.4(a) we plot the evolution of the failure term $F(y)$ (solid) and its on-line approximator $\widehat{F}(y;\widehat{\theta})$ (dashed). It is observed that the on-line approximator (OLA) converges to the actual term in about 2 seconds, as seen also in Figure 6.4(b). The observer output errors (displacement $y_d(t) - \widehat{y}_d(t)$, and velocity $y_v(t) - \widehat{y}_v(t)$) are depicted in Figure 6.5 and both converge to zero. The effects of accommodating the actuator failure are seen in Figure 6.6 where

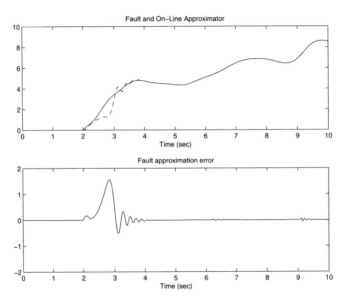

Fig. 6.4. Evolution of (a) failure term $F(y)$ (solid) and its on-line approximator $\widehat{F}(y;\widehat{\theta})$ (dashed), (b) failure error $F(y) - \widehat{F}(y;\widehat{\theta})$.

we compare the difference $y_{ideal} - y_{noaccom}$ between the output of the ideal case (no faults) and the case of no fault accommodation and the difference $y_{ideal} - y_{accom}$ between the ideal case and the case of fault accommodation. This is also the case for the velocity output errors as depicted in Figure 6.6(b).

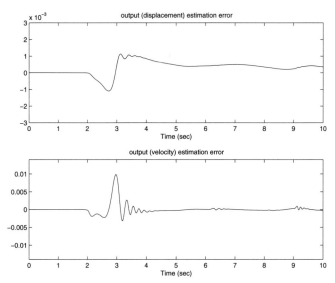

Fig. 6.5. Evolution of (a) displacement estimation error $B^T(x(t) - \widehat{x}(t))$ and (b) velocity estimation error $B^T(\dot{x}(t) - \dot{\widehat{x}}(t))$.

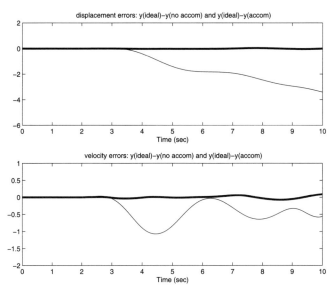

Fig. 6.6. Evolution of (a) displacement errors $(B^T x_{\text{ideal}} - B^T x_{\text{w/o accom}})$ (thin line) and $(B^T x_{\text{ideal}} - B^T x_{\text{with accom}})$ (thick line), (b) velocity errors $(B^T \dot{x}_{\text{ideal}} - B^T \dot{x}_{\text{w/o accom}})$ (thin line) and $(B^T \dot{x}_{\text{ideal}} - B^T \dot{x}_{\text{with accom}})$ (thick line).

It can be observed that when the fault is accommodated via an appropriate control reconfiguration, the closed loop system behaves almost like the case of no faults (healthy closed loop system).

7 Conclusions

We have presented a stable detection and diagnostic observer along with its on-line approximator to assess the nature of actuator faults in dynamical systems. A modification to the (nominal) control law designed for the healthy system was proposed in order to accommodate for the actuator failures. The above were demonstrated via examples wherein it was observed that the detection/diagnostic observer detected the unanticipated occurrence of the actuator failure and diagnosed the nature of the failure (learning). The lack of implementation of the fault accommodation via the appropriate control reconfiguration can have detrimental effects in the overall system performance as was seen in the numerical examples.

References

1. H. T. Banks, R. C. Smith, and Y. Wang, *Smart Material Structures: Modeling, Estimation and Control*. New York: Wiley-Masson, 1996.
2. A. Guran and D. J. Inman, *Smart Structures, Nonlinear Dynamics and Control*. Englewood Cliffs, NJ: Prentice Hall Professional Technical Reference, 1995.
3. M. M. Polycarpou and A. T. Vemuri, Learning methodology for failure detection and accomodation, *Control Systems Magazine, special issue on Intelligent Learning Control*, 15, 16–24, 1995.
4. M. M. Polycarpou and A. J. Helmicki, Automated fault detection and accomodation: A learning systems approach, *IEEE Trans. on Systems, Man and Cybernetics*, 25, 1447–1458, 1995.
5. P. M. Frank, Fault diagnosis in dynamic systems using analytical and knowledge-based redundancy - a survey and some new results, *Automatica*, 26, 459–474, 1990.
6. J. Gertler, Survey of model-based failure detection and isolation in complex plants, *IEEE Control System Magazine*, 8, 3–11, 1988.
7. J. J. Gertler, *Fault Detection and Diagnosis in Engineering Systems*. New York-Basel-Hong Kong: Marcel Dekker, 1998.
8. R. Isermann, Process fault detection based on modeling and estimation methods: a survey, *Automatica*, 20, 387–404, 1984.
9. R. J. Patton, Robust model-based fault diagnosis: the state of the art, in *Proc. of the IFAC Symposium on Fault Detection, Supervision and Safety for Processes (SAFEPROCESS)* (Espoo, Finland), pp. 1–24, June 1994.
10. R. J. Patton, P. M. Frank, and R. N. Clark, *Fault Diagnosis in Dynamical Systems: Theory and Applications*. New York: Prentice Hall, 1989.
11. A. Willsky, A survey of design methods for failure detection in dynamic systems, *Automatica*, 12, 601–611, 1976.

12. A. T. Vemuri and M. M. Polycarpou, On-line approximation based methods for robust fault detection, in *Proceedings of the 13th IFAC World Congress* (San Francisco, CA), pp. 319–324, July 1996.
13. H. Wang, Z. J. Huang, and S. Daley, On the use of adaptive updating rules for actuator and sensor fault diagnosis, *Automatica*, 33, 217–225, 1997.
14. A. T. Vemuri and M. M. Polycarpou, Robust nonlinear fault diagnosis in input-output systems, *International Journal of Control*, vol. 68, 2, 343–360, 1997.
15. H. Wang and S. Daley, Actuator fault diagnosis: An adaptive observer-based technique, *IEEE Trans. Automatic Control*, 41, 1073–1078, 1996.
16. M. A. Demetriou and M. M. Polycarpou, Fault detection, diagnosis and accommodation of dynamical systems with actuator failures via on-line approximators, in *Proceedings of the 1998 American Control Conference* (Philadelphia, PA), June 24-26 1998.
17. M. A. Demetriou and M. M. Polycarpou, Fault accommodation of output induced actuator failures for a flexible beam with collocated input and output, in *Proceedings of the 5th IEEE Mediterranean Conference on Control and Systems*, (Paphos, Cyprus), July 21-23, 1997.
18. M. A. Demetriou and M. M. Polycarpou, Fault diagnosis of output-induced actuator failures for a flexible beam with collocated input and output, in *Proceedings of the IFAC Symposium on Fault Detection, Supervision and Safety for Technical Processes: SAFEPROCESS'97*, (University of Hull, Hull United Kingdom), August 26-28, 1997.
19. B. Anderson and S. Vongpanitlerd, *Network Analysis and Synthesis: A Modern Systems Theory Approach*. Prentice Hall Network Series, Englewood Cliffs, NJ: Prentice Hall, 1973.
20. R. E. Kalman, Lyapunov functions for the problem of lure in automatic control, *Proc. Nat. Acad. Sci. USA*, 49, 201–205, 1963.
21. V. A. Yakubovich, The solution to certain inequalities in automatic control theory, *Dokl. Akad. Nauk. USSR*, 143, 1304–1307, 1962.
22. S. Boyd, L. E. Ghaoui, E. Feron, and V. Balakrishnan, *Linear Matrix Inequalities in System and Control Theory*. SIAM Studies in Applied Mathematics, Philadelphia: Society of Industrial and Applied Mathematics, 1994.
23. H. K. Khalil, *Nonlinear Systems*. New York: Macmillan, 1992.
24. P. A. Ioannou and J. Sun, *Robust Adaptive Control*. Englewood Cliffs, NJ: Prentice Hall, 1995.
25. W. Rudin, *Principles of Mathematical Analysis*, 3rd edn. New York: McGraw-Hill, 1976.
26. W. Kaplan and D. J. Lewis, *Calculus and Linear Algebra*. New York: John Wiley & Sons, Inc, 1971.
27. D. Inman, *Vibration with Control, Measurement, and Stability*. Englewood Cliffs, New Jersey: Prentice Hall, 1989.
28. L. Meirovitch, *Computational Methods in Structural Dynamics*. Alphen aan den Rijn, The Netherlands: Sijthoff and Noordhoff, 1980.
29. L. Meirovitch, *Dynamics and Control of Structures*. New York: Wiley, 1990.
30. J. L. Junkins and Y. Kim, *Introduction to Dynamics and Control of Flexible Structures*. Washington, DC: American Institute of Aeronautics and Astronautics, 1993.
31. R. Skelton, *Dynamic Systems Control: Linear Systems Analysis and Synthesis*. New York: John Wiley & Sons, Inc., 1988.
32. M. J. Balas, Do all linear flexible structures have convergent second-order observers?, *AIAA Journal of Guidance, Control, and Dynamics*, 22, 905–908, December 1999.

33. K. S. Narendra and A. M. Annaswamy, *Stable Adaptive Systems*. Englewood Cliffs, NJ: Prentice Hall, 1989.
34. T. L. Vincent and W. J. Grantham, *Nonlinear and Optimal Control*. New York: John Wiley & Sons, Inc., 1997.
35. F. Verhulst, *Nonlinear Differential Equations and Dynamical Systems*. Berlin: Springer-Verlag, 1990.

Adaptive Control of Systems with Actuator Failures

Gang Tao[1] and Suresh M. Joshi[2]

[1] Department of Electrical and Computer Engineering, University of Virginia, Charlottesville, VA 22903, USA (E-mail: **gt9s@virginia.edu**)
[2] Mail Stop 161, NASA Langley Research Center, Hampton, VA 23681, USA

Abstract

Adaptive control schemes are developed for linear time-invariant plants with actuator failures with characterizations that some of the plant inputs are stuck at some fixed or varying values which cannot be influenced by control action. Control problems of state feedback for state tracking, state feedback for output tracking, and output feedback for output tracking are addressed. Conditions and controller structures for achieving plant-model state or output matching in the presence of actuator failures are derived. Adaptive laws are designed for updating the controller parameters when both the plant parameters and actuator failure parameters are unknown. Closed-loop stability and asymptotic state or output tracking are analyzed.

1. Introduction

Adaptive control systems are capable of accommodating system parametric, structural, and environmental uncertainties caused by payload variation or component aging, component failures, and external disturbances. There have been significant advances in both theory and applications of adaptive control (see, for example, [1, 7, 9, 11–16, 18, 21, 22]). The field of adaptive control continues to develop rapidly with emergence of new challenging problems and their innovative solutions. One such challenging problem is adaptive control of systems with actuator failures, which has many applications such as aircraft and spacecraft flight control systems, process control systems, and power systems.

Actuator failures can be uncertain that it is not known when, how much and how many actuators fail; for example, some unknown inputs may be stuck at some unknown fixed values at unknown time instants. For example, a number of aircraft accidents have been caused by actuator failures, such as the horizontal stabilizer or the rudder being stuck in an unknown position, leading to catastrophic failures. Adaptive control of systems with actuator failures is aimed at compensating such uncertainties with adaptive tuning of controller parameters based on system response errors.

For control of systems with component failures, one class of designs consists of multiple model, switching and tuning schemes [5, 6, 10]. Assuming that the controlled plant belongs to a set of plant models, a multiple model, switching and tuning design has several forms [17]: one based all-fixed plant models, one based on all adaptive plant models, one based on fixed models and one adaptive model, and one based on fixed models with one free-running and one reinitialized adaptive model.

Other types of control designs are adaptive schemes. In [2], an indirect adaptive LQ controller is used to accommodate failures in the pitch control channel or the horizontal stabilizer, leading to performance improvement. In [3], several indirect and direct adaptive control algorithms are presented for control of aircraft with a failure characterized as an aircraft with a locked left horizontal tail surface. An adaptive controller is used to accommodate the system dynamics change caused by such a failure. In [4], an adaptive algorithm is used to control a dynamic system with known dynamics but unknown actuator failures. The control law for the known dynamics is based on a model matching design, while the compensation for actuator failure is based on adaptive tuning of actuation parameter matrices. A fault diagnosis method [27] and an adaptive observer-based method [28, 29] have also been used for control of systems with component failures.

Given the need to handle uncertainties in both system dynamics and component failures which are often uncertain in time, value and pattern, that is, when, how much and which components fail, recently, in [23], and [24], we developed direct adaptive state feedback control schemes for unknown linear time-invariant plants with unknown actuator failures characterized by inputs stuck at some values not influenced by control action. Fundamental issues in adaptive control of systems with actuator failures, such as actuator failure models, controller structure, plant-model matching conditions, error models, adaptive laws, stability analysis and performance evaluation, have been addressed for the considered cases: state feedback for state tracking [23], state feedback for output tracking [24]. Some structural conditions on A and B for state tracking in the presence of actuator failures in the system $\dot{x}(t) = Ax(t) + Bu(t)$ can be relaxed when only output tracking is to be achieved for $y(t) = Cx(t)$. When, as in many cases, the internal state variables of a system are not available for measurement, adaptive output feedback control schemes for output tracking are needed.

In this paper, we formulate the problems of adaptive control of systems with *unknown* parameters and in the presence of *unknown* actuator failures characterized by some of the plant inputs being stuck at some unknown fixed or varying values which cannot be influenced by control action. We present recent results and some new designs and analyses, which provide solutions to some of these adaptive control problems, by using state feedback for state tracking, using state feedback for output tracking, and using output feedback for output tracking. The main goal of this paper is to establish a theoretic foundation for adaptive control of systems with actuator failures, by deriving common actuator failure characterizations, designing some basic adaptive control schemes for plants with such actuator failures, and analyzing their performance. The control objective is to develop adaptive control schemes for unknown linear time-invariant systems to achieve both closed-loop signal boundedness and asymptotic tracking of the state or output of a reference system in the presence of unknown actuator failures.

The paper is organized as follows. In Section 2, we introduce the plant model and a basic actuator failure characterization in which some of the plant inputs are stuck at some fixed values, and formulate the control problems: state feedback for state tracking (to be addressed in Sections 3–6) and state or output feedback for output tracking (to be addressed in Sections 7 and 8). In Section 3, we derive the plant-model state matching conditions and control design when the plant is known and with known actuator failure parameters. In Section 4, using the the plant-model state matching conditions and controller structure of Section 3, we develop an adaptive state tracking control scheme for the case when both the plant parameters and actuator failure parameters are unknown. The adaptive scheme ensures closed-loop signal boundedness and asymptotic state tracking in the presence of actuator failures which are uncertain in time, value and pattern, as well as uncertain plant dynamics. In Section 5, we expand the actuator failure model to include actuator failures when some of the plant inputs are oscillating around some fixed values with known oscillation frequencies but unknown magnitudes, and design and analyze adaptive state tracking control schemes with desired stability and tracking properties. In Section 6, we solve the control problem for more general actuator failures which include plant inputs with arbitrary bounded variations around some fixed values. In Section 7, we consider the output tracking problems by deriving the plant-model output matching conditions and control designs when the plant is known and with known actuator failure parameters, using state feedback or output feedback. In Section 8, we develop adaptive control schemes for the case when both the plant parameters and actuator failure parameters are unknown, and analyze closed-loop signal boundedness and asymptotic output tracking in the presence of unknown actuator failures. Finally in Section 9, we summarize the results developed and some open problems for adaptive control of systems with unknown actuator failures.

2. Problem Statement

Consider a plant described by the differential equation

$$\dot{x}(t) = Ax(t) + Bu(t), \quad y(t) = Cx(t) \qquad (2.1)$$

where $A \in R^{n \times n}$, $B \in R^{n \times m}$, $C \in R^{1 \times n}$ are unknown constant parameter matrices, $x(t) \in R^n$ is the state vector, $u(t) = [u_1, \ldots, u_m]^T \in R^m$ is the input vector whose components may fail during system operation, and $y(t) \in R$ is the plant output.

One type of actuator failure [4] is modeled as

$$u_j(t) = \bar{u}_j, \ t \geq t_j, \ j \in \{1, 2, \ldots, m\} \tag{2.2}$$

where the constant value $\bar{u}_j$ and the failure time instant t_j are unknown. For example, when an aircraft control surface (such as the rudder or an aileron) is stuck at some unknown fixed value.

For the control problems considered in this paper, it is assumed that

(A1) If the plant parameters and the actuator failures (with up to $m-1$ failed actuators) are known, the remaining actuators (controls) can still achieve a desired control objective.

This is the basic existence assumption for a nominal solution. The key task of actuator failure compensation control is to adaptively adjust the remaining controls to achieve the desired performance when there are up to $m-1$ unknown actuator failures, without the knowledge of the plant parameters.

For example, suppose the control surfaces (e.g., ailerons) of an aircraft are divided into several individually actuated segments. If some of the ailerons stop moving and stay at fixed positions, the remaining ailerons may still be able to make a safe landing but they need to be controlled at proper positions. An adaptive controller is expected to automatically adjust the positions of the remaining ailerons to the desired values for a safe landing, without knowing which and how many ailerons have failed, and at what fixed positions.

The input vector u to the plant (2.1) in the presence of actuator failures (2.2) can be described as

$$u(t) = v(t) + \sigma(\bar{u} - v(t)) \tag{2.3}$$

where $v(t)$ is a designed control input, and

$$\bar{u} = [\bar{u}_1, \bar{u}_2, \ldots, \bar{u}_m]^T \tag{2.4}$$

$$\sigma = \text{diag}\{\sigma_1, \sigma_2, \ldots, \sigma_m\} \tag{2.5}$$

$$\sigma_i = \begin{cases} 1 & \text{if the } i\text{th actuator fails, i.e., } u_i = \bar{u}_i \\ 0 & \text{otherwise.} \end{cases} \tag{2.6}$$

Our control objective is to solve the adaptive actuator failure compensation control problems which are stated as follows.

State tracking. Given that the plant dynamics matrices (A, B) are unknown, and so are the actuator failure time t_j, parameters $\bar{u}_j$ and pattern j, design a state feedback control $v(t)$ such that all signals in the closed-loop system are bounded and the state vector $x(t)$ asymptotically tracks a given reference state vector $x_m(t)$ generated from a reference model

$$\dot{x}_m(t) = A_M x_m(t) + b_M r(t), \ x_m(t) \in R^n, \ r(t) \in R \qquad (2.7)$$

where $A_M \in R^{n \times n}$, $b_M \in R^n$ are known constant matrices such that all the eigenvalues of A_M are in the open left-half complex plane, and $r(t)$ is a bounded and piecewise continuous reference input.

For this state tracking problem, more general actuator failure models with time-varying failures than that given in (2.2) will also be considered (see Section 5 and Section 6).

Output tracking. Given that the plant dynamics matrices (A, B, C) are unknown, and so are the actuator failure time t_j, parameters $\bar{u}_j$ and pattern j, design a state or output feedback control $v(t)$ such that all signals in the closed-loop system are bounded and the plant output $y(t)$ asymptotically tracks a given reference output $y_m(t)$ generated from a reference model

$$y_m(t) = W_m(s)[r](t), \ W_m(s) = \frac{1}{P_m(s)} \qquad (2.8)$$

where $P_m(s)$ is a stable monic polynomial in s of degree n^*, and $r(t)$ is a bounded and piecewise continuous reference input[1].

Remark 2.1. In the absence of actuator failures, that is, when $u(t) = v(t)$, a state tracking problem was solved and well-understood when $m = 1$ in (2.1) [16]. Its solution employs a plant-model matching condition which characterizes the class of plants and models admissible to an adaptive state feedback and state tracking design. For issues in actuator failure compensation such as controller parametrization, matching, error model, adaptive law, and stability and tracking analysis, new solutions are needed, which will be developed in the following sections. Similar issues will be addressed for the output tracking problem with actuator failures, which are solvable with less restrictive matching conditions, as in [12, 16, 20] for no failure cases.

3. Plant-Model State Matching Control

To develop adaptive control schemes for the plant (2.1) with unknown parameters and unknown actuator failures, it is useful to understand the desirable controller structure and parameters for the case when the plant parameters and actuator failure pattern and parameters are known, using state feedback for state tracking as addressed in this section (and also in Sections 4–6 for other situations of the plant parameters and actuator failures).

---

[1] The notation $y_m(t) = W_m(s)[r](t)$ denotes the output $y_m(t)$ of a system characterized by the transfer function $W_m(s)$ with input $r(t)$, that is, for $W(s) = \frac{1}{P_m(s)}$, $y_m(t)$ and $r(t)$ are related by $P_m(s)[y_m](t) = r(t)$, where the symbol s denotes, as the case may be, the Laplace transform variable or the differentiation operator: $s[y_m](t) = \dot{y}_m(t)$.

3.1 Basic Plant-Model Matching Conditions

Before addressing the control problem for arbitrary actuator failures, we first derive the existence of controllers for the plant (2.1) in two special cases: when all but one actuator fail, and when no actuator fails, to derive some basic plant-model matching conditions which are useful for controller parametrizations in fixed or adaptive designs for arbitrary actuator failures.

Denoting the ith column of B as $b_i \in R^n$, $i = 1, \ldots, m$: $B = [b_1, \ldots, b_m]$, in view of the actuator failure model (2.2), we see that Assumption (A1) implies that there exist constant vectors $k^*_{s1i} \in R^n$ and non-zero constant scalars $k^*_{s2i} \in R$, $i = 1, \ldots, m$, such that the matching equations

$$A + b_i k^{*T}_{s1i} = A_M, \quad b_i k^*_{s2i} = b_M \qquad (3.1)$$

are satisfied. These conditions are needed for plant-model matching when the special actuator failure pattern: $u_j(t) = \bar{u}_j = 0$, $j = 1, \ldots, i-1, i+1, \ldots, m$, $j \neq i$, occurs. In this case, the ith control input

$$u_i(t) = v_i(t) = v^*_i(t) = k^{*T}_{s1i} x(t) + k^*_{s2i} r(t) \qquad (3.2)$$

leads to the closed-loop system

$$\dot{x}(t) = A_M x(t) + b_M r(t) \qquad (3.3)$$

which matches the reference model (2.7) so that the stated control objective is met: $x(t)$ and $u(t)$ are bounded, $\lim_{t \to \infty}(x(t) - x_M(t)) = 0$ exponentially.

On the other hand, Assumption (A1) also implies that there exist constant vectors $k^*_{1i} \in R^n$ and constant scalars $k^*_{2i} \in R$, $i = 1, \ldots, m$, such that

$$A + BK^{*T}_1 = A + \sum_{i=1}^{m} b_i k^{*T}_{1i} = A_M, \quad Bk^*_2 = \sum_{i=1}^{m} b_i k^*_{2i} = b_M. \qquad (3.4)$$

These equations mean that when no actuator failure is present, that is, $Bu(t) = Bv(t)$, with $K^*_1 = [k^*_{11}, \ldots, k^*_{1m}] \in R^{n \times m}$, $k^*_2 = [k^*_{21}, \ldots, k^*_{2m}]^T \in R^m$, the state feedback control law

$$u(t) = v(t) = v^*(t) = K^{*T}_1 x(t) + k^*_2 r(t) \qquad (3.5)$$

also leads to the closed-loop system (3.3).

Suppose there are p failed actuators, that is, $u_j(t) = \bar{u}_j$, $j = j_1, \ldots, j_p$, $1 \leq p \leq m-1$, but with $\bar{u}_j = 0$, $j = j_1, \ldots, j_p$. In this case, the desired matching conditions clearly are

$$A + \sum_{i \neq j_1, \ldots, j_p} b_i k^{*T}_{1i} = A_M, \quad \sum_{i \neq j_1, \ldots, j_p} b_i k^*_{2i} = b_M. \qquad (3.6)$$

Using $b_i k^*_{s2i} = b_M$ in (3.1), $i = 1, \ldots, m$, we can write $\sum_{i \neq j_1, \ldots, j_p} b_i k^*_{2i} = b_M$ in (3.6) as

$$\sum_{i \neq j_1, \ldots, j_p} \frac{k_{2i}^*}{k_{s2i}^*} = 1. \tag{3.7}$$

Similarly, using $A + b_i k_{s1i}^{*T} = A_M$, $i = 1, \ldots, m$, and $A + \sum_{i \neq j_1, \ldots, j_p} b_i k_{1i}^{*T} = A_M$, we obtain

$$\sum_{i \neq j_1, \ldots, j_p} \frac{k_{1i}^*}{k_{s2i}^*} = \frac{k_{s1j}^*}{k_{s2j}^*}, \ j = 1, \ldots, m. \tag{3.8}$$

Remark 3.1. The parameters k_{s1i}^* and k_{s2i}^* which satisfy (3.1) may be unique, while the matching parameters k_{1i}^* and k_{2i}^* in (3.4) are not unique and can be determined from (3.7) and (3.8) and their existence is ensured by the existence of k_{s1i}^* and k_{s2i}^*. The equalities (3.7) and (3.8) also hold for $p = 0$, that is, when there is no failure. The condition (3.8) implies that $\frac{k_{s1j}^*}{k_{s2j}^*} = \frac{k_{s1i}^*}{k_{s2i}^*} = \bar{k}_{s1}^*$ for some $\bar{k}_{s1}^* \in R^n$ and any i, j. One choice of k_{1i}^* to satisfy (3.8) is $k_{1i}^* = \alpha_i \bar{k}_{s1}^*$, $i \neq j_1, \ldots, j_p$, for some $\alpha_i \in R$ such that $\sum_{i \neq j_1, \ldots, j_p} \frac{\alpha_i}{k_{s2i}^*} = 1$.

Remark 3.2. Condition (3.1) implies condition (3.4), that is, for a special choice: $k_{1i}^* = k_{s1i}^*$, $k_{2i}^* = k_{s2i}^*$, $k_{1j}^* = 0$, $k_{2j}^* = 0$, all $j \neq i$. While condition (3.1) is necessary for solving the formulated actuator failure compensation problems, as shown in the following sections, it is also sufficient for solving the adaptive actuator failure compensation problem. We should note that condition (3.1) is also the plant-model matching condition needed for a solution to the model reference control problem [16], without actuator failures and with or without plant parameter uncertainties when $m = 1$ in the plant (2.1).

In summary, the basic plant-model matching conditions for solving the formulated actuator failure compensation state tracking problem under Assumption (A1) can be given as:

Proposition 3.1. *The necessary and sufficient condition for non-adaptive or adaptive actuator failure compensation to exist is that there exist constant vectors $k_{s1i}^* \in R^n$ and non-zero constant scalars $k_{s2i}^* \in R$, $i = 1, \ldots, m$, such that $A + b_i k_{s1i}^{*T} = A_M$, $b_i k_{s2i}^* = b_M$.*

This condition will be used in Section 3.2 to derive a plant-model state matching controller in the presence of actuator failures modeled in (2.2) with known parameters, and in Sections 4, 5 and 6 for adaptive actuator failure compensation designs using parameter estimates for different failure models: constant, parametrizable or unparametrizable varying failures.

3.2 Actuator Failure Compensation

To solve the control problem with arbitrary actuator failures in a known system, we propose the following state feedback structure

$$v(t) = v^*(t) = K_1^{*T} x(t) + k_2^* r(t) + k^* \tag{3.9}$$

where the parameters $K_1^* \in R^{n \times m}$ and $k_2^* \in R^m$ have been introduced in (3.4) and (3.5), and $k^* = [k_1^*, \ldots, k_m^*]^T \in R^m$ is to be defined.

Suppose there are p failed actuators, that is, $u_j(t) = \bar{u}_j$, $j = j_1, \ldots, j_p$, $1 \leq p \leq m - 1$. In this case, the plant-model matching condition for K_1^* and k_2^* is also that in (3.6), that is,

$$A + B(I - \sigma)K_1^{*T} = A + \sum_{i \neq j_1, \ldots, j_p} b_i k_{1i}^{*T} = A_M,$$

$$B(I - \sigma)k_2^* = \sum_{i \neq j_1, \ldots, j_p} b_i k_{2i}^* = b_M. \tag{3.10}$$

Under this condition, with (2.3), the controller (3.9), when applied to the plant (2.1), leads to the closed-loop system

$$\begin{aligned}
\dot{x}(t) &= Ax(t) + Bv^*(t) + B\sigma(\bar{u} - v^*(t)) \\
&= A_M x(t) + b_M r(t) + B\sigma(K_1^{*T} x(t) + k_2^*(t)) + Bk^* + B\sigma(\bar{u} - v^*(t)) \\
&= A_M x(t) + b_M r(t) + B(I - \sigma)k^* + B\sigma\bar{u}. \tag{3.11}
\end{aligned}$$

For this system to match the reference model (2.7), we need to choose k_i^*, $i \neq j_1, \ldots, j_p$, to make

$$B(I - \sigma)k^* + B\sigma\bar{u} = \sum_{i \neq j_1, \ldots, j_p} b_i k_i^* + \sum_{j = j_1, \ldots, j_p} b_j \bar{u}_j = 0. \tag{3.12}$$

In this case, the choice of k_{1j}^*, k_{2j}^*, and k_j^*, $j = j_1, \ldots, j_p$, is not relevant because their actuators have failed (they are not present in the closed-loop system (3.11)). One nominal choice could be

$$k_{1j}^* = 0, \ k_{2j}^* = 0, \ k_j^* = 0, \ j = j_1, \ldots, j_p. \tag{3.13}$$

Using $b_i = \frac{1}{k_{s2i}^*} b_M$ in (3.1), from (3.12), we have the matching condition

$$\sum_{i \neq j_1, \ldots, j_p} \frac{k_i^*}{k_{s2i}^*} + \sum_{j = j_1, \ldots, j_p} \frac{\bar{u}_j}{k_{s2j}^*} = 0 \tag{3.14}$$

which determines the choice of k_i^*, $i \neq j_1, \ldots, j_p$, for achieving the desired closed-loop system (3.3).

There are two special cases of (3.14). One is when only one $u_j(t)$ fails, and in this case (3.14) becomes

$$\sum_{i \neq j} \frac{k_i^*}{k_{s2i}^*} + \frac{\bar{u}_j}{k_{s2j}^*} = 0. \tag{3.15}$$

Another case is when all $u_j(t)$ but $u_i(t)$ fail, that is, when $u_j(t) = \bar{u}_j$, all $j \neq i$, and $u_i(t) = v_i(t)$. In this case, (3.14) leads to the unique choice

$$k_i^* = -k_{s2i}^* \sum_{j \neq i} \frac{\bar{u}_j}{k_{s2j}^*}. \tag{3.16}$$

For the case when no actuator fails, the parameters K_1^* and k_2^* are defined in Section 3.1, while the parameters in k^* are defined by

$$\sum_{i=1}^{m} \frac{k_i^*}{k_{s2i}^*} = 0. \tag{3.17}$$

One choice is $k_i^* = 0$, $i = 1, \ldots, m$, as that in (3.5).

In all cases, the existence of K_1^*, k_2^* and k^* in the controller (3.9) is ensured to make the closed-loop system (3.11) match the reference model (2.7), resulting in the desired closed-loop system (3.3).

In summary, the plant-model matching parameters K_1^*, k_2^* and k^* are determined by (3.10) and (3.14), or by (3.1), (3.7), (3.8) and (3.14). Those parameters depend on the plant parameters A and B as well as the knowledge of the actuator failures. Whenever the actuator failure pattern changes, the values of K_1^*, k_2^* and k^* will change, according to the conditions in (3.10) and (3.14), or by (3.1), (3.7), (3.8) and (3.14). Therefore, the matching parameters K_1^*, k_2^* and k^* are actually piecewise constant time-varying.

4. Adaptive State Tracking Control

Now we develop an adaptive control scheme for the plant (2.1) with unknown parameters A and B, and with unknown actuator failures (2.2). We propose the adaptive state feedback controller structure

$$v(t) = K_1^T(t)x(t) + k_2(t)r(t) + k(t) \tag{4.1}$$

where $K_1(t) \in R^{n \times m}$, $k_2(t) \in R^m$ and $k(t) \in R^m$ are adaptive estimates of the unknown parameters K_1^*, k_2^* and k^*. To design stable adaptive laws for updating the parameter estimates

$$\begin{aligned} K_1(t) &= [k_{11}(t), \ldots, k_{1m}(t)], \quad k_2(t) = [k_{21}(t), \ldots, k_{2m}(t)]^T, \\ k(t) &= [k_1(t), \ldots, k_m(t)]^T, \end{aligned} \tag{4.2}$$

we assume that

(A2) $\text{sgn}[k_{s2i}^*]$, [2] the sign of k_{s2i}^* in (3.1), is known, for $i = 1, \ldots, m$.

To proceed, we define the parameter errors

$$\tilde{k}_{1i}(t) = k_{1i}(t) - k_{1i}^*, \quad \tilde{k}_{2i}(t) = k_{2i}(t) - k_{2i}^*, \quad \tilde{k}_i(t) = k_i(t) - k_i^* \tag{4.3}$$

---

[2] $\text{sgn}[\chi] = \begin{cases} 1 & \text{if } \chi > 0 \\ -1 & \text{if } \chi < 0 \end{cases}$

for $i = 1, \ldots, m$, and the tracking error $e(t) = x(t) - x_m(t)$.

Let (T_i, T_{i+1}), $i = 0, 1, \ldots, m_0$, with $T_0 = 0$, be the time intervals on which the actuator failure pattern is fixed, that is, actuators only fail at time T_i, $i = 1, \ldots, m_0$. Since there are m actuators, at least one of them does not fail, we have $m_0 < m$ and $T_{m_0+1} = \infty$. Then, at time T_j, $j = 1, \ldots, m_0$, the unknown plant-model matching parameters K_1^*, k_2^* and k^* (that is, k_{1j}^*, k_{2j}^*, k_j^*, $j = 1, \ldots, m$) change their values such that

$$k_{1j}^* = k_{1j(i)}^*, \ k_{2j}^* = k_{2j(i)}^*, \ k_j^* = k_{j(i)}^*, \ t \in (T_i, T_{i+1}) \quad (4.4)$$

for $j = 1, \ldots, m$, $i = 0, 1, \ldots, m_0$.

Suppose there are p failed actuators, that is, $u_j(t) = \bar{u}_j$, $j = j_1, \ldots, j_p$, $1 \leq p \leq m-1$, at time $t \in (T_i, T_{i+1})$ (with $i = i(p) \leq p$ because there may be more than one actuator failing at the same time T_i). Using (2.1), (2.3), (3.10), (3.12), (4.1) and (4.2), we obtain

$$\begin{aligned}\dot{x}(t) &= Ax(t) + B(I-\sigma)(K_1^{*T}x(t) + k_2^*r(t)) + B(I-\sigma)k^* + B\sigma\bar{u} \\&\quad + B(I-\sigma)\left(\tilde{K}_1^T(t)x(t) + \tilde{k}_2(t)r(t) + \tilde{k}(t)\right) \\&= A_M x(t) + b_M r(t) + \sum_{j \neq j_1,\ldots,j_p} b_j \tilde{k}_{1j}^T(t)x(t) \\&\quad + \sum_{j \neq j_1,\ldots,j_p} b_j \tilde{k}_{2j}(t)r(t) + \sum_{j \neq j_1,\ldots,j_p} b_j \tilde{k}_j(t) \\&= A_M x(t) + b_M r(t) + b_M \left(\sum_{j \neq j_1,\ldots,j_p} \frac{1}{k_{s2j}^*} \tilde{k}_{1j}^T(t)x(t) \right. \\&\quad \left. + \sum_{j \neq j_1,\ldots,j_p} \frac{1}{k_{s2j}^*}\tilde{k}_{2j}(t)r(t) + \sum_{j \neq j_1,\ldots,j_p} \frac{1}{k_{s2j}^*}\tilde{k}_j(t)\right). \end{aligned} \quad (4.5)$$

Then using (2.7) and (4.5), we have the tracking error equation

$$\begin{aligned}\dot{e}(t) &= A_M e(t) + b_M \left(\sum_{j \neq j_1,\ldots,j_p} \frac{1}{k_{s2j}^*}\tilde{k}_{1j}^T(t)x(t) \right. \\&\quad \left. + \sum_{j \neq j_1,\ldots,j_p} \frac{1}{k_{s2j}^*}\tilde{k}_{2j}(t)r(t) + \sum_{j \neq j_1,\ldots,j_p} \frac{1}{k_{s2j}^*}\tilde{k}_j(t)\right).\end{aligned} \quad (4.6)$$

Consider the positive definite function

$$\begin{aligned}V_p(e, \tilde{k}_{1j}, \tilde{k}_{2j}, \tilde{k}_j, j \neq j_1, \ldots, j_p) \\= e^T P e + \sum_{j \neq j_1,\ldots,j_p} \frac{1}{|k_{s2j}^*|}\tilde{k}_{1j}^T \Gamma_{1j}^{-1}\tilde{k}_{1j} \\+ \sum_{j \neq j_1,\ldots,j_p} \frac{1}{|k_{s2j}^*|}\tilde{k}_{2j}^2 \gamma_{2j}^{-1} + \sum_{j \neq j_1,\ldots,j_p} \frac{1}{|k_{s2j}^*|}\tilde{k}_j^2 \gamma_j^{-1}\end{aligned} \quad (4.7)$$

where $k_{1j}^* = k_{1j(i)}^*$, $k_{2j}^* = k_{2j(i)}^*$ and $k_j^* = k_{j(i)}^*$ are defined in (4.4) (note that this $V_p(\cdot)$ is defined for all $t \in [0, \infty)$ with the matching parameters $k_{1j(i)}^*$, $k_{2j(i)}^*$ and $k_{j(i)}^*$ for the case when there are p failed actuators over the interval (T_i, T_{i+1}), $i = i(p) \leq p$), $P \in R^{n \times n}$, $P = P^T > 0$ such that

$$PA_M + A_M^T P = -Q \tag{4.8}$$

for any constant $Q \in R^{n \times n}$ such that $Q = Q^T > 0$, $\Gamma_{1j} \in R^{n \times n}$ is constant such that $\Gamma_{1j} = \Gamma_{1j}^T > 0$, $\gamma_{2j} > 0$ and $\gamma_j > 0$ are constant, $j = 1, \ldots, m$.

Choose the adaptive laws for $k_{1j}(t)$, $k_{2j}(t)$ and $k_j(t)$, $j = 1, \ldots, m$, as

$$\dot{k}_{1j}(t) = -\text{sgn}[k_{s2j}^*]\Gamma_{1j}x(t)e^T(t)Pb_M \tag{4.9}$$

$$\dot{k}_{2j}(t) = -\text{sgn}[k_{s2j}^*]\gamma_{2j}r(t)e^T(t)Pb_M \tag{4.10}$$

$$\dot{k}_j(t) = -\text{sgn}[k_{s2j}^*]\gamma_j e^T(t)Pb_M. \tag{4.11}$$

Then, from (4.6)–(4.11) we have

$$\dot{V}_p(t) = -e^T(t)Qe(t) \leq 0, \ t \in (T_{i(p)}, T_{i(p)+1}), \ i(p) \leq p \tag{4.12}$$

where $V_p(t) \triangleq V_p(e(t), \tilde{k}_{1j}(t), \tilde{k}_{2j}(t), \tilde{k}_j(t), j \neq j_1, \ldots, j_p)$.

Since there are only finite times of actuator failures, that is, $i \leq m_0 < m$, and eventually there are $\bar{m}_0$ failed actuators, $m_0 \leq \bar{m}_0 < m$, from (4.12), it follows that $\dot{V}_{\bar{m}_0}(t) = -e^T(t)Qe(t) \leq 0$, $t \in (T_{\bar{m}_0}, \infty)$ so we can conclude that $V_{\bar{m}_0}(t)$ is bounded (it can be shown that $V_{\bar{m}_0}(T_{\bar{m}_0}) \leq \alpha V_0(0) + \beta$ for some constant $\alpha > 0$ and $\beta > 0$; see Example 4.1 below) and so are $x(t)$, $k_{1j}(t)$, $k_{2j}(t)$ and $k_j(t)$, $j \neq j_1, \ldots, j_{\bar{m}_0}$. To show the boundedness of other signals, from (4.9), we have

$$k_{1i}(t) = k_{1i}(0) - \text{sgn}[k_{s2i}^*]\Gamma_{1i}\int_0^t x(\tau)e^T(\tau)Pb_M d\tau, \ i \neq j_1, \ldots, j_{\bar{m}_0} \tag{4.13}$$

$$k_{1j}(t) = k_{1j}(0) - \text{sgn}[k_{s2j}^*]\Gamma_{1j}\int_0^t x(\tau)e^T(\tau)Pb_M d\tau, \ j = j_1, \ldots, j_{\bar{m}_0}. \tag{4.14}$$

From (4.13), we obtain

$$\int_0^t x(\tau)e^T(\tau)Pb_M d\tau = \text{sgn}[k_{s2i}^*]\Gamma_{1i}^{-1}\left(k_{1i}(0) - k_{1i}(t)\right), \ i \neq j_1, \ldots, j_{\bar{m}_0} \tag{4.15}$$

which, together with (4.14), implies

$$k_{1j}(t) = k_{1j}(0) - \text{sgn}[k_{s2j}^*]\Gamma_{1j}\text{sgn}[k_{s2i}^*]\Gamma_{1i}^{-1}\left(k_{1i}(0) - k_{1i}(t)\right), \tag{4.16}$$

for $j = j_1, \ldots, j_{\bar{m}_0}$ and any $i \neq i_1, \ldots, i_{\bar{m}_0}$. Since $k_{1i}(t)$ is bounded, $i \neq j_1, \ldots, j_{\bar{m}_0}$, it follows that $k_{1j}(t)$ is bounded, $j = j_1, \ldots, j_{\bar{m}_0}$. Similarly, $k_{2j}(t)$ and $k_j(t)$ are bounded, $j = j_1, \ldots, j_{\bar{m}_0}$. From (4.1), it follows that $v(t)$ is bounded. Therefore all signals in the closed-loop system are bounded.

Since at least one actuator does not fail, that is, $m_0 < m$ and $T_{m_0+1} = \infty$, over the time interval (T_{m_0}, ∞), the function $V_{\bar{m}_0}(t)$ has a finite initial value and its derivative is $\dot{V}_{\bar{m}_0}(t) = -e^T(t)Qe(t) \leq 0$, which means that $e(t) \in L^2$. From (4.6) and closed-loop signal boundedness, we have that $\dot{e}(t) \in L^\infty$ so that $\lim_{t \to \infty} e(t) = 0$. Furthermore, since $e(t) \in L^2 \cap L^\infty$ and $x(t) \in L^\infty$, we have from (4.9) that $\dot{\tilde{k}}_{1j}(t) \in L^2 \cap L^\infty$, $\dot{\tilde{k}}_{2j}(t) \in L^2 \cap L^\infty$ and $\dot{\tilde{k}}_j(t) \in L^2 \cap L^\infty$, $j = 1, \ldots, m$. It can also be verified that $\ddot{\tilde{k}}_{1j}(t) \in L^\infty$, $\ddot{\tilde{k}}_{2j}(t) \in L^\infty$ and $\ddot{\tilde{k}}_j(t) \in L^\infty$, $j = 1, \ldots, m$, so that $\lim_{t \to \infty} \dot{\tilde{k}}_{1j}(t) = 0$, $\lim_{t \to \infty} \dot{\tilde{k}}_{2j}(t) = 0$, and $\lim_{t \to \infty} \dot{\tilde{k}}_j(t) = 0$, $j = 1, \ldots, m$.

In summary we have the following result.

Theorem 4.1. *The adaptive controller (4.1), with the adaptive laws (4.9)–(4.11), applied to the plant (2.1) with actuator failures (2.2), guarantees that all closed-loop signals are bounded and the tracking error $e(t)$ goes to zero as the time t goes to infinity.*

It should noted that for $t \in (T_i, T_{i+1})$, $i = 0, 1, \ldots, m_0 - 1$, the developed adaptive actuator failure compensation control scheme ensures that $\dot{V}_p(t) = -e^T(t)Qe(t) \leq 0$, a desired performance measure which is the same as that which can be achieved in an adaptive control system without any actuator failures, over the same time interval (T_i, T_{i+1}).

Example 4.1. As an example to illustrate the analysis of the boundedness of $V_p(t)$, we consider the case when $m = 3$, in which the actuator $u_3(t)$ fails at $t = T_1$, the actuator $u_2(t)$ fails at $t = T_2 > T_1$, and the actuator $u_1(t)$ does not fail (that is, $u_1(t) = v_1(t), t \geq 0$). In this case, we have

$$V_0(e, \tilde{k}_{1i}, \tilde{k}_{2i}, \tilde{k}_i, i = 1, 2, 3)$$
$$= e^T P e + \sum_{i=1}^{3} \frac{1}{|k_{s2i}^*|} \left(\tilde{k}_{1i}^T \Gamma_{1i}^{-1} \tilde{k}_{1i} + \tilde{k}_{2i}^2 \gamma_{2i}^{-1} + \tilde{k}_i^2 \gamma_i^{-1} \right) \quad (4.17)$$

$$V_1(e, \tilde{k}_{1i}, \tilde{k}_{2i}, \tilde{k}_i, i = 1, 2)$$
$$= e^T P e + \sum_{i=1}^{2} \frac{1}{|k_{s2i}^*|} \left(\tilde{k}_{1i}^T \Gamma_{1i}^{-1} \tilde{k}_{1i} + \tilde{k}_{2i}^2 \gamma_{2i}^{-1} + \tilde{k}_i^2 \gamma_i^{-1} \right) \quad (4.18)$$

$$V_2(e, \tilde{k}_{11}, \tilde{k}_{21}, \tilde{k}_1)$$
$$= e^T P e + \frac{1}{|k_{s21}^*|} \left(\tilde{k}_{11}^T \Gamma_{11}^{-1} \tilde{k}_{11} + \tilde{k}_{21}^2 \gamma_{21}^{-1} + \tilde{k}_1^2 \gamma_1^{-1} \right), \quad (4.19)$$

that is, $m_0 = \bar{m}_0 = 2$. The time-derivatives of these functions are

$$\dot{V}_0 = -e^T Q e, \ t \in [0, T_1) \quad (4.20)$$

$$\dot{V}_1 = -e^T Q e, \ t \in (T_1, T_2) \quad (4.21)$$

$$\dot{V}_2 = -e^T Q e, \ t \in (T_2, \infty). \quad (4.22)$$

For V_0, V_1 and V_2 as functions of t, it follows from (4.17)–(4.22) that

$$V_0(0) \geq V_0(T_1), \ V_1(T_1) \geq V_1(T_2), \ V_2(T_2) \geq V_2(t), \ t \geq T_2. \tag{4.23}$$

For simplicity of presentation, we let $\Gamma_{1i} = I$, $\gamma_{2i} = \gamma_i = 1$ and $k_{s2i}^* = 1$ in (4.17)–(4.19). Then, with the parameters $k_{1i(j)}^*$, $k_{2i(j)}^*$ and $k_{i(j)}^*$ defined in (4.14), and using the inequality

$$(k_{11}(t) - k_{11(2)}^*)^T (k_{11}(t) - k_{11(2)}^*)$$
$$\leq 2(k_{11}(t) - k_{11(1)}^*)^T (k_{11}(t) - k_{11(1)}^*)$$
$$+ 2(k_{11(2)}^* - k_{11(1)}^*)^T (k_{11(2)}^* - k_{11(1)}^*) \tag{4.24}$$

and similar ones for $(k_{21}(t) - k_{21(2)}^*)^2$ and $(k_1(t) - k_{1(2)}^*)^2$, we obtain

$$V_2 \leq 2V_1 + 2(k_{11(2)}^* - k_{11(1)}^*)^T (k_{11(2)}^* - k_{11(1)}^*)$$
$$+ 2(k_{21(2)}^* - k_{21(1)}^*)^2 + 2(k_{1(2)}^* - k_{1(1)}^*)^2 \tag{4.25}$$

$$V_1 \leq 2V_0 + 2\sum_{i=1}^{2}(k_{1i(2)}^* - k_{1i(1)}^*)^T (k_{1i(2)}^* - k_{1i(1)}^*)$$
$$+ 2\sum_{i=1}^{2}(k_{2i(2)}^* - k_{2i(1)}^*)^2 + 2\sum_{i=1}^{2}(k_{i(2)}^* - k_{i(1)}^*)^2. \tag{4.26}$$

Let ΔV_1 and ΔV_2 be the finite parameter variations in (4.25) and (4.26): $V_1 \leq 2V_0 + \Delta V_1$, $V_2 \leq 2V_2 + \Delta V_2$. Using (4.23), (4.25) and (4.26), we have

$$V_1(T_1) \leq 2V_0(T_1) + \Delta V_1 \leq 2V_0(0) + \Delta V_1 \tag{4.27}$$

$$V_2(T_2) \leq 2V_1(T_2) + \Delta V_2 \leq 2V_1(T_1) + \Delta V_2 \leq 4V_0(0) + 2\Delta V_1 + \Delta V_2. \tag{4.28}$$

Therefore, it follows from (4.23) and (4.28) that

$$V_2(t) \leq V_2(T_2) \leq 4V_0(0) + 2\Delta V_1 + \Delta V_2 \tag{4.29}$$

is bounded. Then, from (4.13)–(4.16), we have that $V_0(t)$ and $V_1(t)$ are also bounded, and from (4.22) we have $e(t) \in L^2$ so that $\lim_{t \to \infty} e(t) = 0$. □

5. Compensation Designs for Parametrizable Time-varying Failures

In this section, we consider a more general type of time-varying actuator failure. Such failures can occur, for example, due to hydraulics failures that can produce unintended movements in the control surfaces of an aircraft. Other examples include orbiting satellites subjected to periodic torques of unknown magnitude due to gravity gradient, Earth's magnetic field, or onboard rotating machinery with variable loads.

We consider the actuator failure model described by

$$u_j(t) = \bar{u}_j + \bar{d}_j(t), \ t \geq t_j \tag{5.1}$$

where $\bar{u}_j$ is an unknown constant, and

$$\bar{d}_j(t) = \sum_{l=1}^{q} \bar{d}_{jl} f_{jl}(t) \tag{5.2}$$

for some unknown scalar constants $\bar{d}_{jl}$ and known scalar signals $f_{jl}(t)$, $j = 1, \ldots, m$, $l = 1, \ldots, q$, $q \geq 1$. With such actuator failures, the actuation vector Bu can be described as

$$Bu(t) = Bv(t) + B\sigma(\bar{u} + \bar{d}(t) - v(t)) \tag{5.3}$$

where $v(t)$ is a designed control input, and

$$\bar{d}(t) = [\bar{d}_1(t), \ldots, \bar{d}_m(t)]^T. \tag{5.4}$$

5.1 Plant-Model Matching Control Design

We first develop a plant-model matching controller of the form

$$v(t) = v^*(t) = K_1^{*T} x(t) + k_2^* r(t) + k^* + g^*(t) \tag{5.5}$$

where $K_1^* = [k_{11}^*, \ldots, k_{1m}^*] \in R^{n \times m}$, $k_2^* = [k_{21}^*, \ldots, k_{2m}^*]^T \in R^m$, and $k^* = [k_1^*, \ldots, k_m^*]^T \in R^m$ are parameters, and $g^*(t) = [g_1^*(t), \ldots, g_m^*(t)]^T \in R^m$ is a vector design signal.

Suppose at time t there are p failed actuators, that is, $u_j(t) = \bar{u}_j + \bar{d}_j(t)$, $j = j_1, \ldots, j_p$, $1 \leq p \leq m - 1$. Then, the parameters K_1^*, k_2^* and k^* are chosen to satisfy (3.10) and (3.12), plus (3.13). With the state feedback controller (5.5), using (5.3) and (2.1), the closed-loop system becomes

$$\begin{aligned}
\dot{x}(t) &= Ax(t) + Bv^*(t) + B\sigma(\bar{u} - v^*(t)) \\
&= A_M x(t) + b_M r(t) + B\sigma(K_1^{*T} x(t) + k_2^* r(t)) \\
&\quad + Bk^* + Bg^*(t) + B\sigma(\bar{u} - v^*(t)) \\
&= A_M x(t) + b_M r(t) + B(I - \sigma)k^* + B\sigma\bar{u} + B(I - \sigma)g^*(t) + B\sigma\bar{d}(t) \\
&= A_M x(t) + b_M r(t) + B(I - \sigma)g^*(t) + B\sigma\bar{d}(t). \tag{5.6}
\end{aligned}$$

To match this system to (2.7), the design task now is to choose $g^*(t)$ to make

$$B(I - \sigma)g^*(t) + B\sigma\bar{d}(t) = \sum_{i \neq j_1, \ldots, j_p} b_i g_i^*(t) + \sum_{j=j_1, \ldots, j_p} b_j \bar{d}_j^*(t) = 0. \tag{5.7}$$

Using (3.1), that is, $b_i = \frac{1}{k^*_{s2i}} b_M$, from (5.7), we have the plant-model matching condition for $g_i^*(t)$, $i \neq j_1, \ldots, j_p$:

$$\sum_{i \neq j_1, \ldots, j_p} \frac{g_i^*(t)}{k^*_{s2i}} + \sum_{j=j_1, \ldots, j_p} \frac{\bar{d}_j(t)}{k^*_{s2j}} = 0. \tag{5.8}$$

Similar to that in (3.13), one nominal choice of the terms $g_j^*(t)$, $j = j_1, \ldots, j_p$, irrelevant to the closed-loop system (5.6), is

$$g_j^*(t) = 0, \ j = j_1, \ldots, j_p. \tag{5.9}$$

However, the condition (5.8) does not give an explicit parametrized form of $g_i^*(t)$. To derive a desirable parametrized form of $g_i^*(t)$, we first consider the case when all $u_j(t)$ but $u_i(t)$ fail, that is, when $u_j(t) = \bar{u}_j + \bar{d}_j(t)$, $j = 1, 2, \ldots, m$, $j \neq i$, and $u_i(t) = v_i(t) = v_i^*(t)$. In this case, (5.8) becomes

$$g_i^*(t) = -k^*_{s2i} \sum_{j \neq i} \frac{\bar{d}_j(t)}{k^*_{s2j}}. \tag{5.10}$$

In view of (5.2), we rewrite $g_i^*(t)$ as

$$g_i^*(t) = \sum_{j \neq i}^{m} \sum_{l=1}^{q} g_{ijl}^* f_{jl}(t) \tag{5.11}$$

where the parameters g_{ijl}^* are

$$g_{ijl}^* = -\frac{k^*_{s2i}}{k^*_{s2j}} d_{jl}^*, \ j = 1, \ldots, m, \ j \neq i, \ l = 1, \ldots, q. \tag{5.12}$$

Although the parametrized form of $g_i^*(t)$ in (5.11) is derived from the special case when all $u_j(t)$ but $u_i(t)$ fail, it is the right form for the general case when there are p failed actuators, that is, $u_j(t) = \bar{u}_j + \bar{d}_j(t)$, $j = j_1, \ldots, j_p$.

Proposition 5.1. *The parametrized form (5.11) of $g_i^*(t)$, $i = 1, \ldots, m$, is necessary and sufficient for the matching condition (5.8).*

Proof: The necessity is clear from (5.10), when all $u_j(t)$ but $u_i(t)$ fail. To prove the sufficiency, we see that the term in (5.8), $\sum_{j=j_1, \ldots, j_p} \frac{\bar{d}_j(t)}{k^*_{s2j}}$, contains $f_{jl}(t)$, $j = j_1, \ldots, j_p$, $l = 1, \ldots, q$. Since $p < m$, in the term $\sum_{i \neq j_1, \ldots, j_p} \frac{g_i^*(t)}{k^*_{s2i}}$, there is at least a $g_{j_p+1}^*$ which, from (5.11), has the form

$$g_{j_p+1}^*(t) = \sum_{j \neq j_p+1}^{m} \sum_{l=1}^{q} g_{j_p+1 jl}^* f_{jl}(t). \tag{5.13}$$

This $g_{j_p+1}^*(t)$ does not contain $f_{j_p+1 l}$, $l = 1, \ldots, q$, but all other $f_{jl}(t)$, $j = 1, \ldots, m$, $j \neq j_p+1$, $l = 1, \ldots, q$, that is, $g_{j_p+1}^*(t)$ also contains $f_{jl}(t)$,

$j = j_1, \ldots, j_p$, $l = 1, \ldots, q$. Therefore, the parametrized form (5.11) of the individual $g_i^*(t)$, $i = 1, \ldots, m$, is sufficient for parametrizing the set of $g_i^*(t)$, $i \neq j_1, \ldots, j_p$, to meet the matching condition (5.8) in which $\sum_{j=j_1,\ldots,j_p} \frac{\bar{d}_j(t)}{k_{s2j}^*}$, with $\bar{d}_j(t)$ in (5.2), is parametrizable by $f_{jl}(t)$, $j = j_1, \ldots, j_p$, $l = 1, \ldots, q$. It should be noted that when there are p failed actuators, that is, $u_j(t) = \bar{u}_j + \bar{d}_j(t)$, $j = j_1, \ldots, j_p$, $g_j^*(t) = 0$ (see (5.9)) is a special case of (5.11) with parameters $g_{jkl}^* = 0$. ∇

For the case when no actuator fails, the parameters K_1^* and k_2^* are defined in (3.4)–(3.8), while the parameters in k^* are defined in (3.17). The signals $g_i^*(t)$, $i = 1, \ldots, m$, are defined by

$$\sum_{i=1}^{m} \frac{g_i^*(t)}{k_{s2i}^*} = 0. \tag{5.14}$$

One choice is that $g_i^*(t) = 0$, $i = 1, \ldots, m$.

It can be easily verified that with the parametrization (5.11), the special case when only one $u_j(t)$ fails is also covered. In this case, the matching condition (5.8) becomes

$$\sum_{i \neq j} \frac{g_i^*(t)}{k_{s2i}^*} + \frac{\bar{d}_j(t)}{k_{s2j}^*} = 0 \tag{5.15}$$

for which there is a lot of freedom in choosing the parameters of $g_i^*(t)$, $i \neq j$.

5.2 Adaptive Control Design

When the plant parameters and actuator failures are unknown, we use an adaptive version of the state feedback controller (5.5), which is

$$v(t) = K_1^T(t)x(t) + k_2(t)r(t) + k(t) + g(t) \tag{5.16}$$

where $K_1(t)$, $k_2(t)$ and $k(t)$ are adaptive estimates of K_1^*, k_2^* and k^*, and

$$g(t) = [g_1(t), \ldots, g_m(t)]^T \tag{5.17}$$

$$g_i(t) = \sum_{j \neq i}^{m} \sum_{l=1}^{q} g_{ijl}(t) f_{jl}(t) \tag{5.18}$$

with $g_{ijl}(t)$ being the estimate of g_{ijl}^* defined in (5.12).

Suppose at time t there are p failed actuators, that is, $u_j(t) = \bar{u}_j + \bar{d}_j(t)$, $j = j_1, \ldots, j_p$, $1 \leq p \leq m - 1$. Our task now is to develop adaptive laws to update the parameter estimates $K_1(t)$, $k_2(t)$, $k(t)$ and $g_{ijl}(t)$. We start with the closed-loop control system with (2.1), (2.3), (3.10), (3.12), (4.2), (5.7) and (5.16), and $\tilde{g}(t) = g(t) - g^*(t)$, that is,

$$\begin{aligned}
\dot{x}(t) &= Ax(t) + B(I-\sigma)(K_1^{*T}x(t) + k_2^*r(t)) + B(I-\sigma)k^* \\
&\quad + B\sigma\bar{u} + B(I-\sigma)g^*(t) + B\sigma\bar{d}(t) \\
&\quad + B(I-\sigma)\left(\tilde{K}_1^T(t)x(t) + \tilde{k}_2(t)r(t) + \tilde{k}(t) + \tilde{g}(t)\right) - B\sigma(v(t) - v^*(t)) \\
&= A_M x(t) + b_M r(t) + \sum_{i\neq j_1,\ldots,j_p} b_i \tilde{k}_{1i}^T(t)x(t) + \sum_{i\neq j_1,\ldots,j_p} b_i \tilde{k}_{2i}(t)r(t) \\
&\quad + \sum_{i\neq j_1,\ldots,j_p} b_i \tilde{k}_i(t) + \sum_{i\neq j_1,\ldots,j_p} b_i \sum_{j\neq i}^{m}\sum_{l=1}^{q} \tilde{g}_{ijl}(t) f_{jl}(t) \\
&= A_M x(t) + b_M r(t) \\
&\quad + b_M \left(\sum_{i\neq j_1,\ldots,j_p} \frac{1}{k_{s2i}^*} \tilde{k}_{1i}^T(t)x(t) + \sum_{i\neq j_1,\ldots,j_p} \frac{1}{k_{s2i}^*} \tilde{k}_{2i}(t)r(t) \right. \\
&\quad \left. + \sum_{i\neq j_1,\ldots,j_p} \frac{1}{k_{s2i}^*} \tilde{k}_i(t) + \sum_{i\neq j_1,\ldots,j_p} \frac{1}{k_{s2i}^*} \sum_{j\neq i}^{m}\sum_{l=1}^{q} \tilde{g}_{ijl}(t) f_{jl}(t) \right). \quad (5.19)
\end{aligned}$$

Then, the error equation for $e(t) = x(t) - x_m(t)$ is

$$\begin{aligned}
\dot{e}(t) &= A_M e(t) + b_M \left(\sum_{i\neq j_1,\ldots,j_p} \frac{1}{k_{s2i}^*} \tilde{k}_{1i}^T(t)x(t) + \sum_{i\neq j_1,\ldots,j_p} \frac{1}{k_{s2i}^*} \tilde{k}_{2i}(t)r(t) \right. \\
&\quad \left. + \sum_{i\neq j_1,\ldots,j_p} \frac{1}{k_{s2i}^*} \tilde{k}_i(t) + \sum_{i\neq j_1,\ldots,j_p} \frac{1}{k_{s2i}^*} \sum_{j\neq i}^{m}\sum_{l=1}^{q} \tilde{g}_{ijl}(t) f_{jl}(t) \right), \quad (5.20)
\end{aligned}$$

similar to (4.7) for the case when $d(t) = 0$ (so that $g(t) = 0$ is chosen), where $\tilde{g}_{ijl}(t) = g_{ijl}(t) - g_{ijl}^*$ and $x_m(t)$ is in (2.7).

Consider the positive definite function

$$\begin{aligned}
&V_p(e, \tilde{k}_{1i}, \tilde{k}_{2i}, \tilde{k}_i, \tilde{g}_{ijl}, i \neq j_1, \ldots, j_p) \\
&= e^T P e + \sum_{i\neq j_1,\ldots,j_p} \frac{1}{|k_{s2i}^*|} \tilde{k}_{1i}^T \Gamma_{1i}^{-1} \tilde{k}_{1i} + \sum_{i\neq j_1,\ldots,j_p} \frac{1}{|k_{s2i}^*|} \tilde{k}_{2i}^2 \gamma_{2i}^{-1} \\
&\quad + \sum_{i\neq j_1,\ldots,j_p} \frac{1}{|k_{s2i}^*|} \tilde{k}_i^2 \gamma_i^{-1} + \sum_{i\neq j_1,\ldots,j_p} \frac{1}{|k_{s2i}^*|} \sum_{j\neq i}^{m}\sum_{l=1}^{q} \tilde{g}_{ijl}^2 \gamma_{ijl}^{-1} \quad (5.21)
\end{aligned}$$

where P is defined in (4.8), $\Gamma_{1i} \in R^{n\times n}$ is a constant matrix such that $\Gamma_{1i} = \Gamma_{1i}^T > 0$, and $\gamma_{2i} > 0$, $\gamma_i > 0$ and $\gamma_{ijl} > 0$ are constants, $i = 1, \ldots, m$, $j \neq i, l = 1, \ldots, q$.

Choose the adaptive laws for $g_{ijl}(t)$ as

$$\dot{g}_{ijl}(t) = -\text{sgn}[k_{s2i}^*] \Gamma_{ijl} f_{jl}(t) e^T(t) P b_M \quad (5.22)$$

for $i = 1, \ldots, m$, $j \neq i$, $l = 1, \ldots, q$, and use the adaptive laws for $k_{1i}(t)$, $k_{2i}(t)$ and $k_i(t)$ as those given in (4.9)–(4.11).

Then, similar to (4.12), we can obtain

$$\dot{V}_p(t) = -e^T(t)Qe(t) \leq 0, \ t \in (T_{i(p)}, T_{i(p)+1}), \ i(p) \leq p \tag{5.23}$$

where, as in Section 4, $i = i(p) \leq p$ is such that there are p failed actuators for $t \in (T_i, T_{i+1})$, that is, actuators only fail at time T_i, (T_i, T_{i+1}), $i = 0, 1, \ldots, m_0 < m$, are the time intervals on which the actuator failure pattern is fixed. Since eventually there are only $\bar{m}_0$ failed actuators, $m_0 \leq \bar{m}_0 < m$, it follows that $\dot{V}_{\bar{m}_0}(t) = -e^T(t)Qe(t) \leq 0, \ t \in (T_{m_0}, \infty)$.

From here on, based on a similar analysis to that used for Theorem 4.1, we can establish the following result.

Theorem 5.1. *The adaptive controller (5.16) (via (2.3)), with the adaptive laws (4.9)–(4.11) and (5.22), applied to the plant (2.1), guarantees that all closed-loop signals are bounded and* $\lim_{t \to \infty}(y(t) - y_m(t)) = 0$.

6. Compensation Designs for Unparametrizable Time-varying Failures

In this section, we consider the even more general actuator failure model:

$$u_j(t) = \bar{u}_j + \bar{d}_j(t) + \bar{\delta}_j(t), \ t \geq t_j \tag{6.1}$$

where $\bar{u}_j$ and $\bar{d}_j(t)$ are defined in (2.2) and (5.2) respectively, and $\bar{\delta}_j(t)$ is an unknown and unparametrizable but bounded term. In this case, we have

$$Bu(t) = Bv(t) + B\sigma(\bar{u} + \bar{d}(t) + \bar{\delta}(t) - v(t)) \tag{6.2}$$

where $v(t)$ is a designed control input, and

$$\bar{\delta}(t) = [\bar{\delta}_1(t), \ldots, \bar{\delta}_m(t)]^T. \tag{6.3}$$

The control objective is to design a feedback control $v(t)$ such that all closed-loop signals are bounded and the state vector $x(t)$ of the plant (2.1) tracks the state vector $x_m(t)$ of the reference model (2.7) as close as possible.

6.1 Stabilizing Control

We first develop a stabilizing controller of the form

$$v(t) = v^*(t) = K_1^{*T}x(t) + k_2^*r(t) + k^* + g^*(t) + v_s(t) \tag{6.4}$$

where $K_1^* = [k_{11}^*, \ldots, k_{1m}^*] \in R^{n \times m}$, $k_2^* = [k_{21}^*, \ldots, k_{2m}^*]^T \in R^m$, and $k^* = [k_1^*, \ldots, k_m^*]^T \in R^m$ are piecewise constant parameters (see Section 3), $g^*(t) = [g_1^*(t), \ldots, g_m^*(t)]^T \in R^m$ is a design signal (see Section 5.1), and

$$v_s(t) = [v_{s1}(t), \ldots, v_{sm}(t)]^T \qquad (6.5)$$

is to be designed to compensate the unknown $\bar{\delta}(t)$ for stability and tracking.

Suppose at time t there are p failed actuators, that is, $u_j(t) = \bar{u}_j + \bar{d}_j(t) + \bar{\delta}_j(t)$, $j = j_1, \ldots, j_p$, $1 \leq p \leq m-1$. Similar to (3.12), (5.7) for k^*, $g^*(t)$, an ideal choice of $v_s(t)$ would be such that

$$B(I - \sigma)v_s(t) + B\sigma\bar{\delta}(t) = \sum_{i \neq j_1, \ldots, j_p} b_i v_{si}(t) + \sum_{j=j_1,\ldots,j_p} b_j \bar{\delta}_j(t) = 0 \qquad (6.6)$$

which, in view of (3.1), would imply

$$\sum_{i \neq j_1,\ldots,j_p} \frac{v_{si}(t)}{k^*_{s2i}} + \sum_{j=j_1,\ldots,j_p} \frac{\bar{\delta}_j(t)}{k^*_{s2j}} = 0 \qquad (6.7)$$

which, however, cannot be parametrized because $\bar{\delta}_j(t)$ is not parametrizable, and thus is not useful in the adaptive control case when $\bar{\delta}_j(t)$ is unknown.

With the choice of K_1^*, k_2^*, k^* and $g^*(t)$ in (3.10), (3.12), (3.13), (5.8) and (5.9), we consider the resulting closed-loop system

$$\dot{x}(t) = A_M x(t) + b_M r(t) + B(I - \sigma)v_s(t) + B\sigma\bar{\delta}(t) \qquad (6.8)$$

which leads to the error system for $e(t) = x(t) - x_m(t)$,

$$\begin{aligned} \dot{e}(t) &= A_M e(t) + B(I - \sigma)v_s(t) + B\sigma\bar{\delta}(t) \\ &= A_M e(t) + \sum_{i \neq j_1,\ldots,j_p} b_i v_{si}(t) + \sum_{j=j_1,\ldots,j_p} b_j \bar{\delta}_j(t) \\ &= A_M e(t) + b_M \left(\sum_{i \neq j_1,\ldots,j_p} \frac{v_{si}(t)}{k^*_{s2i}} + \sum_{j=j_1,\ldots,j_p} \frac{\bar{\delta}_j(t)}{k^*_{s2j}} \right). \end{aligned} \qquad (6.9)$$

We choose the design signal $v_{si}(t)$ as

$$v_{si}(t) = -k^0_{2i}\delta^0 \text{sgn}[e^T(t)Pb_M]\text{sgn}[k^*_{s2i}], \ i \neq j_1,\ldots,j_p \qquad (6.10)$$

where k^0_{2i} is a known upper bound on $|k^*_{s2i}|$: $k^0_{2i} > |k^*_{s2i}|$,

$$\delta^0 \geq \max\{\delta_1,\ldots,\delta_{m-1}\} \qquad (6.11)$$

$$\delta_p = \max_{j_1,\ldots,j_p \in \{1,\ldots,m\}} |\sum_{j=j_1,\ldots,j_p} \frac{\bar{\delta}_j(t)}{k^*_{s2j}}|, \ 1 \leq p \leq m-1 \qquad (6.12)$$

and $P = P^T > 0$ satisfying (4.8).

Consider the positive definite function

$$W(e) = e^T P e \qquad (6.13)$$

and its time-derivative

$$\dot{W} = -e^T Q e - 2|e^T P b_M|\delta^0 \sum_{i\neq j_1,\ldots,j_p} \frac{k_{2i}^0}{|k_{s2i}^*|} + 2e^T P b_M \sum_{j=j_1,\ldots,j_p} \frac{\bar{\delta}_j(t)}{k_{s2j}^*}$$

$$\leq -e^T Q e - 2|e^T P b_M|\delta^0 \sum_{i\neq j_1,\ldots,j_p} \frac{k_{2i}^0}{|k_{s2i}^*|} + 2|e^T P b_M|\delta_p$$

$$\leq -e^T Q e. \tag{6.14}$$

Since the control signal $v_{si}(t)$ in (6.10) is not continuous, although $\dot{W}(t) \leq -e^T(t)Qe(t)$, the solution to the system (6.8) or (6.9) should be understood in the Filippov sense [8, 26].

To avoid system chatterings caused by such discontinuous control laws, one can use the following common approximations for the sgn[·] function:

$$\text{sgn}[\chi] \approx \sigma_{s1}[\chi] = \frac{\chi}{|\chi|+\epsilon}, \quad \epsilon > 0 \tag{6.15}$$

$$\text{sgn}[\chi] \approx \sigma_{s2}[\chi] = \begin{cases} \frac{\chi}{|\chi|} & \text{if } |\chi| \geq \epsilon \\ \frac{\chi}{\epsilon} & \text{if } |\chi| < \epsilon, \end{cases} \quad \epsilon > 0. \tag{6.16}$$

With sgn[·] in (6.10) replaced by $\sigma_{s1}[\cdot]$ in (6.15), we obtain

$$\dot{W} = -e^T Q e - 2\frac{|e^T P b_M|^2}{|e^T P b_M|+\epsilon}\delta^0 \sum_{i\neq j_1,\ldots,j_p} \frac{k_{2i}^0}{|k_{s2i}^*|} + 2e^T P b_M \sum_{j=j_1,\ldots,j_p} \frac{\bar{\delta}_j(t)}{k_{s2j}^*}$$

$$\leq -e^T Q e - 2|e^T P b_M|(\frac{|e^T P b_M|}{|e^T P b_M|+\epsilon}\delta^0 \sum_{i\neq j_1,\ldots,j_p} \frac{k_{2i}^0}{|k_{s2i}^*|} - \delta_p)$$

$$\leq -e^T Q e + 2\delta_0 \frac{|e^T P b_M|\epsilon}{|e^T P b_M|+\epsilon} \sum_{i\neq j_1,\ldots,j_p} \frac{k_{2i}^0}{|k_{s2i}^*|}$$

$$\leq -e^T Q e + 2k_0\epsilon \tag{6.17}$$

where $k_0 = \delta_0 \sum_{i\neq j_1,\ldots,j_p} \frac{k_{2i}^0}{|k_{s2i}^*|}$ is a constant. From (6.17), we have

$$\dot{W} \leq -\|e\|^2 q_1 + 2k_0\epsilon \tag{6.18}$$

where q_1 is the minimum eigenvalue of Q. Hence, $\dot{W} < 0$ whenever $\|e\|^2 > \frac{2k_0\epsilon}{q_1}$, i.e., $\|e\| > \sqrt{\frac{2k_0\epsilon}{q_1}}$, which means that $\|e\|$ decreases to a lower bound proportional to $\sqrt{\epsilon}$. When $\epsilon > 0$ is small, the error $e(t)$ will be small.

With sgn[·] in (6.10) replaced by $\sigma_{s2}[\cdot]$ in (6.16), we obtain

$$\dot{W} \leq \begin{cases} -e^T Q e < 0 & \text{if } |e^T P b_M| \geq \epsilon \\ -e^T Q e - \\ 2|e^T P b_M|\left(\frac{|e^T P b_M|\delta^0}{\epsilon}\sum_{i\neq j_1,\ldots,j_p} \frac{k_{2i}^0}{|k_{s2i}^*|} - \delta_p\right) & \text{if } |e^T P b_M| < \epsilon. \end{cases} \tag{6.19}$$

When $|e^T P b_M| < \epsilon$, we have

$$\dot{W} \leq -e^T Q e - 2|e^T P b_M| \left(\left(\frac{|e^T P b_M|}{\epsilon} - 1 \right) \delta^0 \sum_{i \neq j_1, \ldots, j_p} \frac{k_{2i}^0}{|k_{s2i}^*|} \right)$$

$$\leq -e^T Q e + 2 k_0 \epsilon \tag{6.20}$$

for the same k_0 in (6.17), which means that $\|e\|$ decreases to a lower bound proportional to $\sqrt{\epsilon}$. Therefore, for $\epsilon > 0$ small, the tracking error $e(t)$ also converges to a small set.

6.2 Adaptive Control Design

The adaptive controller structure is

$$v(t) = K_1^T(t) x(t) + k_2(t) r(t) + k(t) + g(t) + v_s(t) \tag{6.21}$$

where $K_1(t)$, $k_2(t)$, $k(t)$ and $g(t)$ are adaptive estimates of K_1^*, k_2^*, k^* and $g^*(t)$, respectively, and $v_s(t)$ is given in (6.10).

Suppose there are p failed actuators, that is, $u_j(t) = \bar{u}_j + \bar{d}_j(t) + \bar{\delta}_j(t)$, $j = j_1, \ldots, j_p$, $1 \leq p \leq m-1$, at time $t \in (T_i, T_{i+1})$, $i = i(p) < p$. To develop adaptive update laws for $K_1(t)$, $k_2(t)$, $k(t)$ and $g(t)$, in view of (5.22) and (6.9), and similar to (4.7) and (5.22), with the controller (6.21), we have the error system for $e(t) = x(t) - x_M(t)$ as

$$\begin{aligned}
\dot{e}(t) &= A_M e(t) + b_M \left(\sum_{i \neq j_1, \ldots, j_p} \frac{1}{k_{s2i}^*} \tilde{k}_{1i}^T(t) x(t) + \sum_{i \neq j_1, \ldots, j_p} \frac{1}{k_{s2i}^*} \tilde{k}_{2i}(t) r(t) \right. \\
&\quad + \sum_{i \neq j_1, \ldots, j_p} \frac{1}{k_{s2i}^*} \tilde{k}_i(t) + \sum_{i \neq j_1, \ldots, j_p} \frac{1}{k_{s2i}^*} \sum_{j \neq i}^m \sum_{l=1}^q \tilde{g}_{ijl}(t) f_{jl}(t) \\
&\quad \left. + \sum_{i \neq j_1, \ldots, j_p} \frac{v_{si}(t)}{k_{s2i}^*} + \sum_{j = j_1, \ldots, j_p} \frac{\bar{\delta}_j(t)}{k_{s2j}^*} \right). \tag{6.22}
\end{aligned}$$

With the adaptive laws (4.9)–(4.11), (5.22) for $K_1(t)$, $k_2(t)$, $k(t)$ and $g(t)$, respectively, and the design signal $v_s(t)$ in (6.10), we also have the time-derivative of $V_p(e, \tilde{k}_{1i}, \tilde{k}_{2i}, \tilde{k}_i, \tilde{g}_{ijl}, i \neq j_1, \ldots, j_p)$ in (5.23) as that given in (6.2), while, due to discontinuities in $v_s(t)$, the system solution should also be understood in the Filippov sense [8, 26].

When the approximations in (6.15) and (6.16) are used for the sgn[·] function in (6.10), the adaptive laws (4.9)–(4.11) and (5.22) ensure

$$\dot{V}_p \leq -e^T Q e + 2 k_0 \epsilon, \ t \in (T_{i(p)}, T_{i(p)+1}), \ i(p) \leq p \tag{6.23}$$

for a design parameter $\epsilon > 0$ (which can be made small) and some constant $k_0 > 0$ independent of ϵ. The inequality (6.23) implies that, with V_p in (5.23), the tracking error $e(t)$ is bounded. However it does not mean that the parameters $K_1(t)$, $k_2(t)$, $k(t)$ and $g(t)$ are bounded (that is, parameter drifts [12] can occur). To enhance robustness of the adaptive control system, we need to modify the individual adaptive law (4.9), (4.10), (4.11) or (5.22) with an additional design signal [12, 15, 22]. A standard parameter projection [22] or a soft parameter projection (the switching-σ modification [12]) for such a design signal ensures the boundedness of the parameter estimates $K_1(t)$, $k_2(t)$, $k(t)$ and $g(t)$, in addition to the desired inequality (6.23) which also implies that $e(t)$ is bounded. Therefore, all closed-loop signals are bounded. Then, using (6.23), we obtain

$$\int_{t_1}^{t_2} \|e(t)\|^2 dt \leq \frac{V_p(t_1) - V_p(t_2)}{q_1} + \frac{2k_0\epsilon}{q_1}(t_2 - t_1), \qquad (6.24)$$

$\forall t_2 > t_1$, $t_2, t_1 \in (T_{i(p)}, T_{i(p)+1})$, where q_1 is the minimum eigenvalue of Q. Since $V_p(t)$ is bounded, we have

$$\frac{1}{t_2 - t_1} \int_{t_1}^{t_2} \|e(t)\|^2 dt \leq \frac{c_0}{t_2 - t_1} + \frac{2k_0\epsilon}{q_1}, \qquad (6.25)$$

$\forall t_2 > t_1$, $t_2, t_1 \in (T_{i(p)}, T_{i(p)+1})$, for some constant $c_0 > 0$. The last inequality indicates that the mean value of the error $\|e(t)\|^2$ can be made small by a choice of small ϵ.

When $t_2 > t_1$ but $t_1 \in [T_i, T_{i+1}]$ and $t_2 \in [T_j, T_{j+1}]$, $j > i$, the error $e(t)$ is bounded for $t \in [t_1, t_2]$ so that the mean value of $\|e(t)\|^2$ is also bounded. After a finite number of actuator failures, $T_{m_0+1} = \infty$ for some m_0 but T_{m_0} is finite. Then, for any $t_2 > t_1$, $t_2, t_1 \in (T_{m_0}, \infty)$, we have (6.25).

In summary we have:

Theorem 6.1. *The adaptive controller (6.21) with the design signal $v_s(t)$ in (6.10) approximated with (6.15) or (6.16), updated by the adaptive laws (4.9)–(4.11) and (5.22) modified by parameter projection, applied to the plant (2.1), guarantees that all closed-loop signals are bounded and the tracking error $e(t)$ satisfies (6.25).*

Thus far, we have solved the adaptive actuator compensation problem: state feedback control of the plant (2.1) to ensure, despite the uncertainties in actuator failures and plant parameters, closed-loop signal boundedness and asymptotical tracking, by $x(t)$, of a reference state $x_m(t)$ from the reference model (2.7). To achieve this control objective, the matching condition (3.1) is necessary and sufficient (even for the non-adaptive control case), which implies that the model matrix A_M and the plant matrix A should be closely related: $A + b_i k_{s1i}^{*T} = A_M$, and that $b_i = \frac{b_M}{k_{s2i}^*}$, $i = 1, 2, \ldots, m$, that is, the columns of B should be parallel to each other. This condition usually can only be met for A in a canonical form, even if b_i and b_j are parallel.

A different control objective for adaptive actuator failure compensation, stated in Section 2 as the *output tracking* problem, is to use state or output feedback control for the plant (2.1) with actuator failures (2.2) under

Assumption (A1), to ensure that all signals in the closed-loop system are bounded and the plant output $y(t)$ asymptotically tracks a given reference output $y_m(t)$ generated from the reference model (2.8). In this case, state dynamics matching is not needed so that conditions on plant matrices A and B could be less restrictive than those stated in (3.1).

Next in Sections 7 and 8, we will derive conditions for plant-model *output* matching in the presence of actuator failures and develop adaptive output tracking control schemes for systems with unknown parameters and unknown actuator failures.

7. Plant-Model Output Matching Control

For the above stated output tracking control objective, before addressing the adaptive control problem for an unknown plant with unknown actuator failures, we first derive the ideal controller structure and parametrization for plant-model output matching, in terms of the plant parameters and actuator failure parameters. For the output tracking problem, we will consider the actuator failure model (2.2).

7.1 State Feedback Designs

Consider the plant (2.1) in the input-output form

$$y(s) = C(sI - A)^{-1}Bu(s) = \frac{Z(s)}{P(s)}u(s) \qquad (7.1)$$

where $Z(s)$ is a polynomial vector and $P(s)$ is a monic polynomial of degree n. In the presence of actuator failures modeled as in (2.2), $u(t)$ can be expressed as in (2.3). The control task is to design a state feedback control for $v(t)$ to ensure closed-loop signal boundedness and asymptotic output tracking of a given reference output $y_m(t)$ generated from the reference model (2.8).

7.1.1 Basic matching conditions. To derive suitable parametrizations for a plant-model output matching controller, we denote

$$B = [b_1, \ldots, b_m], \ b_i \in R^n, i = 1, \ldots, m \qquad (7.2)$$

$$Z(s) = [Z_1(s), \ldots, Z_m(s)] \qquad (7.3)$$

$$Z_j(s) = z_{(n-n^*)j} s^{n-n^*} + \cdots + z_{1j} s + z_{0j}. \qquad (7.4)$$

With the actuator failure model (2.2), we see that Assumption (A1) implies that, for the special actuator failure pattern $u_j(t) = \bar{u}_j = 0, j = 1, \ldots, i - 1, i+1, \ldots, m$, that is, $j \neq i$, there exist constant vectors $k^*_{s1i} \in R^n$ and non-zero constant scalars $k^*_{s2i} \in R$, $i = 1, \ldots, m$, such that the following matching equations are satisfied:

$$C(sI - A - b_i k_{s1i}^{*T})^{-1} b_i k_{s2i}^* = W_m(s), \ i = 1, \ldots, m. \quad (7.5)$$

This condition implies that the ith control input

$$u_i(t) = v_i(t) = k_{s1i}^{*T} x(t) + k_{s2i}^* r(t) \quad (7.6)$$

leads to the desired closed-loop system

$$y(t) = W_m(s)[r](t). \quad (7.7)$$

The existence of such matching parameters k_{s1i}^* and k_{s2i}^* and the system internal stability are ensured by the conditions: (i) (C, A, b_i) is stabilizable and detectable or (A, b_i) is controllable, $i = 1, \ldots, m$; and (ii) $Z_j(s)$ is stable and $\partial Z_j(s) = n - n^*$ (that is, $z_{(n-n^*)j} \neq 0$ in (7.4)), $j = 1, \ldots, m$.

In summary to this case, we have:

Proposition 7.1. *A necessary condition for the plant-model output matching $y(t) = W_m(s)[r](t)$ under (A1) is that there exist constant $k_{s1i}^* \in R^n$ and $k_{s2i}^* \in R$, $i = 1, \ldots, m$, such that the condition (7.5) is satisfied.*

More generally, for the plant (2.1) with the failure model (2.2), we consider the state feedback controller structure

$$v(t) = v^*(t) = K_1^{*T} x(t) + k_2^* r(t) + k^* \quad (7.8)$$

where $K_1^* = [k_{11}^*, \ldots, k_{1m}^*] \in R^{n \times m}$, $k_2^* = [k_{21}^*, \ldots, k_{2m}^*]^T \in R^n$, and $k^* = [k_1^*, \ldots, k_m^*]^T \in R^m$ are piecewise constant parameters for different failure patterns, similar to that in (4.4) for the state tracking case.

Suppose that the plant (2.1) at time t has p failed actuators, that is, $u_j(t) = \bar{u}_j, \ j = j_1, \ldots, j_p, \ 1 \leq p \leq m - 1$. From (7.8) and (2.3), we have

$$\dot{x} = (A + \sum_{i \neq j_1, \ldots, j_p} b_i k_{1i}^{*T}) x + \sum_{i \neq j_1, \ldots, j_p} b_i k_{2i}^* r + \sum_{i \neq j_1, \ldots, j_p} b_i k_i^* + \sum_{i = j_1, \ldots, j_p} b_i \bar{u}_i$$
$$y = Cx. \quad (7.9)$$

For this system to match the desired closed-loop system (7.7), we need to choose K_1^*, k_2^* and k^* to make

$$C(sI - A - \sum_{j \neq j_1, \ldots, j_p} b_j k_{1j}^*)^{-1} \sum_{j \neq j_1, \ldots, j_p} b_j k_{2j}^* = W_m(s) \quad (7.10)$$

$$C(sI - A - \sum_{j \neq j_1, \ldots, j_p} b_j k_{1j}^*)^{-1} (\sum_{j \neq j_1, \ldots, j_p} b_j k_j^* + \sum_{j = j_1, \ldots, j_p} b_j \bar{u}_j) = 0. \quad (7.11)$$

Other controller parameters $k_{1j}^*, k_{2j}^*, k_j^*, \ j = j_1, \ldots, j_p$, do not have an effect on the closed-loop system, which can be taken as any nominal values or zero. Therefore, for this general case, we have the design desired conditions:

Proposition 7.2. *With the controller structure (7.8), actuator failure compensation for the plant (2.1) with the actuator failure pattern $u_j(t) = \bar{u}_j$, $j = j_1, \ldots, j_p$, for any $1 \leq p \leq m - 1$, is solvable if and only if the conditions (7.10) and (7.11) are satisfied.*

To illustrate the conditions (7.10) and (7.11), let us consider the case all $u_j(t)$ but $u_i(t)$ fail, that is, when $u_j(t) = \bar{u}_j$, all $j \neq i$, and $u_i(t) = v_i(t)$. In this case, the matching condition (7.10) is

$$W_m(s) = C(sI - A - b_i k_{1i}^{*T})^{-1} b_i k_{2i}^* \tag{7.12}$$

and the condition (7.11) becomes

$$C(sI - A - b_i k_{1i}^{*T})^{-1}(b_i k_i^* + \sum_{j \neq i} b_j \bar{u}_j) = 0, \tag{7.13}$$

which means that $b_i k_i^* + \sum_{j \neq i} b_j \bar{u}_j$ belongs to the null space of $C(sI - A - b_i k_{1i}^{*T})^{-1}$. To understand this condition, we give an illustrative example.

Example 7.1. Consider the plant (2.1) with

$$A = \begin{bmatrix} 0 & 1 & 0 \\ 0 & 0 & 1 \\ 0 & 0 & 0 \end{bmatrix}, b_1 = \begin{bmatrix} 1 \\ 0 \\ 1 \end{bmatrix}, b_2 = \begin{bmatrix} 0 \\ 1 \\ 1 \end{bmatrix}, b_3 = \begin{bmatrix} 1 \\ 1 \\ 1 \end{bmatrix},$$

$$C = \begin{bmatrix} 2 & 1 & 0 \end{bmatrix}. \tag{7.14}$$

The zero polynomials of $(C, A, B = [b_1, b_2, b_3])$ are

$$Z(s) = [2s^2 + s + 2, s^2 + 3s + 2, 3s^2 + 3s + 2]. \tag{7.15}$$

The plant relative degree is $n^* = 1$. For $W_m(s) = \frac{1}{s+1}$, the matching gains to meet (7.12) and (7.13) are

$$k_{11}^{*T} = -[1, 1.5, 0.5], \ k_{12}^{*T} = -[2, 3, 1], \ k_{13}^{*T} = -[\tfrac{2}{3}, 1, \tfrac{1}{3}]. \tag{7.16}$$

For $k_{11}^{*T} = -[1, 1.5, 0.5]$, we have

$$C(sI - A - b_1 k_{11}^{*T})^{-1}(b_1 k_1^* + b_2 \bar{u}_2 + b_3 \bar{u}_3) = \frac{2k_1^* + \bar{u}_2 + 3\bar{u}_3}{s+1}. \tag{7.17}$$

For any $\bar{u}_2$ and $\bar{u}_3$, to satisfy (7.13), one can choose

$$k_1^* = -0.5\bar{u}_2 - 1.5\bar{u}_3. \tag{7.18}$$

In this case, $C(sI - A - b_2 k_{12}^{*T})^{-1}$ and $C(sI - A - b_3 k_{13}^{*T})^{-1}$ have the same null space as that of $C(sI - A - b_1 k_{11}^{*T})^{-1}$, that is, $C(sI - A - b_2 k_{12}^{*T})^{-1}(\beta_1 w_1 + \beta_2 w_2) = 0$, $\forall s \in C$, for $w_1 = [1, -2, 0]^T$, $w_2 = [0, 0, 1]^T$ and any $\beta_1, \beta_2 \in R$. Furthermore, $C(sI - A - b_1 k_{11}^{*T})^{-1} b_j = \alpha_{1j} W_m(s)$ for some $\alpha_{1j} \in R$, $j = 1, 2, 3$, and $C(sI - A - b_i k_{1i}^{*T})^{-1} b_j = \alpha_{1j} W_m(s)$, $i = 2, 3$, $j = 1, 2, 3$. □

Remark 7.1. When $(C, A + b_i k_{1i}^{*T})$ is observable (which is the case if (A, b_i) is controllable and $Z_i(s) = z_{0i}$, that is, the plant (2.1) does not have finite zero and relative degree of $W_m(s)$ is n), a necessary and sufficient condition for (7.11) is that $b_j = \alpha_j b_i$, for some constants α_j, $j = 1, \ldots, m$, $j \neq i$. This condition in general, that is, when $(C, A + b_i k_{1i}^{*T})$ is not observable, is only sufficient but not necessary for (7.11) (see Example 7.1), which is also sufficient for (7.10). When $C = I \in R^{n \times n}$, the output $y(t)$ becomes the state $x(t)$. In this case, $(C, A + b_i k_{1i}^{*T})$ is always observable so that the matching condition $b_j = \alpha_j b_i$, more restrictive than that in (7.10) and (7.11), is needed for plant-model state matching (see Section 3).

Next we present a necessary and sufficient condition for plant-model output matching in the presence of the actuator failure model (2.2), with up to $m - 1$ failures, that is, for *all* $p = 1, 2, \ldots, m - 1$.

Theorem 7.1. *A necessary and sufficient condition for the actuator failure compensation condition (7.11) and the plant-model output matching condition (7.10), for any $p = 1, 2, \ldots, m - 1$, is*

$$C(sI - A - b_i k_{1i}^{*T})^{-1} b_j \frac{1}{\alpha_{ij}} = W_m(s) \tag{7.19}$$

for some $k_{1i}^* \in R^n$ and $\alpha_{ij} \in R$, $i, j = 1, \ldots, m$.

Proof. Sufficiency: For a given failure pattern $u_j = \bar{u}_j$, $j = j_1, \ldots, j_p$, from (7.19), there exist $k_{1i}^* \in R^n$, $i \neq j_1, \ldots, j_p$, $\alpha_{ii} \in R$, such that

$$C(sI - A - b_i k_{1i}^{*T})^{-1} b_i \frac{1}{\alpha_{ii}} = W_m(s) \tag{7.20}$$

which, with $k_{1j}^* = 0$, $k_{2j}^* = 0$, $j \neq i$, $j \neq j_1, \ldots, j_p$, and $k_{2i}^* = \frac{1}{\alpha_{ii}}$, implies (7.10). To prove (7.11), we see that

$$C(sI - A - \sum_{j \neq j_1, \ldots, j_p} b_j k_{1j}^{*T})^{-1} (\sum_{j \neq j_1, \ldots, j_p} b_j k_j^* + \sum_{j = j_1, \ldots, j_p} b_j \bar{u}_j)$$

$$= C(sI - A - b_i k_{1i}^{*T})^{-1} (\sum_{j \neq j_1, \ldots, j_p} b_j k_j^* + \sum_{j = j_1, \ldots, j_p} b_j \bar{u}_j)$$

$$= W_m(s) (\sum_{j \neq j_1, \ldots, j_p} \alpha_{ij} k_j^* + \sum_{j = j_1, \ldots, j_p} \alpha_{ij} \bar{u}_j). \tag{7.21}$$

For any fixed failure values of $\bar{u}_j$, $j = j_1, \ldots, j_p$, there always exist parameters k_j^*, $j \neq j_1, \ldots, j_p$, such that

$$\sum_{j \neq j_1, \ldots, j_p} \alpha_{ij} k_j^* + \sum_{j = j_1, \ldots, j_p} \alpha_{ij} \bar{u}_j = 0 \tag{7.22}$$

so that (7.11) is satisfied (one choice is with $k_j^* = 0$, $j \neq i$, $j \neq j_1, \ldots, j_p$).

Necessity: From (7.10), in particular, for $p = m - 1$, we have

$$C(sI - A - b_i k_{1i}^{*T})^{-1} b_i k_{2i}^* = W_m(s), \; \forall i \in \{1, \ldots, m\} \quad (7.23)$$

and in this case, from (7.11), it follows that

$$C(sI - A - b_i k_{1i}^{*T})^{-1} (b_i k_i^* + \sum_{j \neq i} b_j \bar{u}_j^*) = 0. \quad (7.24)$$

Substituting (7.23) into (7.24) gives

$$W_m(s) \frac{k_i^*}{k_{2i}^*} + C(sI - A - b_i k_{1i}^{*T})^{-1} \sum_{j \neq i} b_j \bar{u}_j = 0. \quad (7.25)$$

Since $\bar{u}_j$ is arbitrary, we can let one $\bar{u}_j \neq 0$ and all other $\bar{u}_j$ be zero, to get

$$C(sI - A - b_i k_{1i}^{*T})^{-1} b_j (-\frac{k_{2i}^* \bar{u}_j}{k_i^*}) = W_m(s) \quad (7.26)$$

so there exist constants α_{ij} such that (7.19) is satisfied. $\triangledown$

Remark 7.2. We should note that in fact α_{ij} does not depend on $\bar{u}_j$, as (7.26) is only a necessary condition and k_i^* in fact depends on $\bar{u}_j$. It can be verified that $\alpha_{ij} = \alpha_{jj}, i, j = 1, 2, \ldots, m$, and that $\alpha_{ij} = CA^{n^*-1} b_j$, while $CA^k b_j = 0$ for $k = 0, 1, \ldots, n^* - 2$. It turns out that $\alpha_{ij} = z_{(n-n^*)j}$ defined in (7.4), for $j = 1, 2, \ldots, m$, that is, α_{ij} is the high frequency gain of $G_i(s)$.

Remark 7.3. For system internal stability, all zeros of $G_i(s) = C(sI - A)^{-1} b_i$, $i = 1, 2 \ldots, m$, should be stable so that there is no unstable pole-zero cancellation in (7.10).

Conditions in terms of the plant parameters $A, C, b_i, i = 1, 2, \ldots, m$, may be derived for the matching condition (7.19). Such conditions can be useful for constructing a dynamic system suitable for actuator failure compensation using the approach developed in this work.

Proposition 7.3. *Assume (C, A, b_i) is controllable, $i = 1, 2, \ldots, m$. A necessary and sufficient condition for (7.19) is that $G_i(s) = C(sI - A)^{-1} b_i$, $i = 1, 2 \ldots, m$, and $W_m(s)$ have the same relative degree n^*.*

This condition is equivalent to the existence of $k_{s1i}^* \in R^n$ and $k_{s2i}^* \in R$, $i = 1, \ldots, m$, such that the condition (7.5): $C(sI - A - b_i k_{s1i}^{*T})^{-1} b_i k_{s2i}^* = W_m(s)$, $i = 1, \ldots, m$, is satisfied, that is, plant-model output matching by individual inputs. This significantly relaxes the matching condition (3.1): $A + b_i k_{s1i}^{*T} = A_M, b_i k_{s2i}^* = b_M, i = 1, \ldots, m$, for state tracking.

The necessity of the proposition's condition is obvious. The sufficiency proof is to use the condition

$$C(sI - A - b_i k_{1i}^{*T})^{-1} b_i = \alpha_{ii} W_m(s), \; W_m(s) = \frac{1}{P_m(s)} \quad (7.27)$$

with $P_m(s) = s^{n^*} + a^*_{n^*} s^{n^*-1} + a^*_{n^*-1} s^{n^*-2} + \cdots + a^*_2 s + a^*_1$, which is ensured under the proposition's condition, to show that

$$C(sI - A - b_i k_{1i}^{*T})^{-1} b_j = \alpha_{ij} W_m(s), \tag{7.28}$$

for some α_{ij}, all $j \neq i$. Here, we present the proof for the $n = 3$ case with different relative degrees. We will use the resolvent formula

$$(sI - \bar{A})^{-1} = \frac{N(s)}{\det(sI - \bar{A})}, \quad \bar{A} = A + b_i k_{1i}^{*T} \in R^{3 \times 3} \tag{7.29}$$

$$\det(sI - \bar{A}) = s^3 + a_3 s^2 + a_2 s + a_1 \tag{7.30}$$

$$N(s) = s^2 I + s(\bar{A} + a_3 I) + (\bar{A}^2 + a_3 \bar{A} + a_2 I). \tag{7.31}$$

For (C, A, b_i) controllable, without loss of generality, we set $\bar{A}$, C and b_i as

$$\bar{A} = A + b_i k_{1i}^{*T} = \begin{bmatrix} 0 & 1 & 0 \\ 0 & 0 & 1 \\ -a_1 & -a_2 & -a_3 \end{bmatrix},$$

$$C = [c_1, c_2, c_3], \ b_i = [0, 0, 1]^T. \tag{7.32}$$

If $G_i(s) = C(sI - A)^{-1} b_i$ and $G_j(s) = C(sI - A)^{-1} b_j$ have relative degree one, then $Cb_i \neq 0$ (that is, $c_3 \neq 0$) and $Cb_j \neq 0$. Let $c_3 = 1$ so that $\alpha_{ii} = 1$ in (7.27) from which with $C(sI - A - b_i k_{1i}^{*T})^{-1} b_i = \frac{s^2 + c_2 s + c_1}{s^3 + a_3 s^2 + a_2 s + a_1}$, $P_m(s) = s + a_1^*$ and (7.32), it follows that $C(a_1^* I + \bar{A}) = 0$. Set $\alpha_{ij} = Cb_j$ in (7.28) so that $C(b_i - \frac{1}{\alpha_{ij}} b_j) = 0$ as $Cb_i = 1$. Then, using (7.29), we have

$$C(sI - \bar{A})^{-1}(b_i - \frac{1}{\alpha_{ij}} b_j)$$

$$= \frac{C\bar{A}(b_i - \frac{1}{\alpha_{ij}} b_j)s + C(\bar{A}^2 + a_3 \bar{A})(b_i - \frac{1}{\alpha_{ij}} b_j)}{\det(sI - \bar{A})}. \tag{7.33}$$

Combining (7.27) and (7.29), we have

$$(s^2 I + s(C\bar{A}b_i + a_3) + C\bar{A}^2 b_i + Ca_3 \bar{A}b_i + a_2)(s + a_1^*) = s^3 + a_3 s^2 + a_2 s + a_1 \tag{7.34}$$

as $Cb_i = 1$, which implies that

$$C\bar{A}b_i = -a_1^*,$$
$$C\bar{A}^2 b_i = -a_1^* C\bar{A}b_i - a_n(C\bar{A}b_i - a_1^*) = -a_1^* C\bar{A}b_i. \tag{7.35}$$

Using this result we write

$$C\bar{A}(b_i - \frac{1}{\alpha_{ij}} b_j) = -\frac{1}{\alpha_{ij}} C(a_1^* + \bar{A}) b_j = 0,$$

$$C\bar{A}^2(b_i - \frac{1}{\alpha_{ij}} b_j) = -a_1^* C\bar{A}b_i - C\bar{A}^2 \frac{1}{\alpha_{ij}} b_j$$

$$= -\frac{1}{\alpha_{ij}} C(a_1^* I + \bar{A}) \bar{A} b_j = 0. \tag{7.36}$$

Substituting (7.36) into (7.33) leads to $C(sI - \bar{A})^{-1}(b_i - \frac{1}{\alpha_{ij}}b_j) = 0$, which implies (7.28).

If $G_i(s) = C(sI - A)^{-1}b_i$ and $G_j(s) = C(sI - A)^{-1}b_j$ have relative degree two, then $Cb_i = Cb_j = 0$ (so that $C(b_i - \frac{1}{\alpha_{ij}}b_j) = 0$ for any $\alpha_{ij} \neq 0$), $CAb_i \neq 0$, and $CAb_j \neq 0$. In this case, $c_3 = 0$ and $c_2 \neq 0$. Let $c_2 = 1$ so that $CAb_i = 1$, $C\bar{A}b_i = 1$, and $\alpha_{ii} = 1$ in (7.27) from which with $C(sI - A - b_i k_{1i}^{*T})^{-1}b_i = \frac{s+c_1}{s^3+a_3s^2+a_2s+a_1}$, $P_m(s) = s^2 + a_2^* s + a_1^*$ and (7.32), it follows that $C(a_1^* I + a_2^* \bar{A} + \bar{A}^2) = 0$ so that $C(a_2^* \bar{A} + \bar{A}^2)b_j = 0$ as $Cb_j = 0$. Set $\alpha_{ij} = C\bar{A}b_j$ in (7.28) so that $C\bar{A}(b_i - \frac{1}{\alpha_{ij}}b_j) = 0$ as $C\bar{A}b_i = 1$. Then, using (7.29), we have

$$C(sI - \bar{A})^{-1}(b_i - \frac{1}{\alpha_{ij}}b_j) = \frac{C\bar{A}^2(b_i - \frac{1}{\alpha_{ij}}b_j)}{\det(sI - \bar{A})}. \quad (7.37)$$

Combining (7.27) and (7.29), we have

$$(sI + C\bar{A}^2 b_i + a_3)(s^2 + a_2^* s + a_1^*) = s^3 + a_3 s^2 + a_2 s + a_1 \quad (7.38)$$

as $Cb_i = 0$ and $C\bar{A}b_i = 1$, which implies that

$$C\bar{A}^2 b_i = -a_2^*. \quad (7.39)$$

Using this result we write

$$\begin{aligned} C\bar{A}^2(b_i - \frac{1}{\alpha_{ij}}b_j) &= a_2^* - C\bar{A}^2 \frac{1}{\alpha_{ij}} b_j \\ &= -\frac{1}{\alpha_{ij}} C(a_2^* \bar{A} + \bar{A}^2)b_j = 0. \end{aligned} \quad (7.40)$$

Substituting (7.40) into (7.37) leads to $C(sI - \bar{A})^{-1}(b_i - \frac{1}{\alpha_{ij}}b_j) = 0$, which implies (7.28).

If $G_i(s) = C(sI - A)^{-1}b_i$ and $G_j(s) = C(sI - A)^{-1}b_j$ have relative degree three, then $Cb_i = Cb_j = 0 = CAb_i = CAb_j = 0$ (so that $C\bar{A}b_i = C\bar{A}b_j = 0$, and $C(b_i - \frac{1}{\alpha_{ij}}b_j) = C\bar{A}(b_i - \frac{1}{\alpha_{ij}}b_j) = 0$ for any $\alpha_{ij} \neq 0$), $CA^2 b_i \neq 0$, and $CA^2 b_j \neq 0$. In this case, $c_3 = c_2 = 0$ and $c_1 \neq 0$. Let $c_1 = 1$ so that $CA^2 b_i = 1$, $C\bar{A}^2 b_i = 1$, and $\alpha_{ii} = 1$ in (7.27). Set $\alpha_{ij} = C\bar{A}^2 b_j$ in (7.28) so that $C\bar{A}^2(b_i - \frac{1}{\alpha_{ij}}b_j) = 0$ as $C\bar{A}^2 b_i = 1$. Then, using (7.29), we have $C(sI - \bar{A})^{-1}(b_i - \frac{1}{\alpha_{ij}}b_j) = 0$, which implies (7.28). From (7.32) and $Cb_j = CAb_j = 0$, we have $b_j = \alpha_j b_i$ for some non-zero $\alpha_j \in R$, that is, b_i and b_j are parallel to each other (see Remark 7.1).

For the general case, the parameter α_{ij} is taken as $\alpha_{ij} = C\bar{A}^{n^*-1}b_j$ (with $C\bar{A}^{n^*-1}b_1 = 1$), and the equality $C(\bar{A}^{n^*} + a_{n^*}^* \bar{A}^{n^*-1} + a_{n^*-1}^* \bar{A}^{n^*-2} + \cdots + a_2^* \bar{A} + a_1^* I) = 0$, similar to the above $C(a_1^* I + \bar{A}) = 0$ ($n^* = 1$), $C(a_2^* \bar{A} + \bar{A}^2)b_j = 0$ ($n^* = 2$), can be established. Expressions of $C\bar{A}^k b_i$, $k = n^*, n^* + 1, \ldots, n-1$, in terms of the parameters a_i^*, $i = 1, 2, \ldots, n^*$, of $P_m(s)$, and $C\bar{A}^j b_i$, $j < k$, can be derived, similar to those in (7.35) and (7.39).

7.1.2 A compensation scheme.
An alternative design of the actuator failure compensation controller structure (7.8) is

$$v_i(t) = v_i^*(t) = v_0^*(t) = k_{11}^{*T} x(t) + k_{21}^* r(t) + k_1^*, \ i = 1, \ldots, m \quad (7.41)$$

for $v(t) = [v_1(t), \ldots, v_m(t)]^T$, that is, an equal-component design for the control signal $v(t)$ as a special structure of (7.8) with

$$k_{1i}^* = k_{11}^*, \ k_{2i}^* = k_{21}^*, \ k_i^* = k_1^*, \ i = 2, \ldots, m. \quad (7.42)$$

Like other compensation schemes developed so far, the parameters of the controller structure (7.41) are derived based on a set of assumptions on the plant (2.1), which are stated as follows:

(A3a) $(A, \sum_{j \neq j_1, \ldots, j_p} b_j), \ p \in \{0, 1, \ldots, m-1\}$, are controllable;
(A3b) $(C, A, \sum_{j \neq j_1, \ldots, j_p} b_j), \ p \in \{0, 1, \ldots, m-1\}$, have the same relative degree n^*;
(A3c) $(C, A, \sum_{j \neq j_1, \ldots, j_p} b_j), \ p \in \{0, 1, \ldots, m-1\}$, are minimum phase; and
(A3d) $CA^{n^*-1} \sum_{j \neq j_1, \ldots, j_p} b_j, \ p \in \{0, 1, \ldots, m-1\}$, have the same sign.

To show the existence of the plant-model output matching parameters k_{11}^*, k_{21}^* and k_1^* of the controller (7.41), with the choice in (7.41), we express the matching condition (7.10) as

$$C(sI - A - \sum_{j \neq j_1, \ldots, j_p} b_j k_{11}^{*T})^{-1} \sum_{j \neq j_1, \ldots, j_p} b_j k_{21}^* = W_m(s). \quad (7.43)$$

The existence of $k_{11}^* \in R^n$ and $k_{21}^* \in R$ for this equation is indeed ensured by the assumptions (A3a) and (A3b); in particular,

$$k_{21}^* = CA^{n^*-1} \sum_{j \neq j_1, \ldots, j_p} b_j. \quad (7.44)$$

Similarly, the condition (7.11) becomes

$$C(sI - A - \sum_{j \neq j_1, \ldots, j_p} b_j k_{11}^{*T})^{-1} (\sum_{j \neq j_1, \ldots, j_p} b_j k_1^* + \sum_{j=j_1, \ldots, j_p} b_j \bar{u}_j) = 0 \quad (7.45)$$

which may be met asymptotically by some parameter $k_1^* \in R$. To show this, we treat $k_1^* \in R$ as an input and $\bar{u}_j$ as disturbances, and define

$$f_p(t) \triangleq C(sI - A - \sum_{j \neq j_1, \ldots, j_p} b_j k_{11}^{*T})^{-1} [\sum_{j \neq j_1, \ldots, j_p} b_j k_1^* + \sum_{j=j_1, \ldots, j_p} b_j \bar{u}_j](t)$$

$$= W_m(s) [\frac{k_1^*}{k_{21}^*}](t)$$

$$+ C(sI - A - \sum_{j \neq j_1, \ldots, j_p} b_j k_{11}^{*T})^{-1} [\sum_{j=j_1, \ldots, j_p} b_j \bar{u}_j](t) \quad (7.46)$$

whose s-domain expression is

$$F_p(s) = C(sI - A - \sum_{j \neq j_1,\ldots,j_p} b_j k_{11}^{*T})^{-1}(\sum_{j \neq j_1,\ldots,j_p} b_j \frac{k_1^*}{s} + \sum_{j=j_1,\ldots,j_p} b_j \frac{\bar{u}_j}{s})$$

$$= W_m(s)\frac{k_1^*}{k_{21}^* s} + C(sI - A - \sum_{j \neq j_1,\ldots,j_p} b_j k_{11}^{*T})^{-1} \sum_{j=j_1,\ldots,j_p} b_j \frac{\bar{u}_j}{s}. \quad (7.47)$$

Since all zeros of $\det(sI - A - \sum_{j \neq j_1,\ldots,j_p} b_j k_{11}^{*T})$ are stable, there exists a constant parameter k_1^* such that

$$\lim_{s \to 0} sF(s) = 0, \quad (7.48)$$

that is, in the time-domain,

$$\lim_{t \to \infty} f_p(t) = 0 \text{ exponentially.} \quad (7.49)$$

This asymptotic property is crucial for parametrization of the actuator failure compensation control law (7.41) to ensure that

$$\lim_{t \to \infty} (y(t) - y_m(t)) = 0 \text{ exponentially} \quad (7.50)$$

in the presence of up to $m - 1$ actuator failures in the plant (2.1), while the system internal stability is ensured by the assumption (A3c). We note that the assumption (A3d) is for an adaptive update law for estimates of the parameters k_{11}^*, k_{21}^* and k_1^* when they are unknown in the adaptive control case as the plant and failures parameters are unknown.

As in Section 4 for the state tracking problem, we let (T_i, T_{i+1}), $i = 0, 1, \ldots, m_0$, with $T_0 = 0$, be the time intervals on which the actuator failure pattern is fixed, that is, actuators only fail at time T_i, $i = 1, \ldots, m_0$. Since there are m actuators, at least one of them does not fail, we have $m_0 < m$ and $T_{m_0+1} = \infty$. Then, at time T_j, $j = 1, \ldots, m_0$, the plant-model matching parameters k_{11}^*, k_{21}^* and k_1^* change their values such that

$$k_{11}^* = k_{11(i)}^*, \; k_{21}^* = k_{21(i)}^*, \; k_1^* = k_{1(i)}^*, \; t \in (T_i, T_{i+1}) \quad (7.51)$$

for $i = 0, 1, \ldots, m_0$, that is, the plant-model matching parameters k_{11}^*, k_{21}^* and k_1^* are piecewise constant parameters, because the plant has different characterizations under different failure conditions so that the plant-model matching parameters are also different.

For the adaptive control problem (see Section 8.1.2), the plant parameters (A, B, C) and the actuator failure parameters $\bar{u}_j$, $j = j_1, j_2, \ldots, j_p$, are unknown so that the matching parameters k_{11}^*, k_{21}^* and k_1^* are unknown. Adaptive laws are needed to update the estimates of these unknown parameters, which are used in an adaptive version of the controller (7.41).

7.2 Output Feedback Design

As we recall from Section 2, for an output feedback design, the control objective is to find an output feedback control $v(t) = [v_1(t), \ldots, v_m(t)]^T \in R^m$ which ensures closed-loop signal boundedness and output tracking of a reference output $y_m(t)$ generated from the reference model (2.7).

To parametrize a plant-model output matching controller using output feedback, we denote the plant (2.1) in the input-output form

$$y(s) = \frac{\sum_{j=1}^m k_{pj} Z_j(s)}{P(s)} u_j(s) \tag{7.52}$$

where k_{pj} is a scalar, $Z_j(s)$ is a monic polynomial, $j = 1, 2, \cdots m$, and $P(s)$ is a monic polynomial of degree n. We use the controller structure

$$\begin{aligned} v_i(t) &= v_0(t) = v_0^*(t) \\ &= \theta_1^{*T} \omega_1(t) + \theta_2^{*T} \omega_2(t) + \theta_{20}^* y(t) + \theta_3^* r(t) + \theta_4^*, \ i = 1, \ldots, m \end{aligned} \tag{7.53}$$

where $\theta_1^* \in R^{n-1}$, $\theta_2^* \in R^{n-1}$, $\theta_{20}^* \in R$, $\theta_3^* \in R$ are parameters to be defined for plant-model output matching, and $\theta_4^* \in R$ is a constant to be chosen for compensation of the actuation error $u - v = \sigma(\bar{u} - v)$, and

$$\omega_1(t) = \frac{a(s)}{\Lambda(s)}[v_0](t), \ \omega_2(t) = \frac{a(s)}{\Lambda(s)}[y](t) \tag{7.54}$$

with $a(s) = (1, s, \cdots, s^{n-2})^T$, and $\Lambda(s)$ being a monic Hurwitz polynomial of degree $n - 1$. The controller (7.53) uses only the measured system input $u(t)$ and output $y(t)$ plus the given reference input $r(t)$, and not the internal system state variables $x(t)$. The equal-component design (7.53) for the control signal $v(t)$ is similar to that in (7.41) for state tracking but with different dynamic feedback components based on output feedback.

From (7.53), (2.3) and (2.5), when there are p failed actuators, that is, $u_j(t) = \bar{u}_j$, $j = j_1, \cdots, j_p$, we can express the plant (2.1) or (7.52) as

$$y(t) = G(s)[v_0](t) + \bar{y}(t) \tag{7.55}$$

where

$$G(s) = \frac{\sum_{j \neq j_1, \cdots, j_p} k_{pj} Z_j(s)}{P(s)} \triangleq \frac{k_p Z_{uf}(s)}{P(s)} \tag{7.56}$$

for some scalar k_p and monic polynomial $Z_{uf}(s)$, and

$$\bar{y}(t) = \sum_{j=j_1, \cdots, j_p} \frac{k_{pj} Z_j(s)}{P(s)}[\bar{u}_j](t). \tag{7.57}$$

To ensure closed-loop stability, we assumed that $Z_{uf}(s)$ is a monic Hurwitz polynomial with degree $n - n^*$. To derive the closed-loop system, we define

$$F_1(s) = \theta_1^{*T} \frac{a(s)}{\Lambda(s)}, \quad F_2(s) = \theta_2^{*T} \frac{a(s)}{\Lambda(s)}. \tag{7.58}$$

Then the control signal $v_0(t) = v_0^*(t)$ from (7.53) is

$$\begin{aligned} v_0(t) &= F_1(s)[v_0](t) + F_2(s)G(s)[v_0](t) + F_2(s)[\bar{y}](t) \\ &\quad + \theta_{20}^* G(s)[v_0](t) + \theta_{20}^* \bar{y}(t) + \theta_3^* r(t) + \theta_4^* \\ &= (F_1(s) + F_2(s)G(s) + \theta_{20}^* G(s))[v_0](t) \\ &\quad + F_2(s)\bar{y}(t) + \theta_{20}^* \bar{y}(t) + \theta_3^* r(t) + \theta_4^*, \end{aligned} \tag{7.59}$$

which can be further expressed as

$$\begin{aligned} v_0(t) &= (1 - F_1(s) - F_2(s)G(s) - \theta_{20}^* G(s))^{-1} [F_2(s)\bar{y}(t) \\ &\quad + \theta_{20}^* \bar{y} + \theta_3^* r + \theta_4^*](t). \end{aligned} \tag{7.60}$$

From (7.55) and (7.60), the closed-loop system is

$$\begin{aligned} y(t) &= G(s)(1 - F_1(s) - F_2(s)G(s) - \theta_{20}^* G(s))^{-1} [F_2(s)[\bar{y}] \\ &\quad + \theta_{20}^*[\bar{y}] + \theta_3^* r + \theta_4^*](t) + \bar{y}(t). \end{aligned} \tag{7.61}$$

To proceed, we express

$$\begin{aligned} &(1 - F_1(s) - F_2(s)G(s) - \theta_{20}^* G(s))^{-1} \\ &= \left(1 - \frac{\theta_1^{*T} a(s)}{\Lambda(s)} - \theta_2^{*T} \frac{a(s)}{\Lambda(s)} G(s) - \theta_{20} G(s)\right)^{-1} \\ &= \frac{\Lambda(s)P(s)}{\Lambda(s)P(s) - \theta_1^{*T} a(s)P(s) - \theta_2^{*T} a(s)k_p Z_{uf}(s) - \theta_{20}^* k_p Z_{uf}(s)} \end{aligned} \tag{7.62}$$

and determine the parameters θ_1^*, θ_2^*, θ_{20}^*, θ_3^* from the matching condition

$$\begin{aligned} &\theta_1^{*T} a(s)P(s) + (\theta_2^{*T} a(s) + \theta_{20}^* \Lambda(s))k_p Z_{uf}(s) \\ &= \Lambda(s)(P(s) - k_p \theta_3^* Z_{uf}(s)P_m(s)), \quad \theta_3^* = k_p^{-1} \end{aligned} \tag{7.63}$$

which leads to

$$(1 - F_1(s) - F_2(s)G(s) - \theta_{20}^* G(s))^{-1} = \frac{P(s)}{Z_{uf}(s)P_m(s)}. \tag{7.64}$$

Substituting (7.56) and (7.64) into (7.61), we have

$$\begin{aligned} y(t) &= \frac{1}{\theta_3^* P_m(s)}[F_2(s)[\bar{y}] + \theta_{20}^* \bar{y} + \theta_3^* r + \theta_4^*](t) + \bar{y}(t) \\ &= \frac{1}{\theta_3^* P_m(s)}[F_2(s)[\bar{y}] + \theta_{20}^* \bar{y} + \theta_3^* r + \theta_4^* + \theta_3^* P_m(s)[\bar{y}]](t). \end{aligned} \tag{7.65}$$

Using (7.53), we have

$$F_2(s) + \theta_{20}^* + \theta_3^* P_m(s) = \frac{\theta_2^{*T} a(s) + \theta_{20} \Lambda(s) + \theta_3^* P_m(s) \Lambda(s)}{\Lambda(s)}$$

$$= \frac{P(s)}{k_p Z_{uf}(s)} - \frac{\theta_1^{*T} a(s) P(s)}{\Lambda(s) k_p Z_{uf}(s)}. \quad (7.66)$$

Substituting (7.66) into (7.65), we have

$$y(t) = \frac{1}{P_m(s)}[r](t)$$
$$+ \frac{1}{\theta_3^* P_m(s)} \left[\frac{\Lambda(s) - \theta_1^{*T} a(s)}{\Lambda(s) k_p Z_{uf}(s)} \sum_{j=j_1,\cdots,j_p} k_{pj} Z_j(s)[\bar{u}_j] + \theta_4^* \right](t). \quad (7.67)$$

Since $P_m(s), \Lambda(s), Z_{uf}(s)$ are all stable, there exists a constant θ_4^* such that

$$f_p(t) \triangleq \frac{1}{\theta_3^* P_m(s)} \left[\frac{\Lambda(s) - \theta_1^{*T} a(s)}{\Lambda(s) k_p Z_{uf}(s)} \sum_{j=j_1,\cdots,j_p} k_{pj} Z_j(s)[\bar{u}_j] + \theta_4^* \right](t)$$
$$\to 0 \text{ exponentially as } t \to \infty. \quad (7.68)$$

This property is crucial for the actuator failure compensation controller (7.53) to ensure the desired plant-model output matching (7.50). Indeed, with it, the asymptotic matching (7.50) is achieved as from (7.67) and (7.70).

In summary, the above derivation of plant-model output matching is based on the following assumptions:

(A4a) $Z_{uf}(s)$ has degree $n - n^*$, i.e., $(C, A, \sum_{j \neq j_1,\ldots,j_p} b_j), p \in \{0, 1, \ldots, m-1\}$, have the same relative degree n^*; and
(A4b) $Z_{uf}(s)$ is Hurwitz, i.e., $(C, A, \sum_{j \neq j_1,\ldots,j_p} b_j), p \in \{0, 1, \ldots, m-1\}$, are minimum phase.

As before, we let $(T_i, T_{i+1}), i = 0, 1, \ldots, m_0$, with $T_0 = 0$, be the time intervals on which the actuator failure pattern is fixed, that is, actuators only fail at time $T_i, i = 1, \ldots, m_0$. Since there are m actuators, at least one of them does not fail, we have $m_0 < m$ and $T_{m_0+1} = \infty$. Then, at time T_j, $j = 1, \ldots, m_0$, the unknown plant-model matching parameters $\theta_1^*, \theta_2^*, \theta_{20}^*, \theta_3^*$ and θ_4^*, similar to that in (7.51) for state feedback designs, change their values such that

$$\theta_1^* = \theta_{1(i)}^*, \ \theta_2^* = \theta_{2(i)}^*, \ \theta_{20}^* = \theta_{20(i)}^*, \ \theta_3^* = \theta_{3(i)}^*, \ \theta_4^* = \theta_{4(i)}^* \quad (7.69)$$

are constant for $t \in (T_i, T_{i+1}), i = 0, 1, \ldots, m_0$, that is, the plant-model matching parameters $\theta_1^*, \theta_2^*, \theta_{20}^*, \theta_3^*$ and θ_4^* are piecewise constant parameters, because the plant has different characterizations under different failure conditions so that the plant-model matching parameters are also different.

Remark 7.4. The analysis carried out in this section is for the closed-loop system operating at a failure pattern that there are p failed actuators, that is, $u_j(t) = \bar{u}_j$, $j = j_1, \cdots, j_p$. When p or $j_1, \cdots, j_p$, change, the parameters of the equivalent plant transfer function $G(s)$ defined in (7.56) also change, and so do the matching parameters θ_1^*, θ_2^*, θ_{20}^*, θ_3^*, and θ_4^*. This parameter jumping also introduces additional transient response starting from each T_i defined above. Since there are only a finite number of actuator failures, the effect of such transient response is always bounded [25], and since the failure pattern will eventually be fixed (that is, no failure for t after T_{m_0}) so that such transient effect actually converges to zero exponentially with time. Therefore, for stability and tracking performance analysis, this effect can be ignored.

8. Adaptive Output Tracking Control

Now we develop adaptive versions of the ideal actuator failure compensation control schemes (7.8), (7.41) and (7.53), for the case when both the plant dynamic matrices $(A,, B, C)$ in (2.1) and the actuator failure parameters $\bar{u}_j$, j and t_j in (2.2) are unknown.

8.1 State Feedback Designs

With both actuator failure compensation control schemes (7.8) and (7.41) using state feedback for output tracking, the control scheme (7.8) represents a general design framework while the control scheme (7.42) employs a special structure for which a set of design conditions (A3a)–(A3c) were specified.

8.1.1 A general design framework. To develop an adaptive version of the control scheme (7.8) for the plant (2.1) with unknown parameters and unknown actuator failures, we assume that the condition (7.19) is satisfied and that the signs of $\alpha_{ij} = z_{(n-n^*)j}$, $\text{sgn}[\alpha_{ij}]$, $i,j = 1, \ldots, m$, are known.

We use the adaptive controller structure

$$v(t) = K_1^T(t)x(t) + k_2(t)r(t) + k(t) \tag{8.1}$$

where $K_1(t) = [k_{11}(t), \ldots, k_{1m}(t)] \in R^{n \times m}$, $k_2(t) = [k_{21}(t), \ldots, k_{2m}(t)]^T \in R^m$ and $k(t) = [k_1(t), \ldots, k_m(t)]^T \in R^m$ are adaptive estimates of the unknown parameters K_1^*, k_2^* and k^*, and define the parameter errors

$$\tilde{k}_{1j} = k_{1j} - k_{1j}^*, \quad \tilde{k}_{2j} = k_{2j} - k_{2j}^*, \quad \tilde{k}_j = k_j - k_j^*. \tag{8.2}$$

Suppose there are p failed actuators, that is, $u_j(t) = \bar{u}_j$, $j = j_1, \ldots, j_p$, $1 \leq p \leq m-1$. Using (2.3), (7.10), (7.11) and (8.1), we rewrite (2.1) as

$$y(t) = y_m(t) + C(sI - A - \sum_{j \neq j_1, \ldots, j_p} b_j k_{1j}^{*T})^{-1}$$

$$\cdot \sum_{j \neq j_1, \ldots, j_p} b_j [\tilde{k}_{1j}^T x + \tilde{k}_{2j} r + \tilde{k}_j](t). \tag{8.3}$$

Defining $e(t) = y(t) - y_m(t)$, using (7.19), we obtain

$$e(t) = \sum_{j \neq j_1, \ldots, j_p} W_m(s)[\alpha_{ij}(\tilde{k}_{1j}^T x + \tilde{k}_{2j} r + \tilde{k}_j)](t) \tag{8.4}$$

where $i \neq j_1, \ldots, j_p$ (here, we used $C(sI - A - \sum_{j \neq j_1, \ldots, j_p} b_j k_{1j}^{*T})^{-1} b_j = C(sI - A - b_i k_{1i}^{*T})^{-1} b_j = \alpha_{ij} W_m(s)$, with one non-zero k_{1i}^*, $i \neq j_1, \ldots, j_p$).

To derive adaptive laws, we define parameters and their errors

$$\theta_j = [k_{1j}^T, k_{2j}, k_j]^T, \ \theta_j^* = [k_{1j}^{*T}, k_{2j}^*, k_j^*]^T, \ \tilde{\theta}_j = \theta_j - \theta_j^* \tag{8.5}$$

$j = 1, \ldots, m$, and introduce auxiliary signals

$$\omega(t) = [x^T(t), r(t), 1]^T \tag{8.6}$$
$$\zeta(t) = W_m(s)[\omega](t) \tag{8.7}$$
$$\xi_j(t) = \theta_j^T(t)\zeta(t) - W_m(s)[\theta_j^T \omega](t). \tag{8.8}$$

Let $\xi(t) = [\xi_1(t), \ldots, \xi_m(t)]^T$, $\rho_j^* = \alpha_{ij}$, $j \neq j_1, \ldots, j_p$, $\rho_j^* = 0$, $j = j_1, \ldots, j_p$, $\rho^* = [\rho_1^*, \ldots, \rho_m^*]^T$, and $\rho(t) = [\rho_1(t), \ldots, \rho_m(t)]^T$ be the estimate of ρ^*, and define the estimation error

$$\epsilon(t) = e(t) + \rho^T(t)\xi(t). \tag{8.9}$$

Using (8.4)–(8.9), we get

$$\epsilon(t) = \sum_{j \neq j_1, \ldots, j_p} \rho_j^* \tilde{\theta}_j^T(t)\zeta(t) + \tilde{\rho}^T(t)\xi(t) + \epsilon_p(t) \tag{8.10}$$

where $\tilde{\rho} = [\tilde{\rho}_1, \ldots, \tilde{\rho}_m]^T$, $\tilde{\rho}_i(t) = \rho_i(t) - \rho_i^*$, $i = 1, \ldots, m$, and

$$\epsilon_p(t) = \sum_{j \neq j_1, \ldots, j_p} \rho_j^*(\theta_j^{*T}(t)\zeta(t) - W_m(s)[\theta_j^{*T} \omega](t))$$
$$+ \sum_{j \neq j_1, \ldots, j_p} W_m(s)[\rho_j^* \tilde{\theta}_j^T \omega](t) - \rho_j^* W_m(s)[\tilde{\theta}_j^T \omega](t). \tag{8.11}$$

Note that $\rho_j^* = \alpha_{ij}$ is piecewise constant for different failure patterns, $j \not{/} j_1, \ldots, j_p$, and so are the parameters K_1^*, k_2^* and k^*.

Based on (8.10), we choose the adaptive laws

$$\dot{\theta}_j(t) = -\frac{sgn[\alpha_{ij}]\Gamma_j \zeta(t)\epsilon(t)}{1 + \zeta^T(t)\zeta(t) + \xi^T(t)\xi(t)}, \ i = 1, \ldots, m \tag{8.12}$$

$$\dot{\rho}(t) = -\frac{\Gamma_\rho \xi(t)\epsilon(t)}{1 + \zeta^T(t)\zeta(t) + \xi^T(t)\xi(t)} \tag{8.13}$$

where $\Gamma_j = \Gamma_j^T > 0$, $j = 1, \ldots, m$, and $\Gamma_\rho = \Gamma_\rho^T > 0$.

For this adaptive design, we have the following result.

Theorem 8.1. *The adaptive law* (8.12)–(8.13) *is stable in the sense that* $\theta_j(t)$, $j = 1, \ldots, m$, $\rho(t)$ *are bounded, and* $\frac{\epsilon(t)}{N(t)} \in L^2 \cap L^\infty$, $\dot\theta_j(t) \in L^2 \cap L^\infty$, $j = 1, \ldots, m$, $\dot\rho(t) \in L^2 \cap L^\infty$, *where* $N(t) = \sqrt{1 + \zeta^T(t)\zeta(t) + \xi^T(t)\xi(t)}$.

Proof: Consider the positive definite function

$$V_p(\tilde\theta_{j(i)}, \tilde\rho_{(i)}, j \neq j_1, \ldots, j_p)$$
$$= \sum_{j \neq j_1, \ldots, j_p} |\rho^*_{j(i)}| \tilde\theta^T_{j(i)}(t) \Gamma_j^{-1} \tilde\theta_{j(i)}(t) + \tilde\rho^T_{(i)}(t) \Gamma_\rho^{-1} \tilde\rho_{(i)}(t) \quad (8.14)$$

where $\tilde\theta_{j(i)} = \theta_j - \theta^*_{j(i)}$ and $\tilde\rho_{(i)} = \rho - \rho^*_{(i)}$, with $\theta^*_{j(i)}$ being the value of the piecewise constant parameter vector θ^*_j and $\rho^*_{(i)}$ ($\rho^*_{j(i)}$) being the value of the piecewise constant parameter vector ρ^* (scalar ρ^*_j) over (T_i, T_{i+1}), $i = 1, \ldots, m_0$, which are the intervals on which the actuator failure pattern is unchanged. Since there are m actuators and at least one of them does not fail, we have $m_0 < m$ and $T_{m_0+1} = \infty$. Assume there are eventually $\bar m_0$ failed actuators. Then, we have $m_0 \leq \bar m_0 < m$. On each interval (T_i, T_{i+1}), $i = 1, \ldots, m_0$, $\theta^*_j = \theta^*_{j(i)}$ is constant, $j = 1, 2, \ldots, m$, and so is $\rho^*_j = \rho^*_{j(i)}$, so that $\tilde\theta_{j(i)} = \theta_j - \theta^*_{j(i)} = \tilde\theta_j$ and $\tilde\rho_{j(i)} = \rho_j - \rho^*_{j(i)} = \tilde\rho_j$. With its definition in (8.14), $V_p(t)$ is continuous and differentiable.

Before evaluating the time-derivative of $V_p(t)$, we first show that

$$\lim_{t \to \infty} \epsilon_p(t) = 0 \text{ exponentially.} \quad (8.15)$$

Note that $\epsilon_p(t) = 0$ if ρ^*_j, θ^*_j were constant for $t \geq 0$, and that $p = \bar m_0$ for $t \geq T_{m_0}$. Consider the individual term $\epsilon_{\theta^*_j}(t)$ of $\epsilon_p(t)$ in (8.11):

$$\epsilon_{\theta^*_j}(t) = \theta^{*T}_j(t)\zeta(t) - W_m(s)[\theta^{*T}_j \omega](t) \quad (8.16)$$

and let the impulse response function of $W_m(s)$ be $w_m(t)$. Then, for $t \in (T_i, T_{i+1})$, we can express (8.16) as

$$\epsilon_{\theta^*_j}(t) = \theta^{*T}_j(t)\int_0^t w_m(t-\tau)\omega(\tau)d\tau - \int_0^t w_m(t-\tau)\theta^{*T}_j(\tau)\omega(\tau)d\tau$$
$$= \theta^{*T}_j(t)\int_0^{T_i} w_m(t-\tau)\omega(\tau)d\tau - \int_0^{T_i} w_m(t-\tau)\theta^{*T}_j(\tau)\omega(\tau)d\tau$$
$$+ \theta^{*T}_j(t)\int_{T_i}^t w_m(t-\tau)\omega(\tau)d\tau - \int_{T_i}^t w_m(t-\tau)\theta^{*T}_j(\tau)\omega(\tau)d\tau$$
$$= \theta^{*T}_j(t)\int_0^{T_i} w_m(t-\tau)\omega(\tau)d\tau - \int_0^{T_i} w_m(t-\tau)\theta^{*T}_j(\tau)\omega(\tau)d\tau. \quad (8.17)$$

For the third equality of (8.17), we used the fact that $\theta^*_j(t)$ is constant for $t \in (T_i, T_{i+1})$. Therefore, for $t > T_{m_0}$, we have

$$\epsilon_{\theta_j^*}(t) = \int_0^{T_{m_0}} w_m(t-\tau)(\theta_{j(m_0)}^* - \theta_j^*(\tau))^T \omega(\tau)d\tau \qquad (8.18)$$

as $\theta_j^*(t) = \theta_{j(m_0)}^*$ is constant for $t > T_{m_0}$. Since T_{m_0} is finite and $w_m(t)$ is exponentially stable, it follows from (8.18) that $\lim_{t\to\infty} \epsilon_{\theta_j^*}(t) = 0$ exponentially, for $j \neq j_1, \ldots, j_p$ and $p = \bar{m}_0$.

Similarly, $\epsilon_{\rho_j^*}(t) = W_m(s)[\rho_j^* \tilde{\theta}_j^T \omega](t) - \rho_j^* W_m(s)[\tilde{\theta}_j^T \omega](t)$ also converges to zero as $t \to \infty$. Hence, (8.15) is true. Since $\lim_{t\to\infty} \epsilon_p(t) = 0$ exponentially, its existence does not destabilize an adaptive gradient update law like (8.12) (see [19]) so we can ignore it in the stability analysis.

Now we analyze the stability of (8.12)–(8.13). When there are p failed actuators, that is, $u_j(t) = \bar{u}_j$, $j = j_1, \ldots, j_p$, $1 \leq p \leq m-1$, at time $t \in (T_i, T_{i+1})$ (with $i = i(p) \leq p$ because there may be more than one actuator failing at T_i), from (8.12)–(8.14), ignoring the exponentially decaying term $\epsilon_p(t)$ in (8.11), we obtain the time-derivative of $V_p(t)$ as

$$\dot{V}_p(t) = -\frac{2\epsilon^2(t)}{N^2(t)} \leq 0, \ t \in (T_{i(p)}, T_{i(p)+1}). \qquad (8.19)$$

In particular, over the interval (T_{m_0}, ∞) on which there are $p = \bar{m}_0$ failed actuators ($\bar{m}_0 < m$), we have

$$\dot{V}_{\bar{m}_0}(t) = -\frac{\epsilon^2(t)}{N^2(t)} \leq 0, \ t \in (T_{m_0}, \infty) \qquad (8.20)$$

from which it follows that for $j \neq j_1, \ldots, j_{\bar{m}_0}$, $\theta_j(t) \in L^\infty$, $\rho(t) \in L^\infty$, that is, they are bounded parameter estimates, and $\frac{\epsilon(t)}{N(t)} \in L^2 \cap L^\infty$, $\dot{\theta}_j(t) \in L^2 \cap L^\infty$, $\dot{\rho}(t) \in L^2 \cap L^\infty$. Since $\dot{\theta}_i(t) = sgn[\rho_i^*]\Gamma_i \Gamma_j^{-1} \dot{\theta}_j(t)$, we also have $\theta_j(t) \in L^\infty$ and $\dot{\theta}_j(t) \in L^2 \cap L^\infty$, for $j = j_1, \ldots, j_{\bar{m}_0}$. $\triangledown$

Remark 8.1. To implement the adaptive law (8.12)–(8.13), $\epsilon(t)$ is computed from (8.9). The controller (8.1) is adaptively updated starting at $t = 0$ and whenever there is a new failure pattern the parameter adaptation will take into account the new failure pattern to adjust the remaining controls, while there is no explicit failure detection needed.

Remark 8.2. As compared with an adaptive control system without actuator failure, see for example, [12], the estimation error expression (8.10) has the additional term $\epsilon_p(t)$ which represents the unknown effect of actuator failures on the adaptive system. When an actuator failure occurs, it causes the controlled plant to change its structure and parameters, leading to equivalent parameter jumps in system operation. Our analysis indicates that such an effect becomes smaller when the closed-loop system operates with a fixed failure pattern longer. There is a transient in $\epsilon_p(t)$ when the failure pattern changes at time T_i. For any final failure pattern, that is, with any less than m failed actuators at any failure values, the proposed controller parametrization and adaptive design ensures that $\lim_{t\to\infty} \epsilon_p(t) = 0$ exponentially.

8.1.2 An adaptive control scheme.
For the plant (2.1) with unknown parameters A and B, and with unknown actuator failures (2.2), as an adaptive version of the control scheme (7.41) with the parameters k_{11}^*, k_{21}^* and k_1^* unknown, we use the state feedback control law

$$v_1(t) = v_2(t) = \cdots = v_m(t) = k_{11}^T(t)x(t) + k_{21}(t)r(t) + k_1(t) \quad (8.21)$$

where $k_{11}(t) \in R^n$, $k_{21}(t) \in R$, $k_1(t) \in R$ are adaptive estimates of $k_{11}^* \in R^n$, $k_{21}^* \in R$, $k_1^* \in R$. The resulting closed-loop system is

$$\begin{aligned}
\dot{x}(t) &= Ax + Bv + B\sigma(\bar{u} - v) \\
&= Ax + \sum_{j \neq j_1,\ldots,j_p} b_j(k_{11}^T x + k_{21} r + k_1) + \sum_{j=j_1,\ldots,j_p} b_j \bar{u}_j \\
&= (A + \sum_{j \neq j_1,\ldots,j_p} b_j k_{11}^{*T})x + \sum_{j \neq j_1,\ldots,j_p} b_j(k_{21}^* r + k_1^*) + \sum_{j=j_1,\ldots,j_p} b_j \bar{u}_j \\
&\quad + \sum_{j \neq j_1,\ldots,j_p} b_j \tilde{k}_{11}^T x + \sum_{j \neq j_1,\ldots,j_p} b_j \tilde{k}_{21} r + \sum_{j \neq j_1,\ldots,j_p} b_j \tilde{k}_1 \quad (8.22)
\end{aligned}$$

where $\tilde{k}_{11} = k_{11} - k_{11}^*$, $\tilde{k}_{21} = k_{21} - k_{21}^*$, $\tilde{k}_1 = k_1 - k_1^*$. The system output is

$$\begin{aligned}
y(t) &= y_m(t) + Ce^{(A + \sum_{j \neq j_1,\ldots,j_p} b_j k_{11}^{*T})t} x(0) + f_p(t) \\
&\quad + C(sI - A - \sum_{j \neq j_1,\ldots,j_p} b_j k_{11}^{*T})^{-1} \sum_{j \neq j_1,\ldots,j_p} b_j [\tilde{k}_{11}^T x + \tilde{k}_{21} r + \tilde{k}_1](t) \quad (8.23)
\end{aligned}$$

where $f_p(t)$ is defined in (7.46) such that $\lim_{t \to \infty} f_p(t) = 0$ exponentially. From (7.45), stability of the reference model (2.8) and assumption (A3c), it follows that $\lim_{t \to \infty} Ce^{(A + \sum_{j \neq j_1,\ldots,j_p} b_j k_{11}^{*T})t} x(0) = 0$ exponentially. Ignoring these exponentially decaying terms, we have the tracking error equation

$$\begin{aligned}
e(t) &= y(t) - y_m(t) \\
&= C(sI - A - \sum_{j \neq j_1,\ldots,j_p} b_j k_{11}^{*T})^{-1} \sum_{j \neq j_1,\ldots,j_p} b_j [\tilde{k}_{11}^T x + \tilde{k}_{21} r + \tilde{k}_1](t) \\
&= W_m(s) \frac{1}{k_{21}^*} [\tilde{\theta}^T \omega](t) \quad (8.24)
\end{aligned}$$

where $\tilde{\theta}(t) = \theta(t) - \theta^*$ with $\theta(t) = [k_{11}^T, k_{21}, k_1]^T$ being the estimate of the piecewise constant $\theta^* = [k_{11}^{*T}, k_{21}^*, k_1^*]^T$ (see (7.51)), and $\omega = [x^T, r, 1]^T$.

Introducing the auxiliary signals

$$\zeta(t) = W_m[\omega](t) \quad (8.25)$$
$$\xi(t) = \theta^T(t)\zeta(t) - W_m(s)[\theta^T \omega](t) \quad (8.26)$$
$$\epsilon(t) = e(t) + \rho(t)\xi(t) \quad (8.27)$$

where $\rho(t)$ is the estimate of $\rho^* = \frac{1}{k_{21}^*}$, we choose the adaptive laws as

$$\dot{\theta}(t) = -\frac{\mathrm{sgn}[k_{21}^*]\Gamma\zeta(t)\epsilon(t)}{1+\zeta^T\zeta+\xi^2}, \quad \Gamma = \Gamma^T > 0 \tag{8.28}$$

$$\dot{\rho}(t) = -\frac{\gamma\xi(t)\epsilon(t)}{1+\zeta^T\zeta+\xi^2}, \quad \gamma > 0. \tag{8.29}$$

To analyze the stability and tracking performance of the adaptive control system, we define the positive definite function

$$V(\tilde{\theta},\tilde{\rho}) = |\rho^*|\tilde{\theta}^T\Gamma^{-1}\tilde{\theta} + \gamma^{-1}\tilde{\rho}^2, \quad \tilde{\theta} = \theta - \theta^*, \quad \tilde{\rho} = \rho - \rho^* \tag{8.30}$$

and express the estimation error $\epsilon(t)$ in (8.27) as

$$\epsilon(t) = \rho^*\tilde{\theta}^T(t)\zeta(t) + \tilde{\rho}(t)\xi(t) + \epsilon_p(t) \tag{8.31}$$

where

$$\epsilon_p(t) = \rho^*(\theta^{*T}(t)\zeta(t) - W_m(s)[\theta^{*T}\omega](t)) \tag{8.32}$$

which, similar to (8.11), satisfies: $\lim_{t\to\infty}\epsilon_p(t) = 0$ exponentially.

Then, with (8.25)–(8.29), ignoring the effect of the exponentially decaying term $\epsilon_p(t)$ in (8.31), we obtain the time-derivative of V as

$$\dot{V}(t) = -\frac{2\epsilon^2(t)}{1+\zeta^T(t)\zeta(t)+\xi^2(t)} \leq 0, \quad t \in (T_i, T_{i+1}), \quad i = 0,1,\ldots,m_0 \tag{8.33}$$

where $T_0 = 0$, and T_i, $i = 1,\ldots,m_0$, are the time instants at which actuator failure occurs (note that $i = i(p) \leq p$ when there are in total p failed actuators at time T_i, and that $m_0 < m$ because at least one actuator does not fail).

In this case, unlike that in (8.14), $V(\cdot)$ as a function of t is not continuous because $\theta^* = [k_{11}^{*T}, k_{21}^*, k_1^*]^T = [k_{11(i)}^{*T}, k_{21(i)}^*, k_{1(i)}^*]^T$, for $t \in (T_i, T_{i+1})$, is a piecewise constant parameter vector as described in (7.51). Since there are only finite actuator failures in the system, that is, T_{m_0} is finite, and

$$\dot{V}(t) = -\frac{2\epsilon^2(t)}{1+\zeta^T(t)\zeta(t)+\xi^2(t)} \leq 0, \quad t \in (T_{m_0}, \infty) \tag{8.34}$$

we have that $\theta(t), \rho(t) \in L^\infty$, $\frac{\epsilon(t)}{\sqrt{1+\zeta^T(t)\zeta(t)+\xi^2(t)}} \in L^2 \cap L^\infty$, $\dot{\theta}(t) \in L^2 \cap L^\infty$, and $\dot{\rho}(t) \in L^2 \cap L^\infty$.

Based on these desired properties, the following closed-loop stability and asymptotic tracking results can be established.

Theorem 8.2. *The adaptive controller (8.21) with the adaptive law (8.28)–(8.29), applied to the plant (2.1) with actuator failures (2.2), guarantees that all closed-loop signals are bounded and the tracking error $e(t) = y(t) - y_m(t)$ goes to zero as t goes to infinity.*

Similar to (8.1), the controller (8.21) uses state feedback $k_{11}^T(t)x(t) + k_{21}(t)r(t)$ for output tracking (which is simpler compared with output feedback) and feedforward $k_1(t)$ for actuator failure compensation, relaxing the condition (3.1) for state tracking to (A3a)–(A3d) for output matching.

8.2 Output Feedback Design

When only the plant output $y(t)$ not the state variables in $x(t)$ are available for measurement, we need to use an output feedback design for actuator failure compensation. For the plant (2.1) with unknown parameters and unknown actuator failures characterized by (2.2), the parameters θ_1^*, θ_2^*, θ_{20}^*, θ_3^* and θ_4^* in the plant-model output matching controller (7.53) are unknown and need to be adaptively updated. As an adaptive version of the control scheme (7.53), we use the control law

$$v_1(t) = v_2(t) = \cdots = v_m(t)$$
$$= \theta_1^T \omega_1(t) + \theta_2^T \omega_2(t) + \theta_{20} y(t) + \theta_3 r(t) + \theta_4 \quad (8.35)$$

where $\theta_1(t) \in R^{n-1}$, $\theta_2(t) \in R^{n-1}$, $\theta_{20}(t) \in R$, $\theta_3(t) \in R$ and $\theta_4(t) \in R$ are adaptive estimates of the unknown parameters $\theta_1^* \in R^{n-1}$, $\theta_2^* \in R^{n-1}$, $\theta_{20}^* \in R$, $\theta_3^* \in R$ and $\theta_4^* \in R$.

To derive an error equation, we define

$$\theta^* = (\theta_1^{*T}, \theta_2^{*T}, \theta_{20}^*, \theta_3^*, \theta_4^*)^T \in R^{2n+1} \quad (8.36)$$
$$\theta(t) = (\theta_1^T(t), \theta_2^T(t), \theta_{20}(t), \theta_3(t), \theta_4(t))^T \in R^{2n+1} \quad (8.37)$$
$$\omega(t) = (\omega_1^T(t), \omega_2^T(t), y(t), r(t), 1)^T \in R^{2n+1} \quad (8.38)$$
$$\tilde{\theta}(t) = \theta(t) - \theta^*. \quad (8.39)$$

Following the development of Section 7.2, ignoring the exponentially decaying term $f_p(t)$ defined in (7.70), we have the tracking error equation

$$e(t) = y(t) - y_m(t) = \frac{k_p}{P_m(s)}[\tilde{\theta}^T \omega](t). \quad (8.40)$$

To see this, operating both sides of (7.63) on $y(t)$, we have

$$\theta_1^{*T} a(s) P(s)[y](t) + (\theta_2^{*T} a(s) + \theta_{20}^* \Lambda(s)) k_p Z_{uf}(s)[y](t)$$
$$= \Lambda(s)(P(s) - k_p \theta_3^* Z_{uf}(s) P_m(s))[y](t). \quad (8.41)$$

Substituting the plant description (see (7.55))

$$P(s)[y](t) = k_p Z_{uf}[v_0](t) + P(s)[\bar{y}](t) \quad (8.42)$$

in (8.41), we have

$$\theta_1^{*T} a(s) k_p Z_{uf}(s)[v_0](t) + \theta_1^{*T} a(s) P(s)[\bar{y}](t)$$
$$+ (\theta_2^{*T} a(s) + \theta_{20}^* \Lambda(s)) k_p Z_{uf}(s)[y](t)$$
$$= \Lambda(s) k_p Z_{uf}(s)[v_0](t) + \Lambda(s) P(s)[\bar{y}](t)$$
$$- \Lambda(s) k_p \theta_3^* Z_{uf}(s) P_m(s)[y](t). \quad (8.43)$$

Because $\Lambda(s)$ and $Z_{uf}(s)$ are Hurwitz, (8.43) can be expressed as

$$v_0(t) = \theta_1^{*T}\frac{a(s)}{\Lambda(s)}[v_0](t) + \theta_1^{*T}\frac{a(s)}{\Lambda(s)k_p Z_{uf}(s)}[\bar{y}](t) + \theta_2^{*T}\frac{a(s)P(s)}{\Lambda(s)}[y](t)$$

$$+\theta_{20}^*[y](t) + \theta_3^* P_m(s)[y](t) - \frac{P(s)}{k_p Z_{uf}(s)}[\bar{y}](t) + \epsilon_1(t)$$

$$= \theta_1^{*T}\frac{a(s)}{\Lambda(s)}[v_0](t) + \theta_2^{*T}\frac{a(s)}{\Lambda(s)}[y](t) + \theta_{20}^*[y](t) + \theta_3^* P_m(s)[y](t) + \theta_4^*(t)$$

$$-\left(\frac{(\Lambda(s) - \theta_1^{*T}a(s))P(s)}{\Lambda(s)k_p Z_{uf}(s)}[\bar{y}](t) + \theta_4^*\right) + \epsilon_1(t) \qquad (8.44)$$

for some initial condition-related exponentially decaying $\epsilon_1(t)$. Substituting (8.35) in (8.44) and ignoring $\epsilon_1(t)$, we obtain

$$\theta_3^* P_m(s)[e](t) = \tilde{\theta}^T \omega(t) + \left(\frac{(\Lambda(s) - \theta_1^{*T}a(s))P(s)}{\Lambda(s)k_p Z_{uf}(s)}[\bar{y}](t) + \theta_4^*\right), \qquad (8.45)$$

which, with (7.57), can be written as

$$e(t) = \frac{1}{\theta_3^* P_m(s)}[\tilde{\theta}^T \omega](t) + f_p(t) \qquad (8.46)$$

for the exponentially decaying $f_p(t)$ defined in (7.68). With $\theta_3^* = k_p^{-1}$ and $f_p(t)$ ignored, we get (8.40). Note that in this derivation we have neglected the effect of parameter jumping on the closed-loop system response as such an effect is also exponentially decaying, see Remark 7.4.

To develop an adaptive update law for the parameter estimates in $\theta(t)$, we introduce the auxiliary signals

$$\zeta(t) = W_m[\omega](t) \qquad (8.47)$$
$$\xi(t) = \theta^T(t)\zeta(t) - W_m(s)[\theta^T \omega](t) \qquad (8.48)$$
$$\epsilon(t) = e(t) + \rho(t)\xi(t) \qquad (8.49)$$

where $\rho(t)$ is the estimate of $\rho^* = \frac{1}{\theta_3^*}$. Then we choose the adaptive laws as

$$\dot{\theta}(t) = -\frac{\text{sgn}[\theta_3^*]\Gamma \zeta(t)\epsilon(t)}{1 + \zeta^T \zeta + \xi^2}, \quad \Gamma = \Gamma^T > 0 \qquad (8.50)$$

$$\dot{\rho}(t) = -\frac{\gamma \xi(t)\epsilon(t)}{1 + \zeta^T \zeta + \xi^2}, \quad \gamma > 0. \qquad (8.51)$$

To implement (8.50), we need the following condition:

(A4c) $CA^{n^*-1}\sum_{j \neq j_1,\ldots,j_p} b_j$, $p \in \{0, 1, \ldots, m-1\}$, have the same sign:

$$\text{sgn}[\theta_3^*] = \text{sgn}[k_p] = \text{sgn}[CA^{n^*-1}\sum_{j \neq j_1,\ldots,j_p} b_j], \quad k_p = \frac{1}{\theta_3^*}. \qquad (8.52)$$

To analyze the stability and tracking performance of the adaptive control system, we define the positive definite function

$$V(\tilde{\theta}, \tilde{\rho}) = |\rho^*|\tilde{\theta}^T \Gamma^{-1}\tilde{\theta} + \gamma^{-1}\tilde{\rho}^2, \quad \tilde{\theta} = \theta - \theta^*, \quad \tilde{\rho} = \rho - \rho^* \tag{8.53}$$

and express the estimation error $\epsilon(t)$ in (8.49) as

$$\epsilon(t) = \rho^* \tilde{\theta}^T(t)\zeta(t) + \tilde{\rho}(t)\xi(t) + \epsilon_p(t) \tag{8.54}$$

where

$$\epsilon_p(t) = \rho^*(\theta^{*T}(t)\zeta(t) - W_m(s)[\theta^{*T}\omega](t)) \tag{8.55}$$

which, similar to that in the proof of Theorem 8.1, can be shown to have the property: $\lim_{t\to\infty} \epsilon_p(t) = 0$ exponentially.

Then, with (8.47)–(8.51), ignoring the effect of the exponentially decaying term $\epsilon_p(t)$ in (8.54), we obtain the time-derivative of V as

$$\dot{V}(t) = -\frac{2\epsilon^2(t)}{1 + \zeta^T(t)\zeta(t) + \xi^2(t)} \leq 0, \; t \in (T_i, T_{i+1}), \; i = 0, 1, \ldots, m_0 \tag{8.56}$$

where $T_0 = 0$, and T_i, $i = 1, \ldots, m_0 < m$, are the time instants at which actuator failure occurs, that is, $i = i(p) \leq p$ when there are in total p failed actuators at time T_i. Note that $V(\cdot)$ as a function of t is not continuous because $\theta^* = [\theta_1^{*T}, \theta_2^{*T}, \theta_{20}^*, \theta_3^*, \theta_4^*]^T = [\theta_{1(i)}^{*T}, \theta_{2(i)}^{*T}, \theta_{20(i)}^*, \theta_{3(i)}^*, \theta_{4(i)}^*]^T$ is piecewise constant as described in (7.69). Since there are only a finite number of failures in the system, it follows that T_{m_0} is finite and

$$\dot{V}(t) = -\frac{2\epsilon^2(t)}{1 + \zeta^T(t)\zeta(t) + \xi^2(t)} \leq 0, \; t \in (T_{m_0}, \infty) \tag{8.57}$$

which implies that $\theta(t), \rho(t) \in L^\infty$, $\frac{\epsilon(t)}{\sqrt{1+\zeta^T(t)\zeta(t)+\xi^2(t)}} \in L^2 \cap L^\infty$, $\dot{\theta}(t) \in L^2 \cap L^\infty$, and $\dot{\rho}(t) \in L^2 \cap L^\infty$.

Using these desired properties, we can prove the following closed-loop stability and asymptotic tracking results.

Theorem 8.3. *The adaptive controller (8.35), with the adaptive law (8.50)–(8.51), applied to the plant (2.1) with actuator failures (2.2), guarantees that all closed-loop signals are bounded and the tracking error $e(t) = y(t) - y_m(t)$ goes to zero as t goes to infinity.*

9. Concluding Remarks

In this paper we formulated some open problems in adaptive control of systems with unknown actuator failures. We developed solutions to adaptive state feedback control for state tracking in the presence of unknown actuator failures that have fixed, or parametrizable time-varying, or non-parametrizable time-varying failure actions. We also developed solutions to adaptive state feedback control for output tracking, and adaptive output feedback control for output tracking, for fixed actuator failure actions. We derived necessary and sufficient conditions for the existence of such adaptive actuator failure compensation schemes, and developed the controller structures, parametrizations, error models, and adaptive laws for such control systems. We analyzed the stability and tracking performance of the adaptive laws. The presented adaptive designs ensure closed-loop stability and tracking in the presence of actuator failures that occur at uncertain instants of time and are uncertain in value and pattern, and in the presence of parametric uncertainties in the plant dynamics.

The key issue in control of systems with actuator failures is the controller parametrization, even if the plant and failure parameters are known. Certain conditions on the controlled system are needed for a parametrization suitable for meeting a given control objective. For actuator failure compensation, it is desirable to use less restrictive conditions which mean that larger classes of systems can be handled with a fixed or adaptive compensation controller. In this paper, some minimal necessary and sufficient conditions for stable plant-model state or output matching are derived. Once a suitable matching parametrization is developed, stable adaptive schemes can be developed for the case the plant and failure parameters are unknown.

The developed adaptive actuator failure compensation schemes have a controller structure suitably parametrized for any failure pattern (that is, $u_j(t) = \bar{u}_j$, $j = j_1, \ldots, j_p$, $1 \leq p \leq m - 1$, for any $\bar{u}_j$, any $j_1, \ldots, j_p$, and and $p \in \{1, 2, \ldots, m - 1\}$). When the control system operates from one failure pattern to another, it has a transient response as the adaptive controller adjusts itself to fit the new system with different parameters and uncertainties During the time interval when the actuator failure pattern does not change, the plant-model tracking error decreases asymptotically. This is the nominal performance achievable in the presence of unknown actuator failures. When some knowledge about actuator failures is available such knowledge can be used in the compensation controller to improve the system transient performance (this has not been studied except for the extreme case when the actuator failures are known which was considered in Section 3, Section 5.1, and Section 7 for developing desired plant-model matching controllers)

For all adaptive actuator failure compensation control schemes developed in this paper, we have verified their desired stability and tracking performance by extensive simulation results which indicate that for a typical step or sinusoidal reference input $r(t)$, the state or output tracking errors in different cases all converge to small values, and at the moments when an actuator fails, there is a transient response of the control signal $v(t)$, states $x(t)$ and output signal $y(t)$, as expected. The adaptive control schemes developed in Section 8.1 and Section 8.2 have been applied to lateral control of an aircraft model, where the actuator failure model (2.2) is introduced in the rudder servomechanism and adaptively compensated by an adaptive actuator failure compensation control design whose desired performance is verified by simulations. The adaptive designs and analyses have also been extended to the case where the state reference model system has multiple inputs that characterize more freedom in the desired system behavior.

Open problems that are currently under investigation include more general output tracking control designs with relaxed plant-model matching conditions using state feedback or output feedback, adaptive control designs for nonminimum phase systems with actuator failures, adaptive control designs for nonlinear systems with actuator failures, and adaptive control designs for systems with sensor or dynamics failures.

Acknowledgement. This research was supported in part by the US NASA Langley Research Center under grant NCC-1-342. The authors would like to thank Shuhao Chen, Xidong Tang and Xiaoli Ma for their discussion on this research and for their help in writing the manuscript.

References

1. K. J. Astrom and B. Wittenmark, *Adaptive Control*, 2nd edition., Addison-Wesley, Reading, MA, 1995.
2. F. Ahmed-Zaid, P. Ioannou, K. Gousman, and R. Rooney, Accommodation of failures in the F-16 aircraft using adaptive control, *IEEE Control Systems Magazine*, vol. 11, no. 1, pp. 73–78, 1991.
3. M. Bodson and J. E. Groszkiewicz, Multivariable adaptive algorithms for reconfigurable flight control, *IEEE Trans. on Control Systems Technology*, vol. 5, no. 2, pp. 217–229, 1997.
4. J. D. Boskovic, S.-H. Yu, and R. K. Mehra, A stable scheme for automatic control reconfiguration in the presence of actuator failures, *Proceedings of the 1998 ACC*, pp. 2455–2459.
5. J. D. Boskovic, S.-H. Yu, and R. K. Mehra, Stable adaptive fault-tolerant control of overactuated aircraft using multiple models, switching and tuning, *Proceedings of the 1998 AIAA Guidance, Navigation and Control Conference*, vol. 1, pp. 739–749, 1998.
6. J. D. Boskovic and R. K. Mehra, Stable multiple model adaptive flight control for accommodation of a large class of control effector failures, *Proceedings of the 1999 ACC*, pp. 1920–1924.
7. H. F. Chen and L. Guo, *Identification and Stochastic Adaptive Control*, Birkhauser, Boston, MA, 1991.
8. A. F. Filippov, Differential equations with discontinuous right hand sides, *American Mathematical Society Translations*, vol. 42, pp. 199–231, 1964.
9. G. C. Goodwin and K. S. Sin, *Adaptive Filtering Prediction and Control*, Prentice-Hall, Englewood Cliffs, NJ, 1984.

10. M. Gopinathan, J. D. Boskovic, R. K. Mehra, and C. Rago, A multiple model predictive scheme for fault-tolerant flight control design, *Proceedings of the 37th IEEE Conference on Decision and Control*, pp. 1376–1381, 1998.
11. P. A. Ioannou and P. V. Kokotović, *Adaptive Systems with Reduced Models*, Springer-Verlag, Berlin, 1983.
12. P. A. Ioannou and J. Sun, *Robust Adaptive Control*, Prentice-Hall, Englewood Cliffs, NJ, 1995.
13. M. Krstić, I. Kanellakopoulos, P. V. Kokotović, *Nonlinear and Adaptive Control Design*, John Wiley & Sons, New York, 1995.
14. Y. D. Landau, *Adaptive Control: The Model Reference Approaches*, Marcel Dekker, New York, 1979.
15. A. S. Morse, Global stability of parameter adaptive control systems, *IEEE Trans. on Automatic Control*, vol. 25, no. 6, pp. 433–439, 1980.
16. K. S. Narendra and A. M. Annaswamy, *Stable Adaptive Systems*, Prentice-Hall, Englewood Cliffs, NJ, 1989.
17. K. S. Narendra and J. Balakrishnan, Adaptive control using multiple modes, *IEEE Trans. on Automatic Control*, vol. 42, no. 2, pp. 171–187, 1997.
18. S. Sastry and M. Bodson, *Adaptive Control: Stability, Convergence, and Robustness*, Prentice-Hall, Englewood Cliffs, NJ, 1989.
19. G. Tao, Inherent robustness of MRAC schemes, *Systems and Control Letters*, vol. 29, no. 3-11, pp. 165–174, November 1996.
20. G. Tao, Friction compensation in the presence of flexibility, *Proceedings of the 1998 American Control Conference*, pp. 2128–2132, Philadelphia, PA.
21. G. Tao and P. A. Ioannou, Stability and robustness of multivariable model reference adaptive control schemes, in *Advances in Robust Control Systems Techniques and Applications*, Academic Press, edited by C. T. Leondes, vol. 53, pp. 99–123, 1992.
22. G. Tao and P. V. Kokotović, *Adaptive Control of Systems with Actuator and Sensor Nonlinearities*, John Wiley & Sons, 1996.
23. G. Tao, X. L. Ma and S. M. Joshi, Adaptive state feedback control of systems with actuator failures, *Proceedings of 2000 American Control Conference*, pp. 2654–2658, Chicago, IL, June 2000.
24. G. Tao, X. L. Ma and S. M. Joshi, Adaptive output tracking control of systems with actuator failures, *Proceedings of 2000 American Control Conference*, pp. 2669–2673, Chicago, IL, June 2000.
25. K. Tsakalis and P. A. Ioannou, *Linear Time Varying Systems: Control and Application*, Prentice-Hall, Englewood Cliffs, NJ, 1993.
26. V. I. Utkin, *Sliding Modes in Control Optimization*, Springer-Verlag, Berlin, 1991.
27. A. T. Vemuri and M. M. Ploycarpou, Robust nonlinear fault diagnosis in input-output systems, *International Journal of Control*, vol. 68, no. 2, pp. 343–360, 1997.
28. H. Wang and S. Daley, Actuator fault diagnosis: an adaptive observer-based technique, *IEEE Trans. on Automatic Control*, vol. 41, no. 7, pp. 1073–1078, 1996.
29. H. Wang, Z. J. Huang and S. Daley, On the use of adaptive updating rules for actuator and sensor fault diagnosis, *Automatica*, vol. 33, pp. 217–225, 1997.

Multi-mode System Identification

Erik I. Verriest

School of ECE, Georgia Institute of Technology
Atlanta, GA 30332-0250, USA

Abstract

Identification schemes are developed for multi-mode and Markov systems. It is shown how the mode parameters can be identified in parallel, and change points detected, given a persistency of excitation condition. The parameters for the mode switching mechanism can also be estimated. Some issues of the performance of the scheme are analyzed. Applications to identification of hybrid systems, piecewise linear systems, and failure detection and isolation are discussed.

This chapter is dedicated to the memory of Yvette Van Tongel and Cal Atkinson.

1 Introduction

Due to variations in parameters and unmodeled environmental influences most, if not all, systems exhibit variations in the operation regime. For small time variations, robust controller design can smooth these influences, or adaptive mechanisms can be devised to track the parameter variations and readjust the controller on-line. However, in many applications the parameter variations may be so severe that the underlying linear time invariant approach is no longer valid and the design based upon it no longer adequate for estimation and control. When the system exhibits behavior that can no longer be captured by a linear model, nonlinear control and estimation are called for. During the past few decades, great progress has been made in theory as well in practice of nonsmooth nonlinear control.

However many problems involve dynamics with nonsmooth behavior. One example is the control of otherwise linear systems with saturating actuators, and/or states. Backlash, hysteresis and dead-zone are other examples of 'hard' nonlinearities which can be modeled by transitions between different linear operations. The book by Tao and Kokotović [1] on adaptive control in the presence of such sensor and actuator nonlinearities and various other chapters in this volume describe these issues well.

In chemical process control, batch operations may occur at different operating points, for instance to get different fractions of reaction products or in response to different specifications. At each of the operating points, typically a linearized model is used for control design. Similarly, start-up and shutdown of reactors contributes to other modes of operation with their own control strategies. Significant economic loss may result from slow transitions between operating points and therefore it is important to have a model that accurately describes the possible behaviors and identifies change points. A method based on an on-line validation of local models followed by the construction of a time-varying global model via interpolation was proposed by Arkun *et al.* [2].

In aerospace applications, different flight regimes such as take-off and landing and high altitude cruise may be embedded in a full nonlinear model, however the dynamics in the different operating regimes may be simplified to a linear model. Another but not less significant aspect is the detection and isolation of faults in a system. In fact the latter was the problem that initially motivated this study. An increased concern for reliability and safety, be it in aircraft control or operation of a nuclear reactor, has dictated a need to design systems with fast response, or even complete recovery, to unknown and unpredictable faults. In most cases, it is desired to identify the fault as well, if only for maintenance purposes.

Other applications abound in communications technology. Multipath channel characteristics, whether in short range acoustic communication or long range microwave transmissions, may be identified on-line. The time variation may be due to the mobility of either receiver, transmitter or interference.

Model predictive control (MPC) can take care of constraints, but in general requires the solution of an optimization problem [3]. An alternative approach is the multiple-model paradigm. Progress in this active research area can be found in the volume edited by Murray-Smith and Johansen [4]. This multiple model is here combined with a switching mechanism, giving a hybrid model. Either at every sampling time, the probability that another model will be switched on is known, or deterministic (albeit unknown to the observer, e.g., controlled by an independent actor), or estimated empirically from the past data. Some cases of interest are: a one time switch at an unknown time e.g., as in component failure, periodic switching, e.g., modeling seasonal effects in ecological systems, or night and day cycle in thermal and power systems, and random switching. In other applications periodicity may occur but with an *a priori* unknown period, e.g., due to limit cycle behavior. Such models provide an interesting alternative to more conventional models based on superposition of harmonic components. In another class, we allow the switching to be determined by the state of the system to approximate the aforementioned nonlinear system and effects like hysteresis where the model is typically determined by the sign of some functional of the state. In time series analysis such state-dependent multi-modal systems are also known as *self-exciting threshold autoregressive moving average* models (SETARMA) [5].

The point of departure for the adaptive control in this chapter is the indirect one. It is opted to gain knowledge of the various system parameters on-line and that the control aspects (adaptation) are secondary. Hence this correlates with gain scheduling and the self-tuning regulators [6]. The estimated parameters are used in the controller design as if they are the true parameters (certainty equivalence principle). The indirect approach also has the advantage that a measure for the quality (or associated uncertainty) of the parameters may be estimated, and a 'robust' controller may be designed accordingly [7]. In addition, it also allows the user to obtain error bounds, an important step in identification for control [8]. This work differs from the work of Narendra *et al.* who focused on the stability and performance of multiple reference model adaptive control in the deterministic and stochastic case [9, 10]. From a pure identification point of view, the literature on multi-model identification is rather scarce. Modeling of time series with changes in regime occupies one chapter in [11], and a modern martingale approach is found in [12].

In this chapter it is assumed that a system may operate *linearly* in *several possible modes*, as for instance an automobile in different gears. The objective is the identification of the system parameters in each mode. The final goal may be the ability to predict future behavior of the system or to control its behavior. The user should be cautioned that adoption of a multi-mode model as an alternative or simplification to an underlying general nonlinear plant may be ad hoc, although in some simulation studies positive results were obtained [2].

The goal is to combine parameter estimation (identification) with detection of mode transitions. A simple approach uses any of the classical on-line identification methods perhaps followed by validation and change detection. With or without change detection, the identification algorithm will track the parameters, but quite some time may lapse before the new values are identified. This inertia is essentially due to the large amount of memory that is associated with the identification method. Consequently good tracking performance can only be obtained if the system parameters are slowly varying. In hybrid systems, the change is assumed to be very abrupt, and in the case of fast mode switching it may be unreasonable to expect a classical identification algorithm to work. What is proposed in this chapter is a method which enables identification of all modes *in parallel*. Of course this can only be feasible if all modes were switched on during the observation epoch. So one should expect some type of *persistency of excitation* condition.

Change point detection may be performed at the same time, basically using the same operations that are necessary for the mode identification. Thus identification and detection are not inseparable: one improves the other. Preliminary results were reported in [13].

This chapter is organized as follows: section 2 recalls some ideas of total least squares (TLS) based identification and its performance. In section 3 the basic principle in the identification of deterministic ARMA models is explained. In order to set the ideas we focused on a two-mode system. Extensions to a larger number of modes are straightforward in principle, but increase the computational complexity. State space models with discretely varying parameters are discussed in the section 4, where also some general comments on performance are given. A general principle allowing the use of subspace identification methods for modeling finite state Markov chains is discussed in section 5. This and section 3 constitute the main ideas. In addition, the multi mode and Markov identification are combined to a powerful algorithm for identifying hybrid systems. Further applications to fault detection and nonlinear/nonsmooth adaptive control are presented in section 6. Some illustrative examples displaying the feasibility of the proposed method are reported in section 7. The emphasis in this chapter is in obtaining mathematically provable constructs. Indeed, the mathematical simplicity of the class of models discussed allows for a preservation of the rigor, while at the same time being interesting enough from an application point of view.

2 Subspace Identification Degradation

The identification of autoregressive moving average (ARMA) models with fixed parameters is a standard problem described in textbooks on identification and/or adaptive control [7, 14, 15]. In this section, the rationale of the total least squares (TLS) method is explained, and the degradation of this parameter estimation algorithm is analyzed in the case the data is generated by a system whose parameter vector at each step takes one of two possible values. Such a system will be called bi-modal, as it displays two modes of operation. This could be the nominal operation and a faulty operation.

2.1 Total Least Squares Identification

Consider the input–output (or ARMA) representation of the system dynamics

$$y_k = -a_1 y_{k-1} - a_2 y_{k-2} - \cdots - a_n y_{k-n} + \\ + b_1 u_{k-1} + b_2 u_{k-2} + \cdots + b_n u_{k-n}. \tag{2.1}$$

If at step k, a data vector $d'_k = [y_k, y_{k-1}, \ldots, y_{k-n}; u_{k-1}, \ldots, u_{k-n}]$ is collected, the equation (2.1) simplifies to

$$d'_k \theta_k = 0,$$

where θ is the parameter vector, normalized so that its first entry equals 1, thus: $\theta' = [1, -a_1, -a_2, \ldots, -a_n, -b_1, \ldots, -b_n]$. The parameter vector is determined statistically. If the data, d_k, is collected over N consecutive steps, then

$$D_N \theta = \begin{bmatrix} d'_n \\ d'_{n+1} \\ \vdots \\ d'_{n+N-1} \end{bmatrix} \theta = 0.$$

If the ARMA model is exact and minimal (requiring that $a(z)$ and $b(z)$ are coprime), then the overdetermined set of equations has a unique solution, θ_0. Due to measurement inaccuracies, the equations will in general be inconsistent, and a solution is sought in the total least squares sense (TLS) [16]. The TLS is preferred over the well-known least squares (LS) solution. This stems from the fact that the A in the equation $Ax = y$ is corrupted by the same inaccuracies as the known vector y. If the inaccuracies are small, this means that the $n \times n$ 'squared-up' data matrix $D'_N D_N$ will have an eigenvalue close to zero. Its corresponding eigenvector, properly normalized so that its first entry is 1, is the total least squares solution θ_{TLS}. Indeed, replacing the exact y_k and u_k respectively by the observed $y_k + \eta_k$ and $u_k + \nu_k$, assuming that the η_k and ν_k are uncorrelated white noise sequences, gives

$$d'_k \theta = \epsilon'_k \theta, \quad \text{where} \quad \epsilon' = [\eta_1, \ldots, \eta_n; \nu_1, \ldots, \nu_n],$$

and thus, denoting the matrix of *observed* data by $D_{\text{ob}} = D + \mathcal{E}$,

$$D_{\text{ob}} \theta = \mathcal{E} \theta.$$

Premultiplying by the data matrix $(D_{\text{ob}} - \mathcal{E})'$, gives

$$(D_{\text{ob}} - \mathcal{E})' D_{\text{ob}} \theta = (D_{\text{ob}} - \mathcal{E})' \mathcal{E} \theta$$

or

$$D'_{\text{ob}} D_{\text{ob}} \theta = (\mathcal{E}' D_{\text{ob}} + D'_{\text{ob}} \mathcal{E}) \theta - \mathcal{E}' \mathcal{E} \theta.$$

Taking mathematical expectation,

$$\mathbf{E}\, D'_{\text{ob}} D_{\text{ob}} \theta = \mathbf{E}\, \mathcal{E}\mathcal{E}' \theta.$$

From the given assumptions on $\mathcal{E}$, the matrix $\mathbf{E}\, \mathcal{E}\mathcal{E}'$ is proportional to the identity matrix, say $\sigma^2 I$. Thus, invoking ergodicity, $D'_{\text{ob}} D_{\text{ob}} \to D' D \theta + \sigma^2 \theta$. All eigenvalues are shifted by σ^2, thus the new matrix $D'_{\text{ob}} D_{\text{ob}} - \sigma^2 I$ will be closer to the exact squared data matrix. Hence the normed singular vector of $D'_{\text{ob}} D_{\text{ob}} - \sigma^2 I$, which is also the eigenvector corresponding to the smallest eigenvalue of $D'_{\text{ob}} D_{\text{ob}}$, is the TLS-solution, θ_{TLS}.

2.2 Degradation in Bi-modal Systems

In this section, it is assumed that the parameter identification algorithm is run on an ARMA system that actually exhibits switching behavior between two parameter vectors, i.e, the actual system is a bi-modal system, which for simplicity we restrict to be scalar. The observed data consists of a sequence of input/output pairs $h_k = [u_k, y_k]'$. Let the system dynamics be given in ARMA form,

$$y_k = -a_1^{(m)} y_{k-1} - a_2^{(m)} y_{k-2} - \cdots - a_n^{(m)} y_{k-n} + \\ + b_1^{(m)} u_{k-1} + b_2^{(m)} u_{k-2} + \cdots + b_n^{(m)} u_{k-n}. \quad (2.2)$$

In the above equation, the superscript, $m = 1, 2$ refers to the particular mode the system is operating in. It is assumed that the order n is the same for all modes. In order to fix the ideas, suppose that mode 1 switched to mode 2 at time $k = 0$. Thus for $k \leq 0$, the data vectors satisfy

$$d_k \theta_1 = [y_k, y_{k-1}, \cdots, y_{k-n}; u_{k-1}, u_{k-2}, \cdots, u_{k-n}] \theta_1 = 0,$$

with

$$\theta_1 = [1, -a_1^{(1)}, \cdots, -a_n^{(1)}; b_1^{(1)}, \cdots, b_n^{(1)}]'.$$

For $k \geq 1$, the data vectors satisfy

$$d_k \theta_2 = [y_k, y_{k-1}, \cdots, y_{k-n}; u_{k-1}, u_{k-2}, \cdots, u_{k-n}] \theta_2 = 0$$

with

$$\theta_2 = [1, -a_1^{(2)}, \cdots, -a_n^{(2)}; b_1^{(2)}, \cdots, b_n^{(2)}]'.$$

Stacking T consecutive data vectors $d_k, d_{k+1}, \ldots, d_{k+T-1}$ in a data matrix

$$\begin{bmatrix} d_k \\ \vdots \\ d_{k+T-1} \end{bmatrix} \stackrel{\text{def}}{=} D_{k+T-1}^k, \quad (2.3)$$

we find thus

$$D_0^{-N} \theta_1 = 0 \\ D_T^1 \theta_2 = 0. \quad (2.4)$$

Denoting the 'past' data matrix, D_0^{-N}, as D_- and the 'future', D_T^1, (with respect to the switching time), as D_+, it is readily seen that the combination

$$D_T^{-N} \theta = \begin{bmatrix} D_- \\ D_+ \end{bmatrix} \theta = 0 \quad (2.5)$$

cannot be satisfied for either $\theta = \theta_1$ nor $\theta = \theta_2$. As the classical identification algorithms [17] are based on the (approximate) solution of $D\theta = 0$, a relatively long time will be required for data crossover, i.e., before one settles into $\theta \approx \theta_2$ from $\theta = \theta_1$. Obviously, if mode switching occurs frequently the problem will further be compounded.

3 Multi-mode ARMA Identification

Consider a system that may operate in M possible modes, with mode switching at arbitrary times. The system dynamics is given in ARMA form which, for simplicity of notation is restricted to a single input single output

$$y_k = -a_1^{(m)} y_{k-1} - a_2^{(m)} y_{k-2} - \cdots - a_n^{(m)} y_{k-n} +$$
$$+ b_1^{(m)} u_{k-1} + b_2^{(m)} u_{k-2} + \cdots + b_n^{(m)} u_{k-n} \quad (3.1)$$

In the above equation, the superscript, m refers to the particular mode the system is operating in and ranges from 1 to M. Again, it is assumed that the order n is the same for all modes.

Let N consecutive data vectors, $d_k; k = 1, \ldots, N$, be collected. The main idea which makes fast identification of models possible is the observation of the logic ('$\vee$' is the logical 'or')

$$\bigvee_{i=1}^{M} \{ d'_k \theta_i = 0 \} \Leftrightarrow \{ \prod_{i=1}^{M} d'_k \theta_i = 0 \} \quad (3.2)$$

i.e., if several system parameters are possible, then in the noiseless case, at least one of the residual factors must be zero. If the system operated in mode i, we get: $d'_k \theta_i = 0$. Consequently, even if it is not known which mode was operational at time k, the *product* of the residuals must always be zero. The product on the right-hand side of (3.2) leads to homogenous polynomials of degree M in the data, while the unknown parameters are scrambled into symmetric functions of the individual mode parameters. In the remainder of this section, the focus will be on the case $M = 2$. Unless mode switching occurs frequently, this analysis is also adequate in the $M > 2$ case. More will be said about this in section 8. In this case, the right-hand side of (3.2) can be reorganized as

$$d'\theta_1 \cdot d'\theta_2 = \operatorname{Tr} dd' \theta_1 \theta'_2 = [\operatorname{vec}(dd')]' \cdot \operatorname{vec}(\theta_1 \theta'_2) \stackrel{\text{def}}{=} d^{(2)} \theta^{(2)} \quad (3.3)$$

('vec' is the column stacking operator). The datavectors $d^{(2)} = \operatorname{vec}(dd')$ are quadratic in the original data. Stacking these consecutive data vectors $d^{(2)}$ in a datamatrix

$$D^{(2)} \stackrel{\text{def}}{=} \begin{bmatrix} d_1^{(2)} \\ \vdots \\ d_N^{(2)} \end{bmatrix} = \begin{bmatrix} [\operatorname{vec}(d_1 d'_1)]' \\ \vdots \\ [\operatorname{vec}(d_N d'_N)]' \end{bmatrix} \quad (3.4)$$

leads to the system of equations

$$D^{(2)} \theta^{(2)} = 0. \quad (3.5)$$

to be solved for the parameters $\theta^{(2)}$. Hence the problem of bi-modal system identification is reduced to a *linear problem in quadratic data*. The identification problem boils down to a three-step procedure:

(1) Preprocessing of data: this prepares the datamatrix $D^{(2)}$ from consecutive inputs and outputs.
(2) Determination of the right singular vector(s) of $D^{(2)}$, call them $\theta^{(2)}$. This is the problem solved in section 2.
(3) Resolve the individual mode parameters θ_1 and θ_2 from $\theta^{(2)}$.

The identification can be combined with a detection step. Assume that at step k the estimates $\hat{\theta}_1[k]$ and $\hat{\theta}_2[k]$ were obtained. Define the residuals

$$\epsilon_i[k] = d_k \hat{\theta}_i[k] \qquad i = 1, 2. \tag{3.6}$$

With exact data, either $\epsilon_1[k]$ or $\epsilon_2[k]$ must be zero. Hence, the *data association* may, proceed as follows:
Decide that at time k mode 1 was active if $|\epsilon_1| \leq |\epsilon_2|$. For future reference, introduce an extended datavector as the input and output data, together with the estimated mode vector $\hat{\theta}$. This will subsequently be used to estimate the mode switching behavior (see section 5). The generalization to M modes should now be clear: the preprocessor constructs data matrices of order M in the original input–output data. The components of the right singular vectors are combinations (symmetric functions) of the components of the M mode parameters. The mode parameters associated with each mode are determined (see below). Finally the detection associates the mode vector which minimizes the residual to the data at time k. Some fine tuning of the algorithm may be required to deal with particular cases. One such particular case occurs if all observed data are generated from within the same mode, i.e., the modes are not persistently excited. Also it needs to be resolved how the symmetric equations lead to a unique solution. The mathematics behind this solution is exposed below.

3.1 Determination of the Mode Parameters

Consider again equation (3.2), expressing the equivalence for all k, and note that also

$$(d'_k \theta_1)(d'_k \theta_2) = \operatorname{Tr} d_k d'_k \theta_2 \theta'_1 = [\operatorname{vec}(d_k d'_k)]' \operatorname{vec}(\theta_2 \theta'_1) = 0. \tag{3.7}$$

Thus not only is $\operatorname{vec}(\theta_1 \theta'_2)$ in the nullspace of the second-order datamatrix $D^{(2)}$, but also $\operatorname{vec}(\theta_2 \theta'_1)$. In fact, there is more. Because of the redundancy in the Kronecker product construction, $d_k d'_k$, fixed relations exist, *independent of the ARMA parameters* between the columns of the data matrix. These structural properties will be investigated in the first section, where we also put some necessary background material of linear algebraic nature. The resolution of the individual parameters is given in the second section.

3.1.1 Structure of the quadratic data matrix.
Assume that θ_1 and θ_2 are the *exact* parameters in the two modes, and that the system dynamics is truly given by this bi-modal system, i.e., the system is not a reduced model of some higher order model (hence also no delays) and there are no neglected nonlinearities or other time dependencies in the system. In addition, let the system satisfy a *persistency of excitation* condition, in the sense that both modes are active during the observation period. Finally, we shall understand that this setup is not degenerate: the modal parameter vectors θ_1 and θ_2 are not identical. Since the first entry in θ_i is the fixed 1, this implies that at least one of the ARMA parameters must be different in the two modes.

Lemma 3.1. *The vectors $x, y \in \mathbb{R}^n$ are linear independent if and only if the matrices xy' and yx' are linear independent.*

Proof: (i) If $x = \alpha y$, then $xy' = \alpha yy' = yx'$. (ii) Conversely, let $xy' = \alpha yx'$. First assume $\alpha \neq 1$, then $(\alpha - 1)x_i y_i = 0$ for all i. If $x = 0$ or $y = 0$ we are done. Thus assume that $\mathcal{I}$, the set of indices such that $x_i = 0$, and its complement $\mathcal{I}^c$ are not empty. From $(\alpha - 1)x_i y_i = 0$, it follows that $y_i = 0$ if $i \in \mathcal{I}^c$. But then $(xy')_{i,j} = \alpha(yx')_{ij}$ leads to a contradiction if $i \in \mathcal{I}$ and $j \in \mathcal{I}^c$. Hence the only possibility left is $\alpha - 1$, with $x \neq 0$ and $y \neq 0$. Again $xy_i = yx_i$ for all i implies for $i \in \mathcal{I}^c$ that $y = \frac{y_i}{x_i} x$, thus the dependence of x and y. ⋄

Under these conditions, it is clear from the above observation and Lemma 3.1 that the nullspace of $D^{(2)}$ is *at least* two-dimensional. More precisely

$$\text{Span}\{\text{vec}(\theta_1 \theta_2'), \text{vec}(\theta_2 \theta_1')\} \subset \mathcal{N}(D^{(2)}).$$

Also, note that

$$[\text{vec}(d_k d_k')]' \text{vec}(e_i e_j') = d_k' e_i \cdot d_k' e_j$$
$$= (d_k)_i (d_k)_j.$$

Hence, letting $E_{ij} = e_i e_j'$:

$$[\text{vec}(d_k d_k')]' \text{vec}(E_{ij} - E_{ji}) = 0, \quad \forall i, j. \tag{3.8}$$

This relation is nontrivial if $i \neq j$, and there are $n(n-1)/2$ such relations. (the factor $1/2$) occurs because we may assume $i > j$. Thus we arrived at

$$\text{Span}\{\text{vec}(\theta_1 \theta_2'), \text{vec}(\theta_2 \theta_1'), \text{vec}(E_{ij} - E_{ji})_{i>j}\} \subset \mathcal{N}(D^{(2)}).$$

However, this spanning set is not minimal. Indeed, taking the sum and difference of the first two, it is easily seen that the difference is the 'vec' operation on a skew symmetric matrix, and is therefore spanned by the set $\{\text{vec}(E_{ij} - E_{ji})_{i>j}\}$. Hence,

$$\text{Span}\,\{\text{vec}\,(\theta_1\theta_2' + \theta_2\theta_1'), \text{vec}(E_{ij} - E_{ji})_{i>j}\} \subset \mathcal{N}\,(D^{(2)}). \qquad (3.9)$$

If the input is persistently exciting in each mode, which implies that both modes are active, it can also be shown that all vectors in the nullspace are spanned by the left-hand side. In conclusion, the dimension of the nullspace $\mathcal{N}\,(D^{(2)})$ equals $1 + n_1(n_1 - 1)/2$, where n_1 is the dimension of $\theta^{(2)}$, i.e., for a single input single output system of order n this is $n_1 = 2n + 1$.

$$\dim \mathcal{N}(D^{(2)}) = n(2n + 1) + 1.$$

For a first-order system, the nullspace of the quadratic data matrix has dimension 4. In contrast, the quadratic data matrix has $(2n + 1)^2$ columns. Roughly, the redundancy of this setup of the problem is 50%.

The bi-modal identification problem is thus reduced to the unimodal problem

$$D^{(2)}\theta^{(2)} = 0$$

with the additional redundancy

$$D^{(2)}S = 0$$

for a fixed structure S. In solving a unimodal form, the problem reduces to finding the right singular vector of the data matrix. In the bi-modal problem we get a multidimensional right singular subspace, and isolating $\theta^{(2)}$ in it might seem impossible, if it were not for the underlying skew symmetric structure of S. The following lemma provides the key:

Lemma 3.2. *Let v be chosen arbitrarily in the nullspace $\mathcal{N}(D^{(2)})$. then*

$$\text{Mat}[v] + \text{Mat}'[v] = \gamma(\theta_1\theta_2' + \theta_2\theta_1'), \quad \gamma \in \mathbb{R}.$$

Proof: In view of (3.9), v may be expressed as a linear combination

$$v = \xi[\text{vec}(\theta_1\theta_2') + \text{vec}(\theta_2\theta_1')] + \sum_{i>j}\mu_{ij}[\text{vec}(E_{ij} - E_{ji})].$$

Hence, since "Mat" is a left inverse for the "vec" operation

$$\text{Mat}[v] + \text{Mat}'[v] = 2\xi(\theta_1\theta_2' + \theta_2\theta_1'). \quad \diamond$$

In principle, if v is chosen at random in the nullspace, the probability that v will be such that $\text{Mat}[v] + \text{Mat}'[v]$ vanishes is zero. However, an intelligent choice can be made if one exploits the fact that $[\text{vec}(\theta_1\theta_2') + \text{vec}(\theta_2\theta_1')]_{11} = 1$, while $[\text{vec}(E_{ij} - E_{ji})]_{11} = 0$. In practice, perform the singular value decomposition of the 'squared-up' matrix $D^{(2)'}D^{(2)}$, of size $(2n+1)^2$ by $(2n+1)^2$, and isolate the singular vector with nonzero first component. If there are many of these, take the one with largest ratio $|v_1|/\|v\|$ for added accuracy. This problem solves what we call for obvious reasons the *symmetric parameter identification problem*. What is left is the resolution of $\text{vec}(\theta_1\theta_2') + \text{vec}(\theta_2\theta_1')$ into the modal parameter vectors θ_1 and θ_2.

3.2 Resolution of the Mode Parameters

We state first a lemma about the eigenproblem of a symmetrized rank one matrix of form (3.1.1).

Lemma 3.3. *Let x and y be linearly independent vectors in $\mathbb{R}^n$ and $M_2 = xy' + yx'$, then the nonzero eigenvalues of M_2 are*

$$\lambda_{\pm} = x'y \pm \|x\|\|y\|. \tag{3.10}$$

A set of associated eigenvectors is

$$z_{\pm} = \|y\|x \pm \|x\|y. \tag{3.11}$$

Proof. From $(xy' + yx')z = \lambda z$, we get $(y'z)x + (x'z)y = \lambda z$, so that z is a linear combination of the vectors x and y. Thus let

$$z = \alpha x + \beta y.$$

Substituting in the eigenequation, one deduces the system

$$\begin{bmatrix} y'x & y'y \\ x'x & x'y \end{bmatrix} \begin{bmatrix} \alpha \\ \beta \end{bmatrix} = \lambda \begin{bmatrix} \alpha \\ \beta \end{bmatrix}.$$

The characteristic equation is

$$\|x\|^2\|y\|^2 - (x'y - \lambda)^2 = 0$$

from which the eigenvalues follow. With $\lambda_+ = x'y + \|x\|\|y\|$ corresponds

$$[x'x, \ -\|x\|\|y\|] \begin{bmatrix} \alpha_+ \\ \beta_+ \end{bmatrix} = 0,$$

or

$$[\|x\|, \ -\|y\|] \begin{bmatrix} \alpha_+ \\ \beta_+ \end{bmatrix} = 0.$$

With λ_+ corresponds then

$$\begin{bmatrix} \alpha_+ \\ \beta_+ \end{bmatrix} = \begin{bmatrix} \|y\| \\ \|x\| \end{bmatrix},$$

determining the corresponding eigenspace.

Likewise, with $\lambda_- = x'y - \|x\|\|y\|$ correspond the scalars

168   Erik I. Verriest

$$\begin{bmatrix} \alpha_- \\ \beta_- \end{bmatrix} = \begin{bmatrix} \|y\| \\ -\|x\| \end{bmatrix},$$

thus proving the lemma. ◇

Schwarz's inequality applied to (3.10) implies that $\lambda_+ > 0$ and $\lambda_- < 0$. The following sequence of lemmas leads essentially to the uniqueness result.

Lemma 3.4. *The mapping M defined by $M : (x,y) \in \mathbb{R}^n \oplus \mathbb{R}^n \to M(x,y) = xy' + yx' \in \mathbb{R}^{n \times n}$ is invariant under the scaling $\mathbf{S}_{(\alpha,\beta)}(x,y) = (\alpha x, \beta y)$ if $\alpha = 1/\beta$.*

Proof: Obvious. ◇

Lemma 3.5. *If $M(x,y) = 0$, then $x = 0$ and/or $y = 0$.*

Proof: From $x_i y_i = 0$, at least one of the factors must be zero. Say $x_i = 0$, but then $x_i y_j + y_i x_j = 0$ yields $x_j = 0$ for all $j \neq i$. ◇

Lemma 3.6. *Given the linearly independent vectors a and b, the equation $M(x,y) = M(a,b)$, with M defined in Lemma 3.4, has solutions:*

$$\begin{cases} x = \alpha a \\ y = \beta b \end{cases} \text{ or } \begin{cases} x = \alpha b \\ y = \beta a \end{cases} \text{ for arbitrary } \alpha, \beta \in \mathbb{R} \text{ with } \alpha\beta = 1.$$

Proof: For all k: $xy_k + yx_k = ab_k + ba_k$. Obviously $x, y \in \text{Span}\{a, b\}$. Let

$$\begin{aligned} x &= \alpha a + \beta b \\ y &= \gamma a + \delta b \end{aligned}$$

and substitute in the equation to get

$$(\alpha a + \beta b)(\gamma a_k + \delta b_k) + (\gamma a + \delta b)(\alpha a_k + \beta b_k) = ab_k + ba_k.$$

Thus:

$$[\alpha(\gamma a_k + \delta b_k) + \gamma(\alpha a_k + \beta b_k) - b_k]a + [\beta(\gamma a_k + \delta b_k) + \delta(\alpha a_k + \beta b_k) - a_k]b = 0.$$

By the linear independence of a and b,

$$\forall k \begin{cases} \alpha(\gamma a_k + \delta b_k) + \gamma(\alpha a_k + \beta b_k - b_k) &= 0 \\ \beta(\gamma a_k + \delta b_k) + \delta(\alpha a_k + \beta b_k - a_k) &= 0. \end{cases}$$

Thus

$$\begin{cases} 2\alpha\gamma a + (\alpha\delta + \beta\gamma - 1)b &= 0 \\ (\beta\gamma + \delta\alpha - 1)a + 2\beta\delta b &= 0. \end{cases}$$

Invoking the independence of a and b again,

$$\begin{cases} 2\alpha\gamma & = 0 \\ \alpha\delta + \beta\gamma - 1 & = 0 \end{cases} \text{ and } \begin{cases} \beta\gamma + \delta\alpha - 1 & = 0 \\ 2\beta\delta & = 0. \end{cases}$$

If $\gamma = 0$ it follows that $\beta = 0$ and $\alpha\delta = 1$, thus giving

$$\begin{cases} x & = \alpha a \\ y & = \delta b \end{cases} \text{ with } \alpha\delta = 1.$$

If $\alpha = 0$ it follows that $\delta = 0$ and $\beta\gamma = 1$, thus giving

$$\begin{cases} x & = \beta b \\ y & = \gamma a \end{cases} \text{ with } \beta\gamma = 1. \quad \diamond$$

In the problem at hand, we are given a symmetric matrix $M(\theta_1, \theta_2)$. Obviously, its (11)-entry is equal to 2. By the previous lemma, if solutions are sought with first component equal to unity, then apart from an obvious permutation the solution will be *unique*.

Combining these lemmas with the observation preceding them, any singular vector $v \in \mathcal{N}(D^{(2)})$, has the property that $\text{Mat}(v) + (\text{Mat}(v))'$ is of the form in the lemma. This will be the key in the proof of the following theorem, correcting some errors in [18].

Theorem 3.1. *The modal parameters of the bi-modal system are obtained by the following procedure:*

1. *Generate the quadratic data matrix $D^{(2)}$ from the input and output measurements.*
2. *The SVD of $D^{(2)}$ yields the nullspace $\mathcal{N}(D^{(2)})$*

$$D^{(2)} = USV'$$

   where S has corank (nullity) $2n^2 + n + 1$, and contains the singular values on its diagonal, and $V = [V_1, V_2]$, with $\dim \text{Span}(V_2) = 2n^2 + n + 1$.
3. *Select a vector, v, in V_2 with first component equal to one, and generate the matrix $T = \text{Mat}(v) + [\text{Mat}(v)']$. By the constraint imposed on v, it holds that $T_{11} = 2$.*
4. *Solve the eigenproblem for T. Let the normalized eigenvectors corresponding to the nonzero eigenvalues λ_+ and λ_- respectively be $\hat{z}_+$ and $\hat{z}_-$.*
5. *The vectors θ_1 and θ_2 are proportional to*

$$\begin{aligned} \theta_1 & \sim \sqrt{\lambda_+}\hat{z}_+ + \sqrt{|\lambda_-|}\hat{z}_- \\ \theta_2 & \sim \sqrt{\lambda_+}\hat{z}_+ - \sqrt{|\lambda_-|}\hat{z}_-. \end{aligned}$$

The scaling is determined by the conditions $(\theta_i)_1 = 1$, for $i = 1, 2$, i.e., the multiplier of y_k in the ARMA model, is one.

Proof: By Lemma 3.3, the eigenvectors corresponding to the nonzero eigenvalues of T are in the linear span of θ_1 and θ_2. Moreover, $\hat{z}_+ \perp \hat{z}_-$. It follows that

$$\begin{cases} \theta_1 &= a\hat{z}_+ + b\hat{z}_- \\ \theta_2 &= c\hat{z}_+ + d\hat{z}_- \end{cases}$$

for appropriate a, b, c and d in $\mathbb{R}$. Now,

$$\begin{aligned} T &= \theta_1\theta_2' + \theta_2\theta_1' \\ &= 2ac\hat{z}_+\hat{z}_+' + 2bd\hat{z}_-\hat{z}_-' + (ad+bc)(\hat{z}_+\hat{z}_-' + \hat{z}_-\hat{z}_+'). \end{aligned}$$

Since $\hat{z}_- \perp \hat{z}_+$ we get

$$(\hat{z}_+\hat{z}_-' + \hat{z}_-\hat{z}_+')\hat{z}_+ = \hat{z}_- \neq 0.$$

Hence, the matrix $(\hat{z}_+\hat{z}_-' + \hat{z}_-\hat{z}_+') \neq 0$. This matrix is also linearly independent of $\hat{z}_+\hat{z}_+'$ and $\hat{z}_-\hat{z}_-'$. Indeed,

$$\hat{z}_+\hat{z}_-' + \hat{z}_-\hat{z}_+' = \gamma \hat{z}_+\hat{z}_+'$$

would imply, by postmultiplication with $\hat{z}_+$, that $\hat{z}_- = \gamma \hat{z}_-$, which contradicts the orthogonality. Hence the eigen decomposition of T implies that

$$\begin{aligned} 2ac &= \lambda_+ \\ 2bd &= \lambda_- \\ ad + bc &= 0. \end{aligned}$$

This has the solution

$$a = \mu\sqrt{\frac{\lambda_+}{2}} \qquad c = \frac{1}{\mu}\sqrt{\frac{\lambda_+}{2}}$$
$$b = \mu\sqrt{\frac{\lambda_-}{2}} \qquad d = \frac{1}{\mu}\sqrt{\frac{\lambda_-}{2}}.$$

The statement then follows by proper choice of μ. In fact, this μ is not explicitly needed, as the first component of the modal parameter vectors are constraint to one. Hence

$$\begin{cases} \theta_1 &= \dfrac{\sqrt{\lambda_+}\hat{z}_+ + \sqrt{\lambda_-}\hat{z}_-}{\left[\sqrt{\lambda_+}\hat{z}_+ + \sqrt{\lambda_-}\hat{z}_-\right]_1} \\ \theta_2 &= \dfrac{\sqrt{\lambda_+}\hat{z}_+ - \sqrt{\lambda_-}\hat{z}_-}{\left[\sqrt{\lambda_+}\hat{z}_+ - \sqrt{\lambda_-}\hat{z}_-\right]_1}. \end{cases}$$

By Lemma 3.6 this solution is unique, up to permutation. ⋄

It is possible to put the above algorithm in a more explicit form:

Special Isolated Form
Isolate in equation (3.1), the y_k terms from the rest, i.e.,

$$(y_k - \delta_k'\theta_1)(y_k - \delta_k'\theta_2) = 0$$

with some abuse of notation, we used the same symbol θ to denote the $2n$-dimensional parameter vector $(\theta_1, \cdots \theta_{2n})'$ and denote $\delta_k = [y_{k-1}, \cdots, y_{k-n}; u_{k-1}, \cdots, u_{k-n}]'$. Thus

$$y_k^2 - (\delta_k'\theta_1 + \delta_k'\theta_2)y_k + (\delta_k'\theta_1)(\delta_k'\theta_2) = 0.$$

from which

$$y_k^2 - [\delta_k'(\theta_1 + \theta_2)]y_k + \text{Tr}\,(\delta_k\delta_k')(\theta_2\theta_1') = 0,$$

or in Kronecker product notation:

$$y_k^2 - (y_k \otimes \delta_k)'(\theta_1 + \theta_2) + (\delta_k \otimes \delta_k)'(\theta_1 \otimes \theta_2) = 0.$$

Let a right singular vector of the data matrix be partitioned as $[1, -S, \Pi]$. Thus

$$[y_k^2, y_k\delta_k', \text{vec}\,(\delta_k\delta_k')'] \begin{bmatrix} 1 \\ -S \\ \Pi \end{bmatrix} = 0 \tag{3.12}$$

or again in Kronecker product form

$$y_k^2 - (y_k \otimes \delta_k)'S + (\delta_k \otimes \delta_k)'P = 0$$

where $P = \text{Mat}\,\Pi$. Clearly

$$\begin{aligned} S_i &= (\theta_1)_i + (\theta_2)_i \\ P_{ii} &= (\theta_1)_i(\theta_2)_i \end{aligned}$$

leading to decoupled quadratic equations. The solution set for the ith pair is

$$\left\{ \frac{1}{2}\left[S_i + \sqrt{S_i^2 - 4P_{ii}}\right], \frac{1}{2}\left[S_i - \sqrt{S_i^2 - 4P_{ii}}\right] \right\} \tag{3.13}$$

but it still needs to be determined which of the two solutions in (3.13) is associated with mode 1. At this point observe that not all of the available information has yet been used. The ijth component of P yields

$$(\theta_1)_i(\theta_2)_j + (\theta_2)_i(\theta_1)_j = 2P_{ij}. \tag{3.14}$$

Introducing a signature vector σ, and assigning

$$\begin{aligned} (\theta_1)_i &= \frac{1}{2}\left[S_i + \sigma_i\sqrt{S_i^2 - 4P_{ii}}\right] \\ (\theta_2)_i &= \frac{1}{2}\left[S_i - \sigma_i\sqrt{S_i^2 - 4P_{ii}}\right] \end{aligned} \tag{3.15}$$

then (3.15) in (3.14) gives

$$\sigma_i\sigma_j = \frac{S_iS_j - 4P_{ij}}{\sqrt{S_i^2 - 4P_{ii}}\sqrt{S_j^2 - 4P_{jj}}}. \tag{3.16}$$

This matrix determines (up to a global sign) the signature vector σ and hence the association of the mode parameters. The global sign change simply permutes the two modes.

Finally, since this bi-modal identification was reduced to the uni-modal problem it is not surprising that it inherits its properties [14].

Theorem 3.2. *If the input is persistently exciting, and both modes are excited, then the estimates of the mode parameters are unbiased.*

Proof: Let $d_k = \bar{d}_k + \eta_k$ be the observed data, and η_k the uncorrelated measurement noise in inputs and outputs. Then, since the exact parameters and data satisfy $\text{Tr}\,(\bar{d}_k \bar{d}_k')\theta_1\theta_2' = 0$, it follows that

$$\mathbf{E}\,\text{Tr}\,(d_k d_k')\theta_1\theta_2' = \mathbf{E}\,\text{Tr}\,(\eta_k \eta_k')\theta_1\theta_2' = \text{Tr}\,\sigma^2 \theta_1\theta_2' = N\sigma^2 \text{Tr}\,\theta_1\theta_2'.$$

Since the estimate of $\theta^{(2)}$ is the smallest eigenvalue of the datamatrix $D^{(2)}$, the above implies unbiasedness of $\theta^{(2)}$. By the previous theorem the persistency of excitation implies uniqueness of the resolved mode parameters. ⋄

4 Multi-mode State Space Identification

The main idea is to reduce the problem to the one solved earlier. One obtains thus a multi-mode ARMA model, and based on the identified ARMA parameters, one determines with a realization algorithm the corresponding state space models.

To fix the ideas, let the state space equations in mode i be

$$\begin{aligned} x_{k+1} &= A^{(i)} x_k + B^{(i)} u_k \\ y_k &= c^{(i)} x_k. \end{aligned} \quad (4.1)$$

In the following sections it is illustrated first, by way of example, that the associated ARMA parameters do not change instantaneously with the mode transition. Next a new canonical from is exhibited, for which such transition happens in the shortest number of steps, which turns out to be one less than the system order. Finally some suggestions regarding performance improvement are presented.

4.1 Mode Transitions and Transition Modes

To fix the ideas, assume that for $k = 0$ the mode in (4.1) changes from mode $i = 1$ to mode $i = 2$. If one uses the ARMA method, a degradation is to be expected since strictly speaking, the bi-modal state space realization does not lead to a bi-modal ARMA relation in general. The I/O model (ARMA)

exhibits a 'transition period', necessary to flush out the state. To illustrate what is meant, consider the following:
Example: Let the state model in mode 1 be

$$A^{(1)} = \begin{bmatrix} 1 & \\ & -1 \end{bmatrix} \quad b^{(1)} = \begin{bmatrix} 1 \\ 1 \end{bmatrix} \quad c^{(1)} = [\,1\,,\,1\,]$$

and in mode 2,

$$A^{(2)} = \begin{bmatrix} -1 & 1 \\ & 1 \end{bmatrix}, \quad b^{(2)} = \begin{bmatrix} 1 \\ -1 \end{bmatrix} \quad c^{(2)} = [\,1\,,\,0\,].$$

If the initial state at step $k = -3$ is $x_0 = [\xi, \eta]'$, we get the consecutive I/O relations:

ARMA	Mode
$y_{-3} = \xi + \eta$	Initialization
$y_{-2} = \xi - \eta + 2u_{-3}$	"
$y_{-1} = y_{-3} + 2u_{-2}$	Mode 1
$y_0 = y_{-2} + 2u_{-1}$	"
Δ	Modeswitching
$y_1 = -\frac{1}{2}y_0 + \frac{1}{2}y_{-1} + u_0$	Transition
$y_2 = -y_0 + u_1$	Transition
$y_3 = y_1 - 2u_1 + u_2$	Mode 2
...	"

Note that in the transition steps from mode 1 to mode 2, fixed input-output patterns occur. This happened in the example for y_1 and y_2. In principle these modes can be detected as *transition modes*, Likewise transition modes for the reverse transition may be identified. A transition between two modes for an nth order system will in general have an n-step transition period between the ARMA modes associated with the state models. With the reverse transition inducing another fixed n ARMA relations, a bi-modal nth order state space system would require the identification of $2(n+1)$ ARMA modes. This renders the method more complex. If the average time between mode switches is larger than the system order, we expect that a simple bi-modal ARMA identification will still perform sufficiently well, albeit somewhat in a degraded way, because the quadratic model will not be satisfied in the transition steps.

Another feature of this state space form is that not every change in the state space parameters leads to a change in the ARMA model. If the second mode $(A^{(2)}, b^{(2)}, c^{(2)})$ is related to the first by *similarity*, the same input–output model results. Detection would still be possible by the mismatch during the transition times, when the state is flushed out as explained above.

4.2 Identifiability Canonical Form

In this subsection a state space realization is sought for which the number of transition modes as defined above is as small as possible. Consider thus the fixed state space model,

$$x_{k+1} = Ax_k + bu_k, \quad y_k = cx_k$$

with *nonsingular* A matrix. The state equation may be iterated backwards to get

$$x_{k-1} = A^{-1}x_k - A^{-1}bu_{k-1}$$
$$x_{k-2} = A^{-2}x_k - (A^{-2}bu_{k-1} + A^{-1}bu_{k-2})$$
$$\vdots$$
$$x_{k-n+1} = A^{-n+1}x_k - (A^{-n+1}bu_{k-1} + \cdots + A^{-1}bu_{k-n+1}).$$

Letting $\mathcal{Y}_-$ denote the vector $[y_k, y_{k-1}, \ldots, y_{k-n+1}]'$ and likewise: $\mathcal{U}_- = [u_k, u_{k-1}, \ldots, u_{k-n+1}]'$, then,

$$\mathcal{Y}_- = \mathcal{O}(c, A^{-1})x_k - \mathcal{T}_{SE}\mathcal{U}_-$$

where $\mathcal{T}_{SE}$ is the 0-banded Toeplitz matrix ('SE' refers to south-east)

$$\mathcal{T}_{SE} = \begin{bmatrix} 0 & & & 0 \\ 0 & cA^{-1}b & & \\ \vdots & \vdots & \ddots & 0 \\ 0 & cA^{-n+1}b & \cdots & cA^{-1}b \end{bmatrix}.$$

If the pair (c, A^{-1}) is observable, then

$$x_k = \mathcal{O}^{-1}(c, A^{-1})\mathcal{Y}_- + \mathcal{O}^{-1}(c, A^{-1})\mathcal{T}_{SE}\mathcal{U}_-.$$

The state space realization for which the observability matrix $\mathcal{O}(c, A^{-1})$ is identity, i.e., for which the pair (c, A^{-1}) is in the *observability canonical form* [19], has special properties: In the absence of an input, its state at time k consists of the previous and present outputs. If the input was not identically zero, the matrix $\mathcal{T}_{SE}$ shows the dependence of the states on these past inputs. Define this form as the *identifiability canonical form*. Using the identity

$$\begin{bmatrix} 0 & 1 & & \\ & & \ddots & \\ & & & 1 \\ -\alpha_n & -\alpha_{n-1} & \cdots & -\alpha_1 \end{bmatrix}^{-1} = \begin{bmatrix} -\frac{\alpha_{n-1}}{\alpha_n} & \cdots & -\frac{\alpha_1}{\alpha_n} & -\frac{1}{\alpha_n} \\ 1 & & & 0 \\ & \ddots & & \vdots \\ & & 1 & 0 \end{bmatrix}$$

and the invariance of the characteristic polynomial, one deduces that $A_{id} = A_r$, where A_r is in the reachable canonical form (this is called the controllable canonical form in [19]). Hence, the identifiable canonical form is characterized by $A_{id} = A_r$ and $c_{id} = e'_1$. In terms of past inputs and outputs, the state is $x_k = \mathcal{Y}_- + \mathcal{T}_{SE}\mathcal{U}_-$. Since the Markov parameters are invariants of the system (mode), the b_{id}-vector follows from

$$\begin{bmatrix} h_1 \\ \vdots \\ h_n \end{bmatrix} = \mathcal{O}_{id} b_{id}.$$

But, invoking the defining property of the identifiability canonical form,

$$\mathcal{O}_{id} = \begin{bmatrix} c_{id} \\ c_{id} A_{id} \\ \vdots \\ c_{id} A_{id}^{n-1} \end{bmatrix} = \begin{bmatrix} c_{id} A_{id}^{-n+1} \\ c_{id} A_{id}^{-n+2} \\ \vdots \\ c_{id} \end{bmatrix} \quad A_{id}^{n-1} = \tilde{I} A_{id}^{n-1}.$$

The matrix $\tilde{I}$ is the identity matrix with reversed columns, i.e., $\tilde{I} = [e_n, e_{n-1}, \ldots, e_1]$. Consequently,

$$b_{id} = [\tilde{I} A_r^{n-1}]^{-1} \begin{bmatrix} h_1 \\ \vdots \\ h_n \end{bmatrix} = A_r^{-(n-1)} \begin{bmatrix} h_n \\ \vdots \\ h_1 \end{bmatrix}.$$

An alternative form is derived from the factorization of the Hankel matrix in the observability and reachability matrices of the realization

$$\mathcal{O}(c, A^{-1}) \mathcal{R}(A^{-1}, b) = \mathcal{H}(c, A^{-1}, b).$$

Again invoking $\mathcal{O}(c_{id}, A_{id}^{-1}) = I$, it follows that

$$b_{id} = \mathcal{R}(A^{-1}, b) e_1 = \mathcal{H}(c, A^{-1}, b) e_1 = \begin{bmatrix} h_1 \\ h_0 \\ \vdots \\ h_{-(n-2)} \end{bmatrix}.$$

Note also that the Markov parameters $h_{-i} = cA^{-(i+1)}b$, for $i = 0, 1, \ldots$ exist by the assumed nonsingularity of A.

The associated ARMA parameters (coefficients of numerator and denominator polynomials of the transfer function) are $(a_1, a_2, \ldots, a_n; \beta_1, \beta_2, \ldots, \beta_n)$, with

$$\begin{bmatrix} \beta_1 \\ \beta_2 \\ \beta_3 \\ \vdots \\ \beta_n \end{bmatrix} = \begin{bmatrix} 1 & 0 & \cdots & & 0 \\ 0 & -a_2 & -a_3 & \cdots & -a_n \\ 0 & -a_3 & \cdots & -a_n & 0 \\ \vdots & \vdots & -a_n & 0 & \vdots \\ 0 & -a_n & 0 & \cdots & 0 \end{bmatrix} \begin{bmatrix} h_1 \\ h_0 \\ h_{-1} \\ \vdots \\ h_{-n+2} \end{bmatrix}$$

For instance, for a third-order system in identifiability canonical form, the associated ARMA model is given by

$$y_k = -a_1 y_{k-1} - a_2(y_{k-2} + h_0 u_{k-2}) - a_3(y_{k-3} + h_{-1} u_{k-2} + h_0 u_{k-3}) + + h_1 u_{k-1}.$$

What is this canonical from good for? If the system switches ARMA parameters at time 0, then the free parameters (A_{id}, b_{id}) will change abruptly. Hence right after the switch, the new ARMA model is consistent with

$$x_{k+1} = A_{id}^{(2)} x_k + b_{id}^{(2)} u_k; \quad k \geq 0,$$

however, the state vector x_1 is *not* consistent with $\mathcal{Y}_1 - \mathcal{T}^{(2)} \mathcal{U}_1$. The system will need $n-1$ steps to flush out the old state variables before consistency is reached. For instance, in the case $n = 3$, it is easily verified that if Σ_1 has AR parameters (a_1, a_2, a_3), with the Markov parameters h_{-1}, h_0 and h_1, while Σ_2 has AR parameters $(\alpha_1, \alpha_2, \alpha_3)$ $\overline{h}_{-1}, \overline{h}_0$ and $\overline{h}_1$ then the ARMA parameters are:

$$\begin{aligned}
\Sigma_1 &= [a_1, a_2, a_3, h_1, -(a_2 h_0 + a_3 h_{-1}), -a_3 h_0] \\
&= [a_1, a_2, a_3, b_1, b_2, b_3] \\
\Sigma_{12,1} &= [\alpha_1, \alpha_2, \alpha_3, \overline{h}_1, \alpha_2 h_0 + \alpha_3 h_{-1}, \alpha_3 h_0] \\
&= [\alpha_1, \alpha_2, \alpha_3, \beta_1, b_{12}, b_{13}] \\
\Sigma_{12,2} &= [\alpha_1, \alpha_2, \alpha_3, \overline{h}_1, -(\alpha_2 \overline{h}_0 + \alpha_3 \overline{h}_{-1}), -\alpha_3 h_0] \\
&= [\alpha_1, \alpha_2, \alpha_3, \beta_1, \beta_2, -b_{13}] \\
\Sigma_2 &= (\alpha_1, \alpha_2, \alpha_3, \overline{h}_1, -(\alpha_2 \overline{h}_0 + \alpha_3 \overline{h}_{-1}), -\alpha_3 \overline{h}_0] \\
&= [\alpha_1, \alpha_2, \alpha_3, \beta_1, \beta_2, \beta_3].
\end{aligned}$$

For an arbitrary state space realizations more than $n - 1$ steps may be required to reach consistency. This discrepancy between state space and ARMA models, that switch instantaneously, stems from the fact that at a switch the input–output data cannot consistently be compressed to n dimensions. All $2n - 1$ samples of input and output are required. Thus a full $2n - 1$ order realization of the form [19, Exc. 2.4-4, p. 156]

$$\theta_k = [y_k \cdots, y_{k-n+1} | u_{k-1}, \cdots, u_{k-n+1}]$$

$$\theta_{k+1} = \left[\begin{array}{ccc|ccc}
-a_1 & \cdots & -a_n & b_2 & \cdots & b_n \\
1 & & 0 & & & \\
& \ddots & \vdots & & 0 & \\
& 1 & 0 & & & \\
\hline
& & & 0 & \cdots & 0 \\
& & & 1 & & \vdots \\
& 0 & & & \ddots & \\
& & & & 1 & 0
\end{array}\right] \theta_k + \left[\begin{array}{c} b_1 \\ 0 \\ \vdots \\ 0 \\ \hline 1 \\ 0 \\ \vdots \\ 0 \end{array}\right] u_k$$

$$y_k = [1, 0, \cdots, 0]\,\theta_k$$

is needed. For a fixed mode, such a realization is non minimal, as it has $n-1$ hidden modes at zero.

4.3 Performance

As indicated, during a switch from mode Σ_1 to mode Σ_2 in the identifiability canonical form of order n, the induced ARMA model will cycle through $n-1$ intermediate models,

$$\Sigma_1 \to \Sigma_{12,1} \to \Sigma_{12,2} \to \cdots \to \Sigma_{12,n-1} \to \Sigma_2.$$

Note that in fact there are a total of $2n$ different models, $\Sigma_{12,k}$ and $\Sigma_{21,l}$ for $k,l = 0,\ldots,n-1$, where $\Sigma_{12,0} = \Sigma_1$ and $\Sigma_{21,0} = \Sigma_2$. If the average switching interval is of the order of n or less, serious degradation of the bimodal identification is to be expected. On the other hand, if switches occur at an average switching interval much larger than n, it will be expected that the two modes will be well identified. Moreover, if the identification algorithm converges to the two modes Σ_1 and Σ_2, then the transition modes $\Sigma_{12,k}$ and $\Sigma_{21,k}, k = 1,\ldots,n-1$ may be computed. A *matched filter verification* for these modes, i.e., computation of the residuals within these transitions, would alert to a possible mode switch in just one step. In the above example for $n = 3$, the intermediate mode for a transition from Σ_1 to Σ_2 are:

$$\begin{aligned}\Sigma_{12,1} &= [\alpha_1,\alpha_2,\alpha_3,\beta_1,-\alpha_3 b_2/a_3 + (\alpha_3 a_2 - \alpha_2 a_3)b_3/a_3^2, -b_3\alpha_3/a_3],\\ \Sigma_{12,2} &= [\alpha_1,\alpha_2,\alpha_3,\beta_1,\beta_2,b_3\alpha_3/a_3].\end{aligned}$$

Finally, note that there is a dual to the identifiable canonical form.

5 Markov Chain and Hybrid Modeling

In this section it will be assumed that the mode switching follows an underlying stationary Markovian or stationary Hidden Markov Model (HMM) with finite state space. In the first case, the conditional probability that mode i will be switched on next, given the present mode, is independent of the past modes. In the second case, the probability that mode i will be on is a fixed probabilistic function depending on the state of an underlying (hidden) Markov model. A Markov chain is characterized by a countable set of states, $\xi = \{S_i\}$ and its one step transition probability matrix, Π, with $\Pi_{ji} = \Pr(S_j|S_i)$, the probability for a transition to the state S_j, given that the chain is currently in state S_i. We break with the tradition of denoting the probability distribution as a row vector. The n-step transition probability matrix is simply Π^n.

We quote some definitions and results from basic Markov chain theory: If the transition matrix is fixed for all steps, then the Markov chain is said to be *stationary*. A Markov chain is *irreducible*, if there is a nonzero probability to reach any arbitrary state from any given state (not necessarily in one step). If a Markov chain is irreducible, its state transition probability matrix cannot be reduced (by relabeling the states) to a block triangular form

$$\Pi = \begin{bmatrix} P_{11} & P_{12} \\ 0 & P_{22} \end{bmatrix}.$$

A state, S, is *recurrent* if there is a nonzero probability of revisiting S from S.

Finally, let m_x be the average recurrence time (i.e., time between successive visits) of a recurrent state x. If $m_x < \infty$, the state is said to be positive recurrent.

Derman's theorem states that an irreducible positive recurrent and stationary Markov chain has a unique *stationary distribution* π with $\pi(x) = 1/m_x$ satisfying $\Pi \pi = \pi$. If Π has one eigenvalue equal to one and all others inside the unit circle, the Markov chain is said to be *ergodic*. In this case, any arbitrary initial distribution p converges to π, i.e., $\lim_{n \to \infty} \Pi^n p = \pi$. If more than one eigenvalue of Π lies on the unit circle, Π^m may not converge to a fixed limit. An example is

$$\Pi = \begin{bmatrix} 0 & 1 \\ 1 & 0 \end{bmatrix}.$$

In this case the Markov chain is periodic. If C is a finite irreducible set of states, then it is known that all states in C are positive recurrent.

Returning to the multi-mode switching model, we shall assume that the switching satisfies the irreducibility and stationarity conditions, so that a unique stationary distribution may be postulated for the modes. We shall refer to this as the *persistent modal excitation assumption*, which is analogous to the well-known persistent excitation condition in identification and adaptive control [7, 15].

More generally, the state transition matrix may also depend on exogenous variables, this means that they contain no information about the state, These are decision or control variables, which we take here to range in a finite set, the *input alphabet*, $\mathcal{U}$. Likewise, once the chain is in the state S_i, the distribution of symbols in an *output alphabet*, $\mathcal{Y}$ is assumed to depend on S_i. Thus we have in a complete stationary hidden Markov model:

1. Input map: assigning with each $u \in \mathcal{U}$ a state transition probability matrix $\Pi(u)$.
2. Output map: assigning for each $x \in \mathcal{S}$ an output distribution $\eta_x(\cdot)$ on $\mathcal{Y}$.

The pure HMM *realization* problem is the problem of determining the $\Pi(u)$, the prior probabilities on $\mathcal{U}$, and the read-out probabilities $\eta(x)$ on $\mathcal{Y}$ from knowledge of the posterior read-out probabilities. Some progress towards the solution of this problem, based on the theory of positive systems is reported in [20]. An alternative 2D-system approach is presented in [21].

A more realistic problem is the one where these exact probabilities are unknown, and the Markov chain parameters (transition probabilities and output distributions) need to be inferred from an empirical sequence of output symbols. For observations taking values in a continuum, recursive prediction error techniques based on EM (expectation maximization) have been explored [22], and an update method for the square root of the transition probabilities (hence guaranteeing positivity) is shown in [23]. Recursive identification with observations in a finite set is developed in [24].

In the next subsection we present what we believe to be a novel approach to this Markov chain identification problem. The key is the transformation of the nonnumeric data to numerical form, amenable again to a subspace identification method. The second subsection applies this solution to the sequence of estimated (rather than observed) modes of the switched system.

5.1 Hidden Markov Model Identification

In order to illustrate the main idea behind this we give first a simple example. Consider a stochastic automaton with nonnumeric input U, taken from an input alphabet, $\{V, W\}$, and output Y from an output alphabet (also nonnumeric) $\{A, B\}$. Assume that the *conditional output distribution*, given the input, is as follows:

$$U = V \quad \Rightarrow \quad Y \sim \begin{pmatrix} A & B \\ \alpha & 1-\alpha \end{pmatrix}, \tag{5.1}$$

$$U = W \quad \Rightarrow \quad Y \sim \begin{pmatrix} A & B \\ 1-\beta & \beta \end{pmatrix}. \tag{5.2}$$

Here α and β are as yet unknown probabilities. Let also $U = V$ with probability p and $U = W$ with probability $q = 1 - p$.

The key ingredient of the method is to give a *numeric representation* of the data. Let us assign with input V a two-dimensional data vector $[1, 0]$ and with W, the vector $[0, 1]$. Likewise we associate $[1, 0]$ with output A and $[0, 1]$ with B. Thus $U \leftrightarrow [u, \bar{u}]$ with $u = 1$ if $U = V$, and $\bar{u} = 1 - u$. At time k we have the input–output datum $d_k = [u_k, \bar{u}_k; y_k, \bar{y}_k]$. Consecutive data are stacked in a data matrix:

$$D = \begin{bmatrix} d_1 \\ d_2 \\ \vdots \\ d_N \end{bmatrix} = \begin{bmatrix} u_1 & \bar{u}_1 & ; & y_1 & \bar{y}_1 \\ u_2 & \bar{u}_2 & ; & y_2 & \bar{y}_2 \\ \vdots & \vdots & & \vdots & \vdots \\ u_N & \bar{u}_N & ; & y_N & \bar{y}_N \end{bmatrix}. \tag{5.3}$$

Now 'squaring up' yields

$$D'D = \begin{bmatrix} \sum u_k^2 & \sum u_k \bar{u}_k & \sum u_k y_k & \sum u_k \bar{y}_k \\ \sum \bar{u}_k u_k & \sum \bar{u}_k^2 & \sum \bar{u}_k y_k & \sum \bar{u}_k \bar{y}_k \\ \sum y_k u_k & \sum y_k \bar{u}_k & \sum y_k^2 & \sum y_k \bar{y}_k \\ \sum \bar{y}_k u_k & \sum \bar{y}_k \bar{u}_k & \sum \bar{y}_k y_k & \sum \bar{y}_k^2 \end{bmatrix}. \tag{5.4}$$

Invoking the law of large numbers, and letting $\#\mathcal{K}$ denote the cardinality of set $\mathcal{K}$,

$$\sum u_k^2 = N \frac{\#\{U=V\}}{N} \to Np$$

$$\sum \bar{u}_k^2 = N \frac{\#\{U=W\}}{N} \to Nq$$

$$\sum u_k \bar{u}_k = 0$$

$$\sum u_k y_k = \#\{U=V; Y=A\} = N \frac{\#\{Y=A; U=V\}}{\#\{U=V\}} \frac{\#\{U=V\}}{N} \to N\alpha p$$

$$\vdots$$

$$\sum y_k^2 = \#\{Y=A\}$$
$$= \#\{Y=A, U=V\} + \#\{Y=A, U=W\} \to N(\alpha p + (1-\beta)q),$$

the quadratic data matrix converges to

$$D'D = N \left[\begin{array}{cc|cc} p & 0 & \alpha p & (1-\alpha p) \\ 0 & q & (1-\beta)q & \beta q \\ \hline \alpha p & (1-\beta)q & \alpha p + (1-\beta)q & 0 \\ (1-\alpha)p & \beta q & 0 & (1-\alpha)p + \beta q \end{array} \right].$$

Its diagonal bears the relevant information:

$$\operatorname{diag}(D'D) = [p, q, \alpha p + (1-\beta)q, (1-\alpha)p + \beta q].$$

These entries identify the marginal probabilities of the input U and the output Y, and are thus obtained from the data. Nothing is really new here, this is just a nice scheme for establishing the statistics. Define now

$$\mathcal{D} = [\operatorname{diag}(D'D)]^{-1} D'D = \left[\begin{array}{cc|cc} 1 & & \alpha & 1-\alpha \\ & 1 & 1-\beta & \beta \\ \hline \frac{\alpha p}{\alpha p + (1-\beta)q} & \frac{(1-\beta)q}{(1-\alpha)p + \beta q} & 1 & \\ \frac{(1-\alpha)p}{\alpha p + (1-\beta)q} & \frac{\beta q}{(1-\alpha)p + \beta q} & & 1 \end{array} \right]$$
$$\tag{5.5}$$

and note that all entries are well defined for a finite ergodic chain. Moreover, this is exactly

$$\mathcal{D} = \left[\begin{array}{c|c} I & P(Y|U) \\ \hline P(U|Y) & I \end{array} \right], \qquad (5.6)$$

thus identifying the forward and the backward transition probability matrices (between input and output). Recall that by Bayes' theorem,

$$P(U|Y) = \frac{P(Y|U)P(U)}{\sum_i P(Y|u_i)P(u_i)}.$$

We shall now look at a generalization of these ideas. It is desired to model a stochastic automaton with input and output alphabet respectively

$$U \in \{u_1, \ldots, u_m\} \qquad Y \in \{y_1, \ldots, y_p\}. \qquad (5.7)$$

Associate with these m input symbols and p output symbols the vectorial indicator functions

$$\chi_U : U \to [\chi_{u_1}(U), \cdots, \chi_{u_m}(U)] \in \{e'_1, \ldots, e'_m\}, \qquad (5.8)$$
$$\chi_Y : Y \to [\chi_{y_1}(Y), \cdots, \chi_{y_p}(Y)] \in \{e'_1, \ldots, e'_p\}, \qquad (5.9)$$

i.e., $\chi_U = e_j$ if $U = u_j$. For simplicity we consider only pure observation of an uncontrolled first-order Markov chain. There is no input, and the output completely characterizes the state in this case. In anticipation of the application to bi-modal identification, let the output alphabet be $\{0,1\}$. For a first-order chain, statistics of two consecutive outputs are required. Let the superscript '$-$' denote the past, The data vector at time k involves therefore the present, y, and immediate past output, y^- through their numerical representation. Thus

$$d = [\chi_0^y, \chi_1^y | \chi_0^{y^-}, \chi_1^{y^-}].$$

Again the 'squared' data matrix of consecutive ds gives

$$\mathcal{D}'\mathcal{D} = \left[\begin{array}{cc|cc} \#\{y=0\} & 0 & \#\{`00'\} & \#\{`10'\} \\ 0 & \#\{y=1\} & \#\{`01'\} & \#\{`11'\} \\ \hline \#\{`00'\} & \#\{`01'\} & \#\{y^-=0\} & 0 \\ \#\{`10'\} & \#\{`11'\} & 0 & \#\{y^-=1\} \end{array} \right], \qquad (5.10)$$

where $\#\{`01'\}$ denotes the number of times the string, 01, appears. If the chain is ergodic:

$$\operatorname{diag}(\mathcal{D}'\mathcal{D}) = N \left[\begin{array}{cc|cc} P(y=0) & & & \\ & P(y=1) & & \\ \hline & & P(y=0) & \\ & & & P(y=1) \end{array} \right]$$

and it follows that

$$\mathcal{D} = [\text{diag}(D'D)]^{-1} D'D = \begin{bmatrix} I & \Pi(y|y^-) \\ \Pi(y^-|y) & I \end{bmatrix}. \tag{5.11}$$

With a larger data vector

$$d = [\chi^y; \chi^{y^-}; \chi^{y^{--}}; \cdots] \quad , \quad \chi = (\chi_0, \chi_1),$$

we get

$$\mathcal{D} = [\text{diag}(D'D)]^{-1} D'D = \begin{bmatrix} I & \Pi(y|y^-) & \Pi(y|y^{--}) & \cdots \\ \Pi(y^-|y) & I & \Pi(y|y^-) & \cdots \\ \Pi(y^{--}|y) & \Pi(y^-|y) & I & \cdots \\ \vdots & \vdots & \vdots & \ddots \end{bmatrix}. \tag{5.12}$$

For the first-order Markov chain,

$$\Pi(y|y^-) = T_y, \quad \Pi(y|y^{--}) = T_y^2, \quad \Pi(y^-|y) = \overline{T}_y, \quad \text{etc.}$$

where T_y and $\overline{T}_y$ are respectively the forward and the backward transition probability matrix. The right singular vectors of $\mathcal{D}$ determine this transition probability matrix, since for this first order chain

$$\begin{bmatrix} I & T_y \\ \overline{T}_y & I \end{bmatrix} \begin{bmatrix} -T_y \\ I \end{bmatrix} = 0. \tag{5.13}$$

Remark: If a state-dependent output probability map $X \to Y$ is present, we get

$$\begin{aligned} P(y|y^-) &= \sum_x P(y, x|y^-) \\ &= \sum_x P(y|x, y^-) P(x|y^-) \\ &= \sum_x P(y|x) \sum_{x^-} P(x|x^-, y) P(x^-|y^-) \\ &= \sum_x P(y|x) \sum_{x^-} P(x|x^-) P(x^-|y^-) \\ &= ZT\overline{Z}. \end{aligned} \tag{5.14}$$

Note that by Bayes' rule

$$P(x^-|y^-) = \frac{P(y^-|x^-) P(x^-)}{\sum_i P(Y^-|x^- = i) P(x^- = i)}.$$

For an ergodic chain in statistical steady state, the right-hand side is

$$\frac{P(y|i)\pi_i}{\sum_j P(y|j)\pi_j}$$

and for $P(y|y^{--})$, the product $ZT^2\overline{Z}$, etc. Identification of the output probability map, the state transition probability and the reverse output probability map, respectively Z, T and $\overline{Z}$ requires then the solution of a *realization problem*, given $\mathcal{D}$. This is still a very active research problem.

5.2 Hybrid System Modeling

There is no standard definition for what is meant by a hybrid system. In this section, we consider a system to be hybrid, if it has a hierarchical structure, dictated by the time scale. At the lower level exists a class of fixed linear time invariant systems. Each system with its parameters forms a *mode* of the multi-mode system. At the next level, it is assumed that the mode switching itself is governed by a probabilistic law, parameterized by a mode transition probability matrix, and state to output (probabilistic) map, etc. It is conceivable to include more steps in the hierarchy. For instance if the probabilistic parameters for the Markov model are in turn switched at a yet larger time scale chosen from some finite set. Such a situation could, however, be modeled by a two-level hierarchy, by combining the two (or more) Markovian levels into the tensor product Markov system. Thus for this reason, we shall assume that there are only two levels in the hierachy.

The two main procedures discussed in the previous sections can then be combined. We just sketch the steps in identifying a two-mode hybrid model from real input output data:

1. Construct the *quadratic* data matrix $D^{(2)}$ from the observed inputs and outputs.
2. Solve the quadratic equations to obtain the estimates of the *mode parameters* $\hat{\theta}_1$ and $\hat{\theta}_2$.
3. Validation and detection: For each datum construct the residuals $\epsilon_i = d'\theta_i$. If $|\epsilon_1| \leq |\epsilon_2|$, decide that mode parameterized by θ_1 was 'on'. Reverse the decision in the other case and create the augmented datum $d_e = [u, y, \hat{\theta}]$ at each step.
4. Obtain a Markovian model for the mode, based on the sequence $\hat{\theta}$, as discussed in the last section. In particular, the identification yields the *mode state probability transition matrix* T and the *mode state to mode probability* Z.

Comments

(i) By computing at each step k the residuals the active mode at k can be estimated. This yields a *derived* (binary) sequence, from which either the deterministic (e.g., period and dutycycle) or stochastic mode switching parameters (e.g., transition matrix) can be estimated. Preliminary ideas were presented in [25, 13, 18]. However, this involves essentially two passes on the data: one for parameter estimation (and mode estimation), then another to determine the mode switching statistics or model. The familiar expectation maximization (EM) algorithm arises in this context. It is possible to improve on this. The observed data may be used optimally, via change-of-probability measure methods as in [12], to obtain model and mode parameters simultaneously. These methods are recursive. and thus outside the scope of the batch (TLS) methods discussed here.

(ii) As an intermediate between the optimal solution alluded above, and the independent modeling of modes and mode switching, the identified HMM may aid in turn in the estimation of the mode parameters. For instance if there is only a small (ϵ) probability for a transition to mode j, this mode could be eliminated, dropping the degree of the requisite monomials in the data. This greatly reduces the computational complexity. For instance, the augmented data vector for a first-order bi-modal system has dimension 7, whereas the unimodal ARMA form has only two parameters. For a multimode system, the less likely modes could be deleted from the algorithm, leaving only a matched filter (check of residual) as validation.

(iii) A more sophisticated type of HMM is as follows: Start from a linear model for a series $\{x_k\}$ and then allow the parameters of the model to vary according to the values of a finite number of past values of x_k, or a finite number of past values of an associated series $\{y_k\}$. Tong calls these *threshold autoregressive models* and introduced them in river flow analysis. A very interesting feature of these models is that they may give rise to limit cycles. Such models are therefore suitable to describe or approximate 'cyclic' phenomena.

6 Applications

The two basic principles described in the previous sections, namely (i) the identification of the ARMA or state space parameters from a polynomial preprocessing of the available input and output data, and (ii) the identification of the Markovian switching model by a numerical representation of the data can be applied to the adaptive control of nonlinear and nonsmooth systems and the detection and identification (isolation) of a system failure.

6.1 Adaptive Control of Nonlinear and Nonsmooth Systems

The class of *piecewise linear* systems is a special class of nonlinear systems. The state manifold of such a system can be partitioned into subsets, $S_i; i = 1, \ldots, N$, such that if the state x_t belongs to S_i, then the next state and output are given by a *linear* update $x_{t+1} = A_i x_t + B_i u_t$

and read out $y_t = C_i x_t$. In the ARMA framework, the model is $y_t = -a_1(S)y_{t-1} - \cdots - a_n(S)y_{t-n} + b_1(S)u_{t-1} + \cdots b_n(S)u_{t-n}$, with parameters $a_1(S), \ldots a_n(S), b_1(S), \ldots, b_n(S)$ dependent on S, which in turn depends on the past data vector $(y_{t-1}, \ldots, y_{t-n}, u_{t-1}, \ldots, u_{t-n})$. Consider the case where the nonlinearity is not known *a priori*. Modeling the system as an N-mode system, the identification algorithm estimates as before the parameters θ_i in each mode. After the initial learning period, the parameters may be reasonably well known. The residuals $\epsilon_i(t) = y_t - \hat{y}_{t|t-1}(\hat{\theta}_i)$ are computed and one decides $\hat{\theta}(t) = \theta_i$ if $\epsilon_i(t)$ is minimal among all residuals at time t. Weights may be introduced to reflect a bias on ignorance or importance, but we shall not clutter our analysis with this greater generality here.

At this point, instead of identifying an underlying Markovian behavior, the estimated mode may be post processed with the data vectors in order to identify the *separatrices*, i.e., the submanifolds dividing the state manifold into the locally linear regions. This subproblem is a classical problem of classification with or without supervision (or learning pattern recognition systems), found in Tsypkin's books [26, 27]. We shall here look at the simplest possible model. Assume that there are two linear regions, and that these are separated by the *discriminant function*, $f(x, c)$, where x is the state of the system (equivalently the data vector d in an ARMA model). (Assume that this function is known up to a parameter vector c, for instance $f(x, c) = c'\phi(x) = \sum_{i=1}^{N} c_i \phi_i(x)$, where the $\phi_i(x)$ are linearly independent functions. Assume that there exists a $\bar{c}$ for which the discriminant function is exact. This means that on S_1, $f(\cdot, \bar{c})$ is positive, and negative on its complement. As penalty function, select a convex function, F, of the difference between $g(\hat{\theta})$ and $f(x, c)$, where $g(\hat{\theta}_1) = +1$ and $g(\hat{\theta}_2) = -1$. The empirical loss is then $L_t = \sum_0^t F(\hat{\theta}(t) - f(x(t), c))$, which leads to a parameter update algorithm of the form

$$c_t = c_{t-1} + \Gamma_t F'(\hat{\theta}(t-1) - c'_{t-1}\phi(x_t))\phi(x_t).$$

The matrix Γ_t may depend on c_t or c_{t-1} and is selected to guarantee convergence of c_t to the exact value $\bar{c}$.

In this application the underlying system is purely deterministic. Given sufficient data, the parameters are in principle exactly identifiable. Slow parameter variations in the discriminant function as well as in the linear submodels will be tracked.

6.2 Fault Detection and Isolation

The need for fault diagnostics is dictated by an ever increasing demand for safety and reliability, be it in aircraft control, traffic control, or operation of a nuclear reactor. As a result, the problem of detecting changes in the parameters of a dynamic system has been studied via different methods during the past decades. The main ideas have been collected in the books [28–30], and the surveys [31–34]. As discussed in recent papers [35, 36], many problems can be categorized under the detection framework. Fault detection and model validation are closely related. In fault detection one checks whether all parameters (of sensors, actuators and process dynamics) coincide with their nominal values, whereas in model validation, one wants to determine if a preconceived model is consistent with the observed input output data. It leaves blank the modeling of the nominal behavior itself. The multi-mode identification framework that was established in the preceding sections lends itself very well to this problem. The potential failure is viewed as a second but unknown mode of operation of the system. This way, it cannot only detect when a failure occurred but in fact goes beyond by allowing the identification of the system parameters once the fault occurred.

Moreover, it suffices to use a bi-modal model, since in this application, one is interested in a single mode transition. Changes of the parameter vector in a particular direction may or may not be anticipated. In the first case, the nominal residuals of the identification algorithm are closely monitored. The appearance of a second mode (the failure) will manifest itself by wild fluctuations in these nominal residuals, $y_t - d'_t \theta_{\text{nom}}$, whereas the failure residuals $y_t - d'_t \theta_{\text{fail}}$, will become small. Threshold detection of this residual behavior facilitates the change point detection, and the new parameters, determining the behavior of the system in failure mode are available. This of course relates to the classical approach based on whitening the data based on the identified model and testing the residuals for deviation from the white noise hypothesis.

This two-model structure may also be combined with nonlinear filtering. A recent approach combines this two-model structure with a nonlinear filtering approach. It assumes that the model corresponding to faulty operation manifests itself through a random drift (Brownian motion) of the parameters [35]. By introducing one additional parameter, θ, multiplying the driving noise for the drift model, the nonfault case corresponds to $\theta = 0$ and the faulty one by $\theta = 1$. In principle this model is a two-mode stochastic system, and is therefore a special case of our general method.

7 Examples

Some simple tests were run illustrating the feasibility of the proposed method. First for deterministic periodic mode switching, input–output data were generated by a second-order system switching mode every tenth step (thus the period of the modal behavior is 20). As input signal, white noise was chosen of unit variance. Then both input and output signal were corrupted by additive Gaussian noise with RMS σ_u and σ_y. These corrupted signals were fed into the identification algorithm. As performance measure of the algorithm, we choose $J_{100} = \|\theta - \hat{\theta}_{100}\|/\|\theta\|$, where $\theta = [a_1^{(1)}, a_2^{(1)}, b_1^{(1)}, b_2^{(2)}]$ is the vector of ARMA parameters. The estimated parameter vector is the one obtained after the first 100 samples. Two different bi-modal systems were considered:

Example 1: Both modes stable, with transfer function respectively $H_1(z) = \frac{z-0.5}{z^2-z+0.5}$ and $H_2(z) = \frac{z-0.5}{z^2-z+0.75}$. The first mode has complex poles at $1/2 \pm 1/2j$, the second has real poles at $1/4$ and $3/4$. Good parameter tracking behavior was found despite the fact that the parameter vectors of both modes are relatively close. The dependence on input and output noise variance is

σ_u	σ_y	J
0	0	0.020
0.05	0	0.026
0	0.05	0.093
0.05	0.05	0.109
0.1	0.1	0.158

Example 2: Both modes stable, with transfer function respectively $H_1(z) = \frac{z}{z^2-z+0.5}$ and $H_2(z) = \frac{0.5z-0.5}{z^2-0.75z+0.75}$. The first mode has complex poles at $1/2 \pm 1/2j$ and zero at $z = 0$, the second has complex poles at $0.375 \pm 0.78j$. The parameter vectors for the two modes are more separated. This gave better parameter tracking behavior than in the first example. The dependence on input and output noise variance was found to be

σ_u	σ_y	J
0	0	0.0004
0.1	0	0.0171
0	0.1	0.0816
0.1	0.1	0.1039

With purely random switching, faster convergence for the two modes was observed.

8 Conclusions

The benefits of a more refined knowledge of the structure of a system in estimation, prediction and control are well known. The book [37] gives a nice overview of the problems associated with jump linear systems. What we discussed in this paper is that such a hybrid structure can be identified from input and output data of the system by extending the known principles. The guiding principles were twofold:

The first is to form an M-fold product of input output data, essentially one (linear) factor per mode. One may argue that the complexity of the method increases quickly as the number of modes is increased. However, noting that a *change in mode* involves only two modes (the present mode and a 'guard' mode) at a time, there is strictly speaking no need to continuously track all modes. A bi-modal system as described in section 3 may suffice if the frequency of mode switching is sufficiently low. The identifier memorizes the estimated parameter vectors for each mode, while the parameter estimates for the current mode are updated as well as a 'guard'-mode. When the latter is identified as being associated with one of the parameter vectors in memory, the guard mode becomes the current mode, and a new guard mode may be initialized. The former current mode, i.e., the parameter vector update, is then combined with the previously memorized parameter vector and returned to memory. This provides a low computational alternative to tracking all modes (in case they would be slowly varying).

On the other hand for truly bi-modal systems, we indicated that it might be useful to use higher mode identifiers for quick detection of changes in state space parameters. This in order to quickly capture the transition from one ARMA model to the other.

The second principle deals with the identification of Markovian finite state systems from nonnumeric inputs and outputs. Here the key was the substitution of the nonnumeric data by its vector valued indicator function, thus making it numeric and receptive to a subspace method which automatically generates the relevant statistics for identification. The two steps, augmented with a (mode)-detection, at little additional computational cost leads to identification of hybrid systems. Of course there remains a lot of room for fine tuning and adapting to specific situations, following the richness of adaptive methods in the well-documented single-mode case. For instance bi-modal identification was also proposed as a viable alternative for the combined fault-detection/fault-estimation problem. As long as the system is working in its nominal mode, the two identified modes will be nearly identical. The differences are due to the unavoidable noise, and since the system tries to identify the noise as part of the model, these fluctuations will give closely spaced modes. However, as soon as a fault occurs, one of the identified modes will wander off to a new average. This way not only is a fault detected, but also identified (as the second-mode parameters).

Acknowledgement. The author is indebted to Professor Manfred Deistler (TU Wien, Austria) for helpful discussions and references. The support of the NSF-CNRS collaborative grant INT-9818312 is gratefully acknowledged.

References

1. G. Tao and P. V. Kokotović, *Adaptive Control of Systems with Actuator and Sensor Nonlinearities*, Wiley, 1996.
2. Y. Arkun, A. Banerjee and N.M. Lakshmanan, Self scheduling MPC using LPV models, in *Nonlinear Model Based Control*, R. Berber, ed, NATO ASI Series, Kluwer Academic Publishers, 1998.
3. R.R. Bitmead, M. Gevers and V. Wertz, *Adaptive Optimal Control: The Thinking Man's GPC*, Prentice Hall, 1990.
4. R. Murray-Smith and T.A. Johansen, eds, *Multiple Model Approaches to Nonlinear Modeling and Control*, Taylor & Frances, London, 1997.
5. H. Tong, *Non-linear Time Series: A Dynamical System Approach*, Oxford University Press, 1990.
6. J.S. Schamma and M. Athans, Analysis of gain scheduled control for nonlinear plants, *IEEE Transactions on Automatic Control*, 35, No. 8, 898-907, 1990.
7. K.J. Aström and B. Wittenmark, *Adaptive Control*, 2nd edn., Addison Wesley, 1995.
8. M. Gevers, Towards a Joint Design of Identification and Control?, in *Essays on Control: Perspectives on the Theory and Its Applications*, H.L. Trentelman, J.C. Willems, eds, Birkhäuser, 1993.
9. K.S. Narendra and J. Balakrishnan, Adaptive control using multiple models, *IEEE Transactions on Automatic Control*, 42, No. 2, 171-187.
10. K.S. Narendra and C. Xiang, Adaptive control of discrete time systems using multiple models, *IEEE Transactions on Automatic Control*, 45, No. 9, 1669-1686, 2000.
11. J.D. Hamilton, *Time Series Analysis*, Princeton University Press, 1994.
12. R.J. Elliott, L. Aggoun and J.B. Moore, *Hidden Markov Models: Estimation and Control*, New York, Springer Verlag, 1994.
13. E.I. Verriest, "Subspace algorithm for identifying multi-mode (hybrid) systems and applications in failure detection," presented at the *International Linear Algebra Symposium*, Winnipeg, Manitoba, CND, June 1997.
14. G. Goodwin and K.S. Sin, *Adaptive Filtering, Prediction and Control*, Prentice Hall, 1984.
15. T. Söderström and P. Stoica, *System Identification*, Prentice Hall, 1989.
16. S. Van Huffel and J. Vandewalle, *The Total Least Squares Problem: Computational Aspects and Analysis*, SIAM, 1991.
17. P. Van Overschee and B. De Moor, *Subspace Identification for Linear Systems: Theory, Implementation, Applications*, Kluwer Academic Publishers, 1996.
18. E.I. Verriest and B. De Moor, Multi-mode system identification, *Proceedings of the 1999 European Control Conference*, Karlsrühe, Germany, September 1999.
19. T.Kailath, *Linear Systems*, Prentice Hall, 1980.
20. B.D.O. Anderson, New Developments in the Theory of Positive Systems, in *Systems and Control in the Twenty-First Century*, C.I Byrnes, B.N. Datta, D.S. Gilliam and C.F. Martin, eds, Birkhäuser, 1997, pp. 17-36.

21. J. Ramos and E.I. Verriest, A 2-D realization theory for Markov chains," *Proceedings of the 29th Conference on Decision and Control*, Honolulu, HI, December 1990, pp. 853-858.
22. V. Krishnamurthy, and J.B. Moore, On-line estimation of hidden Markov model parameters based on the Kullback-Leibler information measure, *IEEE Transactions on Signal Processing*, 41, No. 8, 2557-2573, 1993.
23. I.B. Collings, V. Krishnamurthy, and J.B. Moore, On-line identification of hidden Markov models via recursive prediction error techniques, *IEEE Transactions on Signal Processing*, 42, No. 12, 3535-3539, 1994.
24. F. LeGland and L. Mével, Recursive identification of HMM's with observations in a finite set," *Proc. 34th Conference on Decision and Control*, New Orleans, LA, pp. 216-221, 1995.
25. E.I. Verriest, Subspace algorithms for identifying multi-mode (hybrid) systems and applications, presented at the SIAM 45th Anniversary Meeting, Stanford, CA, July 14-18, 1997.
26. Ja. Zypkin, *Adaption und Lernen in Kybernetischen Systemen*, VEB Verlag Technik, Berlin, 1970.
27. Ya.Z. Tsypkin, *Foundations of the Theory of Learning Systems*, Academic Press, 1973.
28. M. Basseville and A. Benveniste, eds, *Detection of Abrupt Changes in Signals and Dynamical Systems*, Lecture Notes in Control and Information Sciences, No. 77, Springer Verlag, Berlin, 1986.
29. R. Patton, P. Frank and R. Clark, *Fault Diagnosis in Dynamic Systems, Theory and Applications*, Prentice Hall, New York, 1989.
30. R.J. Patton, P.M. Frank and R.N. Clark, eds, *Issues of Fault Diagnosis for Dynamic Systems*, Springer Verlag, 2001.
31. A.S. Willsky, A survey of design methods for failure detection in dynamic systems, *Automatica*, **12**, 601-611, 1976.
32. M. Basseville, Detecting changes in signals and systems - a survey, *Automatica*, 24(3), 309-326, 1988.
33. R. Isermann, Process fault detection based on modeling and estimation methods - a survey, *Automatica*, 20(4), 387-404, 1984.
34. R. Isermann, Fault diagnosis of machines via parameter estimation and knowledge processing - tutorial paper, *Automatica*, 29(4), 815-835, 1993.
35. M.L. Tyler and M. Morari, Change detection using non-linear filtering and likelihood ratio testing, *Proceedings of the ECC-97*, Brussels, Belgium, TH-M J3.
36. W.H. Chung and J.L. Speyer, A game theoretic fault detection filter, *IEEE Transactions on Automatic Control*, **43**, No. 2, 143-161, 1998.
37. M. Mariton *Jump Linear Systems in Automatic Control*, Marcel Dekker, 1990.

On Feedback Control of Processes with 'Hard' Nonlinearities

Bernard Friedland

Department of Electrical and Computer Engineering
New Jersey Institute of Technology
Newark, NJ 07102 USA
e-mail: bf@njit.edu

Abstract

It is frequently necessary to control processes with 'hard' nonlinearities, such as friction, backlash, and saturation, or with state-variable constraints, such as physical limit stops or regions in state space into which the system should not venture. Design of control systems for such processes has eluded treatment by some of the currently popular methods, such as feedback linearization, that require differentiable ('smooth') nonlinearities. The recently-developed state-dependent algebraic Riccati equation (SDARE) method, also known as extended linearization, however, does not suffer from this limitation and may be useful for design of control systems that are not amenable to design by other methods.
This chapter reviews the SDARE method and discusses modeling issues related to the application of the method. The application of the method is demonstrated in processes with friction, backlash, and state-variable constraints. Performance is evaluated by simulation.

1 Introduction

Dynamic processes with so called 'hard' nonlinearitites occur frequently in practical control problems. Such nonlinearities include 'limit stops' (i.e., physical bumpers or other devices intended to prevent displacement of a physical quantity outside boundaries defined by these devices), backlash, friction, saturation and the like. The common feature of these phenomena is that they are characterized mathematically by functions that often lack continuity, have sharp corners, or other nasty properties.

It is sometimes possible to ignore the presence of hard nonlinearities in the design of a control system. Doing this, however, may entail selection of components with performance characteristics that exceed the requirements of the project. To avoid the effects of friction, for example, bearings of exceptional quality (and high cost) might be specified that could be unnecessary if the effects of friction in ordinary bearings could be dealt with effectively;

to avoid the effects of saturation, larger actuators than necessary might be employed.

Control system design methods that are capable of dealing with hard nonlinearities have obvious practical advantages and have been receiving increasing attention in recent years, as evidenced, for example by the contents of this book. The present article presents one such design method, namely the 'state-dependent algebraic Riccati equation (SDARE) method, which was originally suggested by Friedland [1] and developed in a number of papers by Cloutier et al. [2]. The method has been shown to be a practical method for solving the problem of designing controllers for nonlinear processes in many applications. The present work shows how the method can be applied to processes with hard nonlinearities.

2 The State-dependent Algebraic Riccati Equation Method

Underlying the SDARE method is the requirement that the process dynamics be expressed in the form

$$\dot{x} = A(x)x + B(x)u \qquad (1)$$
$$y = C(x)x \qquad (2)$$

where x is the process state vector, u is the control vector, and y is the observation vector. Most practical processes can be expressed in the form of (1) and (2), but the matrices $A(x)$ and $B(x)$ are not unique. (The choice of these matrices can influence the response of the system. This is the so-called 'parameterization problem' which is currently receiving attention.)

With the dynamics expressed by (1) and (2), the control law is given by

$$u = -G(\hat{x})\hat{x} \qquad (3)$$

with the control gain matrix given by

$$G(x) = R^{-1}(x)B'(x)M(x) \qquad (4)$$

where $M(x)$ is the solution to the algebraic Riccati equation

$$0 = M(x)A(x) + A'(x)M(x) - M(x)B(x)R^{-1}(x)B'(x)M(x) + Q(x) \qquad (5)$$

In the 'classical' linear, quadratic control theory, the matrices $Q(x)$ and $R(x)$ are the state weighting and control weighting matrices, respectively, in the quadratic performance integral to be minimized. In the present application, however, these matrices rigorously cannot be associated with a performance integral to be minimized. Rather, these matrices should be regarded as design matrices that can be chosen by the user of the SDARE method as a possible means to achieving a desired performance goal. In particular, as will be shown below, the matrix $Q(x)$ can include a (nonlinear) penalty for state excursions outside a 'forbidden' region.

The estimated state $\hat{x}$ is obtained by means of an observer in the form of a Kalman filter:

$$\dot{\hat{x}} = A(\hat{x})\hat{x} + K(\hat{x})(y - C(\hat{x})\hat{x}) \qquad (6)$$

with the observer gain matrix given by

$$K(x) = P(x)C'(x)W^{-1}(x) \qquad (7)$$

where $P(x)$ is the solution of the algebraic Riccati equation

$$0 = A(x)P(x) + P(x)A'(x) - P(x)C'(x)W^{-1}(x)C(x)P(x) + V(x) \qquad (8)$$

With the classical Kalman filter, the matrices V and W are the covariance matrices of the (assumed white) process noise and observation noise, respectively. Here, however, as with the matrices Q and R, these matrices are regarded as design parameters to be chosen by the user of the method.

Instead of (6)–(8), a reduced-order observer, in which the gain matrix is obtained using a reduced-order state-dependent algebraic Riccati equation can be used.

For implementation the state-dependent algebraic equations (5) and (8) are solved *on-line*. It is assumed that the matrices $A(x), B(x)$ and $C(x)$ possess *at each state* x (in the region of state space for which a solution is required) the requisite controllability and observability conditions to guarantee the existence of solutions of (5) and (8) or, the equivalent algebraic Riccati equation for a reduced-order observer.

It should be noted that the rigorous theory of the state-dependent Riccati equation method is still in development. But dozens of applications of the method have provide ample empirical evidence of its efficacy. The present article adds to this evidence.

3 Modeling Hard Nonlinearities

The presence of a physical constraint (i.e., 'limit stops') on a state variable implies that the state variable cannot exceed a fixed bound. Including this fact in the description of the process is problematic and has engendered the development of a special theory [3] for such systems. The approach proposed in the present investigation is based on the premise that a bounded state variable *can* exceed a fixed bound if a large enough force is available. (There are no immovable objects in nature.) Hence the physical bound is represented in the dynamic model by the use of a highly nonlinear spring which has negligible effect when the state is within the bound, but produces a very large restoring force when the bound is exceeded by even a small amount.

For a symmetric bound, $|x| < B$, one possible model for the restoring force is

$$\psi_1(x) = \begin{cases} k_1 x, & |x| < B \\ C + k_2 x, & x < -B \\ -C + k_2 x, & x > B \end{cases} \quad (9)$$

where k_2 is a very large number, say 10^9, k_1 is a very small number, say 10^{-9} and

$$C = \frac{k_2 - k_1}{B}$$

An alternative representation of the restoring force is

$$\psi_2(x) = (x/B)^{2N+1} \quad (10)$$

where N is a large positive integer (say $N = 10$ or greater). This function has derivatives of all order which may make it more suitable for use with methods, such as feedback linearization, that require this property.

The parameters k_1, k_2 of (9) or N of (10) can be adjusted in accordance with the true physical nature of the limit stops.

The models of (9) or (10) are also useful in the modeling of backlash, as discussed below.

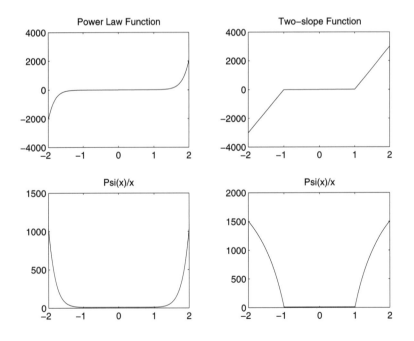

Fig. 1. Representation of hard limits.

When physical limits are not present in the system, there is scant justification for attempting to model the constraints in the dynamics of the process. Instead, it is necessary to account for the constraints in the design of the control law. Optimization of control system performance in the presence of state variable constraints has a long history in the trajectory optimization field; see [4], for example. But it is usually not easy to express the result in a form suitable for implementation in a feedback control system. An alternate approach is to approximate the constraints by including in the performance criterion a very severe penalty for violating them. This alternate approach is adopted here, and the penalty function has the form of the nonlinear functions $\psi(x)$ given by (9) or (10)

The nonlinear functions given in (9) and (10) are appropriate for use in the SDARE method, since

$$\psi_1(x) = c_1(x)x$$

with

$$c_2(x) = (x/B)^{2N}$$

and

$$\phi_2(x) = c_2(x)x$$

with

$$c_2(x) = \begin{cases} k_2 + c_1/x, & x < -B \\ k_1, & |x| \leq B \\ k_2 - c_1/x, & x > B \end{cases}$$

4 Controlling Systems with State Variable Constraints

4.1 Illustrative Examples

The application of the SDARE method using the proposed nonlinear modeling approach is illustrated by means of an inverted pendulum with position constraints.

The dynamics of the system, in normalized coordinates, are given by

$$\dot{x}_1 = x_2 \tag{11}$$
$$\dot{x}_2 = a(x_1)x_1 + u \tag{12}$$

where

$$a(x_1) = 1 - \phi(x_1)$$

in which $\phi(\ )$ is the nonlinear function used to represent the hard constraint, if it is physically present. Otherwise $\phi = 0$.

The SDARE control law for this process is given by

$$u = -g_1(x)x_1 - g_2(x)x_2$$

The state-dependent gain matrix

$$G(x) = [g_1(x), g_2(x)]$$

are obtained from (4) using the solution of the SDARE (5) with

$$A(x) = \begin{bmatrix} 0 & -1 \\ a(x_1) & 0 \end{bmatrix}, \quad B(x) = \begin{bmatrix} 0 \\ 1 \end{bmatrix},$$

$$Q(x) = \begin{bmatrix} q(x_1) & 0 \\ 0 & 0 \end{bmatrix}$$

where

$$q(x_1) = C + x_1^{2N}$$

includes the penalty for exceeding the constraint.

The SDARE in this case is readily solved with pencil-and-paper. The result is

$$g_1(x) = a(x_1) + \sqrt{a^2(x_1) + q(x_1)} \qquad (13)$$
$$g_2(x) = \sqrt{2g_1} \qquad (14)$$

A reduced-order observer for estimating the velocity in the event it cannot be measured directly is given by

$$\hat{x}_2 = Fx_1 + z$$

where

$$\dot{z} = -F\hat{x}_2 + a(x_1)x_1 + u$$

In these equations F is an arbitrary positive constant chosen to yield satisfactory observer behavior.

It can be shown that the error in estimating x_2 by this observer converges to zero.

Simulations showing the performance of the control laws were simulated for both the case in which the control law must prevent the state from entering the forbidden region and the case in which the presence of limit stops prevent the state from entering the forbidden region. The results are shown below.

'Forbidden region' in state space The performance goal in this example is to keep the position x_1 within the bound $|x_1| < 1$. Performance is simulated with an initial velocity for which a linear control law generates a trajectory that violates the constraint.

For an initial velocity $x_2 = 500$ and an initial position of zero, a linear control law, with $C = 10^5$, produces the response shown in Figure 2. It is seen that the position reaches a maximum value of about 18, which represents a serious constraint violation.

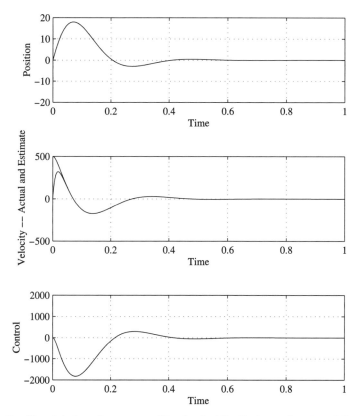

Fig. 2. Simulated performance with physical limit stops absent and linear control ($N = 0$).

In contrast, for the same initial conditions, but using the penalty function defined by $N = 10$, the results are as shown in Figure 3. It is seen that the control law generates a response in which the maximum position is about 3. Although the excursion of the position still exceeds the bound of 1, the improvement over the linear control is a factor of over 6.

Still further improvement is achieved by use of a higher value of N in the penalty function. With $N = 100$ in the penalty function, the response is as shown in Figure 4. It is seen that the excursion is reduced to about 1.4 which represents a moderate improvement over the previous case and over a tenfold improvement over the linear control.

It should be noted that the peak value of the control signal with the nonlinear control law is not greater than its value with the linear control, but it is applied soon enough to limit the position excursion into the forbidden region.

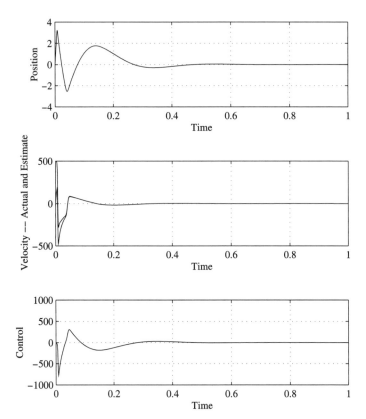

Fig. 3. Simulated performance with physical limit stops absent and a penalty function with $N = 10$.

Limit Stops Present When limit stops are physically present, no penalty function is needed in the control law, but the dynamic model must include a representation of the physical stops. This representation is used in the design of the control law, both in the determination of the nonlinear gains $g_1(x)$ and $g_2(x)$ and in the design of the observer. (The observer is linear when the physical constraints are absent as in the previous example.)

With the control nonlinearities 'matched' to the representation of the physical limit stops, i.e., the same power law representation in each, and $N = 100$, the simulation results are as shown in Figure 5. It is observed that the physical stops prevent the state from exceeding the bound. The pendulum strikes the limit stops several times on each side and then returns smoothly to the origin.

The physical presence of the limit stops eases the burden on design of the control law. In particular, a linear control law with a matched representation

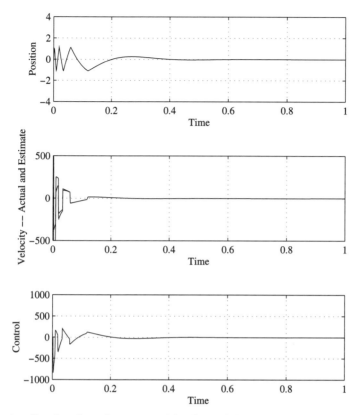

Fig. 4. Simulated performance with physical limit stops absent and a penalty function with $N = 100$.

of the limit stops in the observer produces a transient response that is scarcely distinguishable from that shown in Figure 5.

Ignoring the limit stops in the observer, however, results in unacceptable behavior. The pendulum hits the limit stops repeatedly producing a trajectory that is all but a limit cycle, as shown in Figure 6.

5 Friction and Backlash Control

5.1 Illustrative Example

The application of these techniques is illustrated by means of a mechanical system comprising a d-c motor driving a load through a compliant timing belt as illustrated schematically in Figure 7. In practice a system like this can have many parasitic effects. To exemplify the techniques of this paper, however, attention is focused on two effects: friction, the only source of which

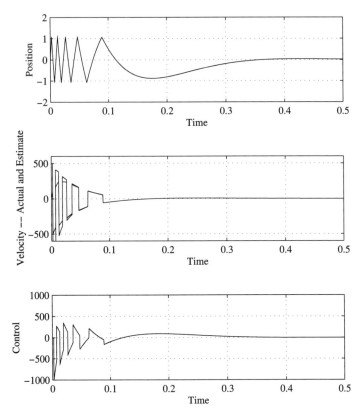

Fig. 5. Simulated performance with physical limit stops present, using nonlinear control law and nonlinear observer.

is assumed to exist between the load and its pivot, and backlash, which is assumed to result from unwanted motion in the timing belt.

In many treatments friction is approximated as pure Coulomb friction, in which case the friction function $F(v) = \text{sgn}(v)$. The backlash function $N(\beta)$ is approximated by a very soft spring near the origin and a very stiff spring outside the backlash region. The nature of these functions are discussed in greater detail in the respective sections dealing with these effects.

In a practical application friction will be present on the drive side and on the load side (and in the drive itself). Friction present at the drive side, if known, can be cancelled directly by a component of the control signal. The friction torque on the load side, however, cannot be cancelled directly, and hence its presence is more problematic. To focus on this problem, it is assumed that the only source of friction present in the system is on the load side.

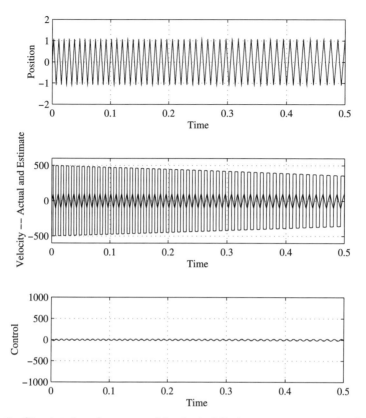

Fig. 6. Simulated performance with physical limit stops present, using linear control law and linear observer.

With friction present only at the load, it is convenient to use the 'backlash angle'

$$\beta = \phi_M - \phi_L \tag{15}$$

in place of the motor angle. In terms of the load-shaft angle and the backlash angle the system dynamics become

$$\ddot{\phi}_L = N(\beta)/\bar{J}_L - \theta F(\dot{\phi}_L) \tag{16}$$
$$\ddot{\beta} = -N(\beta)/\bar{J} - D(\dot{\phi}_L + \dot{\beta}) + B_M e \tag{17}$$

where J_L is the moment of inertia of the load wheel, J_M is the moment of inertia of the drive wheel, $\bar{J} = J_M J_L/(J_M + J_L)$, $D = K^2/RJ_M$ in which K is the motor's torque constant, R its armature resistance, $B_M = K/RJ_M$, and θ is the 'strength' of the friction.

A block diagram representation of the above dynamics is shown in Figure 8.

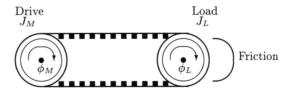

Fig. 7. Belt-driven load illustrates control of system with friction and backlash.

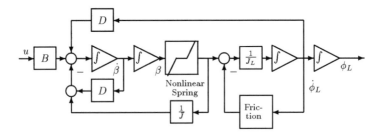

Fig. 8. Block diagram of drive system with friction and backlash.

If friction and the deadzone in the backlash are absent, the system becomes linear and is governed by

$$\dot{x} = Ax + Bu \tag{18}$$

With the state defined by

$$x = \begin{bmatrix} \phi_L & \dot{\phi}_L & \beta & \dot{\beta} \end{bmatrix}' \tag{19}$$

the matrices that define the dynamics are given by

$$A = \begin{bmatrix} 0 & 1 & 0 & 0 \\ 0 & 0 & a_{23} & 0 \\ 0 & 0 & 0 & 1 \\ 0 & a_{42} & a_{43} & a_{44} \end{bmatrix}, \quad B = \begin{bmatrix} 0 \\ 0 \\ 0 \\ B_M \end{bmatrix}, \tag{20}$$

where
$$a_{23} = N/\bar{J}_L, \quad a_{42} = -K^2/J_M R = a_{44}, \quad a_{43} = -N/\bar{J}$$

In the absence of friction and backlash, and with a stiff driving belt the system is in a sense ideal. The dynamics are linear and a full state feeedback control law is:

$$u = G_1(\phi_R - \phi_L) - G_2\dot{\phi}_L - G_3\beta - G_4\dot{\beta} \tag{21}$$

This is considered the reference case. For a plant with the following numerical values: $J_L = J_M = 0.3, K^2/R = 25., N = 60000., B_M = 1.$ the gains $G_1 = 1000., G_2 = 17.9, G_3 = 1126, G_4 = 12.4$ produce the transient response shown in Figure 9, which can be regarded as a basis of comparison with the performance using the other control laws to be considered subsequently.

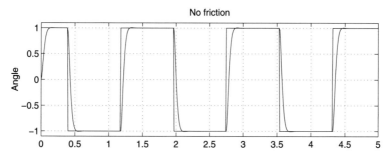

Fig. 9. Transient response in absence of friction. and backlash.

Theory A plant that is linear except for the presence of friction can be described by:

$$\dot{x} = Ax + Bu + F(x)\theta \tag{22}$$

where θ = const. is a parameter (or vector of parameters) that is needed to define the friction and $F(x)$ is a known matrix function, which is not necessarily smooth.

The design goal is to find a control law which minimizes the quadratic performance criterion

$$\lim_{T\to\infty} \int_t^T (x'Qx + u'Ru)d\tau \tag{23}$$

If the matrix F in (22) were not a function of x, θ would be an exogenous variable and the control law that minimizes the performance criterion of (23) can be shown to be given by

$$u = -G_x x - G_\theta \theta \tag{24}$$

where
$$\mathbf{G} = \begin{bmatrix} G_x & G_\theta \end{bmatrix} = R^{-1}\mathbf{B}'\mathbf{M} \tag{25}$$

where
$$\mathbf{M} = \begin{bmatrix} M_x & M_{x\theta} \\ M'_{x\theta} & M_\theta \end{bmatrix} \tag{26}$$

satisfies the algebraic Riccati equation
$$0 = \mathbf{M}\mathbf{A} + \mathbf{A}'\mathbf{M} - \mathbf{M}\mathbf{B}R^{-1}\mathbf{B}'\mathbf{M} + \mathbf{Q} \tag{27}$$

with
$$\mathbf{Q} = \begin{bmatrix} Q & 0 \\ 0 & 0 \end{bmatrix} \tag{28}$$

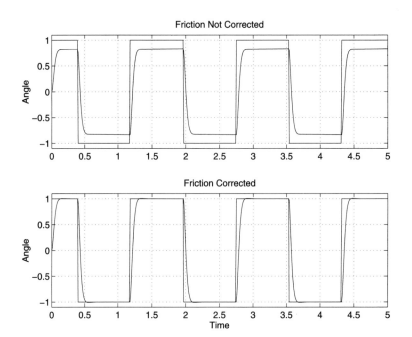

Fig. 10. Simulated performance of control system without and with friction compensation, with stiff drive belt.

As shown in [5],
$$G_x = R^{-1}B'M_x \tag{29}$$

where M_x satisfies the ARE
$$0 = M_x A + A' M_x - M_x B R^{-1} B' M_x + Q \tag{30}$$

and
$$G_\theta = -R^{-1}B'(A'_c)^{-1}M_x F \qquad (31)$$
where
$$A_c = A - BG = \text{closed-loop dynamics matrix}$$

In the present case, however, the matrix F depends on x. In order to obtain a control law we use the principle of the state-dependent algebraic Riccati equation, which in this case, is simply to use $F(x)$ in place of the constant matrix F in (31).

The control law thus obtained is not optimum, of course. In fact, if θ is large, it may not even result in a stable closed-loop system. But, if we are dealing with a parasitic phenomenon, we can legitimately assume that θ is small, and hence that the resulting control law is approximately optimum.

Performance The performance of the control system with friction compensated in the manner described was simulated for several different values of belt 'stiffness'. The result of two cases are shown in Figures 10 and 11. Figure 10 illustrates the behavior for a relatively stiff belt, having a resonance frequency $\omega = \sqrt{N/J_{\text{tot}}} = 316$. Without compensation there is a substantial hangoff which is all but eliminated upon the use of friction compensation. The behavior is similar for a 'softer' belt with a resonance frequency of 3.16. Without compensation the hangoff in this case is even larger, but the hangoff is removed in the steady state through the use of friction compensation.

5.2 Backlash Control

Theory The phenomenon of backlash is inherently more complicated than friction and cannot readily be represented in the form given in (7), because as the dead space $\delta \to 0$ the system approaches a linear fourth-order system with a stiff spring.

Nevertheless, backlash can be treated using the SDARE method. The dynamics matrix has the form of (20) with

$$a_{23} = N(\beta)/\bar{J}_L$$
$$a_{42} = -K^2/J_M R = a_{44}$$
$$a_{43} = -N(\beta)/\bar{J}$$

where $N(\beta) = \psi(\beta)/\beta$ in which $\psi(\beta)$ is the 'backlash function'

The nonlinear functions $\psi(\ )$ used in the representation of backlash is one of the nonlinear functions defined in Section 3, above.

The issue of backlash modeling has been addressed by several prior authors [6,7] and can be quite complex. To use the SDARE method, it is important that the model have a nonzero slope at the origin. If the slope in the model is zero the system is not locally controllable and many design methods,

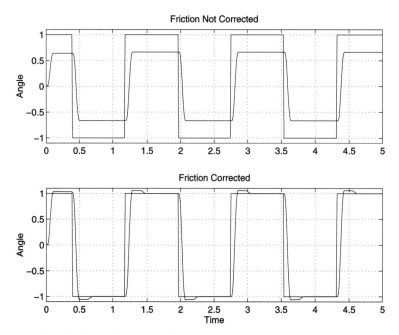

Fig. 11. Simulated performance of control system without and with friction compensation, with soft drive belt.

including the SDARE method, fail. One can argue that the physics of backlash support the assumption of nonzero slope; nevertheless, it is not entirely impossible to design a control system without this assumption [8].

From the standpoint of implementation the two-slope model is preferable, since the nonlinear terms in the A matrix can be approximated by

$$N(\beta) = \begin{cases} k_1, & |\beta| < \delta \\ k_2, & |\beta| \geq \delta \end{cases} \quad (32)$$

which leads to a two-gain control law with switching:

$$u = \begin{cases} -g_{\text{soft}} x, & |\beta| < \delta \\ -g_{\text{hard}} x, & |\beta| \geq \delta \end{cases} \quad (33)$$

where the gains $g_{\text{soft}}, g_{\text{hard}}$ are computed by solving the ARE with k_1 and k_2, respectively.

Performance Performance of the system with control law (33) was evaluated by simulation, for $\delta = 0.25$. For plant numerical values given in section 2, and

$$Q = \text{diag}\begin{bmatrix} 10000 & 0 & 0 & 0 \end{bmatrix}$$

the gain sets are

$$g_{soft} = \begin{bmatrix} 316. & 76.3 & 293. & 5.01 \end{bmatrix}$$
$$g_{hard} = \begin{bmatrix} 316. & 6.71 & 483. & 6.86 \end{bmatrix}$$

Note that only one of the gains changes significantly.

The results and comparisons with other control options is shown in Figure 12. For case shown in the top of the figure, the backlash state variables β and $\dot{\beta}$ were simply ignored. The result is the limit cycle shown. For the simulation shown in the middle of the figure, the backlash states are fed back, but the gains are not switched; the gain set g_{hard} is used for all values of β. For the simulation shown in the lower part of the figure the switched gain control was used. Although the linear control that uses the backlash states eliminates the limit cycle, the overshoot is rather large and the response is oscillatory. The switched gain control law, on the other hand, achieves a rapid response essentially without overshoot.

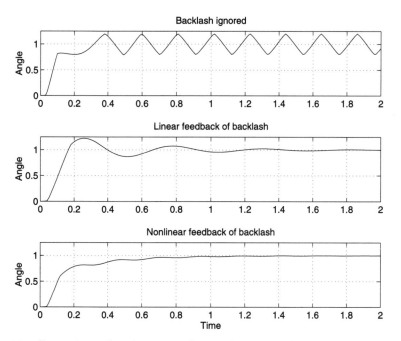

Fig. 12. Comparison of performance of control system without and with backlash compensation

Estimation of backlash states In principle it is possible to measure the backlash angle (possibly by means of a calibrated load cell), but the measurement is not easy to make. To avoid the necessity of measuring this angle and its derivative, an observer can be used. It is important to incorporate the nonlinear dynamics into the observer, however.

To evaluate the performance of a control system with estimated backlash states, a reduced-order observer was included in the control system. In accordance with 'standard' theory [1], a suitable observer is given by

$$\hat{\beta} = k_{11}\theta_L + k_{12}\dot{\theta}_L + z_1 \tag{34}$$

$$\dot{\hat{\beta}} = k_{21}\theta_L + k_{22}\dot{\theta}_L + z_2 \tag{35}$$

where

$$\dot{z}_1 = \frac{-k_{21}}{J_L}\psi(\hat{\beta}) + \dot{\hat{\beta}} - k_{11}\dot{\theta} \tag{36}$$

$$\dot{z}_2 = -(\frac{1}{J} + \frac{k_{22}}{J_L})\psi(\hat{\beta}) - D\dot{\hat{\beta}} - (D + k_{22})\dot{\theta} + \frac{B}{J_M}u \tag{37}$$

The performance of the control system using this observer, with observer gains given by:

$$k_{11} = 0, \quad k_{12} = 10., \quad k_{21} = 0., \quad k_{22} = 100.$$

is shown in Figure 13. Except for the high frequency oscillation, the performance is not very different from that exhibited when the backlash states are directly measured.

6 Conclusions

The control system designer must often contend with hard nonlinearities, which are often difficult to deal with by use of standard methods. As the foregoing examples have shown, the state-dependent algebraic Riccati equation (SDARE) method might serve as an appropriate tool for dealing with such problems.

Acknowledgement. This article is based on two papers by the author [9] and [10] originally presented at the American Control Conference.

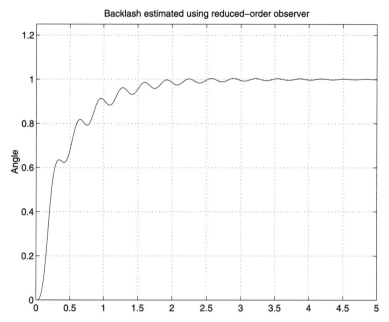

Fig. 13. Performance of backlash compensated control system with observer used to estimate backlash states.

References

1. B. Friedland, *Advanced Control System Design*, Prentice-Hall, Upper Saddle River, NJ, 1996.
2. J.R. Cloutier, State-dependent Riccati equation techniques: an overview,*Proc. American Control Conference*, Albuquerque, NM, June 1997.
3. B. Broglioto, S.-I. Niculescu, and P. Orhant, On the control of finite-dimensional mechanical systems with unilateral constraints,*IEEE Trans. on Automatic Control*, 42, No. 2, 200–215, 1997.
4. A.E. Bryson, Jr. and Y.-C. Ho, *Applied Optimal Control*, Blaisdell Publishing Co., Waltham, MA, 1969.
5. B. Friedland, *Control System Design*, McGraw-Hill, New York, 1986.
6. G. Brandenburg and U. Schafer, Influence and adaptive compensation of simultaneously acting backlash and coulomb friction in elastic two-mass systems of robots and machine tools, *Proc. IEEE Int. Conf. on Control and Applicatons*, Jerusalem, Israel, 1989.
7. M. Nordin, J.Galic, and P.-O. Gutman, New models for backlash and gear play, *Int. J. of Adaptive Control and Signal Processing*, 11, 49–63, 1997.
8. R. Boneh and O. Yaniv, Control of an elastic two-mass system with large backlash, *J. Dynamic Systems, Measurement, and Control*, 121, No. 2, 278–284, 1999.

9. B. Friedland, Feeedback control of systems with parasitic effects, *Proc. American Control Conference*, Albuquerque, NM, June 4–6, 1997.
10. B. Friedland, On controlling systems with state-variable constraints, *Proc. 1998 American Control Conf.*, Philadelphia, PA, June 4–6, 1998.

Adaptive Friction Compensation for Servo Mechanisms

J. Wang, S. S. Ge, and T. H. Lee

Department of Electrical and Computer Engineering,
National University of Singapore,
Singapore 117576
Email: elegesz@nus.edu.sg

Abstract

Friction exists in all machines having relative motion, and plays an important role in many servo mechanisms and simple pneumatic or hydraulic systems. In order to achieve high precision motion control, accurate friction modeling and effective compensation techniques have to be investigated. In this chapter, we shall present a systematic treatment of adaptive friction compensation techniques from both engineering and theoretical aspects. Firstly, a comprehensive list of the commonly used classical friction models and dynamic friction models is presented for comparison and controller design. Then, by considering the position and velocity tracking control of a servo mechanism with friction, adaptive friction compensation schemes are given based on the LIP static friction model and dynamic LuGre model respectively. Finally, extensive simulation comparison studies are presented to verify the effectiveness of the proposed methods.

1 Introduction

Friction exists in all machines having relative motion, and plays an important role in many servo mechanisms and simple pneumatic or hydraulic systems. It is a natural phenomenon that is hard to model, if not impossible. As friction does not readily yield to rigorous mathematical treatment, it is often simply ignored for the lack of control tools available or regarded as a phenomenon unworthy of discussion. In reality, friction can lead to tracking errors, limit cycles, and undesired stick-slip motion. Engineers have to deal with the undesirable effects of friction though lack of effective tools to make it easier to handle.

All surfaces are irregular at the microscopic level, and in contact at a few asperity junctions. These asperities behave like springs, and can deform either elastically or plastically when subjected to a shear force. Thus, friction will act in the direction opposite to motion and will prevent true sliding from taking place as long as the tangential force is below a certain stiction limit at which the springs become deformed plastically. At the macro level, many factors affect friction such as lubrication, velocity, temperature, force orthogonal to the relative motion and even the history of motion. In an effort to deal with the undesired effects of friction effectively, many friction models have been presented in the literature relevant to friction modeling and control.

Friction can be classified into two categories: static and dynamic. Conventionally, friction is modeled as a static function of velocity and the externally applied force. The notorious components in friction are stiction and Coulomb friction forces, which are highly nonlinear functions of velocity although bounded, and cannot be handled by linear control theory. Traditionally, friction is treated as a bounded disturbance, and the standard PID (Proportional-Integral-Derivative) algorithm is used in motion control. However, the integral control action may cause limit cycles around a target position and result in large tracking errors. If only PD control is employed, then friction will cause a finite steady-state error. Though high gain PID can reduce the steady–state position error due to friction, it often causes system instability when the drive train is compliant [18]. Thus, in order to achieve high precision motion control, friction must be appropriately compensated for. Friction compensation can be achieved based on a reasonable accurate model for friction. However, it is difficult to model friction because it depends on many factors such as velocity, position, temperature, lubrication and even the history of motion. Direct compensation of friction is desirable and effective in motion control. However, it is difficult to realize in practice because of the difficulty in obtaining a true representative parametric model. For controller design, the parametric model should be simple enough for analysis, and complex enough to capture the dominant dynamics of the system. If the model used is too simple, such as the simple Coulomb friction and viscous friction model, then there is the possibility of overcompensation resulting from estimation inaccuracies [12]. Adaptive friction compensation schemes have been proposed to compensate for nonlinear friction in a variety of mechanisms [1, 12], but these are usually based on the linearized model or a model which is linear-in-the-parameters for the problems under study. Each model captures only the dominate friction phenomena of the system and may exhibit discrepancies when used for other systems where other friction phenomena appear.

Although static-model-based adaptive friction compensation techniques have been proposed in the literature [4, 13, 15, 19], the results are not always satisfactory in applications with high precision positioning and low velocity tracking. Several interesting properties observed in systems with friction cannot be explained by static models because the internal dynamics of friction are not considered. Examples of the dynamic properties include: stick-slip motion, presliding displacement, Dahl effect and friction lag. Accordingly, dynamic friction models are preferred for the development of advanced friction compensation schemes. A notable dynamic friction model which captures all the static and dynamic characteristics of friction is the dynamic LuGre model proposed in [13]. The necessary and sufficient conditions for passivity of the LuGre model was recently given in [6].

When using the dynamic LuGre friction model, controller design becomes difficult because (i) the friction parameters appear in a nonlinear fashion, and (ii) the system's internal state z, which depends on unknown parameters, is not measurable. Based on the LuGre model [13], several nice model-based controllers have been developed. Under the assumption of known system parameters and functions, a model-based controller was presented in [13] with the unmeasurable friction state being estimated by an observer which is driven by the tracking error. Two globally stable model-based adaptive friction compensation schemes were presented in [15] for structured variations by assuming that changes in friction are mainly due to either changes in the normal force that only affects proportionally the static friction characteristics, or temperature variations affecting uniformly both static and dynamic parameters. Accordingly, the resulting schemes adapt only one single parameter. In [32], using a dual-observer structure for the internal friction state estimation, two elegant controllers were presented for the following two cases: (i) unknown friction coefficients with known inertia parameter and known friction characteristic function, and (ii) known friction characteristic function with unknown multiplying coefficient for handling nonuniform variation in the friction force and normal force variations. Case (ii) is an alternative to the solution given in [15].

Due to its uniform approximation ability to any continuous nonlinear functions, there has been considerable research interest in neural network (NN) control of nonlinear systems, and some have been applied to systems with friction, which are usually difficult to model. In [14], an adaptive NN controller was proposed, where the NN was used to parameterize the nonlinear characteristic function of the dynamic friction model which may be a function of both position and velocity with known system parameters. In [25], a reinforcement learning based NN adaptive control scheme was applied to compensate for stick-slip friction for tracking control of a 1-DOF mechanical system with guaranteed high precision and smoothness of motion. In [21], both model-based and NN-based adaptive controllers were investigated for dynamic friction, where the friction model for controller design is given in an easy-to-use linear-in-the-parameters (LIP) form. By considering the dynamic LuGre model, two NN-based adaptive friction compensation schemes have recently been proposed in [22]. One is for the unknown nonlinear friction characteristic function in the LuGre model, and the other is for completely unknown parameters in the LuGre model.

Though adaptive control of systems with smooth nonlinearities has made remarkable progress in recent decades, they are not directly applicable to systems with frictions, which is a nonsmooth nonlinearity. Control of such nonlinearities has been a tough challenge academically and practically. In this chapter, we shall present a systematic treatment of adaptive friction compensation techniques from both engineering and theoretical aspects.

Firstly, we will give a detailed account of the commonly used friction models including their history and properties. Several classical static friction models and dynamic friction models are described. As a result, we shall present a simple linear-in-the-parameters (LIP) friction model that captures most of the observed static friction phenomena of velocity, and is easy to use for controller design. As the space of primitive functions increases, the model becomes more complete and representative, and thus reduces the possibility of friction overcompensation resulting from estimation inaccuracies caused by simplified friction model structures, such as simple Coulomb friction and viscous friction models. The properties of the dynamic LuGre model are also discussed in detail for controller design. The LuGre model used in our design accounts for both the position and velocity dependence of the friction force.

Secondly, by considering the position and velocity tracking control of a servo mechanism with friction, adaptive friction compensation schemes are presented based on the static friction model and dynamic LuGre model. For static friction model based compensation, the emphasis is on a unified adaptive controller based on our proposed LIP friction model and the study of effects in augmenting the space of basis/primitive functions. For the dynamic LuGre model, in order to provide good control performance, new adaptive controllers are presented, combining NN parameterization, dual-observer for state estimation/stability and adaptive control techniques. As in [14], the nonlinear characteristic function of the dynamic friction model is assumed to be a function of both position and velocity, which covers a larger class of systems. Three cases are considered based on the uncertainty levels of the system: 1. unknown system parameters including all friction coefficients and the inertia parameter; 2. unknown characteristic function $\alpha(x, \dot{x})$ of the dynamic friction model; and 3. full set of unknown parameters. For Case 1, a model-based adaptive controller is presented. For Case 2, an adaptive NN controller is proposed based on a neural network parameterization of the unknown $\alpha(x, \dot{x})$. Note that in designing each of these two adaptive controllers, two nonlinear observers are used operating in parallel for the purpose of (a) estimating the unmeasurable friction state and (b) canceling the cross-coupling terms in the derivative of the Lyapunov function for closed-loop stability. For Case 3, by dividing the friction model into two portions: (i) viscous friction with unknown constant coefficient, and (ii) unknown dynamic friction which is a function of the unmeasurable internal friction state $z(t)$, and bounded by a function which is independent of $z(t)$, an adaptive NN controller is developed based on NN parameterization of the unknown dynamic friction bounding function. Using Lyapunov synthesis, adaptation algorithms are designed to achieve globally asymptotic tracking of the desired trajectory and to guarantee the boundedness of all signals in the closed loop. Both position tracking and velocity tracking are realized at the same time by the utilization of a filtered error signal.

Finally, extensive numerical comparison studies are presented. Some conclusions are drawn and comparisons made concerning applications for practical engineers.

2 Friction Models

It is well known that friction depends on both velocity and position, but its structure is not well defined, especially at low velocity. For ease of analysis and simulation, it is important to have a mathematical model of friction. A friction model should be able to accurately predict the observed friction characteristics, and be simple enough for friction compensation.

Friction is a multifaceted phenomenon, and exhibits the well-known classical Coulomb and viscous friction, nonlinearity at low velocity, and elasticity of contact surfaces. In any given circumstance, some features may dominate over others and some features may not be detectable with the available sensing technology. But all these phenomena are present all the time. The use of a more complete friction model will extend the applicability of analytical results and resolve discrepancies that arise in different investigations. While the classical friction models give only the static relationships between velocity and friction force, the most recent friction model, the so-called LuGre model, is a dynamic model with an unmeasureable internal state. In the following, we shall give a list of commonly used friction models that enable simple controller design and computer simulation.

2.1 Static Models

Static friction models refer to those models that are functions of velocity/position only, without any internal dynamics. In this subsection, we will not only list the commonly known classical static friction components, such as stiction, Coulomb etc., and we will also present a neural network friction model which is easy to use to treat the complexity and difficulty in modelling friction.

Static friction (Stiction)
At zero velocity, the static friction opposes all motion as long as the torque is smaller in magnitude than the maximum stiction force f_s, and is usually described by

$$F = \begin{cases} u, & |u| < f_s \\ f_s \delta(\dot{x}) \text{sgn}(u), & |u| \geq f_s \end{cases} \tag{2.1}$$

where

$$\delta(\dot{x}) = \begin{cases} 1, & \dot{x} = 0 \\ 0, & \dot{x} \neq 0 \end{cases}, \quad \text{sgn}(u) = \begin{cases} +1, & u > 0 \\ -1, & u < 0 \end{cases} \tag{2.2}$$

In actual (numerical) implementation, the impulse function can be approximated differently, such as triangular or rectangular as in Karnopp's version of stiction.

In fact, stiction is not truly a force of friction, but a force of constraint in presliding and behaves like a spring. For small motion, the elasticity of asperities suggests that the applied force is approximately proportional to the presliding displacement

$$F = f_t x \delta(\dot{x}), \qquad (2.3)$$

where f_t is the tangential stiffness of the contact, x is the displacement away from equilibrium position and $\delta(\dot{x})$ is used to describe the fact that stiction occurs only when at rest. At a critical force, breakaway occurs and true sliding begins. Breakaway has been observed to occur at the order of 2 – 5 microns in steel junctions and millimetre motions in robots, where the arms act as levers to amplify the micron motion at the gear teeth. Presliding displacement is of interest to the control community in extremely high precision positioning. If sensors are not sensitive enough, we are able to observe only common stiction (2.1) and Coulomb friction (2.4).

Coulomb friction (dry friction)
Independent of the area of contact, Coulomb friction always opposes relative motion and is proportional to the normal force of contact. Coulomb friction is described by

$$F = f_c \text{sgn}(\dot{x}) \qquad (2.4)$$

where $f_c = \mu |f_n|$ with μ being the coefficient of friction, and f_n the normal force. Constant f_c is independent of the magnitude of relative velocity.

Viscous friction
Viscous friction corresponds to the well-lubricated situation, and is proportional to velocity. It obeys the linear relationship

$$F = f_v \dot{x} \qquad (2.5)$$

Drag friction
Drag friction is caused by resistance to a body moving through a fluid (e.g. wind resistance). It is proportional to the square of velocity as described by

$$F = f_d |\dot{x}| \dot{x} \qquad (2.6)$$

When the speed of travel is small, this term is neglectable. This term cannot be neglected in the control of hard disk drives because of the high speed rotation of spindle motors.

Classical friction models have different combinations of static, Coulomb and viscous friction as their basic building blocks.

Exponential model

In [7], after reviewing several existing models, an exponential model incorporating Coulomb and viscous frictions is given as

$$F(\dot{x}) = f_c \mathrm{sgn}(\dot{x}) + (f_s - f_c)e^{-(\dot{x}/\dot{x}_s)^\delta} + f_v \dot{x} \qquad (2.7)$$

where $\dot{x}_s$ and δ are empirical parameters, f_c is the Coulomb friction level, f_s is the level of the stiction force and f_v is the viscous coefficient. By choosing different parameters, different frictions can be realized [4]. While the range of δ may be large, $\delta = 2$ gives the Gaussian exponential model [3] which is nearly equivalent to the Lorentzian model (2.13). Gaussian models have the following different forms:

Gaussian exponential with one break:

$$F(\dot{x}) = f_c \mathrm{sgn}(\dot{x}) + (f_s - f_c)e^{-(\dot{x}/\dot{x}_s)^2} + f_v \dot{x} \qquad (2.8)$$

Gaussian exponential with two breaks:

$$F(\dot{x}) = f_c \mathrm{sgn}(\dot{x}) + f_v \dot{x} + f_{s_1} e^{-(\dot{x}/\dot{x}_{s_1})^2} + F_{s_2} e^{-(\dot{x}/\dot{x}_{s_2})^2} \qquad (2.9)$$

Gaussian exponential with two breaks and offsets:

$$F(\dot{x}) = f_c \mathrm{sgn}(\dot{x}) + f_v \dot{x} + f_{s_1} e^{-(\dot{x}-\dot{x}_{10})^2/\dot{x}_{s_1}^2} + F_{s_2} e^{-(\dot{x}-\dot{x}_{20})^2/\dot{x}_{s_2}^2} \qquad (2.10)$$

where x_{s1} and x_{s2} are empirical parameters; f_{s1} and f_{s2} are static friction constants; $\dot{x}_{10}$ and $\dot{x}_{20}$ are offset points of breaks.

On the other hand, $\delta = 1$ gives Tustin's model [35] as described by

$$F(\dot{x}) = f_c \mathrm{sgn}(\dot{x}) + (f_s - f_c)e^{-(\dot{x}/\dot{x}_s)} + f_v \dot{x} \qquad (2.11)$$

Tustin's model is one of the best models describing friction force at a velocity close to zero. It includes a decaying exponential term in the friction model which explains the microscopic limit cycle behavior that, after a breakaway point at $\dot{x}$, has a negative exponential characterization. Experimental work has shown that this model can approximate real friction forces with a precision of 90% [2, 11].

Because of the nonlinearity in unknown parameter $\dot{x}_s$ in Tustin's model and the difficulty in dealing with nonlinear parameters, the following simple linear-in-the-parameters (LIP) friction model was proposed [12]:

$$F(\dot{x}) = f_c \mathrm{sgn}(\dot{x}) + f_r \sqrt{|\dot{x}|}\mathrm{sgn}(\dot{x}) + f_v \dot{x} \qquad (2.12)$$

where constants f_i, $(i = c, r, v)$ are not unique and depend on the operating velocity. The simple LIP model has the following advantages: (i) it captures the downward bends and possible asymmetrics; (ii) the unknown parameters are linear and thus suitable for on-line identification; (iii) these parameters can accommodate parametric changes due to environmental variations; (iv) this type of model structure reduces the possibility of friction overcompensation resulting from estimation inaccuracies caused by simplified friction model structures such as Coulomb friction and viscous friction models.

Lorentzian model

In [23], a model of the following form has been employed:

$$F(\dot{x}) = f_c \text{sgn}(\dot{x}) + (f_s - f_c)\frac{1}{1 + (\dot{x}/\dot{x}_s)^2} + f_v \dot{x} \quad (2.13)$$

which shows a systematic dependence of $\dot{x}_s$ and f_v on the lubricant and loading parameters. Similar to the case of the Gaussian model, Lorentzian models also have forms with one break, two breaks or two breaks with offsets.

Remark 2.1. Based on the above discussion, a more complete model may consist of the following components: stiction, Coulomb, viscous and drag friction, and square root friction:

$$F(x, \dot{x}) = f_t x \delta(\dot{x}) + f_c \text{sgn}(\dot{x}) + f_v \dot{x} + f_d \dot{x}|\dot{x}| + f_r \sqrt{|\dot{x}|}\text{sgn}(\dot{x}) \quad (2.14)$$

which can be conveniently expressed in LIP form as

$$F(x, \dot{x}) = S^T(x, \dot{x})P \quad (2.15)$$

where

$$S(x, \dot{x}) = [x\delta(\dot{x}), \text{sgn}(\dot{x}), \dot{x}, \dot{x}|\dot{x}|, \sqrt{|\dot{x}|}\text{sgn}(\dot{x})]^T \quad (2.16)$$
$$P = [f_t, f_c, f_v, f_d, f_r]^T \quad (2.17)$$

where $S(x, \dot{x})$ is a vector of known basis functions, and P is a vector of unknown parameters. Though the LIP form is very desirable for model-based friction compensation as will be shown later, it is in no sense complete, but is a more complete representation. If other nonlinear components, such as the nonlinear exponential term $e^{-(\dot{x}/\dot{x}_s)^\delta}$ and Lorentzian term $1/(1 + (\dot{x}/\dot{x}_s)^2)$ under the assumption of known $\dot{x}_s$ and δ, exist in the friction model, the space of the regressor function can be simply increased by including them. If $\dot{x}_s$ and δ are not known, they can be approximated using the primitives explored in [15].

Remark 2.2. It is generally considered that friction has two different manifestations, i.e. presliding friction and sliding friction [4]. In the presliding stage, which is usually in the range of less than 10^{-5}m, friction is dominated by the elasticity of the contacting asperity of surfaces as described by equation (2.3). It not only depends on both position and velocity of motion, but also exhibits nonlinear dynamic behavior such as hysteresis characteristics with respect to position and velocity as observed by many researchers. In the sliding stage, friction is dominated by lubrication of the contacting surfaces and introduces damping into system. It is usually represented by various functions of velocity. Thus, we can conclude that *friction is continuous* though it may be highly nonlinear, and depends on both position and velocity. The discontinuities modelled by stiction (2.1) and Coulomb friction (2.4) are actually observations at the macro level. Thus, friction can be approximated by neural networks as explained below.

Neural network friction model
Neural networks offer a possible tool for nonlinear mapping approximation. Neural networks can approximate any continuous function to arbitrarily any accuracy over a compact set if the size of the network is large enough [16, 30].

Because of the complexity and difficulty in modelling friction, neural networks may be used to generate input/output maps using the property that a multilayer neural network can approximate any function, under mild assumptions, with any desired accuracy. It is well known that any sufficiently smooth function can be approximated by a suitably large network using various activation functions, $\sigma(.)$, based on the Stone–Weierstrass theorem. Typical choices for $\sigma(.)$ include the sigmoid, hyperbolic tangent, radial basis functions, etc. It has been proven that any continuous functions, not necessarily infinitely smooth, can be uniformly approximated by a linear combinations of Gaussian radial basis functions (RBF). The Gaussian RBF neural network is a particular network architecture which uses l Gaussian functions of the form

$$s_i(x, \dot{x}) = \exp\left(-\frac{(x - \mu_{1i})^2 + (\dot{x} - \mu_{2i})^2}{\sigma^2}\right), \quad i = 1, \cdots, l \qquad (2.18)$$

where $x, \dot{x}$ are the input variables, σ^2 is the variance and μ_{1i}, μ_{2i} are the centres. A Gaussian RBF neural network can be mathematically expressed as

$$F_{nn}(x, \dot{x}) = W^T S(x, \dot{x}) \qquad (2.19)$$

where $S(x, \dot{x}) = [s_1, s_2, ..., s_l]^T \in R^l$ is the basis function vector, and $W \in R^l$ is the corresponding weight vector. A general friction model $F(x, \dot{x})$ can then be written as

$$F(x, \dot{x}) = F_{nn}(x, \dot{x}) + \epsilon(x, \dot{x}) \qquad (2.20)$$

where $F_{nn}(x, \dot{x})$ is given in (2.19), and $\epsilon(x, \dot{x})$ is the bounded neural network functional reconstruction error. If there exist integer l and constant weight W such that $\epsilon = 0$, $F(x, \dot{x})$ is said to be in the functional range of the neural network.

For ease of analysis and controller design later, we present only an LIP NN-based friction model: nonlinear multilayer NN-based friction models can also be investigated following the method of treating multilayer NN in [26, 36]. Accordingly, we have the following LIP friction model in general form:

$$F(x, \dot{x}) = F_m(x, \dot{x}) + \epsilon(x, \dot{x}) \tag{2.21}$$

where $F_m(x, \dot{x}) = S^T(x, \dot{x})P$ is the LIP model for friction and ϵ is the residue modelling error. If $S(x, \dot{x})$ consists of the classical model basis functions listed in (2.16), P is the corresponding coefficient vector. If $S(x, \dot{x})$ is the basis function vector of the neural network model (2.19), P is the neural network weight vector.

2.2 Dynamic Models

The classical models cannot describe all the dynamic effects of friction, such as the presliding displacement, the frictional lag, the Stribeck effect, which all occur in the so-called low-velocity and pre-sliding regions. Driven by applications with high precision positioning and with low velocity tracking, there has been significant interest in dynamic friction models. In this subsection, we present two dynamic models from the literature.

Dahl model

The Dahl model was introduced in [17]. Several experiments were conducted on friction in servo systems with ball bearings. A comparatively simple model was developed to simulate systems with ball bearing friction. Let x be the displacement, F the friction force, and f_c the Coulomb friction force. Then Dahl's model has the form

$$\dot{F} = \sigma(1 - \frac{F}{f_c}\operatorname{sgn}(\dot{x}))^\alpha \dot{x} \tag{2.22}$$

where σ is the stiffness coefficient and α is a parameter that determines the shape of the stress–strain curve. The value $\alpha = 1$ is most commonly used [28]. For the case $\alpha = 1$ the Dahl model (2.22) becomes

$$\dot{F} = \sigma\dot{x} - \frac{F}{f_c}|\dot{x}| \tag{2.23}$$

Introducing $F = \sigma z$, the model (2.23) can be further written as

$$\dot{z} = \dot{x} - \frac{\sigma|\dot{x}|}{f_c}z \tag{2.24}$$

$$F = \sigma z \tag{2.25}$$

Remark 2.3. The Dahl model is a generalization of ordinary Coulomb friction. It captures neither the Stribeck effect nor stiction. Recently, the Dahl model was extended to the LuGre model in [13].

LuGre model

All of the observed static and dynamic characteristics of friction can be accurately captured by the dynamic LuGre model proposed in [13]. The LuGre model considers the dynamic effects of friction to arise from the defection of bristles which model the asperities between two contacting surfaces, and is given by

$$F = \sigma_0 z + \sigma_1 \dot{z} + \sigma_2 \dot{x} \quad (2.26)$$
$$\dot{z} = \dot{x} - \alpha(\dot{x})|\dot{x}|z \quad (2.27)$$

where F is the friction force, z denotes the average deflection of the bristles, which is not measurable, $\sigma_0, \sigma_1, \sigma_2$ are friction force parameters that can be physically explained as the stiffness of bristles, damping coefficient, and viscous coefficient, and the nonlinear friction characteristic function $\alpha(\dot{x})$ is a finite positive function which can be chosen to describe different friction effects. One parameterization of $\alpha(\dot{x})$ to characterize the Stribeck effect is given in [13]

$$\alpha(\dot{x}) = \frac{\sigma_0}{f_c + (f_s - f_c)e^{-(\dot{x}/\dot{x}_s)^2}} \quad (2.28)$$

where f_c is the Coulomb friction level, f_s is the level of the stiction force and $\dot{x}_s$ is the constant Stribeck velocity.

Remark 2.4. In the above model, there are no terms which explicitly account for position dependence of the friction force. However, there may exist some applications where the function $\alpha(\cdot)$ in the LuGre model also depends on the actual position, or on a more complex combination of position and velocity. The dependency on position can also be visualized in the following scenarios as discussed in [14]. It is known that $\alpha(\cdot)$ depends on the normal force [15] between two surfaces. For a robot in operation, the normal forces on the shafts are not only functions of velocities, but also functions of positions and accelerations. In gear boxes, friction will vary as a function of the effective surface area of the gear's teeth in contact. Two–dimensional rolling and spinning friction in ball bearings causes the frictional torque to be dependent on both position and velocity. Therefore, we assume that $\alpha(x, \dot{x})$ is an upper and lower bounded positive smooth function of x and $\dot{x}$, and adopt the LuGre model in the following form:

$$F = \sigma_0 z + \sigma_1 \dot{z} + \sigma_2 \dot{x} \quad (2.29)$$
$$\dot{z} = -\alpha(x, \dot{x})|\dot{x}|z + \dot{x} \quad (2.30)$$

Assumption 2.1. *There exist positive constants α_{min} and α_m such that $0 < \alpha_{min} \leq \alpha(x, \dot{x}) \leq \alpha_m, \forall (x, \dot{x}) \in R^2$.*

Lemma 2.1. *Under Assumption 2.1, if $|z(0)| \leq 1/\alpha_{min}$ then $|z(t)| \leq 1/\alpha_{min}, \forall t \geq 0$ [13].*
Proof Define $V_z = \frac{1}{2} z^2$. Then the time derivative of V_z along (2.30) is

$$\dot{V}_z = z(\dot{x} - \alpha(x,\dot{x})|\dot{x}|z) = -|z||\dot{x}|(\alpha(x,\dot{x})|z| - \text{sgn}(\dot{x})\text{sgn}(z)) \quad (2.31)$$

Thus, $\dot{V}_z \leq 0$ when $|z| \geq 1/\alpha_{min}$. From $|z(0)| \leq 1/\alpha_{min}$, we always have $z(t) \leq 1/\alpha_{min}$. $\diamond$

3 Adaptive Friction Compensation

3.1 Problem Statement

The servo mechanism under study is modeled as a simple mass system with dynamic friction, and is described by

$$m\ddot{x} + F = u \quad (3.1)$$

where m is the mass, x is the displacement, u is the control force, and F is the friction force.

Assumption 3.1. *States x and $\dot{x}$ are measurable for feedback controller design.*

Assumption 3.2. *The desired trajectory x_d, and its first and second derivatives $\dot{x}_d, \ddot{x}_d$ are continuous and bounded.*

The control objective is to design an adaptive controller to track the given desired trajectory. In the literature, many control techniques have been investigated for friction compensation: they include high-gain PID [5], feedforward compensation [8], robust friction compensation [9, 34], adaptive friction compensation [10, 12], neural network control [14, 21, 25]. In this section, we present adaptive controller design methods based on the LIP static friction model and the dynamic LuGre friction model.

Define the tracking errors as

$$e = x - x_d \quad (3.2)$$
$$\dot{x}_r = \dot{x}_d - \lambda e \quad (3.3)$$
$$r = \dot{e} + \lambda e \quad (3.4)$$

where $\lambda > 0$ and r is the filtered tracking error. Then the tracking error dynamics is transformed into

$$m\dot{r} = u - F - m\ddot{x}_r \quad (3.5)$$

3.2 Static Friction Model Based Adaptive Controller Design

Suppose that the friction model is given by the static LIP model (2.21), i.e.,

$$F(x,\dot{x}) = S^T(x,\dot{x})P^* + \epsilon \quad (3.6)$$

where P^* is the optimal coefficient vector and ϵ is the bounded modelling error. Then the adaptive controller can be designed based on the following description.

Substituting LIP friction model (3.6) into (3.5), we have

$$m\dot{r} = u - S^T(x,\dot{x})P^* - \epsilon - m\ddot{x}_r \qquad (3.7)$$

Consider the adaptive controller given by

$$u = -k_1 r + \hat{m}\ddot{x}_r + \hat{F}_m(x,\dot{x}) - k_i \int_0^t r d\tau - u_r \qquad (3.8)$$

where $k_1 > 0$, $\hat{F}_m(x,\dot{x}) = S^T(x,\dot{x})\hat{P}$, and u_r is a robust control term for suppressing any modelling uncertainty. For now, let us consider $u_r = k_2 \text{sgn}(r)$ with $k_2 \geq \epsilon$.

Substituting (3.8) into (3.7), we have the closed-loop error dynamics

$$\begin{aligned} m\dot{r} &= -k_1 r - u_r - k_i \int_0^t r d\tau - \tilde{m}\ddot{x}_r - S^T(x,\dot{x})\tilde{P} + \epsilon \\ &= -k_1 r - u_r - k_i \int_0^t r d\tau - \psi^T \tilde{\theta} + \epsilon \end{aligned} \qquad (3.9)$$

where $(\tilde{*}) = (*) - (\hat{*})$, $\psi^T = [\ddot{x}_r, S^T(x,\dot{x})]$ and $\tilde{\theta} = [\tilde{m}, \tilde{P}]^T$.

The closed-loop stability properties are then summarized in Theorem 3.1.

Theorem 3.1. *Consider the closed–loop system consisting of (3.1) with LIP static friction given by (2.21) and adaptive controller (3.8). If the parameters $\hat{\theta}$ are updated by*

$$\dot{\hat{\theta}} = -\Gamma \psi r \qquad (3.10)$$

where $\Gamma^T = \Gamma > 0$, then the tracking error converges to zero and all the signals in the closed loop are bounded.

Proof Consider the Lyapunov function candidate

$$V = \frac{1}{2}mr^2 + \frac{1}{2}\tilde{\theta}^T \Gamma^{-1}\tilde{\theta} + \frac{1}{2}k_i \left(\int_0^t r d\tau\right)^2 \qquad (3.11)$$

Its time derivative is given by

$$\dot{V} = mr\dot{r} + \tilde{\theta}^T \Gamma^{-1}\dot{\tilde{\theta}} + rk_i \int_0^t r d\tau \qquad (3.12)$$

Substituting (3.9) into (3.12) leads to

$$\begin{aligned} \dot{V} &= r(\epsilon - \psi^T\tilde{\theta} - k_1 r - u_r) + \tilde{\theta}^T \Gamma^{-1}\dot{\tilde{\theta}} \\ &= -k_1 r^2 + r(\epsilon - u_r) - r\psi^T\tilde{\theta} + \tilde{\theta}^T\Gamma^{-1}\dot{\tilde{\theta}} \end{aligned} \qquad (3.13)$$

Noting that $(\dot{\tilde{*}}) = -(\dot{\hat{*}})$ and substituting adaptation law (3.10) into (3.13), we have

$$\dot{V} = -k_1 r^2 + r(\epsilon - u_r) \qquad (3.14)$$

Since $u_r = k_2 \text{sgn}(r)$ and $k_2 \geq |\epsilon|$, we have $\dot{V} = -k_1 r^2 \leq 0$. It follows that $0 \leq V(t) \leq V(0)$, $\forall t \geq 0$. Hence $V(t) \in L_\infty$, which implies that $\tilde{\theta}$ is bounded. In other words, $\hat{\theta}$ is bounded for θ constant although unknown. Since $r \in L_2$, $e \in L_2 \cap L_\infty$, e is continuous and $e \to 0$ as $t \to \infty$, and $\dot{e} \in L_2$. By noting that $r \in L_2$, $x_d, \dot{x}_d, \ddot{x}_d \in L_\infty$, and ψ is of bounded functions, it is concluded that $\dot{r} \in L_\infty$ from equation (3.9). Using the fact that $r \in L_2$ and $\dot{r} \in L_\infty$, thus $r \to 0$ as $t \to \infty$. Hence $\dot{e} \to 0$ as $t \to \infty$. ◇

With regard to the implementation issues, we make the following remarks.

Remark 3.1. The presence of sgn(·) function in the sliding mode control inevitably introduces chattering, which is undesirable as it may excite mechanical resonance and cause mechanical wear and tear. To alleviate this problem, many approximation mechanisms have been used, such as boundary layer, saturation functions [20], and hyperbolic tangent function, tanh(·), which has the following property [31]

$$0 \leq |\alpha| - \alpha \tanh(\frac{\alpha}{\varepsilon}) \leq 0.2785\varepsilon, \quad \forall \alpha \in R \tag{3.15}$$

By smoothing the sgn(·) function, although asymptotic stability can no longer be guaranteed, the closed-loop system is still stable but with a small residue error. For example, if $u_r = k_2 \tanh(r/\epsilon_r)$, where $\epsilon_r > 0$ is a constant, and $k_2 \geq |\epsilon|$, then (3.14) becomes

$$\begin{aligned}
\dot{V} &= -k_1 r^2 + r(\epsilon - u_r) \\
&\leq -k_1 r^2 + |r||\epsilon| - rk_2 \tanh(\frac{r}{\epsilon_r}) \\
&\leq -k_1 r^2 + |r|k_2 - rk_2 \tanh(\frac{r}{\epsilon_r})
\end{aligned} \tag{3.16}$$

Using (3.15), (3.16) can be further simplified to

$$\dot{V} \leq -k_1 r^2 + 0.2785 \epsilon_r k_2 \tag{3.17}$$

Obviously, $\dot{V} \leq 0$ whenever r is outside the compact set

$$D = \{r | r^2 \leq \frac{0.2785 \epsilon_r k_2}{k_1}\} \tag{3.18}$$

Thus, we can conclude that the closed-loop system is stable and the tracking error will converge to a small neighborhood of zero, whose size is adjustable by the design parameters k_1 and ϵ_r.

It should be mentioned that these modification may cause the estimated parameters to grow unboundedly because asymptotic tracking cannot be guaranteed. To deal with this problem, the σ-modification scheme or e-modification, and among others [24], can be used to modify the adaptive laws to guarantee the robustness of the closed-loop system in the presence of approximation errors. For example, $\hat{\theta}$ can be adaptively tuned by

$$\dot{\hat{\theta}} = -\Gamma \phi r - \sigma \hat{\theta} \tag{3.19}$$

where $\sigma > 0$. The additional σ term in (3.19) ensures the boundedness of $\hat{\theta}$ when the system is subject to bounded disturbances without any additional prior information about the plant. The drawback is that tracking errors may only be made arbitrarily small rather than zero.

Remark 3.2. In this paper, only Gaussian RBF neural networks are discussed. If fact, other neural networks can also be used without any difficulty, and include other RBF neural networks, high order neural networks, and multilayer neural networks [26, 36].

3.3 LuGre Friction Model Based Adaptive Controller Design

In this subsection, the dynamic LuGre friction model based adaptive controller design will be presented. For the dynamic friction model (2.29-2.30), the uncertainties can be divided into two groups: (i) the unknown nonlinear friction characteristic function $\alpha(x, \dot{x})$; and (ii) the friction parameters σ_0, σ_1 and σ_2 which are assumed to change nonuniformly. Since $\alpha(x, \dot{x})$ is very hard to model, neural networks (NN) offer a possible tool for nonlinear mapping approximation. For simplicity, RBF NNs shall be used for function approximation though other neural networks can also be used without any difficulty.

In the following, we shall present three novel adaptive compensation schemes for dynamic friction under different situations: (i) unknown friction parameters $\sigma_0, \sigma_1, \sigma_2$ and unknown inertia parameter m; (ii) unknown nonlinear function $\alpha(x, \dot{x})$; and (iii) unknown full set of related parameters and nonlinear function.

Substituting friction dynamics (2.29) into (3.5), we have the following tracking error dynamics

$$m\dot{r} = u - (\sigma_1 + \sigma_2)\dot{x} - \sigma_0 z + \alpha(x, \dot{x})\sigma_1|\dot{x}|z - m\ddot{x}_r \tag{3.20}$$

3.3.1 Controller design for unknown system parameters

Firstly, let us investigate the case where the system parameter m and dynamic friction parameters $\sigma_0, \sigma_1, \sigma_2$ are unknown while nonlinear function $\alpha(x, \dot{x})$ is known.

Motivated by the work in [32], the following dual-observer is adopted to estimate the unmeasurable friction state z:

$$\dot{\hat{z}}_0 = \dot{x} - \alpha(x, \dot{x})|\dot{x}|\hat{z}_0 - r \tag{3.21}$$
$$\dot{\hat{z}}_1 = \dot{x} - \alpha(x, \dot{x})|\dot{x}|\hat{z}_1 + \alpha(x, \dot{x})|\dot{x}|r \tag{3.22}$$

where $\hat{z}_0$ and $\hat{z}_1$ are estimates of the friction state z. Combining with equation (2.30), the corresponding observation errors are given by

$$\dot{\hat{z}}_0 = -\alpha(x,\dot{x})|\dot{x}|\hat{z}_0 + r \quad (3.23)$$
$$\dot{\hat{z}}_1 = -\alpha(x,\dot{x})|\dot{x}|\hat{z}_1 - \alpha(x,\dot{x})|\dot{x}|r \quad (3.24)$$

where $\tilde{z}_0 = z - \hat{z}_0$ and $\tilde{z}_1 = z - \hat{z}_1$.

Remark 3.3. The two friction state observers are introduced here for different purposes. Observers (3.21) and (3.22) driven by $-r$ and $\alpha(x,\dot{x})|\dot{x}|r$ are introduced to cancel the cross-coupling terms $-\sigma_0 r\tilde{z}_0$ and $\alpha(x,\dot{x})\sigma_1|\dot{x}|r\tilde{z}_1$ respectively in (3.34), the derivative of the Lyapunov function candidate, for closed-loop stability. Due to the fact that the filtered tracking error r asymptotically converges to zero as will be proven in Theorem 3.2, it is then expected that the output $\hat{z}_0$ of observer (3.21) will reveal the actual friction state z. Though we can prove that $\hat{z}_1$ is bounded in Theorem 3.2, we could not claim that it is small. In fact, since $\alpha(x,\dot{x})$ can take a very large value, a small error in r will result in a large driving value in (3.24), accordingly $\hat{z}_1$ may become very large in comparison with z as will be verified by a simulation study later.

Let $\hat{\sigma}_0, \hat{\sigma}_1, \hat{\sigma}_2$ and $\hat{m}$ be the estimates of unknown friction parameters $\sigma_0, \sigma_1, \sigma_2$ and inertia parameter m. Consider the adaptive controller u given by

$$u = -c_1 r + \hat{\sigma}_1 \dot{x} + \hat{\sigma}_2 \dot{x} + \hat{\sigma}_0 \hat{z}_0 - \alpha(x,\dot{x})\hat{\sigma}_1|\dot{x}|\hat{z}_1 + \hat{m}\ddot{x}_r \quad (3.25)$$

where c_1 is a positive design constant.

Substituting (3.25) into (3.20), we have the closed-loop error dynamics

$$m\dot{r} = -c_1 r - \tilde{\sigma}_1 \dot{x} - \tilde{\sigma}_2 \dot{x} - \sigma_0 \tilde{z}_0 - \tilde{\sigma}_0 \hat{z}_0 + \alpha(x,\dot{x})\sigma_1|\dot{x}|\tilde{z}_1$$
$$+\alpha(x,\dot{x})\tilde{\sigma}_1|\dot{x}|\hat{z}_1 - \tilde{m}\ddot{x}_r \quad (3.26)$$

where $(\tilde{*}) = (*) - (\hat{*})$ are the parameter estimates errors.

Theorem 3.2. *Consider the closed system consisting of system (3.1) with dynamic friction given by (2.29) and (2.30), adaptive controller (3.25) and dual-observer given by (3.21) and (3.22). If the parameters $\hat{\sigma}_0, \hat{\sigma}_1, \hat{\sigma}_2$ and $\hat{m}$ are updated by*

$$\dot{\hat{\sigma}}_0 = -\eta_0 \hat{z}_0 r \quad (3.27)$$
$$\dot{\hat{\sigma}}_1 = -\eta_1 (r\dot{x} - \alpha(x,\dot{x})|\dot{x}|r\hat{z}_1) \quad (3.28)$$
$$\dot{\hat{\sigma}}_2 = -\eta_2 \dot{x} r \quad (3.29)$$
$$\dot{\hat{m}} = -\eta_m \ddot{x}_r r \quad (3.30)$$

where η_0, η_1, η_2 and η_m are positive constants, then the tracking error converges to zero and all the signals in the closed loop are bounded.

Proof Consider the Lyapunov function candidate

$$V = \frac{1}{2}mr^2 + \frac{1}{2}\sigma_0 \tilde{z}_0^2 + \frac{1}{2}\sigma_1 \tilde{z}_1^2 + \frac{1}{2\eta_0}\tilde{\sigma}_0^2 + \frac{1}{2\eta_1}\tilde{\sigma}_1^2 + \frac{1}{2\eta_2}\tilde{\sigma}_2^2 + \frac{1}{2\eta_m}\tilde{m}^2 \quad (3.31)$$

where $\eta_0, \eta_1, \eta_2, \eta_m$ are positive design constants.

Evaluating the time derivative of V along (3.26), we obtain

$$\begin{aligned}\dot{V} &= mr\dot{r} + \sigma_0 \tilde{z}_0 \dot{\tilde{z}}_0 + \sigma_1 \tilde{z}_1 \dot{\tilde{z}}_1 + \frac{1}{\eta_0}\tilde{\sigma}_0 \dot{\tilde{\sigma}}_0 + \frac{1}{\eta_1}\tilde{\sigma}_1 \dot{\tilde{\sigma}}_1 + \frac{1}{\eta_2}\tilde{\sigma}_2 \dot{\tilde{\sigma}}_2 + \frac{1}{\eta_m}\tilde{m}\dot{\tilde{m}} \\ &= -c_1 r^2 - \tilde{\sigma}_1 r\dot{x} - \tilde{\sigma}_2 r\dot{x} - \sigma_0 r\hat{z}_0 - \tilde{\sigma}_0 r\hat{z}_0 \\ &\quad + \alpha(x,\dot{x})\sigma_1|\dot{x}|r\tilde{z}_1 + \alpha(x,\dot{x})\tilde{\sigma}_1|\dot{x}|r\hat{z}_1 - \tilde{m}r\ddot{x}_r + \sigma_0 \tilde{z}_0 \dot{\tilde{z}}_0 \\ &\quad + \sigma_1 \tilde{z}_1 \dot{\tilde{z}}_1 + \frac{1}{\eta_0}\tilde{\sigma}_0 \dot{\tilde{\sigma}}_0 + \frac{1}{\eta_1}\tilde{\sigma}_1 \dot{\tilde{\sigma}}_1 + \frac{1}{\eta_2}\tilde{\sigma}_2 \dot{\tilde{\sigma}}_2 + \frac{1}{\eta_m}\tilde{m}\dot{\tilde{m}} \end{aligned} \quad (3.32)$$

Re-arranging the related items in (3.32), we have

$$\begin{aligned}\dot{V} &= -c_1 r^2 - \tilde{\sigma}_0(r\hat{z}_0 - \frac{1}{\eta_0}\dot{\tilde{\sigma}}_0) - \tilde{\sigma}_1(r\dot{x} - \alpha(x,\dot{x})|\dot{x}|r\hat{z}_1 - \frac{1}{\eta_1}\dot{\tilde{\sigma}}_1) \\ &\quad - \tilde{\sigma}_2(r\dot{x} - \frac{1}{\eta_2}\dot{\tilde{\sigma}}_2) - \tilde{m}(r\ddot{x}_r - \frac{1}{\eta_m}\dot{\tilde{m}}) - \sigma_0 r\tilde{z}_0 \\ &\quad + \alpha(x,\dot{x})\sigma_1|\dot{x}|r\tilde{z}_1 + \sigma_0 \tilde{z}_0 \dot{\tilde{z}}_0 + \sigma_1 \tilde{z}_1 \dot{\tilde{z}}_1 \end{aligned} \quad (3.33)$$

Noting that $(\dot{\tilde{*}}) = -(\dot{\hat{*}})$ and substituting adaptation laws (3.27-3.30) into (3.33) yields

$$\dot{V} = -c_1 r^2 - \sigma_0 r\tilde{z}_0 + \alpha(x,\dot{x})\sigma_1|\dot{x}|r\tilde{z}_1 + \sigma_0 \tilde{z}_0 \dot{\tilde{z}}_0 + \sigma_1 \tilde{z}_1 \dot{\tilde{z}}_1 \quad (3.34)$$

In order to cancel the terms $-\sigma_0 r\tilde{z}_0$ and $\alpha(x,\dot{x})\sigma_1|\dot{x}|r\tilde{z}_1$ in (3.34), substituting the observer error dynamics (3.23) and (3.24) into (3.34), we obtain

$$\begin{aligned}\dot{V} &= -c_1 r^2 - \sigma_0 r\tilde{z}_0 + \alpha(x,\dot{x})\sigma_1|\dot{x}|r\tilde{z}_1 + \sigma_0 \tilde{z}_0(-\alpha(x,\dot{x})|\dot{x}|\tilde{z}_0 + r) \\ &\quad + \sigma_1 \tilde{z}_1(-\alpha(x,\dot{x})|\dot{x}|\tilde{z}_1 - \alpha(x,\dot{x})|\dot{x}|r) \\ &= -c_1 r^2 - \sigma_0 \alpha(x,\dot{x})|\dot{x}|\tilde{z}_0^2 - \sigma_1 \alpha(x,\dot{x})|\dot{x}|\tilde{z}_1^2 \end{aligned} \quad (3.35)$$

Obviously, the following inequality holds

$$\dot{V} = -c_1 r^2 - \sigma_0 \alpha(x,\dot{x})|\dot{x}|\tilde{z}_0^2 - \sigma_1 \alpha(x,\dot{x})|\dot{x}|\tilde{z}_1^2 \leq 0 \quad (3.36)$$

for c_1, σ_0, σ_1 and $\alpha(\cdot)$ are positive.

From the definition of Lyapunov function V in (3.31) and $\dot{V} \leq 0$, the global uniform boundedness of the tracking error r, the observer errors $\tilde{z}_0, \tilde{z}_1$, and the parameter estimation errors $\tilde{\sigma}_0, \tilde{\sigma}_1, \tilde{\sigma}_2, \tilde{m}$ are guaranteed. Obviously, the estimates $\hat{\sigma}_0, \hat{\sigma}_1, \hat{\sigma}_2$ and $\hat{m}$ are bounded. From the definition of r and Assumption 3.2, it can also be concluded that the tracking error e is bounded. The boundedness of the control u is apparent from (3.25). Since $r \in L_2, e \in L_2 \cap L_\infty$, e is continuous and $e \to 0$ as $t \to \infty$, and $\dot{e} \in L_2$. By noting that $r \in L_2$ and $x_d, \dot{x}_d, \ddot{x}_d \in L_\infty$, it is concluded that $\dot{r} \in L_\infty$ from equation (3.26). Using the fact that $r \in L_2$ and $\dot{r} \in L_\infty$, thus $r \to 0$ as $t \to 0$. Hence $\dot{e} \to 0$ as $t \to 0$. $\diamondsuit$

3.3.2 Controller design for unknown $\alpha(x, \dot{x})$

Assuming that the parameters $m, \sigma_0, \sigma_1, \sigma_2$ are known while the nonlinear function $\alpha(x, \dot{x})$ is unknown. Note that the parameterization given in (2.28) is not exclusive. Rather than finding another analytical description of $\alpha(x, \dot{x})$ for better or for worse, we shall use the RBF NN $\hat{\alpha}(x, \dot{x}) = W^T S(x, \dot{x})$ to approximate the unknown $\alpha(x, \dot{x})$. Since $\alpha(x, \dot{x})$ is a continuous function, according to the general approximation ability of neural networks [20], the following function approximation holds over a compact set $\Omega \subset R^2$

$$\alpha(x, \dot{x}) = W^{*T} S(x, \dot{x}) + \epsilon, \quad \forall (x, \dot{x}) \in \Omega \tag{3.37}$$

where W^* is the optimal weight vector.

Assumption 3.3 *The neural network approximation error ϵ is bounded over the compact set Ω, i.e., $|\epsilon| \leq \epsilon_b$, where ϵ_b is a small positive constant.*

In this case, since $\alpha(x, \dot{x})$ is unknown, observers (3.21) and (3.22) cannot be used. Therefore, an auxiliary variable $\delta = z + \frac{m}{\sigma_1} r$ is defined to implement the estimate for the internal friction state z. Its derivative is then given by

$$\dot{\delta} = \dot{z} + \frac{m}{\sigma_1}\dot{r} = (\dot{x} - \alpha(x, \dot{x})|\dot{x}|z) + \frac{m}{\sigma_1}\dot{r} \tag{3.38}$$

Substituting (3.20) and $z = \delta - \frac{m}{\sigma_1} r$ into (3.38) yields

$$\dot{\delta} = \dot{x} - \frac{\sigma_0}{\sigma_1}\delta + \frac{m\sigma_0}{\sigma_1^2}r + \frac{1}{\sigma_1}(u - (\sigma_1 + \sigma_2)\dot{x} - m\ddot{x}_r) \tag{3.39}$$

To estimate the unmeasurable friction state δ, consider the following dual-observer

$$\dot{\hat{\delta}}_1 = \dot{x} - \frac{\sigma_0}{\sigma_1}\hat{\delta}_1 + \frac{m\sigma_0}{\sigma_1^2}r + \frac{1}{\sigma_1}(u - (\sigma_1 + \sigma_2)\dot{x} - m\ddot{x}_r) - r \tag{3.40}$$

$$\dot{\hat{\delta}}_2 = \dot{x} - \frac{\sigma_0}{\sigma_1}\hat{\delta}_2 + \frac{m\sigma_0}{\sigma_1^2}r + \frac{1}{\sigma_1}(u - (\sigma_1 + \sigma_2)\dot{x} - m\ddot{x}_r) + \sigma_1 r \dot{x} \tag{3.41}$$

where $\hat{\delta}_1$ and $\hat{\delta}_2$ are the estimates of friction state δ. Combining (3.39) with the dual-observer, we can obtain the internal friction state error dynamics as follows:

$$\dot{\tilde{\delta}}_1 = -\frac{\sigma_0}{\sigma_1}\tilde{\delta}_1 + r \tag{3.42}$$

$$\dot{\tilde{\delta}}_2 = -\frac{\sigma_0}{\sigma_2}\tilde{\delta}_2 - \sigma_1 r \dot{x} \tag{3.43}$$

where $\tilde{\delta}_1 = \delta - \hat{\delta}_1$ and $\tilde{\delta}_2 = \delta - \hat{\delta}_2$.

Remark 3.4. Similarly, two friction state observers are introduced here for different purposes. Observers (3.40) and (3.41) driven by $-r$ and $\sigma_1 r \dot{x}$ are introduced to cancel the cross-coupling terms $-\sigma_0 r \tilde{\delta}_1$ and $\sigma_1 r \alpha(x, \dot{x})|\dot{x}|\tilde{\delta}_2$ respectively in (3.50), the derivative of the Lyapunov function candidate, for closed-loop stability. Due to the fact that the filtered tracking error r asymptotically converges to zero as will be proven in Theorem 3.3, it is then expected that the output $\hat{\delta}_1$ of observer (3.40) will reveal the actual friction state δ. Though we can prove that $\hat{\delta}_2$ is bounded in Theorem 3.3, we could not claim that it is small. In fact, since σ_1 can take a large value, a small error in r will result in a large driving value in (3.43), accordingly $\hat{\delta}_2$ may become large in comparison with δ as will be verified by a simulation study later.

Using $z = \delta - \frac{m}{\sigma_1} r$, equation (3.20) can be rewritten as

$$m\dot{r} = -\sigma_0 \delta + \sigma_1 \alpha(x, \dot{x})|\dot{x}|\delta + \frac{m\sigma_0}{\sigma_1} r - m\alpha(x, \dot{x})|\dot{x}|r$$
$$+ u - (\sigma_1 + \sigma_2)\dot{x} - m\ddot{x}_r \qquad (3.44)$$

Consider the control law

$$u = (\sigma_1 + \sigma_2)\dot{x} + m\ddot{x}_r + u_{ar} \qquad (3.45)$$

where u_{ar} is the adaptive robust control term given by

$$u_{ar} = -c_1 r + \sigma_0 \hat{\delta}_1 - \sigma_1 \hat{\alpha}(x, \dot{x})|\dot{x}|\hat{\delta}_2 - \frac{m\sigma_0}{\sigma_1} r - \frac{\sigma_1^2 \sigma_2 \alpha_m r \dot{x}^2}{\sigma_0}$$
$$- k\sigma_1 |\dot{x}||\hat{\delta}_2|\operatorname{sgn}(r) \qquad (3.46)$$

with $\hat{\alpha}(x, \dot{x}) = W^T S(x, \dot{x})$, constants $c_1 > 0$ and $k \geq \epsilon_b$.

Substituting (3.46) into (3.44), we have

$$m\dot{r} = -c_1 r - \sigma_0 \tilde{\delta}_1 + \sigma_1 \alpha(x, \dot{x})|\dot{x}|\tilde{\delta}_2 + \sigma_1 (\tilde{W}^T S(x, \dot{x}) + \epsilon)|\dot{x}|\hat{\delta}_2$$
$$- m\alpha(x, \dot{x})|\dot{x}|r - \frac{\sigma_1^2 \sigma_2 \alpha_m r \dot{x}^2}{\sigma_0} - k\sigma_1 |\dot{x}||\hat{\delta}_2|\operatorname{sgn}(r) \qquad (3.47)$$

where $\tilde{W} = W^* - W$.

Remark 3.5. The presence of k is for robust closed-loop stability because of the existence of $\epsilon \neq 0$. The particular choice of the dynamic gain, $k\sigma_1 |\hat{\delta}_2||\dot{x}|$, has the following advantages: (i) the value of k needs only to be large enough to suppress the bounded approximation error ϵ, and (ii) the gain is a function of $|\hat{\delta}_2||\dot{x}|$ so that it is zero whenever $|\hat{\delta}_2| = 0$ and/or $|\dot{x}| = 0$, and decreases as $|\hat{\delta}_2||\dot{x}|$ diminishes.

Theorem 3.3. *Consider the closed-loop system consisting of system (3.1) with dynamic friction given by (2.29) and (2.30), adaptive controller (3.45) and (3.46), and dual-observer (3.40) and (3.41). If the weights of NN approximation to $\alpha(x, \dot{x})$ are updated by*

$$\dot{W} = \Gamma S(x, \dot{x}) \sigma_1 |\dot{x}| \hat{\delta}_2 r \qquad (3.48)$$

where $\Gamma = \Gamma^T > 0$ is a dimensionally compatible constant matrix, then the tracking error converges to zero and all the signals in the closed loop are bounded.

Proof Consider the Lyapunov function candidate

$$V = \frac{1}{2}mr^2 + \frac{1}{2}\sigma_0\tilde{\delta}_1^2 + \frac{1}{2}\alpha_m\tilde{\delta}_2^2 + \frac{1}{2}\tilde{W}^T\Gamma^{-1}\tilde{W} \tag{3.49}$$

The time derivative of V along (3.47) is

$$\begin{aligned}
\dot{V} &= mr\dot{r} + \sigma_0\tilde{\delta}_1\dot{\tilde{\delta}}_1 + \alpha_m\tilde{\delta}_2\dot{\tilde{\delta}}_2 + \tilde{W}\Gamma^{-1}\dot{\tilde{W}} \\
&= -c_1r^2 - \sigma_0 r\tilde{\delta}_1 + \sigma_1 r\alpha(x,\dot{x})|\dot{x}|\tilde{\delta}_2 + \sigma_1 r(\tilde{W}^T S(x,\dot{x}) + \epsilon)|\dot{x}|\hat{\delta}_2 \\
&\quad -m\alpha(x,\dot{x})|\dot{x}|r^2 - \frac{\sigma_1^2\sigma_2\alpha_m r^2\dot{x}^2}{\sigma_0} - k\sigma_1|\dot{x}||\hat{\delta}_2|\mathrm{rsgn}(r) + \sigma_0\tilde{\delta}_1\dot{\hat{\delta}}_1 \\
&\quad +\alpha_m\tilde{\delta}_2\dot{\hat{\delta}}_2 + \tilde{W}^T\Gamma^{-1}\dot{\tilde{W}} \tag{3.50}
\end{aligned}$$

Substituting the observer error dynamics (3.42) and (3.43) into (3.50) and noting $\alpha(x,\dot{x}) \leq \alpha_m$, we have

$$\begin{aligned}
\dot{V} &= -c_1r^2 - m\alpha(x,\dot{x})|\dot{x}|r^2 - \frac{\sigma_0^2}{\sigma_1}\tilde{\delta}_1^2 - \frac{\alpha_m\sigma_0}{\sigma_2}\tilde{\delta}_2^2 + \sigma_1\alpha(x,\dot{x})r|\dot{x}|\tilde{\delta}_2 \\
&\quad -\sigma_1\alpha_m r\dot{x}\tilde{\delta}_2 - \frac{\sigma_1^2\sigma_2\alpha_m r^2\dot{x}^2}{\sigma_0} + \sigma_1\epsilon|\dot{x}|r\hat{\delta}_2 - k\sigma_1|\dot{x}||\hat{\delta}_2|\mathrm{rsgn}(r) \\
&\quad +\tilde{W}^T(\sigma_1 rS(x,\dot{x})|\dot{x}|\hat{\delta}_2 + \Gamma^{-1}\dot{\tilde{W}}) \\
&\leq -c_1r^2 - m\alpha(x,\dot{x})|\dot{x}|r^2 - \frac{\sigma_0^2}{\sigma_1}\tilde{\delta}_1^2 - \frac{\alpha_m\sigma_0}{\sigma_2}\tilde{\delta}_2^2 \\
&\quad +2\sigma_1\alpha_m|r||\dot{x}||\tilde{\delta}_2| - \frac{\sigma_1^2\sigma_2\alpha_m r^2\dot{x}^2}{\sigma_0} + \sigma_1|\epsilon||\dot{x}||r||\hat{\delta}_2| \\
&\quad -k\sigma_1|\dot{x}||\hat{\delta}_2|\mathrm{rsgn}(r) + \tilde{W}^T(\sigma_1 rS(x,\dot{x})|\dot{x}|\hat{\delta}_2 + \Gamma^{-1}\dot{\tilde{W}}) \tag{3.51}
\end{aligned}$$

Using the fact that

$$\begin{aligned}
2\sigma_1\alpha_m|r||\dot{x}||\tilde{\delta}_2| &\leq \frac{(2\sigma_1\alpha_m|r||\dot{x}|)^2}{4\alpha_m\sigma_0/\sigma_2} + \frac{\alpha_m\sigma_0}{\sigma_2}\tilde{\delta}_2^2 \\
&= \frac{\sigma_1^2\sigma_2\alpha_m r^2\dot{x}^2}{\sigma_0} + \frac{\alpha_m\sigma_0}{\sigma_2}\tilde{\delta}_2^2 \tag{3.52}
\end{aligned}$$

and substituting adaptation law (3.48) into (3.51), and noting $|\epsilon| \leq k$, we have the following inequality

$$\dot{V} \leq -c_1r^2 - m\alpha(x,\dot{x})|\dot{x}|r^2 - \frac{\sigma_0^2}{\sigma_1}\tilde{\delta}_1^2 \leq 0 \tag{3.53}$$

for c_1, σ_0, σ_1 and $\alpha(\cdot)$ are positive.

From the definition of Lyapunov function V in (3.49) and $\dot{V} \leq 0$, the global uniform boundedness of the tracking error r, the observer errors $\tilde{\delta}_1, \tilde{\delta}_2$, and the parameter estimation errors $\tilde{W}$ are guaranteed. From the definition of r and Assumption 3.2, it can also be concluded that the tracking error e is bounded. The boundedness of control u is apparent from (3.45).

Since $r \in L_2, e \in L_2 \cap L_\infty$, e is continuous and $e \to 0$ as $t \to \infty$, and $\dot{e} \in L_2$. By noting that $r \in L_2$ and $x_d, \dot{x}_d, \ddot{x}_d \in L_\infty$, it is concluded that $\dot{r} \in L_\infty$ from equation (3.47). Using the fact that $r \in L_2$ and $\dot{r} \in L_\infty$, thus $r \to 0$ as $t \to 0$. Hence $\dot{e} \to 0$ as $t \to 0$. $\diamond$

Remark 3.6. It is well known that when using dynamic friction models, controller design becomes difficult because (i) the friction parameters appear in a nonlinear fashion, and (ii) the systems's internal state z, which depends on the unknown parameters, is not measurable. One of the main contributions here is the introduction of an easily implementable approach to cope with the completely unknown $\alpha(x, \dot{x})$ by using the concept of dual-observer and neural network parameterization.

3.3.3 Controller design for full set of unknown parameters

In practice, friction compensation becomes more challenging when the related parameters and functions of the system are unknown. In [21], neural network parameterization was presented for the completely unknown static friction model. Through intensive computer simulation study, it was found that the proposed controller is effective in handling dynamic friction. By separating the dynamic friction into a static function of velocity with a dynamic perturbation in friction, and using a convex/concave parameterization, an adaptive nonlinear control scheme was presented in [27] to achieve dynamic friction compensation. In [29], using the boundedness property of the internal friction state $z(t)$, a simple linearly parameterized friction model was first presented and then adaptive control was developed based on the simplified model. In this section, motivated by the work in [21, 29], the dynamic friction is first separated into two parts: (i) the viscous friction with unknown constant coefficient, and (ii) the unknown dynamic friction which is a function of the unmeasurable internal friction state $z(t)$ and is bounded by a function which is independent of $z(t)$. Then an RBF NN is applied to approximate this unknown bounding function. Based on Lyapunov synthesis, adaptation algorithms for both the NN weights and the unknown system and friction parameters are presented.

In particular, the dynamic friction (2.29) can be further written as

$$\begin{aligned} F &= \sigma_1 \dot{x} + \sigma_2 \dot{x} + \sigma_0 z - \sigma_1 \alpha(x, \dot{x}) |\dot{x}| z \\ &= \theta \dot{x} + F_z(x, \dot{x}, z) \end{aligned} \quad (3.54)$$

where $\theta = \sigma_1 + \sigma_2$, $\theta \dot{x}$ represents the viscous friction force and $F_z(x, \dot{x}, z) = \sigma_0 z - \sigma_1 \alpha(x, \dot{x}) |\dot{x}| z$ is the dynamic friction force which depends on z.

From Lemma 2.1, we know that F_z is bounded by

$$|F_z(x,\dot{x},z)| = |(\sigma_0 - \sigma_1\alpha(x,\dot{x}))||z(t)| \leq \frac{\sigma_0 + \sigma_1\alpha(x,\dot{x})}{\alpha_{min}} = F_{zm}(x,\dot{x}) \quad (3.55)$$

where $F_{zm}(x,\dot{x})$ is the bounding function of $F_z(x,\dot{x},z)$ and is independent of the unmeasurable internal friction state z. An RBF NN can be applied to approximate $F_{zm}(x,\dot{x})$, and similarly there exists the following function approximation

$$F_{zm}(x,\dot{x}) = W^{*T}S(x,\dot{x}) + \epsilon \quad (3.56)$$

with W^* being the optimal weight vector, and the NN approximation error ϵ being bounded by a small positive constant ϵ_d, i.e., $|\epsilon| \leq \epsilon_d$.

System tracking error dynamics (3.20) can be rewritten as

$$\begin{aligned} m\dot{r} &= u - (\sigma_1 + \sigma_2)\dot{x} - \sigma_0 z + \alpha(x,\dot{x})\sigma_1|\dot{x}|z - m\ddot{x}_r \\ &= u - \theta\dot{x} - F_z(x,\dot{x},z) - m\ddot{x}_r \quad (3.57) \end{aligned}$$

Consider the following controller

$$u = -c_1 r + \hat{\theta}\dot{x} + \hat{m}\ddot{x}_r - \hat{F}_{zm}(x,\dot{x})\text{sgn}(r) - k\text{sgn}(r) \quad (3.58)$$

where constant $c_1 > 0$, $\hat{\theta}$ and $\hat{m}$ are the estimates of unknown θ and m respectively, $\hat{F}_{zm}(x,\dot{x}) = W^T S(x,\dot{x})$ is the RBF NN approximation of function bound $F_{zm}(x,\dot{x})$, and $k > \epsilon_d$.

Substituting (3.58) into (3.57) yields

$$m\dot{r} = -c_1 r - \tilde{\theta}\dot{x} - \tilde{m}\ddot{x}_r - \hat{F}_{zm}(x,\dot{x})\text{sgn}(r) - k\text{sgn}(r) - F_z(x,\dot{x},z) \quad (3.59)$$

where $(\tilde{*}) = (*) - (\hat{*})$ denotes the unknown parameter estimation errors.

Adding and subtracting $F_{zm}(x,\dot{x})\text{sgn}(r)$ in (3.59) and noting equation (3.56), we have

$$\begin{aligned} m\dot{r} &= -c_1 r - \tilde{\theta}\dot{x} - \tilde{m}\ddot{x}_r + (\tilde{W}^T S(x,\dot{x}) + \epsilon)\text{sgn}(r) - F_z(x,\dot{x},z) \\ &\quad - F_{zm}(x,\dot{x})\text{sgn}(r) - k\text{sgn}(r) \\ &= -c_1 r - \tilde{\theta}\dot{x} - \tilde{m}\ddot{x}_r + \tilde{W}^T S(x,\dot{x})\text{sgn}(r) - F_z(x,\dot{x},z) \\ &\quad - F_{zm}(x,\dot{x})\text{sgn}(r) - (k - \epsilon)\text{sgn}(r) \quad (3.60) \end{aligned}$$

where $\tilde{W} = W^* - W$.

Theorem 3.4 *Consider the closed-loop system consisting of system (3.1) with dynamic friction given by (2.29) and (2.30), and adaptive controller (3.58). If the parameters $\hat{\theta}, \hat{m}$ and NN weight W are updated by*

$$\dot{\hat{\theta}} = -\eta_\theta \dot{x} r \quad (3.61)$$
$$\dot{\hat{m}} = -\eta_m \ddot{x}_r r \quad (3.62)$$
$$\dot{W} = \Gamma S(x,\dot{x})|r| \quad (3.63)$$

where η_θ and η_m are positive constants, $\Gamma = \Gamma^T > 0$ is a dimensionally compatible constant matrix, then the tracking error converges to zero and all the signals in the closed loop are bounded.

Proof Consider the candidate Lyapunov function candidate

$$V = \frac{1}{2}mr^2 + \frac{1}{2\eta_\theta}\tilde{\theta}^2 + \frac{1}{2\eta_m}\tilde{m}^2 + \frac{1}{2}\tilde{W}^T\Gamma^{-1}\tilde{W} \qquad (3.64)$$

The time derivative of V along (3.60) is

$$\begin{aligned}
\dot{V} &= m\dot{r}r + \frac{1}{\eta_\theta}\tilde{\theta}\dot{\tilde{\theta}} + \frac{1}{\eta_m}\tilde{m}\dot{\tilde{m}} + \tilde{W}^T\Gamma^{-1}\dot{\tilde{W}} \\
&= -c_1r^2 - \tilde{\theta}\dot{x}r - \tilde{m}\ddot{x}_r r + \tilde{W}^T S(x,\dot{x})r\text{sgn}(r) + \epsilon r\text{sgn}(r) \\
&\quad - F_z(x,\dot{x},z)r - F_{zm}(x,\dot{x})r\text{sgn}(r) - kr\text{sgn}(r) \\
&\quad + \frac{1}{\eta_\theta}\tilde{\theta}\dot{\tilde{\theta}} + \frac{1}{\eta_m}\tilde{m}\dot{\tilde{m}} + \tilde{W}^T\Gamma^{-1}\dot{\tilde{W}} \qquad (3.65)
\end{aligned}$$

Re-arranging the related items in (3.65) yields

$$\begin{aligned}
\dot{V} &= -c_1r^2 - \tilde{\theta}(\dot{x}r - \frac{1}{\eta_\theta}\dot{\tilde{\theta}}) - \tilde{m}(\ddot{x}_r r - \frac{1}{\eta_m}\dot{\tilde{m}}) + \tilde{W}^T(S(x,\dot{x})r\text{sgn}(r) \\
&\quad + \Gamma^{-1}\dot{\tilde{W}}) + \epsilon|r| - k|r| - F_z(x,\dot{x},z)r - F_{zm}(x,\dot{x})|r| \qquad (3.66)
\end{aligned}$$

Noting that $(\dot{\tilde{*}}) = -(\dot{\hat{*}})$ and substituting adaptation laws (3.61-3.63) into (3.66), we have

$$\dot{V} = -c_1r^2 + \epsilon|r| - k|r| - F_z(x,\dot{x},z)r - F_{zm}(x,\dot{x})|r| \qquad (3.67)$$

Because $|\epsilon| \leq k$ and $|F_z(x,\dot{x},z)| \leq F_{zm}(x,\dot{x})$, (3.67) can be further simplified to

$$\dot{V} \leq -c_1r^2 \leq 0 \qquad (3.68)$$

From the definition of Lyapunov function V in (3.64) and $\dot{V} \leq 0$, the global uniform boundedness of the tracking error r, the parameter estimation errors $\tilde{\theta}, \tilde{m}$, and the NN weight estimation error $\tilde{W}$ are guaranteed. Obviously, the estimates $\hat{\theta}, \hat{m}$ and weight W are bounded. From the definition of r and Assumption 3.2, it can also be concluded that the tracking error e is bounded. The boundedness of the control u is apparent from (3.58).

Since $r \in L_2, e \in L_2 \cap L_\infty$, e is continuous and $e \to 0$ as $t \to \infty$, and $\dot{e} \in L_2$. By noting that $r \in L_2$ and $x_d, \dot{x}_d, \ddot{x}_d \in L_\infty$, it is concluded that $\dot{r} \in L_\infty$ from equation (3.60). Using the fact that $r \in L_2$ and $\dot{r} \in L_\infty$, thus $r \to 0$ as $t \to 0$. Hence $\dot{e} \to 0$ as $t \to 0$. $\Diamond$

4 Simulation Studies

In this section, simulation studies are presented to illustrate the tracking performance of the proposed adaptive friction compensators under different cases respectively. For simplicity, the system parameter is chosen as $m = 1$. The dynamic friction model given by (2.27-2.28) is used in the simulation, and the parameters are chosen as [21, 27]

$$\sigma_0 = 10^5, \sigma_1 = \sqrt{10^5}, \sigma_2 = 0.4, \dot{x}_s = 0.001, f_c = 1 \text{ and } f_s = 1.5 \quad (4.1)$$

The control objective is to make the output $x(t)$ track a desired trajectory $x_d(t) = 0.5\sin(2\pi t)$. The initial states are $[x(0), \dot{x}(0)]^T = [0.1, 0]^T$. The filtered tracking error $r = \dot{e} + \lambda e$, $\lambda = 5$, $e = x - x_d$, and $\ddot{x}_r = \ddot{x}_d - \lambda\dot{e}$.

In the following, we shall show the tracking results for the case of LIP friction model based design and the case of dynamic LuGre friction model based design. First, we consider the case where the complex friction behavior is approximated by the LIP static friction model presented in Section 2.1, and present the performance of the corresponding model based adaptive control algorithms proposed in Section 3.2. Both model based and neural network based adaptive friction compensation results are comparatively illustrated. Secondly, based on the uncertainty levels with the dynamic LuGre model (2.29-2.30), the proposed adaptive friction compensation controllers in Section 3.3 are verified by the simulation results. Finally, some concluding remarks are given on the performance of the adaptive friction compensation schemes proposed in this chapter.

4.1 Static Model Based Adaptive Control

In this subsection, we shall study the performance of the static model based adaptive control. As shown in Section 2.1, basic friction behavior can be captured by conventional LIP static friction model (2.15) or NN based LIP static friction model (2.20).

4.1.1 Conventional LIP model based adaptive control
For analysis, the conventional LIP friction models used in the simulation are in the form $F(x, \dot{x}) = S^T(\dot{x})P$ with the following different choice of $S(\dot{x})$ and P:

(1) $S(\dot{x}) = \text{sgn}(\dot{x})$, $P = f_c$;
(2) $S(\dot{x}) = [\text{sgn}(\dot{x}), \dot{x}]^T$, $P = [f_c, f_v]^T$;
(3) $S(\dot{x}) = [\text{sgn}(\dot{x}), \dot{x}, |\dot{x}|\dot{x}]^T$, $P = [f_c, f_v, f_d]^T$;
(4) $S(\dot{x}) = [\text{sgn}(\dot{x}), \dot{x}, |\dot{x}|\dot{x}, |\dot{x}|^{1/2}\text{sgn}(\dot{x})]^T$, $P = [f_c, f_v, f_d, f_r]^T$.

The corresponding adaptive controller is given by (3.8), i.e.,

$$u = -k_1 r + \hat{m}\ddot{x}_r + S^T(\dot{x})P - k_i \int_0^t r d\tau - u_r \quad (4.2)$$

where $k_i = 0$, $u_r = k_2 \text{sgn}(r)$ with $k_2 = 2$, and $k_1 = 10$ for cases 1 and 2 while $k_1 = 80$ for cases 3 and 4. The adaptation gain is chosen as $\Gamma = \text{diag}[0.1]$ for cases 1 and 2, and $\Gamma = \text{diag}[10]$ for cases 3 and 4. The tracking performance is shown in Figure 4.1, while the control signals are shown in Figure 4.2.

It was found that the system becomes unstable for high-gain feedback because the friction models are too simple to sufficiently approximate the LuGre model in the plant. The allowable feedback gain is $k_1 = 10$ for cases 1 and 2, and large tracking errors exist. Because of the existence of large tracking errors, the gain of adaptation cannot be chosen too large either. If a high adaptive gain is chosen, the system may become unstable.

For cases 3 and 4, as the friction models become complex and capture the dominate dynamics behaviors, it was found the high feedback gains can be used and the speed of adaptation can also be increased. For comparison studies, the tracking performance is also shown in Figure 4.1, while the control signals are shown in Figure 4.2 when $k_1 = 80$ and $\Gamma = \text{diag}[10]$. It can be seen that tracking performance become much better. The boundedness of the adaptive parameters are shown in Figure 4.3.

4.1.2 NN LIP model based adaptive control

RBF NN $F(x, \dot{x}) = S^T(x, \dot{x})W$ is used as the friction model, and the corresponding adaptive controller is also given by (3.8), i.e.,

$$u = -k_1 r + \hat{m}\ddot{x}_r + S^T(\dot{x})W - u_r - k_i \int_0^t r d\tau \qquad (4.3)$$

where $k_1 = 10$, $k_i = 0$, $u_r = k_2 \text{sgn}(r)$ with $k_2 = 2$. The adaptation gain is chosen as $\Gamma = \text{diag}[100]$.

To show the effectiveness of NN-based adaptive control, a Gaussian RBF neural network of 100 nodes is chosen to approximate friction with $\sigma^2 = 10.0$.

The tracking performance is shown in Figure 4.4, while the control signals and neural network weights are shown in Figure 4.5 and Figure 4.6, respectively.

It can be seen that a NN LIP model based adaptive controller can produce good tracking performance and guarantees the boundedness of all the closed-loop signals because the neural network friction model can capture the dynamic behavior of the LuGre model in the plant.

4.2 Dynamic LuGre Friction Model Based Adaptive Control

In this subsection, we study the performance of the adaptive friction compensation schemes proposed in Section 3.2 based on different uncertainty levels of dynamic LuGre friction.

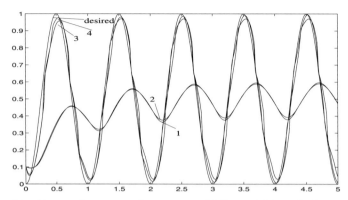

Fig. 4.1. Tracking performance of conventional LIP model based adaptive control.

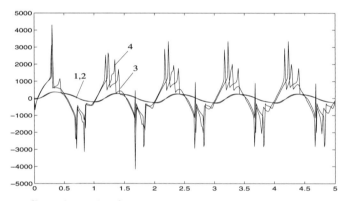

Fig. 4.2. Control signals of conventional LIP model based adaptive control.

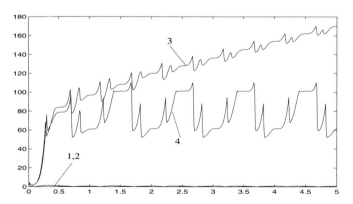

Fig. 4.3. Variations in parameters.

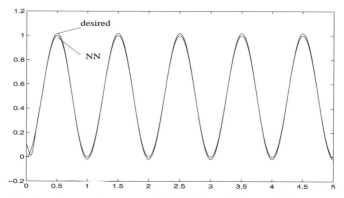

Fig. 4.4. Tracking performance of NN LIP model based adaptive control.

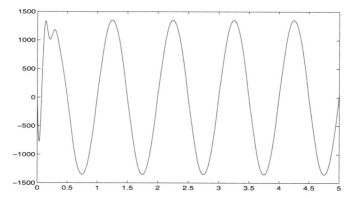

Fig. 4.5. Control signals of NN LIP model based adaptive control.

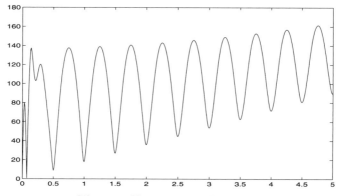

Fig. 4.6. Variations in NN weights.

4.2.1 Adaptive control for unknown system parameters

The system parameter m and dynamic friction parameters $\sigma_0, \sigma_1, \sigma_2$ are assumed unknown while nonlinear function $\alpha(x, \dot{x})$ is given by (2.28). The corresponding adaptive controller is given by (3.25), i.e.,

$$u = -c_1 r + \hat{\sigma}_1 \dot{x} + \hat{\sigma}_2 \dot{x} + \hat{\sigma}_0 \hat{z}_0 - \alpha(x, \dot{x}) \hat{\sigma}_1 |\dot{x}| \hat{z}_1 + \hat{m} \ddot{x}_r \qquad (4.4)$$

where $c_1 = 50$, $\hat{z}_0$ and $\hat{z}_1$ are the outputs of the observers (3.23) and (3.24) respectively. The adaptation gains are chosen as $\eta_0 = 10, \eta_1 = 10, \eta_2 = 10$ and $\eta_m = 10$. Figure 4.7 and Figure 4.8 show that both the actual position and velocity outputs asymptotically track the desired system trajectory and velocity outputs. The bounded dual-observer outputs and the actual internal dynamic friction state z are shown in Figure 4.10. It can be seen from Figure 4.10 that (i) the output $\hat{z}_0$ of the first observer is roughly the same as the actual internal friction state z, and (ii) the output $\hat{z}_1$ of the second observer is not at the same scale with z though bounded. This is due to the different purposes of the two observers. The first one is mainly for friction state estimation and the second one can be regarded as an auxiliary signal used to cancel the cross-coupling terms in the derivative of the Lyapunov function for closed-loop stability. Figure 4.9 indicates the bounded control signal u. The boundedness of the estimated parameters is illustrated in Figure 4.11.

4.2.2 Adaptive control for unknown $\alpha(x, \dot{x})$

The parameters m, σ_0, σ_1 and σ_2 are known while the nonlinear function $\alpha(x, \dot{x})$ is unknown. The RBF NN $\hat{\alpha}(x, \dot{x}) = W^T S(x, \dot{x})$ is used to approximate the unknown $\alpha(x, \dot{x})$. The NN is chosen of size $l = 49$, variance $\sigma = 1$, centers $\mu_i = [\mu_{i1}, \mu_{i2}]^T, i = 1, 2, \cdots, l$ covering all the 49 combinations of $\mu_{i1} = \{-1.5, -1, -0.5, 0, 0.5, 1, 1.5\}$ and $\mu_{i2} = \{-10, -5, -2.5, 0, 2.5, 5, 10\}$. The initial conditions for NN are $W(0) = 0$. The corresponding adaptive controller is given by (3.45-3.46), i.e.,

$$u = (\sigma_1 + \sigma_2) \dot{x} + m \ddot{x}_r + u_{ar} \qquad (4.5)$$

with

$$u_{ar} = -c_1 r + \sigma_0 \hat{\delta}_1 - \sigma_1 \hat{\alpha}(x, \dot{x}) |\dot{x}| \hat{\delta}_2 - \frac{m \sigma_0}{\sigma_1} r - \frac{\sigma_1^2 \sigma_2 \alpha_m r \dot{x}^2}{\sigma_0}$$
$$- k \sigma_1 |\dot{x}| |\hat{\delta}_2| \mathrm{sgn}(r) \qquad (4.6)$$

where the control parameters are chosen as $c_1 = 30, k = 10$. From the parameterization of $\alpha(x, \dot{x})$ in equation (2.28), we have $0 < \alpha(\dot{x}) \leq \frac{g_0}{f_c} = 10^5$. Accordingly, we choose $\alpha_m = \frac{g_0}{f_c} = 10^5$ in the simulation. The adaptation gain for NN weights is chosen as $\Gamma = 10$. Figure 4.12 and Figure 4.13 show that good position and velocity tracking performances are achieved. Figure 4.14 indicates the bounded control signal u. The bounded dual-observer outputs and the actual value of the auxiliary variable $\delta = z + \frac{m}{\sigma_1} r$ are illustrated in

Figure 4.15. It can be seen from Figure 4.15 that (i) the output $\hat{\delta}_1$ of the first observer is roughly the same as the actual friction state δ, and (ii) the output $\hat{\delta}_2$ of the second observer is not at the same scale as δ though bounded. This is due to the fact that the first observer is for friction state estimation and the second one is an auxiliary signal for the cancelation of the cross-coupling term in the derivative of the Lyapunov function. Figure 4.16 presents the boundedness of NN weights.

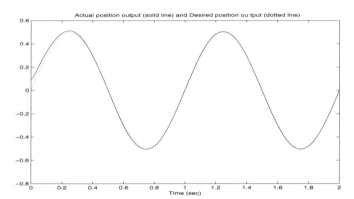

Fig. 4.7. Position tracking performance of Case 1.

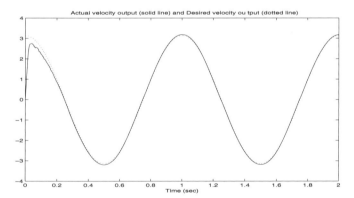

Fig. 4.8. Velocity tracking performance of Case 1.

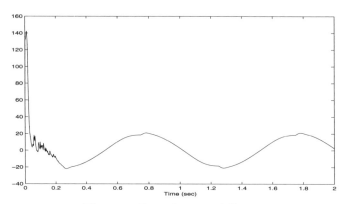

Fig. 4.9. Control input of Case 1.

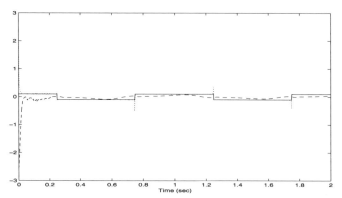

Fig. 4.10. Dual-observer outputs of Case 1 ($z*10^4$: solid line; $\hat{z}_0*10^4$: dotted line; $\hat{z}_1$: dashed line).

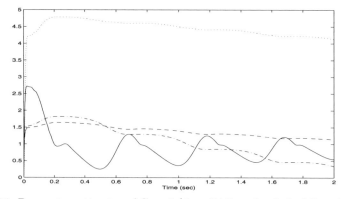

Fig. 4.11. Parameter estimates of Case 1 ($\hat{m}$: solid line; $\hat{\sigma}_0$: dashed line; $\hat{\sigma}_1$: dotted line; $\hat{\sigma}_2$: dashdot line).

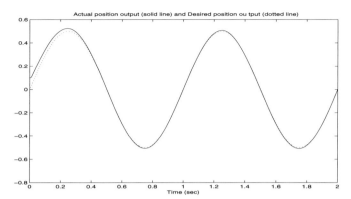

Fig. 4.12. Position tracking performance of Case 2.

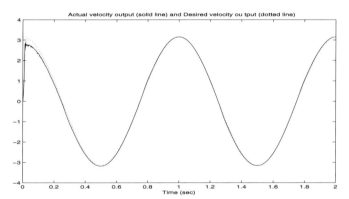

Fig. 4.13. Velocity tracking performance of Case 2.

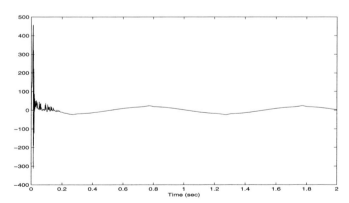

Fig. 4.14. Control input of Case 2.

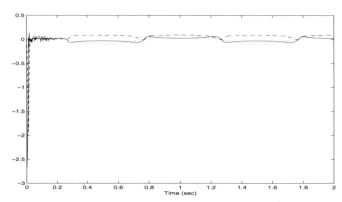

Fig. 4.15. Dual-observer outputs of Case 2 ($\delta*10$: solid line; $\hat{\delta}_1*10$: dotted line; $\hat{\delta}_2$: dashed line).

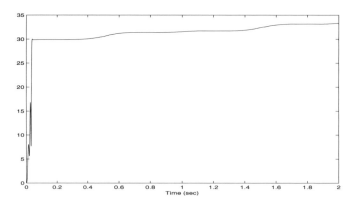

Fig. 4.16. Norm of estimated weights $\|W\|$ of Case 2.

4.2.3 Adaptive control for full set of unknown parameters

In this case, the full set of system and friction model parameters are assumed unknown. The RBF NN $\hat{F}_{zm}(x,\dot{x}) = W^T S(x,\dot{x})$ is used to approximate the function bound $F_{zm}(x,\dot{x})$. The NN is chosen of size $l = 25$, variance $\sigma = 1$, centers $\mu_i = [\mu_{i1}, \mu_{i2}]^T, i = 1, 2, \cdots, l$ covering all the 25 combinations of $\mu_{ij} = \{-1, -0.5, 0, 0.5, 1\}, j = 1, 2$. The initial conditions for NN are $W(0) = 0$. The corresponding adaptive controller is given by (3.58), i.e.,

$$u = -c_1 r + \hat{\theta}\dot{x} + \hat{m}\ddot{x}_r - \hat{F}_{zm}(x,\dot{x})\text{sgn}(r) - k\text{sgn}(r) \qquad (4.7)$$

where the control parameters are chosen as $c_1 = 50, k = 10$. The adaptation gains are chosen as $\Gamma = 10, \eta_\theta = 10$ and $\eta_m = 10$. Figure 4.17 and Figure 4.18 respectively show the ideal position and velocity tracking performance. Figure 4.19 indicates the bounded control signal u. Figure 4.20 presents the boundedness of NN weights, and the bounded parameter estimation results are illustrated in Figure 4.21.

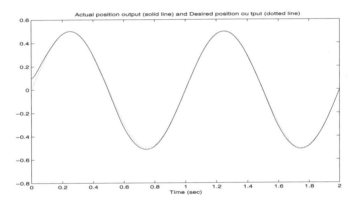

Fig. 4.17. Position tracking performance of Case 3.

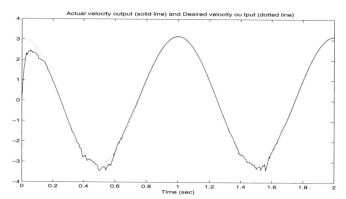

Fig. 4.18. Velocity tracking performance of Case 3.

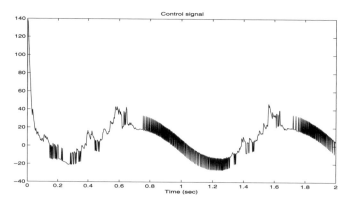

Fig. 4.19. Control input of Case 3.

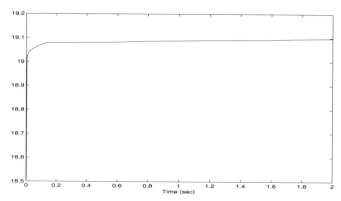

Fig. 4.20. Norm of estimated weights $\|W\|$ of Case 3.

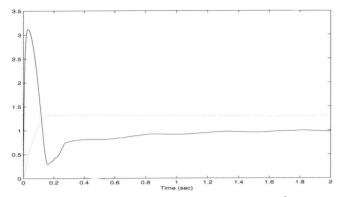

Fig. 4.21. Parameter estimates of Case 3 ($\hat{m}$: solid line; $\hat{\theta}$: dotted line).

Remark 4.1. From the simulation results in Section 4.1 and Section 4.2, it can easily be found that the performance of dynamic LuGre model based adaptive control is better than the results of LIP model based adaptive control. This is because the dynamic LuGre friction model more accurately captures the complex static and dynamic characteristics of friction. High precision motion control can be achieved by using the dynamic LuGre model based adaptive control schemes proposed in Section 3.2.

Remark 4.2. For conventional LIP model based adaptive control, we found from Figure 4.1 that by augmenting the basis function space to capture the dynamics of friction more accurately, better performance can be obtained. In fact, systems with simpler friction models may become unstable for high-gain feedback, and the gain of adaptation cannot be chosen too large either, whereas for systems with complex friction models it was found that high feedback gains can be used and the speed of adaptation can also be increased.

Remark 4.3. For NN LIP model based adaptive control, though better performance is obtained compared with conventional LIP model base adaptive control results, its performances is worse than that of the dynamic LuGre model based adaptive control. The reason is that the complex friction structure uncertainty is not revealed by direct NN approximation.

Remark 4.4. All of the results for dynamic LuGre model based adaptive control are ideal as shown in Section 4.2. However, the first two methods need the assumption of partially known friction model parameters, which may be hard to meet in practice. These assumptions are relaxed in the third approach without any sacrifice in performance.

5 Conclusion

In this chapter, a systematic treatment of adaptive friction compensation techniques has been presented. First, the commonly used static friction models and dynamic friction models were detailed using the corresponding model features analysis. Then, by considering the position and velocity tracking control of a servo mechanism with friction, adaptive friction compensation schemes are given based on the proposed LIP static friction model and dynamic LuGre friction model respectively. Using Lyapunov synthesis, adaptation algorithms were designed to achieve asymptotic tracking of the desired trajectory and guarantee the boundedness of all signals in the closed loop. Both position tracking and velocity tracking were realized at the same time by utilization of a filtered error signal. Extensive numerical simulation results have shown that the proposed approaches are effective for friction compensation in servo mechanisms.

References

1. A. M. Annaswamy, F. P. Skantze and A. P. Loh, Adaptive control of continuous time systems with convex/concave parameterization, *Automatica*, 34, no. 1, 33-49, 1998.
2. B. Armstrong, Friction experimental determination, modeling and compensation, *IEEE Int. Conf. Robotics and Automation*, pp. 1422-1427, 1988.
3. B. Armstrong, Stick-slip arising from stribeck friction, *IEEE Int. Conf. Robotics and Automation*, Cincinnati, pp. 1377-1382, 1990.
4. B. Armstrong-Helouvry, P. Dupont and C. Canudas de Wit, A survey of analysis tools and compensation methods for the control of machines with friction, *Automatica*, 30, no. 7, 1083-1138, 1994.
5. B. Armstrong and B. Amin, PID control in the presence of static friction: a comparison of algebraic and describing function analysis, *Automatica*, 32, no. 5, 679-692, 1996.
6. N. Barabanov and R. Ortega, Necessary and sufficient conditions for passivity of the LuGre friction model, *Proc. of IFAC 14 World Congress*, Beijing, pp. 399-402, 1999.
7. L. C. Bo and D. Pavelescu, The friction-speed relation and its influence on the critical velocity of the stick-slip motion, *Wear*, 3, 277-289, 1982.
8. L. Cai and G. Song, A smooth robust nonlinear controller for robot manipulators with joint job stick-slip friction, *Proc. IEEE Int. Conf. Robotics Automation*, Atlanta, GA, pp. 449-454, 1993.
9. L. Cai and G. Song, Joint stick-slip friction compensation of robot manipulators by using smooth robust controller, *J. of Robotic Systems*, 11, no. 6, 451-470, 1994.
10. C. Canudas de Wit, K. J. Åström and K. Brawn, Adaptive friction compensation in DC motor drives, *IEEE J. of Robotics and Automation*, 3, no. 6, 681-685, 1987.

11. C. Canudas De Wit and J. Carillo, A modified EW-RLS algorithm for system with bounded disturbance, *Automatica*, 26, no. 4, 599-606, 1990.
12. C. Canudas de Wit, C. P. Noel, A. Auban and B. Brogliato, Adaptive friction compensation in robot manipulators: low velocities, *Int. J. of Robotics Research*, 10, no. 3, 189-199, 1991.
13. C. Canudas de Wit, H. Olsson, K. J. Åström and P. Lischinsky, A new model for control of systems with friction, *IEEE Trans. on Automatic control*, 40, no. 3, 419-425, 1995.
14. C. Canudas de Wit and S. S. Ge, Adaptive friction compensation for systems with generalized velocity/position friction dependency, *Proceedings of the 36th IEEE Conf. on Decision and Control*, San Diego, CA, pp. 2465-2470, 1997.
15. C. Canudas de Wit and P. Lischinsky, Adaptive friction compensation with partially known dynamic friction model, *Int. J. of Adaptive Control & Signal Processing*, 11, 65-80, 1997.
16. T. P. Chen and H. Chen, Approximation capability to functions of several variables, nonlinear functionals, and operators by radial basis function neural networks, *IEEE Trans. Neural Networks*, 6, no. 4, 904-910, 1995.
17. P. R. Dahl, A solid friction model, *Report AFO 4695-67-C-0158*, Aerospace Corporation, EI Segundo, California, 1968.
18. P. E. Dupont, Avoiding stick-slip through PD control, *IEEE Trans. on Automatic Control*, 39, no. 5, 1094-1097, 1994.
19. B. Friedland and Y. J. Park, On adaptive friction compensation, *IEEE Trans. on Automatic Control*, 37, no. 10, 1609-1612, 1992.
20. S. S. Ge, T. H. Lee and C. J. Harris, Adaptive Neural Network Control of Robot Manipulators, World Scientific, River Edge, NJ, 1998.
21. S. S. Ge, T. H. Lee and S. X. Ren, Adaptive friction compensation of servo mechanisms, *Int. J. of Systems Sciences*, to appear, 2001.
22. S. S. Ge, T. H. Lee and J. Wang, Adaptive NN Control of Dynamic Systems with Unknown Dynamic Friction, *Proceedings of the 39th IEEE Conf. on Decision and Control*, Sydney, Dec. 12-15, 2000.
23. D. P. Hess and A. Soom, Friction at a lubricatal line contact operating at oscillating sliding velocity, *J. Tribology*, 112, no. 1, 147-152, 1990.
24. P. A. Ioannou and J. Sun, *Robust Adaptive Control*, Prentice-Hall, Upper Saddle River, NJ, 1996.
25. Y. H. Kim and F. L. Lewis, Reinforcement adaptive learning neural-net-based friction compensation control for high speed and precision, *IEEE Trans. on Control Systems Technology*, 8, no. 1, 118-126, 2000.
26. F. L. Lewis and K. Liu, Multilayer neural-net robot controller with guaranteed tracking performance, *IEEE Trans. on Neural Network*, 7, 188-198, 1996.
27. K. M. Misovec and A. M. Annaswamy, Friction compensation using adaptive nonlinear control with persistent excitation, *Int. J. Control*, 72, no. 5, 457-479, 1999.
28. H. Olsson, K. J. Åström, C. Canudas de wit, M. Gafvert and P. Lischinsky, Friction models and friction compensation, *European Journal of Control*, 4, 176-195, 1998.
29. R. Ortega, A. Loria, P. J. Nicklasson and H. Sira-Ramirez, Passivity-based Control of Euler-Lagrange Systems, Springer-Verlag, London, 1998

30. T. Poggio and F. Girosi, Networks for approximation and learning. *Proc. of IEEE*, 78, 1481-1497, 1990.
31. M. M. Polycarpou and P. A. Ioannou, A robust adaptive nonlinear control design, *Proceedings of ACC*, San Francisco, CA, pp. 1365-1369, 1993.
32. Y. Tan and I. Kanellakopoulos, Adaptive nonlinear friction compensation with parametric uncertainties, *Proceedings of ACC*, San Diego, CA, pp. 2511-2515, 1999.
33. A. Tustin, The effects of backlash and of speed-dependent friction on the stability of closed-cycle control system, *J. Institution of Electrical Engineers*, 94, no. 2, 143-151, 1947.
34. S. Ward, S. C. Radcliff and C. J. MacCluer, Robust nonlinear stick-slip friction compensation, *ASME Journal of Dynamic Systems, Measurement, and Control*, 113, 639-645, 1991.
35. A. Tustin, The effects of backlash and of speed-dependent friction on the stability of closed-cycle control system, *J. Institution of Electrical Engineers*, 94, no. 2, 143-151, 1947.
36. T. Zhang, S. S. Ge and C. C. Hang, "Design and performance analysis of a direct adaptive controller for nonlinear systems", *Automatica*, 35, no. 11, 1809-1817, 1999.

Relaxed Controls and a Class of Active Material Actuator Models

A. Kurdila

Department of Aerospace Engineering,
Mechanics and Engineering Science
University of Florida
Gainesville, FL32611-6250
ajk@aero.ufl.edu

Abstract

Previous research has demonstrated that rigorous modeling and identification theory can be derived for structural dynamical models that incorporate control influence operators that are *static* Krasnoselskii-Pokrovskii integral hysteresis operators. Experimental evidence likewise has shown that some *dynamic* hysteresis models provide more accurate representations of a class of structural systems actuated by some active materials including shape memory alloys and some piezoceramics. In this paper, we show that the representation of control influence operators via static hysteresis operators can be interpreted in terms of a homogeneous Young's measure. Within this framework, we subsequently derive dynamic hysteresis operators represented in terms of Young's measures that are parameterized in time. We show that the resulting integrodifferential equations are similar to the class of relaxed controls discussed by Warga [10],Gamkrelidze [24], and Roubicek [25]. The formulation presented herein differs from that studied in [10], [24] and [25] in that the kernel of the hysteresis operator is a history dependent functional, as opposed to Caratheodory integral satisfying a growth condition. The theory presented provides representations of *dynamic* hysteresis operators that have provided good agreement with experimental behavior in some active materials. The convergence of finite dimensional approximations of the governing equations is also discussed.

1 Introduction

It has been noted in [2,3] that the derivation of accurate, low-dimensional models of hysteresis that are accurate for active material actuators remains a formidable task. There is overwhelming empirical and anecdotal evidence that the nonlinear hysteresis effects in actuators based on piezoelectrics, shape memory alloys, magnetorheological fluids, magnetostrictives, electrostrictives, ferro-electrics and electrorheological fluids must be accommodated in analysis and control models. The reader is referred to [9,22] for discussions of hysteresis observed in PZT-based actuators. Experimental and analytical characterizations of hysteresis in magnetostrictives and ferroelectrics are studied in [6,8].

While hysteretically nonlinear behavior is evident in all of these materials, we consider specific research related to both (1) shape memory alloys and (2) relaxed control theory that helps to focus attention on the methodology described in this paper. There is quite literally an enormous literature that describes various aspects of hysteresis in shape memory alloys. Research studying this complex structural response tends to be polarized in three disparate classes, (1) microstructural models, (2) continuum models and (3) dynamics and control models. For some time, researchers have been intrigued by the nonconvex nature of the martensitic solid–solid phase transformation in shape memory alloys. A number of researchers, including for example Ball and James in [13–15], James and Kinderlehrer in [23], have utilized parameterized, or Young's, measures to achieve relaxations of nonconvex energy functionals. These models rely on rather abstract constructions, weak topologies on functions spaces that have values in regular countably additive measures. They seek to represent structural response on an extremely fine scale, as small as the $O(10)$ microns, for example [25]. It should be emphasized that these formulations predominantly treat static equilibria. In one notable exception, reference [25] studies the well-posedness of dynamical systems that incorporate Young's measures.

Other reseachers have sought to characterize the structural response of solid–solid phase transformations with models that represent length scales at a somewhat coarser scale, on a length scale of $O(100)$–$O(1000)$ microns. Bo and Lagoudas et al. in [26] utilize thermodynamical formalisms in conjunction with choices of internal hysteresis loop switching laws, to achieve accurate constitutive models for uniaxial cyclic load testing of wires. While most continuum models treat the SMA transformation as homogeneous for uniformly loaded specimens, Abeyaratne and Knowles [20] have proposed a continuum thermodynamic framework for the kinetics of inhomogeneous phase transformations in a 1-D setting. Shaw, in [11] and [12], provides clear experimental evidence of intricate dynamics of phase transformations in uniaxial loaded SMA specimens. This research is significant for several reasons. By using full-field experimental measurement methods, phase front dynamics in the transitional length scales between microstructure and the full structural length can be observed. Furthermore, he shows that fine structure effects manifest at larger length scales and can be represented through continuum models.

A significant complementary effort has been pursued by researchers who seek to use SMA to actuate structural systems, including the work of [7] and [16] in which structural shape control is studied. Brokate and Sprekels in [18] discuss the mathematical foundations of hysteresis in Falk's phenomenological model of SMA constitutive response. Banks, Kurdila and Webb in [2] and [3] derive convergent approximation methods for identification of measure-valued representations of SMA actuators. In a series of papers, Kurdila and Webb [4,5] derive a systematic methodology for the control of linear-nonlinear cascaded systems. The technique uses the identification theory derived in [2,3], and subsequently proves the closed loop stability of the model reference and adaptive control methods.

In summary, there is some stratification in the approaches used to model shape memory alloys. A given microstructural model may have little, if any, bearing on a different continuum model, phenomenological model or control model. This fact stands in stark contrast to many other fields in structural mechanics wherein there is a significant interplay between modeling efforts at various length scales. This issue has far-reaching implications and it is anticipated that it will require several man-years to resolve to any reasonable level of completion. Techniques to propagate classes of models for solid–solid phase transformations across length scales are generally not available today.

It is significant that researchers seeking to characterize microstructural effects in shape memory alloys have sometimes employed Young measures, or parameterized measures, to obtain generalized representations of strain equilibrium. One approach replaces the conventional strain energy density functional, which is not convex or quasiconvex, for shape memory alloys with one that is. An alternative representation of the structural energy is expressed in terms of a parameterized probability measure. This measure, roughly speaking, characterizes the mixture of martensite and austenite in static equilibrium configurations. For some time, the same researchers studying phase transformations in mechanics have understood that parameterized measures can be traced, historically speaking, to relaxation methods for optimal control problems. However, they have not exploited this relationship to derive consistent control methodologies for shape memory alloys. The methods of measure-valued controls studied by Warga [10] and Gamkrelidze [24] are typical of these control theoretic techniques.

In [2] and [3] the authors of this paper have formulated system level structural evolution equations incorporating hysteretically nonlinear control influence operators. These models were motivated by observations of system response at the structural scale. In subsequent experimental tests [4] and [5], the models proved to be effective in providing low dimensional approximations of the hysteretic nonlinear response. We emphasize that these models were selected to be *static* hysteresis operators having the form

$$\int_S k_v(t,s)\mu(ds) \tag{1.1}$$

Where $k_v(t,s)$ is a history dependent kernel, v is the control input, t is time and μ is a probability measure defined over a compact metric space S. The form of the integral hysteresis operator is a specific case of the generalized play studied in Visintin [1], or studied earlier by Krasnoselskii and Pokrovskii in [19].

The performance in tracking control methods derived for this class of operator proved to be best for extremely low bandwidth response. However, researchers and developers have steadily increased the bandwidth of SMA based actuation. For example, convective cooling effects in [5] lead to tracking of signals having a frequency of roughly 10 hertz. Some commercially available microvalves cite operating frequencies of 20 hertz. Thin film SMA actuators attain operating frequencies nearing 1000 hertz. At this frequency range, the static hysteresis models and associated control methodologies proved increasingly ineffective. Careful study at the control structure showed that the deterioration in performance could be associated with temporal evolution of the current-to-strain hysteresis cycle in the SMA actuator. Subsequently, experimental studies by the authors have shown that the assumption that the control influence operator is a *dynamic* hysteresis operator can lead to the development of adaptive control strategies that yield superior performance in contrast to the static hysteresis operator assumed in (1.1). While this strategy has been effective in experiment, the development of this approach in [18,19] is formal at best.

Consequently, this paper seeks to provide a rigorous foundation for the formal, but effective, methodologies employed in [5]. In this paper, we show that the formulation derived in [2,3] can in fact be viewed as employing a homogeneous parameterized measure, or Young's measure, to represent the hysteretic, history dependent control influence operator. We then establish the well-posedness of an infinite dimensional dynamical system driven by a hysteretic, history dependent control influence operator that is represented in terms of a time-parameterized Young's measure. We show that consistent discretizations of these equations yield the effective models discussed in [5]. Finally, we wish to emphasize that while the motivation for the derived methodology has been based on some historical considerations of shape memory alloys, the control methodology is applicable to a large class of systems. For example, the technique has been studied in [9] for the control of aeroelastic systems actuated by PZT bimorphs. This system is shown in Figure 2. Thus, the methodology is another practical realization of the generalized or relaxed controls of Warga [10] or Gamkrelidze [24], albeit for history dependent kernels.

2 The Governing Dynamical Equations

In this section, we discuss the class of nonlinear structural systems considered in this paper. We will restrict attention to systems in which we can identify a distinct, isolated hysteretic active material actuator that induces control on a linearly elastic structure. For example, in Figure 2, we depict such a system in the form of a helicopter rotor blade with a trailing edge flap that

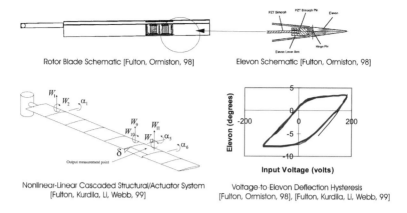

Fig. 2.1. Trailing edge flap and rotor.

is actuated by a PZT bimorph. Because of the aerodynamic loads imposed on the trailing edge flap, the voltage-to-flap deflection mapping is strongly hysteretic, as is expected for PZT at high voltage levels. A system model can be constructed that is comprised of an isolated hysteretically nonlinear actuator and a linear structural sybsystem that represents blade deformation. We consider the representation of the structural subsystem and control actuator separately.

2.1 Infinite Dimensional Structural Systems

In this section, we review the class of structural systems studied in this paper, based on models derived in [2], [3], [4] and [5]. In [2], [3] and [4] a linear structural system model is employed that is assumed to be driven by a static integral hysteresis operator. To formulate the governing equations, we employ the Gelfand triple

$$V \hookrightarrow H \approx H^* \hookrightarrow V^* \tag{2.1}$$

where, the the real separable Banach space V is densely and continuously embedded in real Hilbert space H. The Hilbert space is identified with its dual H^* via the Riesz mapping, i.e., it is a pivot space. The governing equations take the form

$$\ddot{w}(t) + A_1 \dot{w}(t) + A_0 w(t) = (\mathcal{B}v)(t) \quad \text{in} \quad V^* \tag{2.2}$$
$$w(0) = w_0 \tag{2.3}$$
$$\dot{w}(0) = w_1 \tag{2.4}$$

where V^* is the topological dual of the separable Banach space V. The bounded linear mappings A_0 and A_1 are defined in terms of bilinear forms $a_0(\cdot,\cdot)$ and $a_1(\cdot,\cdot)$ as

$$\langle A_0 u, v \rangle_{V^* \times V} = a_0(u,v) \qquad \forall u,v \in V$$

$$\langle A_1 u, v \rangle_{V^* \times V} = a_1(u,v) \qquad \forall u,v \in V$$

As usual, it is assumed that the bilinear form $a_0(\cdot,\cdot) : V \times V \to \mathbb{R}$ is symmetric, bounded and V-coercive

$$a_0(u,v) = a_0(v,u) \qquad \forall u,v \in V \qquad (2.5)$$
$$a_0(u,v) \leq c_1 \|u\|_V \|v\|_V \qquad \forall u,v \in V \qquad (2.6)$$
$$a_0(u,v) \geq c_2 \|u\|_V^2 \qquad \forall u \in V \qquad (2.7)$$

where $c_1 \geq 0$, $c_2 > 0$. Similarly, the damping bilinear form $a_1(\cdot,\cdot) : V \times V \to \mathbb{R}$ is bounded and statisfies a Garding inequality

$$a_1(u,v) \leq c_1 \|u\|_V \|v\|_V \qquad \forall u,v \in V \qquad (2.8)$$
$$a_1(u,v) + \lambda \|u\|_H^2 \geq c_2 \|u\|_V^2 \qquad \forall u \in V \qquad (2.9)$$

where $\lambda \geq 0$, $c_1 \geq 0$ and $c_2 > 0$. It is well known that these assumptions are appropriate for a host of linear structural systems including assemblies of linearly elastic plates, shells, beams and rods. The existence and uniqueness of solutions to the governing equations (2.2), subject to the assumptions (2.6)–(2.8), is discussed for example in [17,21,22]. Suppose $\tau = [0,T]$. As an example, the theory (see p.98 Theorem 4.1 [22]) guarantees the existence of solutions $(w,\dot{w}) \in L^2(\tau;V) \times L^2(\tau;V)$ provided $(\mathcal{B}v) \in L^2(\tau;V^*)$, $w_0 \in V$, $w_1 \in H$ for the particular assumptions considered in this paper. More general cases can certainly be considered.

We emphasize that the primary consideration of this paper is the study of the nonlinear (hysteretic) control influence operator to be discussed in Section 3. Consequently, we note that the evolution equation embodied in equations (1.1) through (2.2) is, strictly speaking, simpler in form than the systems studied in [2,3]. The evolution equations corresponding to the structural system in [2] and [3] make provision for a wider class of damping operators by introducing a space V_2 intermediate between V and H. In addition, inasmuch as the goal of [2] and [3] was to derive identification methods, the operators A_0, A_1 in [2] and [3] depend there on a parameter selected from some compact metric space. The purpose of this paper is to interpret a class of approximations to equations (1.1) through (2.2) in terms of the relaxed controls of Warga [10]. In this context, the damping space V_2 does not play an essential role, and neither does the parametric dependence of the operators A_0, A_1. By limiting our discussion to the form of equations presented above in (1.1)–(2.2) significant simplification in the presentation is achieved. Generalizations to alternative damping operators and parametrically dependent infinitisemal generators will be considered in future work.

2.2 The Control Influence Operator

2.2.1 Static hysteresis operator. A static control influence operator is defined in terms of a hysteretically nonlinear integral operator having the separable form

$$(\mathcal{B}_S v)(t) \equiv \int_S k_v(t,s) \mu(ds) \cdot g$$

In this equation, $\mathcal{S}$ is the compact subset of $\mathbb{R}^2$

$$\mathcal{S} = \{(s_1, s_2) \in \mathbb{R}^2 : \underline{s} \leq s_1 \leq s_2 \leq \bar{s}\}$$

The symbols $\underline{s}$, $\bar{s}$ denote fixed real numbers, $\mu(\cdot)$ is a probability measure on $\mathcal{S}$, $g \in V^*$ and the kernel

$$k.(\cdot,\cdot) : C[0,T] \times [0,T] \times \mathcal{S} \to \mathbb{R}$$

In this paper, we focus on the action of the kernel for a fixed control input $v \in [0,T]$. Notationally, this is represented as the mapping

$$k_v : [0,T] \times \mathcal{S} \to \mathbb{R}$$

The kernel k_v is derived in several steps and follows standard techniques discussed in [1,19]. Let $(s_1, s_2) \in \mathcal{S}$ and define the translates r_{s_1}, r_{s_2}

$$r_{s_1}(x) = r(x - s_1) \qquad (2.10)$$
$$r_{s_2}(x) = r(x - s_2) \qquad (2.11)$$

where $r(x)$ is a Lipschitz continuous ridge function. If $v_m(t)$ is a monotone, continuous function, a monotone output operator $\mathcal{M}$ is defined for each $s = (s_1, s_2) \in \mathcal{S}$ and $\xi \in \text{range}(r)$ as follows:

$$\mathcal{M}(v_m, s, \xi)(t) = \begin{cases} \max\{\xi, r(v_m(t) - s_2)\} & \text{if } v_m \text{ is nondecreasing} \\ \min\{\xi, r(v_m(t) - s_1)\} & \text{if } v_m \text{ is nonincreasing} \end{cases}$$

We next extend this definition to piecewise linear splines. Let $\mathcal{S}_{1,j}[0,T]$ be the set of piecewise linear splines on the uniform mesh with knots $\{t_k\}_{k=0}^{j}$. The action of the kernel on $\mathcal{S}_{1,j}[0,T]$ is defined inductively. Set $\mathcal{M}_0 = \xi$. We define

$$k_{v_p}(t,s) = \begin{cases} \mathcal{M}(v_p, s, \mathcal{M}_{k-1})(t) & t \in [t_{k-1}, t_k] \\ \mathcal{M}_k = \mathcal{M}(v_p, s, \mathcal{M}_{k-1})(t_k) & k = 1, \ldots, j \end{cases}$$

where the piecewise linear function $v_p \in \mathcal{S}_{1,j}[0,T]$. Since the set of piecewise linear splines is dense in $C[0,T]$, the definition of $k.(t,s)$ can be extended by continuity to act on $C[0,T]$.

In summary, we state the governing equations modeling linear structural systems actuated by static hysteresis operators considered in this paper. We seek to find $(w, \dot{w}) \in L^2(\tau; V) \times L^2(\tau; V)$ such that

$$\left\langle \frac{\partial^2 w}{\partial t^2}(t),\phi \right\rangle + a_1 \left(\frac{\partial w}{\partial t}(t),\phi \right) + a_0(w(t),\phi) = \langle (\mathcal{B}_\mathcal{S} v)(t),\phi \rangle \qquad \forall \phi \in V \tag{2.12}$$

subject to the initial conditions

$$w(0) = w_0 \tag{2.13}$$
$$\dot{w}(0) = w_1 \tag{2.14}$$

In these equations, the control influence operator is defined to be

$$(\mathcal{B}_\mathcal{S} v)(t) = \int_\mathcal{S} k_v(t,s)\mu(ds) \cdot g \tag{2.15}$$
$$\equiv (\mathcal{B}_\mathcal{S} v)(t) \cdot g \tag{2.16}$$

2.2.2 Dynamic hysteresis operator. Moreover, based on experimental data, the static model of hysteresis above was modified to achieve a representation of the hysteresis operator that evolved over time in [5]. While the model has proven to be quite effective for representing structural response and designing controls, its theoretical foundation remained in question. At least formally, the hysteresis operator employed in [5] can be expressed as a integral hysteretic contol influence operator

$$(\mathcal{B}_D v)(t) = \int_\mathcal{S} k_v(t,s)\mu_t(ds) \cdot g = (Bv)(t) \cdot g \tag{2.17}$$

In this equation, carefully note the appearence of a parameterized probability measure μ_t. The purpose of this paper, then, is to determine in what sense we can interprete the governing equations in (2.12) where the control influence operator is given by (2.17).

Duality in $L^1(\tau; C(\mathcal{S}))$ and Young measures To make rigorous the introduction of a time-parameterized hysteresis operator in equation (2.17), we review an equivalent realization of the Banach space $L^1(\tau; C(\mathcal{S}))$ of Bochner integrable functions taking values in $C(\mathcal{S})$. Recall that $\tau = [0, T]$ and $\mathcal{S}$ are compact metric spaces, and τ is endowed with the Lebesque measure. It is well known that the set of functions

$$g : \tau \times \mathcal{S} \to \mathbb{R}$$

such that $g(\cdot, s)$ is measurable, $g(t, \cdot) \in C(\mathcal{S})$ and $\sup_{s \in \mathcal{S}} |g(t,s)|$ is integrable defines a Banach space G with norm

$$\|g\|_G \equiv \int_\tau \sup_{s \in \mathcal{S}} |g(t,s)| dt$$

In fact the Banach space $(G, \|\quad\|_G)$ is isometrically isomorphic to $L^1(\tau; C(\mathcal{S}))$ [p.265 Warga]. If we define

$$U \equiv \{f : \tau \to \mathcal{S}, \quad f \text{ is measurable}\}$$

it is possible to imbed U in $L^1(\tau; C(\mathcal{S}))^*$. For any $u \in U$ and any $g \in G \approx L^1(\tau; C(\mathcal{S}))$ we define an imbedding i

$$i : U \to (L^1(\tau; C(\mathcal{S})))^*$$

via

$$\langle i(u), g \rangle = \int_T g(t, u(t)) dt$$

We subsequently define the family of Young measures Y as the weak* closure of $i(U)$ in $L^1(\tau; C(\mathcal{S}))^*$

$$Y = \{\xi \in L^1(\tau : C(\mathcal{S}))^* : \exists \{u_d\}_{d \in D} \subset U : i(u_d) \to \xi \quad \text{weakly*}\} \quad (2.18)$$

Still another characterization of Young measures will be required for the analysis in this paper. If Z is Banach space, a weakly measurable function $f : \tau \to Z^*$ is one such that

$$t \mapsto \langle f(t), h \rangle_{Z^* \times Z}$$

is measurable for each $h \in Z$. The vector space of weakly measurable mappings is a Banach space with norm

$$\|f\|_{L_w^\infty(\tau, Z^*)} \equiv esssup_{t \in \tau} \|f(t)\|_{Z^*}$$

and is denoted $L_w^\infty(\tau, Z^*)$. A fundamental result of Dunford and Pettis is that the spaces $L^1(\tau, Z)^*$ and $L_w^\infty(\tau, Z^*)$ are isometrically isomorphic. Of particular interest in this paper will be the choice $Z = C(\mathcal{S})$. In this case, the norm dual $C(\mathcal{S})^*$ is the collection of scalar-valued, regular, countably additive set functions on $\mathcal{S}$, $rca(\mathcal{S})$. The isometric isomorphism between $L_w^\infty(\tau, rca(\mathcal{S}))$ and $L^1(\tau; C(\mathcal{S}))$ is realized by the mapping

$$\psi : L_w^\infty(\tau; rca(\mathcal{S})) \to L^1(\tau; C(\mathcal{S}))^*$$

$$\psi : \mu \mapsto \xi$$

$$\langle \xi, g \rangle = \int_T \int_S g(t, s) \mu_t(ds) dt$$

Now we obtain our alternative useful characterization of the set of Young measures Y defined in equation (2.18). Suppose $\mathcal{P}(\mathcal{S})$ is the set of all probability measures over $\mathcal{S}$.

$$Y = \{\xi \in L^1(\tau; C(\mathcal{S}))^* : \exists \mu \in L_w^\infty(\tau; rca(\mathcal{S})) \quad (2.19)$$
$$\text{s.t.} \quad \psi(\mu) = \xi, \mu_t \in \mathcal{P}(\mathcal{S}) \quad \text{a.e.} \quad t \in \tau\} \quad (2.20)$$

In other words, the set of Young measures are time-parameterized probability measures almost everywhere $t \in \tau$. Figure (2.2) depicts the relationship of these spaces schematically.

In summary then, for the dynamic hysteresis operator, we seek to find $(w, \dot{w}) \in L^2(\tau; V) \times L^2(\tau; V)$ such that

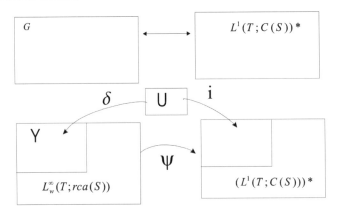

Fig. 2.2. Schematic of function spaces.

$$\left\langle \frac{\partial^2 w}{\partial t^2}(t), \phi \right\rangle + a_1\left(\frac{\partial w}{\partial t}(t), \phi\right) + a_0(w(t), \phi) = \langle (\mathcal{B}_D v)(t), \phi \rangle \quad \forall \phi \in V \tag{2.21}$$

subject to the initial conditions

$$w(0) = w_0 \tag{2.22}$$
$$\dot{w}(0) = w_1 \tag{2.23}$$

where

$$(\mathcal{B}_D v)(t) = \int_{\mathcal{S}} k_v(t,s)\mu_t(ds) \cdot g \tag{2.24}$$
$$\equiv (B_D v)(t) \cdot g \tag{2.25}$$

and

$$\mu \in L_w^\infty(\tau, rca(\mathcal{S})) \tag{2.26}$$

2.3 Existence and Uniqueness of Solutions

The study of the existence of solutions for the case of a static hysteresis control operator is given in [3]. In this paper it is shown that there is a solution $(w, \dot{w})$

$$w \in C(\tau, V) \subset L_2(\tau; V)$$
$$\dot{w} \in C(\tau, H) \cap L_2(\tau; V)$$

to equations (2.12)–(2.15) provided that $w_0 \in V$, $w_1 \in H$ and the bilinear forms $a_0(\cdot, \cdot)$, $a_1(\cdot, \cdot)$ satisfy the conditions in equations (2.6) –(2.8), respectively. We have the following theorem that guarantees the well-posedness of the history dependent, nonlinear dynamical system in equations (2.21)–(2.26) associated with the dynamic hysteresis control influence operator.

Theorem 2.1. *Suppose the bilinear forms $a_0(\cdot,\cdot)$ and $a_1(\cdot,\cdot)$ satisfy the conditions (2.6)–(2.8), and the control influence operator*

$$(\mathcal{B}_D v)(t) = \int_\mathcal{S} k_v(t,s)\mu_t(ds) \cdot g$$

where $\mu \in Y \subset L^\infty_w(\tau, rca(\mathcal{S}))$, and $k_v(\cdot,\cdot)$ is defined in Section 2.2.1. If $w_o \in V$ and $w_1 \in H$, there is a solution $(w, \dot{w})$ of equation (2.21) such that

$$w \in C(\tau, V) \subset L^2(\tau; V)$$

$$\dot{w} \in C(\tau, H) \subseteq L^2(\tau; V)$$

Proof. The proof of this theorem follows immediately from Theorem (4.1) p.98 of [22] provided that we can show

$$\mathcal{B}_D v \in L^2(\tau; V^*)$$

Since

$$(\mathcal{B}_D v)(t) = \int_\mathcal{S} k_v(t,s)\mu_t(ds) \cdot g$$

where $g \in H$, it suffices to show that

$$(\mathcal{B}_D v)(t) = \int_\mathcal{S} k_v(t,s)\mu_t(ds) \tag{2.27}$$

is in $L^2(\tau)$. Since we know only that

$$\mu \in L^\infty_w(\tau; rca(\mathcal{S})) \approx (L^1(\tau; C(\mathcal{S})))^*$$

it is not clear that $\mathcal{B}_D v \in L^2(\tau)$. The following lemma shows that this is indeed the case.

Lemma 2.1. *The kernel derived in Section 2.2.1 has several notable continuity properties:*

1. $k_v(\cdot,\cdot) \in C(\tau \times \mathcal{S})$ for each $v \in C(\tau)$
2. $k_v(\cdot,\cdot) \in L^1(\tau, C(\mathcal{S}))$
3. $\mathcal{B}_D v \in L^2(\tau)$

Proof. Properties (1) and (2) above are derived in [2,3], and the interested reader is referred to these references for the details. There are several methods for deriving the third claim above. We follow the methodology outlined in Roubicek [25 p. 161].

Choose any $\mu \in Y \subset L^\infty_w(\tau, C(\mathcal{S}))$. By definition there is a net

$$i(u_d) \to \psi(\mu) \text{ weakly* in } (L^1(\tau, C(\mathcal{S})))^*$$

Of course, each $u_d \in U$ can be identified with the Young measure $\psi(\delta_{u_d}) = \psi(\mu_d) = i(u_d)$. Consider the net

$$\{\langle i(u_d), k_v\rangle\}_{d\in D} = \{k_v(t, u_d(t))\}_{d\in D} = \left\{\int_S k_v(t,s)\delta_{u_d(t)}(ds)\right\}_{d\in D} = \{f_d\}_{d\in D}$$

By construction of k_v, this net is bounded in $L^\infty(\tau)$. This net must have a subnet that converges weakly* in $L^\infty(\tau)$ to a function in $L^\infty(\tau)$. On the other hand, we claim that for any $\mu \in Y$, the map

$$f_d(g) \equiv \int_\tau g(t) \int_S k_v(t,s)\delta_{u_d(t)}(ds)dt$$

defines a bounded linear functional on $C(\tau)$. Since linearity is clear, we must show that $|l_d(g)| < c\|g\|_{C(\tau)}$.

$$\left|\int_\tau g(t)\int_S k_v(t,s)\delta_{u_d(t)}(ds)dt\right| \leq c\|g\|_{C(\tau)} \sup_t |k_v(t, u_d(t))| \quad (2.28)$$
$$\leq c'\|g\|_{C(\tau)} \quad (2.29)$$

If $g \in C(S)$, then $g(t)k_v(t,s) \in L^1(\tau, C(S))$

$$\langle i(u_d), g \cdot k_v\rangle = l_d(g) \to \int_\tau g(t)\int_S k_v(t,s)\mu_t(ds)dt$$

The functional

$$l(g) \equiv \int_\tau \int_S g(t)k_v(t,s)\mu_t(ds)dt$$

is also a bounded linear functional on $C(\tau)$. Thus, we have shown that the sequence of functionals

$$\{f_d\}_{d\in D} = \left\{\int_S k_v(t,s)\delta_{u_d(t)}(ds)\right\}_{d\in D} \quad (2.30)$$

converges to

$$\int_S k_v(t,s)\mu_t(ds) \quad (2.31)$$

in $C(\tau)^*$. Recall that homeomorphical embedding

$$C(\tau) \subset L^\infty(\tau)$$

induces the inclusion

$$L^\infty(\tau) \subset rca(S) = C(\tau)^*$$

We have also shown that $\{f_d\}_{d\in D}$ has a subnet that converges weakly* to an element in $L^\infty(\tau)$. We conclude that

$$\int_S k_v(t,s)\mu_t(ds) \in L^\infty(\tau) \quad (2.32)$$

3 Convergent Approximations

In this section, we derive a framework for the convergent approximation of the governing dynamical equations. Again, the case in which the control influence is realized by a static hysteresis operator is studied in [4]. We focus on the extension of these results to the dynamic hysteresis operators considered in equations (2.21)–(2.26).

3.1 Finite Dimensional Approximation of Structural System

We choose conventional Galerkin approximation of the infinite dimensional equations appearing in equation (2.2). That is, we select a family of finite dimesional spaces V_n and H-orthogonal projections $P_n : H \to V_n$ such that

$$V_n \subset V \subset H \quad \forall \; n \in \mathbb{N}$$

$$P_n f \to f \text{ in } V - \text{norm as } n \to \infty$$

If $\{\phi_{n,k}\}_{k=1}^{dim(V_n)}$ is a basis for V_n, ie $V_n = span\{\phi_{n,k}\}_{k=1}^{dim(V_n)}$, we set

$$w_n(t,x) = \sum_{k=1}^{dim(V_n)} w_{n,k}(t)\phi_{n,k}(x) \quad t \in \mathbb{R}^+, \quad x \in \mathbb{R}^N$$

The introduction of the approximation above into equation (2.2) yields the familiar representation

$$\sum_{l=1}^{dim(V_n)} (\phi_{n,k},\phi_{n,l})\ddot{w}_{n,l}(t) + \sum_{l=1}^{dim(V_n)} a_1(\phi_{n,k},\phi_{n,l})\dot{w}_{n,l}(t) \quad (3.1)$$

$$+ \sum_{l=1}^{dim(V_n)} a_0(\phi_{n,k},\phi_{n,l})w_{n,l}(t) = \langle(\mathcal{B}_D v)(t), \phi_{n,k}\rangle \quad (3.2)$$

or

$$[M^n]\ddot{\mathbb{W}}^n(t) + [C^n]\dot{\mathbb{W}}^n(t) + [K^n]\mathbb{W}^n(t) = f^n(t)$$

where

$$M^n_{i,j} = (\phi_{n,i},\phi_{n,j}) \quad (3.3)$$
$$C^n_{i,j} = a_1(\phi_{n,i},\phi_{n,j}) \quad (3.4)$$
$$K^n_{i,j} = a_0(\phi_{n,i},\phi_{n,j}) \quad (3.5)$$
$$f^n_i(t) = \langle(\mathcal{B}_D v)(t), \phi_{n,i}\rangle$$

The first-order form of the finite dimensional governing equations are written as

$$\dot{X}^n(t) = [A^n]X^n(t) + F^n(t)$$
$$X^n(0) = X^n_0$$

where

$$[A^n] = \begin{Bmatrix} 0 & [I^n] \\ -[M^n]^{-1}[K^n] & -[M^n]^{-1}[C^n] \end{Bmatrix} \quad F^n(t) = \begin{Bmatrix} 0 \\ [M^n]^{-1} f^n(t) \end{Bmatrix} \quad (3.6)$$

$$X^n(t) = \begin{Bmatrix} W(t) \\ \dot{W}(t) \end{Bmatrix} \quad (3.7)$$

An approximation of the structural subsystem by a Galerkin method then yields

$$\dot{X}^n(t) = [A^n]X^n(t) + B^n \int_S k_v(t,s)\mu_t(ds) \quad (3.8)$$

subject to the initial conditions

$$X^n(0) = X_0^n \quad (3.9)$$

where $X^n, [A^N]$ are defined in equations (3.6) and (3.7), and $B_i^n = \langle g, \phi_{n,i} \rangle$.

In our first result, we fix a specific subspace V_n for the approximation of the structural system. We show that for any selection of the Young's measure characterizing the dynamic hysteresis operator, there is a solution to equation (3.8).

Theorem 3.2. *For any $\mu \in Y$ there is a solution to equation (3.8). Moreover, if $\psi(\mu) = i(u)$ then the solutions of equations (3.8) coincides with the solution of the equation*

$$\dot{X}^n(t) = [A^n]X^n(t) + B^n k_v(t, u(t)) \quad (3.10)$$

subject to the initial conditions

$$X^n(0) = X_0^n \quad (3.11)$$

Proof. If $\mu \in Y$, then by construction there is a net $\{u_d\}_{d \in D} \subseteq L^\infty(\tau; \mathcal{S}) \subseteq L^\infty(\tau; \mathbb{R})$ such that

$$i(u_d) \to \eta \quad \text{weakly* in} \quad (L^1(\tau; C(\mathcal{S})))^*$$

where $\eta = \psi(\mu)$, and ψ is the isometric isomorphism between $L_w^\infty(\tau; rca(\mathcal{S}))$ and $(L^1(\tau; C(\mathcal{S})))^*$. For each $d \in D$ consider the approximate governing equations

$$\dot{X}_d^n(t) = [A^n]X_d^n(t) + B^n k_v(t, u_d(t)) \quad (3.12)$$
$$X_d^n(0) = X_0^n \quad (3.13)$$

For the remainder of this proof, we will suppress the superscript n that emphasizes that we consider a fixed discretization of the structural subsystem throughout

$$\dot{X}_d(t) = [A]X_d(t) + Bk_v(t, u_d(t))$$
$$X_d(0) = X_0$$

We claim that this equation has a solution for each $d \in D$. Clearly

$$\|X_d(t)\| = \|X_0 + \int_0^t \{[A]X_d(\tau) + Bk_v(\tau, u_d(\tau))\}d\tau\| \quad (3.14)$$

$$\leq \|X_0\| + \int_0^t \{c_1\|B\| + c_2\|X_d(\tau)\|\}d\tau \quad (3.15)$$

By proposition 1.6.1 Roubicek [25] p.50, there is a unique solution $X_d \in W^{1,2}(0,T;\mathbb{R}^n)$. By the Gronwall inequality $\{X_d\}_{d \in D}$ is uniformly bounded in $L^\infty(0,T;\mathbb{R}^n)$. Since

$$\|\dot{X}_d(t)\| \leq c\|X_d(t)\| + c_1\|B\|$$

the net $\{\dot{X}_d\}$ is likewise bounded in $L^\infty(0,T;\mathbb{R}^n)$. Thus, $\{X_d\}_{d \in D}$ is bounded in $W^{1,2}(0,T;\mathbb{R}^N)$. There is a subnet

$$X_d \to X_0 \quad \text{weakly in} \quad W^{1,2}(0,T;\mathbb{R}^n)$$

Now, by definition,

$$i(u_d) \to \eta = \psi(\mu) \quad \text{weakly * in} \quad (L^1(\tau; C(\mathcal{S})))^*$$

so that

$$\langle i(u_d), k_v \rangle = \int_0^T k_v(t, u_d(t)) dt \to \int_0^T \int_S k_v(t, \xi) \mu_t(d\xi) dt = \langle \eta, k_v \rangle = \langle \psi(\mu), k_v \rangle$$

But in particular, it is easy to show that if $\psi \in C_0^\infty(0,T)$

$$(t,s) \mapsto \psi(t) k_v(t,s)$$

is in $L^1(\tau; C(\mathcal{S}))$. By the definition of weak * convergence in $L^1(\tau; C(\mathcal{S}))$ we have

$$\langle i(u_d), \psi k_v \rangle = \int_0^T \phi(t) k_v(t, u_d(t)) dt$$

$$\to \int_0^T \int_S \phi(t) k_v(t,s) \mu_t(ds) dt = \langle \psi(\mu), \phi k_v \rangle$$

We can multiply equation (2.12) by $\phi(t) \in C_0^\infty(0,T)$ and conclude that

$$\int_0^T \left(\dot{X}_0(t) - [A]X_0(t) - B \int_S k_v(t,s) \mu_t(ds) \right) \phi(t) dt = 0$$

Hence, X_0 is a solution to equation (3.8).

3.2 Approximation of the Dynamic Hysteresis Operator

We have seen that a rigorous interpretation of the dynamic hysteresis operator empirically studied in [5].

$$(\mathcal{B}_D v)(t) \equiv \int_S k_v(t,s) \mu_t(ds) \cdot g$$

where

$$(B_D v)(t) \equiv \int_S k_v(t,s) \mu_t(ds)$$

is achieved when we interpret μ as a Young measure

$$\mu \in Y \subseteq L_w^\infty(\tau; rca(\mathcal{S})) \approx (L^1(\tau; C(\mathcal{S})))^*$$

In this section we define a consistent approximation scheme for this dynamic hysteresis operator. Suppose that the compact set $\mathcal{S}$ is the union of C^0-conforming finite elements $\Delta_{j,k}$

$$\mathcal{S} = \cup_k \Delta_{j,k} \tag{3.16}$$

where h_j is the mesh parameter, i.e.,

$$\text{diam}(\Delta_{j,k}) < h_j \tag{3.17}$$

$\mathcal{S}_{j,k}$ are the associated nodes, and $\{N_{j,k}\}$ are the associated finite element basis functions. For any $\mu \in Y$ we define the mapping

$$Q_j : Y \to Y \tag{3.18}$$

$$Q_j : \mu \mapsto \mu_j \tag{3.19}$$

by requiring that

$$\mu_{j,t} = \sum_k \alpha_{j,k}(t) \delta_{\mathcal{S}_{j,k}} \tag{3.20}$$

where

$$\alpha_{j,k}(t) \equiv \int_S N_{j,k}(s) \mu_t(ds) \tag{3.21}$$

The operator Q_j embodies the approximation used for the Young's measure that represents the dynamic hysteresis operator $\mathcal{B}$. The convergence of this approximation relies on the following proposition.

Proposition 3.1. For any $\mu \in Y$,

$$\mu_j = Q_j \mu \to \mu \quad \text{weakly} * \text{in} \quad (L^1(\tau; C(\mathcal{S})))^*$$

Proof. Pick some $g \in L^1(\tau; C(\mathcal{S}))$. We can write

$$|\langle \psi(\mu) - \psi(\mu_j), g \rangle| = \left| \int_\tau \int_\mathcal{S} g(t,s) \mu_t(ds) dt - \int_\tau \int_\mathcal{S} g(t,s) \mu_{j,t}(ds) dt \right|$$

Since

$$\mu_{j,t} = \sum_k \alpha_{j,t}(t) \delta_{S_{j,k}} = \sum_k \int_S N_{j,k}(\xi) \mu_t(d\xi) \delta_{S_{j,k}}$$

We have for any $g \in L^1(\tau; C(S))$

$$\int_\tau \int_S g(t,s) \mu_{j,t}(ds) dt \tag{3.22}$$

$$= \int_\tau \int_S g(t,s) \left(\sum_k \int_S N_{j,k}(\xi) \mu_t(d\xi) \right) \delta_{S_{j,k}}(ds) \tag{3.23}$$

$$= \int_\tau \sum_k \int_S N_{j,k}(\xi) \mu_t(d\xi) g(t, S_{j,k}) dt \tag{3.24}$$

$$= \int_\tau \int_S \sum_k g(t, S_{j,k}) N_{j,k}(\xi) \mu_t(d\xi) dt \tag{3.25}$$

Finally we conclude that

$$|\langle \psi(\mu) - \psi(\mu_j), g \rangle| = \left| \int_\tau \int_S (g(t,s) - \sum_k g(t, S_{j,k}) N_{j,k}(s)) \mu_t(ds) dt \right|$$

Suppose, in particular, that $g = u \otimes v \in L^1(\tau) \otimes C(S)$. Then, we have

$$|\langle \psi(\mu) - \psi(\mu_j), g \rangle| \tag{3.26}$$

$$= \left| \int_\tau u(t) \int_S (v(s) - \sum_k v(S_{j,k}) N_{j,k}(s)) \mu_t(ds) dt \right| \tag{3.27}$$

$$\leq \| v(s) - \sum_k v(S_{j,k}) N_{j,k}(s) \|_\infty \tag{3.28}$$

The right-hand side of the above inequality is expressed in terms of the error induced via interpolation of $v \in C(s)$ by C^0-conforming linear finite elements. This error approaches zero as $j \to \infty$. The proposition then follows by the density of $L^1(\tau) \otimes C(S)$ in $L^1(\tau; C(S))$.

Now we can express the finite dimensional governing equation that incorporate approximations of the structural state and the dynamic control influence operator. We seek to find

$$w_n(t) \equiv \sum_{k=1}^n w_{n,k}(t) \phi_{n,k}$$

such that

$$(\ddot{w}_n(t),\xi) + a_1(\dot{w}_n(t),\xi) + a_0(w_n(t),\xi) = ((\mathcal{B}_{D,n}v)(t),\xi) \qquad \forall \xi \in V_n$$

where

$$(\mathcal{B}_{D,n}v)(t) = \int_S k_v(t,s)(Q_n\mu)_t(ds) \cdot g$$

Recall that the approximation of the structural subsystem in equation (2.12) resulted in the equations (3.8)

$$\dot{X}^n(t) = [A^n]X^n(t) + B^n \int_S k_v(t,s)\mu_t(ds) \qquad (3.29)$$

subject to the initial conditions

$$X^n(0) = X_0^n \qquad (3.30)$$

If we use the operator Q_n to approximate the control influence operator, the finite dimensional equations in 3.2

$$\dot{X}^n(t) = [A^n]X^n(t) + B^n \int_S k_v(t,s)(Q_n\mu)_t(ds) \qquad (3.31)$$

and

$$X^n(0) = X_0^n \qquad (3.32)$$

But by definition,

$$\int_S k_v(t,s)(Q_n\mu)_t(ds) = \int_S k_v(t,s) \sum_k \alpha_{n,k}(t)\delta_{s_{n,k}}(ds) \qquad (3.33)$$

$$= \sum_k \alpha_{n,k}(t)k_v(t,s_{n,k}) \qquad (3.34)$$

where

$$\alpha_{n,k}(t) = \int_S N_{n,k}(s)\mu_t(ds)$$

That is, we have the finite dimensional equations with the approximated control influence operator

$$\dot{X}^n(t) = [A^n]X^n(t) + B^n \sum_k \alpha_{n,k}(t)k_v(t,s_{n,k}) \qquad (3.35)$$

and the initial conditions

$$X^n(0) = X_0^n \qquad (3.36)$$

But these equations have precisely the form used in [5] to obtain adaptive control methodologies for hysteretically nonlinear systems.

Now it is straightforward to establish the convergence of the solutions of the finite dimensional approximations in equations (3.8) to the solution of equation (2.12).

Theorem 3.3. *Fix $\mu \in Y \subseteq L_w^\infty(\tau; C(\mathcal{S}))$, and suppose that $(w, \dot{w})$ is the solution of the infinite dimensional equations in (2.12) generated by the $(\mathcal{B}_D v)(t)$. Suppose that*

1. *$\{V_k\}_{k=1}^\infty$ is a sequence of subspaces satisfying the approximation hypotheses (2.12) and (3.8).*
2. *The hysteresis operator is approximated by*

$$(\mathcal{B}_D^n v)(t) = \int_{\mathcal{S}} k_v(t, s)(Q_n \mu)_t(ds)$$

   where Q_n is defined in equations (3.16)–(3.20).
3. *$(w_n, \dot{w}_n)$ defined in equation (2.12) is the solution of the finite dimensional equations in (2.12) (or equivalently (3.8)) generated by $(\mathcal{B}_D^n v)(t)$.*

where $(w, \dot{w})$ is the solution of the infinite dimensional equations guaranteed by Theorem (3.1). Then there is a subsequence s.t.

$$w_n \to w \quad \text{weakly in} \quad L^2(\tau, V)$$
$$\dot{w}_n \to \dot{w} \quad \text{weakly in} \quad L^2(\tau, V)$$

Proof. We note that the proof of this theorem would follow directly from Theorem (5.2) p.126 of [22], if we could show that the function

$$\mu \mapsto \int_{\mathcal{S}} k_v(t, s) \mu_t(ds)$$

acting from $Y \subset L_w^\infty(\tau; rca(s))$ to $L^2(\tau)$ is (weakly*-strongly) continuous. However, we know only that this map is (weakly*-weakly) continuous under the present set of hypothesis. Suppose $(w_n, \dot{w}_n)$ is a solution of the finite dimensional equations. If we substitute $\xi = \dot{w}_{n,k}(t) \phi_{n,k}$, and sum the resulting equations, we obtain

$$(\ddot{w}_n(t), \dot{w}_n(t))_H + a_1(\dot{w}_n(t), \dot{w}_n(t)) + a_0(w_n(t), \dot{w}_n(t)) \quad (3.37)$$
$$= ((\mathcal{B}_{D,n} v)(t), \dot{w}_n(t))_{V^* \times V} \quad (3.38)$$

Since the first two terms are comprised of finite sums, we can write

$$\frac{1}{2} \frac{d}{dt} \left\{ (\dot{w}_n(t), \dot{w}_n(t))_H + a_0(w_n(t), w_n(t)) \right\} + a_1(\dot{w}_n(t), \dot{w}_n(t)) \quad (3.39)$$
$$= ((\mathcal{B}_{D,n} v)(t), \dot{w}_n(t))_{V^* \times V} \quad (3.40)$$

If we integrate in time, we obtain

$$\|\dot{w}_n(t)\|_H^2 + a_0(w_n(t), w_n(t)) + 2 \int_0^t a_1(\dot{w}_n(s), \dot{w}_n(s)) ds \quad (3.41)$$
$$= \|\dot{w}_n(0)\|_H^2 + a_0(w_n(0), w_n(0)) \quad (3.42)$$
$$+ 2 \int_0^t ((\mathcal{B}_{D,n} v)(s), \dot{w}_n(s))_{V^* \times V} ds \quad (3.43)$$

From the inequality
$$|ab| \leq \frac{1}{2}a^2 + \frac{1}{2}b^2$$
we can write

$$|\langle(\mathcal{B}_{D,n}v)(s), \dot{w}_n(s)\rangle|_{V^* \times V} \tag{3.44}$$
$$\leq \|(\mathcal{B}_{D,n}v)(s)\|_{V^*} \|\dot{w}_n(s)\|_V \tag{3.45}$$
$$\leq \frac{1}{2}\left[\left(\frac{1}{\sqrt{2\varepsilon}}\|(\mathcal{B}_{D,n}v)(s)\|_{V^*}\right)^2 + \left(\sqrt{2\varepsilon}\|\dot{w}_n(s)\|_V\right)^2\right] \tag{3.46}$$
$$= \frac{1}{4\varepsilon}\|(\mathcal{B}_{D,n}v)(s)\|_{V^*}^2 + \varepsilon\|\dot{w}_n(s)\|_V^2 \tag{3.47}$$

for any $\varepsilon > 0$. Recall that $a_1(\cdot,\cdot)$ satisfies a Garding inequality

$$a_1(u,v) + \lambda\|u\|_H^2 \geq c\|u\|_V^2 \tag{3.48}$$

We can combine (3.48), (3.44) and (3.41) to obtain

$$\|\dot{w}_n(t)\|_H^2 + c\|w_n(t)\|_V^2 + 2\int_0^t (c\|\dot{w}_n(s)\|_V^2 - \lambda\|\dot{w}_n(s)\|_H^2)ds \tag{3.49}$$
$$\leq \|\dot{w}_n(0)\|_H^2 + c\|w_n(0)\|_V^2 \tag{3.50}$$
$$+ 2\int_0^t \left\{\frac{1}{4\varepsilon}\|(\mathcal{B}_{D,n}v)(s)\|_{V^*}^2 + \varepsilon\|\dot{w}_n(s)\|_V^2\right\}ds \tag{3.51}$$

When we introduce

$$k_n(t) \equiv \|\dot{w}_n(0)\|_H^2 + c\|w_n(0)\|_V^2 + \frac{1}{2\varepsilon}\int_0^t \|(\mathcal{B}_{D,n}v)(s)\|_{V^*}^2 ds$$

equation (3.49) can be re-written as

$$\|\dot{w}_n(t)\|_H^2 + c\|w_n(t)\|_V^2 + 2\int_0^t (c-\varepsilon)\|\dot{w}_n(s)\|_V^2 ds \tag{3.52}$$
$$\leq k_n(t) + 2\lambda \int_0^t \|\dot{w}_n(s)\|_H^2 ds \tag{3.53}$$

But $k_n(t)$ is uniformly bounded, $|k_n(t)| < c$, since

$$\dot{w}_n(0) \to w_1 \quad \text{in} \quad H$$
$$w_n(0) \to w_0 \quad \text{in} \quad V$$

and by construction $\mathcal{B}_{D,n}v$ is uniformly bounded

$$\|\mathcal{B}_{D,n}v\|_{V^*}^2 = \left|\int_S k_v(t,s)(Q_n\mu)_t(ds)\right|^2 \|g\|_{V^*}^2 \leq \text{constant}$$

Via Gronwall's inequality we can conclude that

$$\|\dot{w}_n\|_{L^2(0,T;H)} < constant$$

and subsequently

$$\|w_n\|_{L^2(0,T;V)} < constant$$

$$\|\dot{w}_n\|_{L^2(0,T;V)} < constant$$

There is a subsequence such that

$$w_{n_k} \to w \quad \text{weakly in} \quad L^2(0,T;V)$$

and

$$\dot{w}_{n_k} \to \hat{w} \quad \text{weakly in} \quad L^2(0,T;V)$$

But by definition, we have

$$\int_0^T \dot{w}_{n_k}(t)\phi(t)dt = -\int_0^T w_{n_k}(t)\dot{\phi}(t)dt$$

for all $\phi \in C_0^\infty(0,T)$. Choose any $v^* \in V^*$. Then

$$\langle v^*, \int_0^T \dot{w}_{n_k}(t)\phi(t)dt \rangle = \langle v^*, -\int_0^T w_{n_k}(t)\dot{\phi}(t)dt \rangle \tag{3.54}$$

But the left-hand side can be rewritten as

$$\langle v^*, \int_0^T \dot{w}_{n_k}(t)\phi(t)dt \rangle = \int_0^T \langle \phi(t)v^*, \dot{w}_{n_k}(t) \rangle dt$$

Since $\phi(t)v^* \in L^2(0,T;V^*)$, we conclude that

$$\langle v^*, \int_0^T \dot{w}_{n_k}(t)\phi(t)dt \rangle \to \int_0^T \langle \phi(t)v^*, \tilde{w}(t) \rangle dt \tag{3.55}$$

$$= \int_0^T \langle v^*, \phi(t)\tilde{w}(t) \rangle dt \tag{3.56}$$

$$= \langle v^*, \int_0^T \phi(t)\tilde{w}(t)dt \rangle \tag{3.57}$$

Similarly, the right-hand side of equation (3.54) can be rewritten as

$$\langle v^*, \int_0^T w_{n_k}(t)\dot{\phi}(t)dt \rangle = \int_0^T \langle \dot{\phi}(t)v^*, w_{n_k}(t) \rangle dt \tag{3.58}$$

$$\to \int_0^T \langle \dot{\phi}(t)v^*, w(t) \rangle dt \tag{3.59}$$

$$= \langle v^*, \int_0^T \dot{\phi}(t)w(t)dt \rangle \tag{3.60}$$

Therefore, we have

$$\langle v^*, \int_0^T \dot\phi(t)w(t)dt + \int_0^T \phi(t)\tilde w(t)dt\rangle = 0$$

for all $v \in V^*$, or

$$\int_0^T \phi(t)\tilde w(t)dt = -\int_0^T \dot\phi(t)w(t)dt$$

This is simply the definition of the distributional derivative of $w(t)$

$$\tilde w(t) \equiv \dot w$$

The only remaining task is to show that the pair $(w, \dot w) \in L^2(0,T;V) \times L^2(0,T;V)$ satisfies the governing equations. If we substitute $\xi = \psi(t)\eta$, where $\psi \in C_0^\infty(0,T)$ and $q \in V$, into equation (3.54) and integrate over time we get

$$\int_0^T \{(\ddot w_n(t),\eta)_H + a_1(\dot w_n(t),\eta) + a_0(w_n(t),\eta)\}\psi(t)dt \qquad (3.61)$$

$$= \int_0^T ((\mathcal{B}_{D,n}v)(t),\eta)\psi(t)dt \qquad (3.62)$$

We can integrate the first term by parts to get

$$(\dot w_n(t),\eta)_H \psi\Big|_0^T - \int_0^T (\dot w_n(t),\eta)_H \dot\psi(t)dt$$

It is straightforward to show that each of the maps

$$f \mapsto \int_0^T (f(t),\eta)_H \dot\psi(t)dt$$

$$f \mapsto \int_0^T a_1(f(t),\eta)_H \psi(t)dt$$

$$f \mapsto \int_0^T a_0(f(t)\eta)_H \psi(t)dt$$

defines a bounded linear functional on $L^2(0,T;V)$. Since $(w_n, \dot w_n) \in L^2(0,T;V) \times L^2(0,T;V)$ are weakly convergent, we conclude that the left-hand side of equation (3.61) converges to

$$\int_0^T \Big\{-(\dot w_n(t),\eta)_H \dot\psi(t) + (a_1(\dot w_n(t),\eta) + a_0(w_n(t),\eta))\psi(t)\Big\}dt$$

If we expand the right-hand side of equation (3.61), we obtain

$$\int_0^T ((\mathcal{B}_{D,n}v)(t),\eta)\psi(t)dt = \int_0^T \int_S \psi(t)k_v(t,s)(Q_n\mu)_t(ds)dt\langle g,\eta\rangle$$

However, since $\psi(t)k_v(t,s) \in L^1(\tau, C(\mathcal{S}))$ and

$$Q_n \mu \mapsto \mu \quad \text{weakly* in} \quad (L^1(\tau, C(\mathcal{S})))^*$$

we conclude that

$$\int_0^T ((\mathcal{B}_{D,n}v)(t), \eta)\psi(t)dt \to \int_0^T ((\mathcal{B}_D v)(t), \eta)\psi(t)dt$$

Combining equations (3.44) and (3.41), we have

$$\int_0^T \left\{ \frac{d}{dt}(\dot{w}(t), \eta)_H + a_1(\dot{w}(t), \eta) + a_0(w(t), \eta) - ((\mathcal{B}_D v)(t), \eta) \right\} \psi(t) dt = 0$$

for any $\psi \in C_0^\infty(0,T)$, $\forall \eta \in V$. We conclude that

$$\langle \ddot{w}(t), \eta \rangle_{V^* \times V} + a_1(\dot{w}(t), \eta) + a_0(w(t), \eta) = \langle (\mathcal{B}_D v)(t), \eta \rangle_{V^* \times V} \quad \forall \eta \in V$$

and the theorem is proved.

References

1. Visintin A. *Differential Models of Hysteresis*. Springer-Verlag, 1991.
2. H.T. Banks, A.J. Kurdila and G. Webb. Identification and hysteretic control influence operators representing smart actuators, Part (II): Convergent approximations. *Journal of Intelligent Material Systems and Structures*, 8:536–550, June 1997.
3. H.T. Banks, A.J. Kurdila and G. Webb. Identification and hysteretic control influence operators representing smart actuators, Part (I):Formulation. *Mathematical Problems in Engineering*, 3:287–328, 1997.
4. A.J. Kurdila and G. Webb. Compensation for distributed hysteresis operators in active structural systems. *Journal of Guidance, Control and Dynamics*, 20(6), November 1997.
5. G. Webb, A.J. Kurdila and D.L. Lagoudas. Hysteresis modeling of SMA actuators for control applications. *Journal of Intelligent Material Systems and Structures, accepted for publication*, 1998.
6. Q. Su, J. Cherif, Y. Wen, C. Bailly and M. Wuttig. Magneto-mechanical properties of Terfenol-D thin films. In *Proceedings of the SPIE, Smart Structures and Material, 1997, Mathematics and Control in Smart Structures*, p. 234–240, San Diego, CA, March 1997.
7. C.A. Rogers. Active vibration and structural acoustic contol of shape memory alloy hybrid composites: experimental results. *Journal of the Acoustical Society of America*, 88:2803–2811, 1990.
8. M. Kamlah, U. Bohle, D. Munz and C. Tsakmakis. Macroscopic description of the nonlinear electromechanical coupling in ferroelectrics. In *Proceedings of the SPIE, Smart Structures and Material, 1997, Matehimatics and Control in Smart Structures*, p. 144–155, San Diego, CA, March 1997.
9. A.J. Kurdila, J. Li and M. Fulton. Nonlinear control of PZT actuated trailing edge flaps. In *presented at Smart Structures and Materials*, Newport Beach, CA, March 1999.

10. J. Warga. *Optimal Control of Differential and Functional Equations*. Academic Press, New York, 1972.
11. J.A. Shaw and S. Kyriakides. On the nucleation and propagation of phase transformation fronts in NiTi alloy. *Acta Mater.*, 45(2):683–700, 1997.
12. J.A. Shaw and S. Kyriakides. Initiation and propagation of localized deformation in elasto-plastic strips under uniaxial tension. *International Journal of Plasticity*, 13(10):837–871, 1999.
13. J.M. Ball. Convexity conditions and existence theorems in nonlinear elasticity. *Archive Rat. Mech. Anal.*, 63:337–403, 1977.
14. J.M. Ball and R.D. James. Fine phase mixtures as minimizers of energy. *Archive Rat. Mech. Anal.*, 100:13–52, 1988.
15. J.M. Ball and R.D. James. Proposed experimental tests of a theory of fine microstructrue and the two well problem. *Phil. Trans. Royal Soc. London, Series A*, 338:389–450, 1992.
16. A. Baz, K. Imam and J. McCoy. Active vibration control of flexible beams using shape memory actuators. *Journal of Sound and Vibration*, 140:437–456, 1990.
17. A. Bensoussan, G. DaPrato, M. Delfour and S. Mitter. *Representation and Control of Infinite Dimensional Systems, Volume I*. Birkhauser, Boston, 1992.
18. M. Brokate and J. Sprekels. Optimal control of thermomechanical phase transitions in shape memory alloys: Necessary conditions of optimality. *Math. Methods in Appl. Sci*, 14:265–280, 1993.
19. M. Krasnoselskii A. Pokrovskii. *Systems with Hysteresis*. Springer-Verlag, Berlin, 1989.
20. R. Abeyaratne and J.K. Knowles. A continuum model of a thermoelastic solid capable of undergoing phase transitions. *Journal of the Mechanics and Physics of Solids*, 41:541–571, 1993.
21. R. Temam. *Infinite Dimensional Dynamical Systems in Mechanics and Physics*. Springer-Verlag, New York, 1988.
22. H.T. Banks, R.C. Smith and Y. Wang. *Smart Material Structures: Models, Estimation and Control*. John Wiley & Sons, Masson, Paris, 1996.
23. R.D. James and D. Kinderlehrer. Mathematical Approaches to the Study of Smart Materials. *Mathematics of Smart Materials and Strctures*, 1919:2–18, 1993.
24. R.V. Gamkrelidze. *Principles of Optimal Control Theory*. Plenum Press, New York, 1978.
25. T. Roubicek. *Relaxation in optimization Theory and Variational Calculus*. Walter de Gruyter Publisher, New York, 1997.
26. Z. Bo and D.L. Lagoudas. Thermomechnical modeling of polycrystalline SMA's under cyclic loading, Part(IV): Modeling of minor hysteresis loops. *International Journal of Engineering Science*, 1998.

Robust Adaptive Control of Nonlinear Systems with Dynamic Backlash-like Hysteresis

Chun-Yi Su[1], Masahiro Oya[2], and Xinkai Chen[3]

[1] Department of Mechanical Engineering, Concordia University, Montreal, Quebec, H3G 1M8, Canada
[2] Department of Control Engineering, Kyushu Institute of Technology, 1-1 Sensui, Tobata, Kitakyushu, 804-8550, Japan
[3] Department of Information Sciences, Tokyo Denki University, Hatoyama-machi, hiki-gun, Saitama 350-0394, Japan

Abstract

This paper deals with adaptive control of nonlinear dynamic systems preceded by unknown backlash-like hysteresis nonlinearities, where the hysteresis is modeled by a dynamic equation. By utilizing this dynamic model and by combining a universal function approximator with adaptive control techniques, a stable robust adaptive control algorithm is developed without constructing a hysteresis inverse. The stability of the close-loop system is shown using Lyapunov arguments. The effectiveness of the proposed method is demonstrated through simulations.

1 Introduction

Some mechanical actuators exhibit a backlash-like hysteresis in their characteristics. In a typical hydraulic actuator, as one example, the backlash is caused by the gears controlling the valve which controls the actual input to the tank. The spacing between gears results in a hysteresis-type function that relates the input and the output angles. If backlash spacing in gears is not accounted for, it will result in the degradation of system performance, reducing positioning accuracy and even may lead to instability [15]. Generally speaking, hysteresis characteristics are nondifferentiable nonlinearities for which traditional control techniques are insufficient and control of a system preceded by a hysteresis is typically challenging.

The development of control techniques to mitigate effects of backlash hystereses has been addressed in several papers [1, 2, 4, 7, 8, 14-16]. A common feature of those schemes is that they rely on the construction of an inverse hysteresis to mitigate the effects of the hysteresis. These results, especially [15, 16], provide a theoretic framework which can serve as a base for future research.

Inspired by the above research, this paper will address the control of nonlinear systems preceded by unknown backlash-like hysteresis nonlinearities.

The novelty is that a dynamic hysteresis model is defined in this paper to pattern a backlash-like hysteresis. Rather than constructing an inverse hysteresis nonlinearity to mitigate the effects of the hysteresis, we propose a new approach for controller synthesis by using the properties of the hysteresis model. To deal with the nonlinearities in the plant, a universal function approximator is adopted for compensation of plant nonlinearities, where parameters of the approximator are adapted using a Lyapunov-based design. The new control law ensures global stability of the adaptive system. Simulations performed on a nonlinear system illustrate and clarify the approach.

2 Problem Statement

The controlled system consists of a nonlinear plant preceded by a backlash-like hysteresis actuator, that is, the hysteresis is present as an input of the nonlinear plant. It is a challenging task of major practical interests to develop a control scheme for unknown backlash-like hysteresis. The development of such a control scheme will now be pursued.

A backlash-like hysteresis nonlinearity can be denoted as an operator

$$w(t) = P[v](t) \tag{2.1}$$

with $v(t)$ as input $v(t)$ and $w(t)$ as output $w(t)$. The notation $[\cdot](t)$ represents the fact that the operator in $[\cdot]$ is dependent on the trajectory, $v \in C^o[0, t]$, not an instantaneous value $v(t)$. The operator $P(v(t))$ will be discussed in details in the subsequent section. The nonlinear dynamic system being preceded by the above hysteresis is described in the canonical form,

$$x^{(n)}(t) + f(x(t), \dot{x}(t), ..., x^{(n-1)}(t)) = bw(t) \tag{2.2}$$

where f is an unknown continuous nonlinear function, and control gain b is unknown but constant. It is a common assumption that the sign of b is known. From this point onward, without losing generality, we shall assume $b > 0$. It should be noted that more general classes of nonlinear systems can be transformed into this structure [5].

The control objective is to design a control law for v(t) in (2.1), to force the plant state vector, $\mathbf{x} = [x, \dot{x}, ..., x^{(n-1)}]^T$, to follow a specified desired trajectory, $\mathbf{x}_d = [x_d, \dot{x}_d, ..., x_d^{(n-1)}]^T$, i.e., $\mathbf{x} \to \mathbf{x}_d$ as $t \to \infty$. The following assumption regarding the desired trajectory $\mathbf{x}_d$ is required in the paper.

Assumption 1: The desired trajectory, $\mathbf{x}_d = [x_d, \dot{x}_d, ..., x_d^{(n-1)}]^T$ is continuous and available. Furthermore $[\mathbf{x}_d^T, x_d^{(n)}]^T \in \Omega_d \subset R^{n+1}$ with Ω_d a compact set.

3 Backlash-like Hysteresis Model and its Properties

Traditionally, a backlash hysteresis nonlinearity can be described by,

$$w(t) = P(v(t)) = \begin{cases} c(v(t) - B) & \text{if } \dot{v}(t) > 0 \text{ and } w(t) = c(v(t) - B) \\ c(v(t) + B) & \text{if } \dot{v}(t) < 0 \text{ and } w(t) = c(v(t) + B) \\ w(t_-) & \text{otherwise} \end{cases} \quad (3.1)$$

where $c > 0$ is the slope of the lines and $B > 0$ is the backlash distance. This model is itself discontinuous and may not be amenable to controller design for the nonlinear systems (2.2).

Instead of using the above model, in this paper we define a continuous-time dynamic model to describe a class of backlash-like hysteresis, as given by,

$$\frac{dw}{dt} = \alpha \left| \frac{dv}{dt} \right| (cv - w) + B_1 \frac{dv}{dt} \quad (3.2)$$

where α, c, and B_1 are constants, satisfying $c > B_1$.

Remark: Generally, modelling hysteresis nonlinearities is still a research topic and the reader may refer to [6] for a recent review.

We shall now examine the solution properties of the dynamic model (3.2) and explain the corresponding switching mechanism, which is crucial for design of the controller. The equation (3.2) can be solved explicitly for v piecewise monotone,

$$w(t) = cv(t) + d(v) \quad (3.3)$$

with $\quad d(v) = [w_o - cv_o]e^{-\alpha(v-v_o)sgn\dot{v}} + e^{-\alpha v sgn\dot{v}} \int_{v_o}^{v} [B_1 - c]e^{\alpha \zeta(sgn\dot{v})} d\zeta$

for $\dot{v}$ constant and $w(v_o) = w_o$. Analyzing the solution (3.3), we see that it is composed of a line with the slope c, together with a term $d(v)$. For $d(v)$, it can be easily shown that if $w(v; v_o, w_o)$ is the solution of (3.3) with initial values (v_o, w_o), then, if $\dot{v} > 0$ ($\dot{v} < 0$) and $v \to +\infty$ ($-\infty$), one has

$$\lim_{v \to \infty} d(v) = \lim_{v \to \infty} [w(v; v_o, w_o) - f(v)] = -\frac{c - B_1}{\alpha} \quad (3.4)$$

$$(\lim_{v \to -\infty} d(v) = \lim_{v \to -\infty} [w(v; v_o, w_o) - f(v)] = \frac{c - B_1}{\alpha}) \quad (3.5)$$

It should be noted that the above convergence is exponential at the rate of α. Solution (3.3) and properties (3.4) and (3.5) show that $w(t)$ eventually satisfies the first and second conditions of (3.1). Furthermore, setting $\dot{v} = 0$ results in $\dot{w} = 0$ which satisfies the last condition of (3.1). This implies that the dynamic equation (3.2) can be used to model a class of backlash-like hystereses and is an approximation of backlash hysteresis (3.1).

Let us use an example for specified initial data to show the switching mechanism for the dynamic model (3.2) when $\dot{v}$ changes direction. We note that when $\dot{v} > 0$ on $w(0) = 0$ and $v(0) = 0$, the solution (3.3) gives

$$w(t) = cv(t) - \frac{c - B_1}{\alpha}(1 - e^{-\alpha v(t)}) \quad \text{for} \quad v(t) \geq 0 \text{ and } \dot{v} > 0 \tag{3.6}$$

Let v_s be a positive value of v and consider now a specimen such that v is increasing along the initial curve (3.6) until a time t_s at which v reaches the level v_s. Suppose now that from the time t_s, the signal v is decreased. In this case, w is given by

$$w(t) = cv(t) + \frac{c - B_1}{\alpha}[1 - (2e^{-\alpha v_s} - e^{-2\alpha v_s})e^{\alpha v(t)}] \quad \text{for} \quad \dot{v} < 0 \tag{3.7}$$

where $v < v_s$. Equations (3.6) and (3.7) indeed show that w switches exponentially from the line $cv(t) - \frac{c-B_1}{\alpha}$ to $cv(t) + \frac{c-B_1}{\alpha}$ to generate backlash-like hysteresis curves.

To confirm the above analysis, the solutions of (3.2) can be obtained by numerical integration with v as the independent variable. Figure 7.1 shows that model (3.2) indeed generates backlash-like hysteresis curves, which confirms the above analysis. The details are described in the section of simulation studies. It should be mentioned that the parameter α determines the rate at which $w(t)$ switches between $-\frac{c-B_1}{\alpha}$ and $\frac{c-B_1}{\alpha}$. The larger the parameter α is, the faster the transition in $w(t)$ is going to be. However, the backlash distance is determined by $\frac{c-B_1}{\alpha}$ and the parameter must satisfy $c > B_1$. Therefore, the parameter α cannot be chosen freely. A compromise should be made in choosing a suitable parameter set $\{\alpha, c, B_1\}$ to model the required shape of backlash-like hysteresis. If the values of the backlash slope and distance are not known implicitly, then adaptations will be used to estimate them. This topic will be clarified shortly.

4 Lyapunov-based Control Structure

From the solution structure (3.3) of the model (3.2) we see that the signal $w(t)$ is expressed as a linear function of input signal $v(t)$ plus a bounded term. In this section, we shall propose a Lyapunov-based control structure for plants of the form described by (2.2), preceded by the hysteresis described in (3.2). The proposed controller will lead to global stability and yields desired tracking.

Using the solution expression (3.3), the system (2.2) becomes,

$$x^{(n)}(t) + f(x(t), \dot{x}(t), ..., x^{(n-1)}(t)) = bcv(t) + bd(v(t)) \tag{4.1}$$

which results in a linear relation to the input signal $v(t)$. It is very important to note that equations (3.4) or (3.5) imply that there exists a uniform bound ρ such that

$$\|d(v)\| \leq \rho \tag{4.2}$$

In presenting the Lyapunov-based control structure, the following definition is required:

$$\tilde{\mathbf{x}} = \mathbf{x} - \mathbf{x}_d \tag{4.3}$$

where $\tilde{\mathbf{x}}$ represents the tracking error vector.

A filtered tracking error is defined as

$$s(t) = (\frac{d}{dt} + \lambda)^{(n-1)} \tilde{x}(t) \quad with \quad \lambda > 0 \tag{4.4}$$

Remark: It has been shown in [11] that the equation $s(t) = 0$ defines a time-varying hyperplane in R^n on which the tracking error vector $\tilde{\mathbf{x}}(t)$ decays exponentially to zero.

Define a continuous function,

$$sat(s) = \begin{cases} 1 - exp(-s/\gamma), & if\ s > 0 \\ -1 + exp(s/\gamma) & if\ s \leq 0 \end{cases} \tag{4.5}$$

with γ being any small positive constant. As $\gamma \to 0$, $sat(s)$ approaches a step transition from -1 at $s = 0^-$ to 1 at $s = 0^+$ continuously [3].

With the above in mind, we have the following lemma to establish the existence of an ideal control, v^*, that leads to $\mathbf{x} \to \mathbf{x}_d$ as $t \to \infty$.

Lemma 1 *For the plant in equation (2.2) with the hysteresis (3.2) at the input, all the closed-loop signals are bounded and the state vector $\mathbf{x}(t) \to \mathbf{x}_d(t)$ as $t \to \infty$ with a desired Lyapunov-based controller*

$$v^*(t) = -ks(t) + v_f(\mathbf{x}(t)) + \frac{1}{bc}v_{fd}(t) - k_d sat(s) \tag{4.6}$$

where $v_f(\mathbf{x}(t)) = \frac{1}{bc}f(\mathbf{x}(t))$, $v_{fd}(t) = x_d^{(n)}(t) - \Lambda_v^T \tilde{\mathbf{x}}(t)$ with $\Lambda_v^T = [0, \lambda^{(n-1)}, (n-1)\lambda^{(n-2)}, ..., (n-1)\lambda]$, k is a constant, and k_d satisfies $k_d \geq \rho/c$.

Proof: Equation (4.4) can be rewritten as $s(t) = \Lambda^T \tilde{\mathbf{x}}(t)$ with $\Lambda^T = [\lambda^{(n-1)}, (n-1)\lambda^{(n-2)}, ..., 1]$. The derivative of the error metric (4.4) can be written as:

$$\begin{aligned} \dot{s}(t) &= -x_d^{(n)}(t) + \Lambda_v^T \tilde{\mathbf{x}}(t) - f(\mathbf{x}(t)) + bcv^*(t) + bd(v^*(t)) \\ &= -v_{fd}(t) - f(\mathbf{x}(t)) + bcv^*(t) + bd(v^*(t)) \\ &= bc(-ks(t) - k_d sat(s)) + bd(v^*(t)) \end{aligned} \tag{4.7}$$

Define a Lyapunov Function candidate $V(t) = \frac{1}{2bc}s^2$. Differentiating V along the system (4.7) yields

$$\begin{aligned} \dot{V}(t) &= -ks^2 - k_d s \cdot sat(s) + \frac{1}{c}d(v^*(t))s \\ &\leq -ks^2 \end{aligned} \tag{4.8}$$

where the fact that $s \cdot sat(s) \geq 0$ has been used. This shows that V is a Lyapunov function which leads to global boundedness of s. To complete the proof and establish asymptotic convergence of the tracking error, it is necessary to show that $s \to 0$ as $t \to \infty$. This is accomplished by applying Barbalat's Lemma [9] to the continuous, non-negative function:

$$V_1(t) = V(t) - \int_0^t (\dot{V}(\tau) + ks^2(\tau))d\tau \text{ with}$$
$$\dot{V}_1(t) = -ks^2(t) \qquad (4.9)$$

It can easily be shown that every term in (4.7) is bounded, hence $\dot{s}$. This implies that $\dot{V}_1(t)$ is a uniformly continuous function of time. Since V_1 is bounded below by 0, and $\dot{V}_1(t) \leq 0$ for all t, use of Barbalat's lemma proves that $\dot{V}_1(t) \to 0$. Therefore, from (4.9) it can be demonstrated that $s(t) \to 0$ as $t \to \infty$. The remark following equation (4.4) indicates that $\tilde{\mathbf{x}}(t) \to 0$ as $t \to \infty$.

If the parameters b and c as well as the nonlinear function $f(\mathbf{x}(t))$ are unknown, the control law described above cannot be directly applied. Therefore, the challenge addressed in the sequent sections is the development of an adaptive controller to deal with unknown nonlinear function $f(\mathbf{x}(t))$ as well as the unknown parameters b and c.

5 Function Approximation Using Gaussian Functions

When the nonlinearity $f(\mathbf{x}(t))$ as well as the constants b and c are unknown, the desired controller $v^*(t)$ given in (4.6) is not applicable. In order to develop a stable adaptive control law, a parameterized approximator shall be used to approximate the unknown nonlinearity $v_f(\mathbf{x}(t)) = \frac{1}{bc} f(\mathbf{x}(t))$. Here, it should be emphasized that in this paper we are addressing the situation where an explicit linear parameterization of the nonlinear function $f(x(t), \dot{x}(t), ..., x^{(n-1)}(t))$ is either unknown or impossible. Otherwise, construction of a parameterized approximator is not necessary [10, 13].

In literature, several function approximators can be applied for this purpose [10, 13]. In this paper, we shall employs Gaussian as base functions to approximate $v_f(\cdot)$ over a compact region of the state space. The general form of the function approximators can be expressed as,

$$h(\mathbf{x}(t)) = \sum_{j=1}^{N} \omega_j(t) g_j \left(\sigma_j(t) \| \mathbf{x}(t) - \xi_j(t) \| \right) \qquad (5.1)$$

where $h: U \subset R^n \to R$, $\omega_j(t)$ is the connection weight; $g_j \left(\sigma_j(t) \| \mathbf{x}(t) - \xi_j(t) \| \right) = \prod_{i=1}^{n} \mu_{A_j^i}(x_i(t))$ and $\mu_{A_j^i}(x_i(t))$ is the *Gaussian membership function*, defined by

$$\mu_{A_j^i}(x_i(t)) = \exp\left(-(\sigma_j^i(t)(x_i(t) - \xi_j^i(t)))^2\right) \qquad (5.2)$$

where $\sigma_j^i(t)$ and $\xi_j^i(t)$ are real-valued parameters, $\sigma_j(t) = [\sigma_j^1(t), \sigma_j^2(t), \ldots, \sigma_j^n(t)]^T$ and $\xi_j(t) = [\xi_j^1(t), \xi_j^2(t), \cdots, \xi_j^n(t)]^T$. Notice that contrary to the traditional notation, in this paper we use $1/\sigma_j^i(t)$ to represent the variance just for the convenience of later development. Note that the above function approximator can be implemented by either neural networks [10] or fuzzy logic [13].

Then, the function approximator (5.1) can be re-written as

$$h(\mathbf{x}(t)) = W^T(t) \cdot G(\mathbf{x}(t), \xi(t), \sigma(t)) \tag{5.3}$$

where $G(\mathbf{x}(t), \xi(t), \sigma(t)) = [g_1(\sigma_1(t)\|\mathbf{x}(t) - \xi_1(t)\|), g_2(\sigma_2(t)\|\mathbf{x}(t) - \xi_2(t)\|), \cdots, g_N(\sigma_N(t)\|\mathbf{x}(t) - \xi_N(t)\|)]^T$, $W(t) = [\omega_1(t), \omega_2(t), \cdots, \omega_N(t)]^T$, $\xi(t) = [\xi_1(t), \xi_2(t), \cdots, \xi_N(t)]^T$, $\sigma(t) = [\sigma_1(t), \sigma_2(t), \cdots, \sigma_N(t)]^T$.

The ability of the approximator (5.3) to uniformly approximate smooth functions over a compact set is well documented in the literature (see [10] for a reference). In particular, it has been shown that given a smooth function $v_f(\mathbf{x}(t)) : U \mapsto R$, where $U \in R^n$ is a compact set and $\varepsilon_h > 0$, there exists a Gaussuian function vector $G(\mathbf{x}, \xi^*, \sigma^*)$ and a weight vector W^* such that $\sup_{\mathbf{x} \in U} |v_f(\mathbf{x}(t)) - W^{*T} G(\mathbf{x}, \xi^*, \sigma^*)| < \varepsilon_h$.

To construct $v_f^* = W^{*T} G(\mathbf{x}, \xi^*, \sigma^*)$, the values of the parameter vectors W^*, σ^*, and ξ^* are required. Unfortunately, they are usually unavailable. Normally, the unknown parameter vectors W^*, σ^*, and ξ^* are replaced by their estimates $\hat{W}$, $\hat{\sigma}$, and $\hat{\xi}$. Then, the estimation function $\hat{v}_f = \hat{W}^T G\left(\mathbf{x}, \hat{\xi}, \hat{\sigma}\right)$ is used instead of v_f^* to approximate the unknown function v_f. Using the estimation function $\hat{v}_f$, the approximation error between $v_f(\mathbf{x}(t))$ and $\hat{v}_f(\mathbf{x}(t))$ can be established as follows.

Lemma 2 *Define the estimation errors of the parameter vectors as*

$$\tilde{W}(t) = W^* - \hat{W}; \quad \tilde{\xi}(t) = \xi^* - \hat{\xi}; \quad \tilde{\sigma}(t) = \sigma^* - \hat{\sigma} \tag{5.4}$$

The estimation error function $\tilde{v}_f(t) = v_f(\mathbf{x}(t)) - \hat{v}_f(\mathbf{x}(t))$ is

$$\tilde{v}_f(t) = \tilde{W}^T(t) \cdot (\hat{G}(t) - G'_\xi \hat{\xi}(t) - G'_\sigma \hat{\sigma}(t)) + \hat{W}^T(t) \cdot (G'_\xi \tilde{\xi}(t) + G'_\sigma \tilde{\sigma}(t)) + d_f(t) \tag{5.5}$$

where $G'_\xi \in R^{N \times (Nn)}$ and $G'_\sigma \in R^{N \times (Nn)}$ are derivatives of $G(\mathbf{x}(t), \xi^*, \sigma^*)$ with respect to ξ^* and σ^* at $(\hat{\xi}(t), \hat{\sigma}(t))$, respectively, $d_f(t)$ is a residual term. Moreover, $d_f(t)$ satisfies,

$$|d_f(t)| < \theta_f^{*T} \cdot Y_f(t) \tag{5.6}$$

and $\theta_f^* \in R^{4 \times 1}$ is an unknown constant vector, being composed of optimal weight matrices and some bounded constants; and $Y_f(t) = [1, \|\hat{W}(t)\|, \|\hat{\xi}(t)\|, \|\hat{\sigma}(t)\|]^T$ is a known function vector.

Proof: See Appendix A

Remarks:

1. The role of Lemma 2 is that through the first Taylor's expansion of $v_f^*(\mathbf{x}(t))$ near $(\hat{\xi}(t), \hat{\sigma}(t))$, the function approximation error $\tilde{v}_f(t)$ in (5.5) has been expressed in a linearly-parameterizable form with respect to $\tilde{\xi}$ and $\tilde{\sigma}$, which makes the updates of $(\hat{\xi}(t), \hat{\sigma}(t))$ possible. Moreover, the residual term is bounded by a linear expression with a known function vector as in (5.6). Thus, adaptive control techniques can be applied to deal with this residual term.
2. It should be noted that no explicit expressions for ξ^*, σ^*, W^*, θ_f^* are required since these values can be learned by using the adaptive algorithm developed in the following section.

6 Adaptive Controller Design

We are now ready to develop an adaptive control law to achieve the control objective for the plant described by (2.2), preceded by the hysteresis described in (3.2) with unknown nonlinear function $f(\mathbf{x}(t))$ and b as well as c. Before proposing an adaptive control law, the following assumptions regarding the plant and hysteresis are required.

(A1) There exist known constants b_{min} and b_{max} such that the control gain b in (2.2) satisfies $b \in [b_{min}, b_{max}]$.

(A2) There exist known constants c_{min} and c_{max} such that the slope c in (3.1) satisfies $c \in [c_{min}, c_{max}]$.

(A3) The bound ρ for the relation $\|d(v)\| \leq \rho$ is known.

Remark: Assumption (A1) is common for the nonlinear system described by (2.2) [11]. Assumption (A2) assumes the slope range of a backlash hysteresis nonlinearity, which is reasonable. Assumption (A3) requires knowledge of the upper bound of the hysteresis loop, which is quite reasonable and practical.

In presenting the developed robust adaptive control law, the following definition is required:

$$\tilde{\phi} = \phi - \hat{\phi} \tag{6.1}$$

where $\hat{\phi}$ is an estimate of ϕ, which is defined as $\phi \triangleq (bc)^{-1}$.

Given the plant in (2.2), hysteresis model (3.2), and the function approximation (5.5), subject to the assumption described above, the following control and adaptation laws are presented:

$$v(t) = -ks(t) + \hat{\phi} v_{fd}(t) + \hat{W}^T G\left(\mathbf{x}, \hat{\xi}, \hat{\sigma}\right) - (\hat{\theta}_f^T Y_f + k^*) sat(s) \tag{6.2}$$

$$u_{fd}(t) = x_d^{(n)}(t) - \Lambda_v^T \tilde{\mathbf{x}}(t) \tag{6.3}$$

$$\dot{\hat{W}} = -\Gamma_1 s(t)(\hat{G} - G'_\xi \hat{\xi} - G'_\sigma \hat{\sigma}) \tag{6.4}$$

$$\dot{\hat{\xi}} = -\Gamma_2 s(t)(\hat{W}^T G'_\xi)^T \tag{6.5}$$

$$\dot{\hat{\sigma}} = -\Gamma_3 s(t)(\hat{W}^T G'_\sigma)^T \tag{6.6}$$

$$\dot{\hat{\theta}}_f = \Gamma_4 |s(t)| Y_f \tag{6.7}$$

$$\dot{\hat{\phi}} = Proj(\hat{\phi}, -\eta v_{fd} s) \tag{6.8}$$

where $\Lambda_v^T = [0, \lambda^{(n-1)}, (n-1)\lambda^{(n-2)}, ..., (n-1)\lambda]$; k^* is a control gain, satisfying $k^* \geq \rho/c_{min}$, whereby, ρ is defined in (4.2). In addition, the parameters η and Γ_i ($i = 1, \cdots, 4$) are positive constants determining the rates of adaptations, and $Proj(\cdot, \cdot)$ is a projection operator, which is formulated as follows:

$$Proj(\hat{\phi}, -\eta v_{fd} s) = \begin{cases} 0 & \text{if } \hat{\phi}_i = \phi_{max} \text{ and } \eta v_{fd} s < 0 \\ -\eta v_{fd} s & \text{if } [\phi_{min} < \hat{\phi} < \phi_{max}] \\ & \text{or } [\hat{\phi} = \phi_{max} \text{ and } \eta v_{fd} s \geq 0] \\ & \text{or } [\hat{\phi} = \phi_{min} \text{ and } \eta v_{fd} s \leq 0] \\ 0 & \text{if } \hat{\phi} = \phi_{min} \text{ and } \eta v_{fd} s > 0 \end{cases} \tag{6.9}$$

Remarks:

1. In the above control law, a projection operator has been introduced. It can be easily proved that the projection operator for $\hat{\phi}$ has the following properties: (i) $\hat{\phi}(t) \in \Omega_\phi$ if $\hat{\phi}(0) \in \Omega_\phi$; (ii) $\|Proj(p, y)\| \leq \|y\|$; and (iii) $(p^* - p)^T \Lambda Proj(p, y) \geq (p^* - p)^T \Lambda y$, where Λ is a positive definite symmetric matrix.

2. The projection operator requires knowledge of the parameters ϕ_{min} and ϕ_{max}. These represent the upper and lower bounds of ϕ, respectively. Assumptions (A1) and (A2) are fundamental to this end. However, it should be noted that these parameters are only used to specify the range of parameter changes for the projection operator. With regards to this paper, such a range is not restricted as long as the estimated parameter is bounded (required for the stability proof); hence one can always choose suitable ϕ_{min} and ϕ_{max}, although such a choice may be conservative.

The stability of the closed-loop system described by (2.2), (3.2) and (6.2)–(6.9) is established in the following theorem.

Theorem 6.1. *For the plant in equation (2.2) with the hysteresis (3.2) at the input subject to assumptions (A1) – (A3), the robust adaptive controller specified by equations (6.2) – (6.9) ensures that if $\hat{\phi}(t_0) \in \Omega_\phi$, all the closed-loop signals are bounded and the state vector $\mathbf{x}(t)$ converges to $\mathbf{x}_d$ as $t \to \infty$.*

Proof Using the expression (4.1), the time derivative of the filtered error (4.4) can be written as:

$$\dot{s}(t) = -v_{fd}(t) - f(\mathbf{x}(t)) + bcv(t) + bd(v) \tag{6.10}$$

Using the control law (6.2) – (6.9), the above equation can be rewritten as:

$$\begin{aligned}\dot{s} &= -v_{fd} - f(\mathbf{x}) + bc[-ks + \hat{\phi}v_{fd} + \hat{W}^T G\left(X, \hat{\xi}, \hat{\sigma}\right)\\ &\quad - (\hat{\theta}_f^T Y_f + k^*)sat(s)] + bd(v)\end{aligned} \tag{6.11}$$

To establish global boundedness, we define a Lyapunov function candidate,

$$V(t) = \frac{1}{2}[\frac{1}{bc}s^2 + \frac{1}{\eta}(\phi - \hat{\phi})^2 + \tilde{W}^T \Gamma_1^{-1}\tilde{W} + \tilde{\xi}^T \Gamma_2^{-1}\tilde{\xi} + \tilde{\sigma}^T \Gamma_3^{-1}\tilde{\sigma} + \tilde{\theta}_f^T \Gamma_4^{-1}\tilde{\theta}_f] \tag{6.12}$$

The derivative of $\dot{V}$ along (6.11) leads to

$$\begin{aligned}\dot{V}(t) &= \frac{1}{bc}s\dot{s} - \frac{1}{\eta}(\phi - \hat{\phi})\dot{\hat{\phi}} - \tilde{W}^T \Gamma_1^{-1}\dot{\tilde{W}} - \tilde{\xi}^T \Gamma_2^{-1}\dot{\tilde{\xi}} - \tilde{\sigma}^T \Gamma_3^{-1}\dot{\tilde{\sigma}} - \tilde{\theta}_f^T \Gamma_4^{-1}\dot{\tilde{\theta}}_f \\ &= -ks^2 + s[\hat{\phi}v_{fd}(t) + \hat{W}^T G\left(X, \hat{\xi}, \hat{\sigma}\right) - (\hat{\theta}_f^T Y_f + k^*)sat(s)] \\ &\quad + \frac{1}{bc}s[-v_{fd} - f(\mathbf{x}) + bd(v)] - \frac{1}{\eta}(\phi - \hat{\phi})\dot{\hat{\phi}} \\ &\quad - \tilde{W}^T \Gamma_1^{-1}\dot{\tilde{W}} - \tilde{\xi}^T \Gamma_2^{-1}\dot{\tilde{\xi}} - \tilde{\sigma}^T \Gamma_3^{-1}\dot{\tilde{\sigma}} - \tilde{\theta}_f^T \Gamma_4^{-1}\dot{\tilde{\theta}}_f \\ &= -ks^2 + s[\hat{\phi}v_{fd}(t) + \hat{W}^T G\left(X, \hat{\xi}, \hat{\sigma}\right) - (\hat{\theta}_f^T Y_f + k^*)sat(s)] \\ &\quad + s[-\phi v_{fd} - v_f(\mathbf{x}) + \frac{d}{c}] - \frac{1}{\eta}(\phi - \hat{\phi})\dot{\hat{\phi}} \\ &\quad - \tilde{W}^T \Gamma_1^{-1}\dot{\tilde{W}} - \tilde{\xi}^T \Gamma_2^{-1}\dot{\tilde{\xi}} - \tilde{\sigma}^T \Gamma_3^{-1}\dot{\tilde{\sigma}} - \tilde{\theta}_f^T \Gamma_4^{-1}\dot{\tilde{\theta}}_f \end{aligned} \tag{6.13}$$

Using Lemma 2, one has

$$\begin{aligned}v_f(\mathbf{x}) - \hat{W}^T G\left(X, \hat{\xi}, \hat{\sigma}\right) &= \frac{1}{bc}f(\mathbf{x}) - \hat{W}^T G\left(X, \hat{\xi}, \hat{\sigma}\right) \\ &= \tilde{W}^T(t) \cdot (\hat{G}(t) - G'_\xi \hat{\xi}(t) - G'_\sigma \hat{\sigma}(t)) \\ &\quad + \hat{W}^T(t) \cdot (G'_\xi \tilde{\xi}(t) + G'_\sigma \tilde{\sigma}(t)) + d_f(t) \end{aligned} \tag{6.14}$$

The equation (6.13) can then be expressed as

$$\begin{aligned}\dot{V}(t) &= -ks^2 + s[\hat{\phi}v_{fd}(t) - (\hat{\theta}_f^T Y_f + k^*)sat(s)] \\ &\quad - s[\tilde{W}^T(t) \cdot (\hat{G}(t) - G'_\xi \hat{\xi}(t) - G'_\sigma \hat{\sigma}(t)) + \hat{W}^T(t) \cdot (G'_\xi \tilde{\xi}(t) + G'_\sigma \tilde{\sigma}(t)) \\ &\quad + d_f(t)] + s[-\phi v_{fd} + \frac{d}{c}] - \frac{1}{\eta}(\phi - \hat{\phi})\dot{\hat{\phi}} - \tilde{W}^T \Gamma_1^{-1}\dot{\tilde{W}} - \tilde{\xi}^T \Gamma_2^{-1}\dot{\tilde{\xi}} \\ &\quad - \tilde{\sigma}^T \Gamma_3^{-1}\dot{\tilde{\sigma}} - \tilde{\theta}_f^T \Gamma_4^{-1}\dot{\tilde{\theta}}_f \end{aligned} \tag{6.15}$$

Since $|d_f| \leq \theta_f^{*T} Y_f$, the above becomes

$$\begin{aligned}
\dot{V}(t) \leq\ & -ks^2 + s[\hat{\phi} v_{fd}(t) - (\hat{\theta}_f^T Y_f + k^*) sat(s)] - s\tilde{W}^T(t)(\hat{G}(t) - G'_\xi \hat{\xi}(t) \\
& - G'_\sigma \hat{\sigma}(t)) - s\hat{W}^T(t)(G'_\xi \tilde{\xi}(t) + G'_\sigma \tilde{\sigma}(t)) + |s|\theta_f^{*T} Y_f + s[-\phi v_{fd} + \frac{d}{c}] \\
& - \frac{1}{\eta}(\phi - \hat{\phi})\dot{\hat{\phi}} - \tilde{W}^T \Gamma_1^{-1} \dot{\hat{W}} - \tilde{\xi}^T \Gamma_2^{-1} \dot{\hat{\xi}} - \tilde{\sigma}^T \Gamma_3^{-1} \dot{\hat{\sigma}} - \tilde{\theta}_f^T \Gamma_4^{-1} \dot{\hat{\theta}}_f \quad (6.16)
\end{aligned}$$

By using adaptive law (6.8) and the property $-\frac{1}{\eta}(\phi - \hat{\phi}) Proj(\hat{\phi}, -\eta v_{fd}s) \leq (\phi - \hat{\phi}) v_{fd}s$, one obtains

$$\begin{aligned}
\dot{V}(t) \leq\ & ks^2 + s[\hat{\phi} v_{fd}(t) - (\hat{\theta}_f^T Y_f + k^*) sat(s)] - s\tilde{W}^T(t) \cdot (\hat{G}(t) - G'_\xi \hat{\xi}(t) \\
& - G'_\sigma \hat{\sigma}(t)) - s\tilde{W}^T(t)(G'_\xi \tilde{\xi}(t) + G'_\sigma \tilde{\sigma}(t)) + |s|\theta_f^{*T} Y_f - s\phi v_{fd} \\
& + |s| \frac{\rho}{c_{min}} + (\phi - \hat{\phi}) v_{fd}s - \tilde{W}^T \Gamma_1^{-1} \dot{\hat{W}} - \tilde{\xi}^T \Gamma_2^{-1} \dot{\hat{\xi}} \\
& - \tilde{\sigma}^T \Gamma_3^{-1} \dot{\hat{\sigma}} - \tilde{\theta}_f^T \Gamma_4^{-1} \dot{\hat{\theta}}_f \\
\leq\ & -ks^2 - |s|\hat{\theta}_f^T Y_f + (-k^* + \frac{\rho}{c_{min}})|s| - s\tilde{W}^T(t) \cdot (\hat{G}(t) - G'_\xi \hat{\xi}(t) \\
& - G'_\sigma \hat{\sigma}(t)) - s\tilde{W}^T(t)(G'_\xi \tilde{\xi}(t) + G'_\sigma \tilde{\sigma}(t)) + |s|\theta_f^{*T} Y_f \\
& - \tilde{W}^T \Gamma_1^{-1} \dot{\hat{W}} - \tilde{\xi}^T \Gamma_2^{-1} \dot{\hat{\xi}} - \tilde{\sigma}^T \Gamma_3^{-1} \dot{\hat{\sigma}} - \tilde{\theta}_f^T \Gamma_4^{-1} \dot{\hat{\theta}}_f \\
\leq\ & -ks^2 \quad (6.17)
\end{aligned}$$

Equation (6.17) implies that V is a Lyapunov function which leads to global boundedness of s and $(\hat{\phi} - \phi)$, as well as $\hat{W}$, $\hat{\xi}$, $\hat{\sigma}$, $\hat{\theta}_f$. It is easily shown that if $\tilde{x}(0)$ is bounded, then $\tilde{x}(t)$ is also bounded for all t, and since $\mathbf{x}_d(t)$ is bounded by design, $\mathbf{x}(t)$ must also be bounded. To complete the proof and establish asymptotic convergence of the tracking error, it is necessary to show that $s \to 0$ as $t \to \infty$. This is accomplished by applying Barbalat's Lemma [9] to the continuous, non-negative function:

$$\begin{aligned}
V_1(t) &= V(t) - \int_0^t (\dot{V}(\tau) + ks^2(\tau)) d\tau \text{ with} \\
\dot{V}_1(t) &= -ks^2(t) \quad (6.18)
\end{aligned}$$

It can easily be shown that every term in (6.11) is bounded, hence $\dot{s}$ is bounded. This implies that $\dot{V}_1(t)$ is a uniformly continuous function of time. Since V_1 is bounded below by 0, and $\dot{V}_1(t) \leq 0$ for all t, use of Barbalat's Lemma proves that $\dot{V}_1(t) \to 0$. Therefore, from (6.18) it can be demonstrated that $s(t) \to 0$ as $t \to \infty$. The remark following equation (4.4) indicates that $\mathbf{x}(t)$ will converge to $\mathbf{x}_d(t)$.

7 Simulation Studies

In this section, we illustrate the above methodology on a simple nonlinear system described by

$$\dot{x} = a\frac{1 - e^{-x(t)}}{1 + e^{-x(t)}} + bw(t) \tag{7.1}$$

where $w(t)$ represents an output of hysteresis. The actual parameter values are $b = 1$ and $a = 1$. Without control, i.e., $w(t) = 0$, the system (7.1) is unstable, because $\dot{x} = \frac{1-e^{-x(t)}}{1+e^{-x(t)}} > 0$ for $x > 0$, and $\dot{x} = \frac{1-e^{-x(t)}}{1+e^{-x(t)}} < 0$ for $x < 0$. The objective is to control the system state x to follow a desired trajectory x_d, which will be specified later.

The backlash-like hysteresis is described by

$$\frac{dw}{dt} = \alpha \left|\frac{dv}{dt}\right|[cv - w] + \frac{dv}{dt}B_1 \tag{7.2}$$

with parameters $\alpha = 1$, $c = 3.1635$, and $B_1 = 0.345$. Using input signal $v(t) = k\sin(2.3t)$ with k=2.5, 3.5, 4.5, 5.5, 6.5, the responses of this dynamic equation with the initial condition $w(0) = 0$ are shown in Figure 7.1. We should mention that when using a variety of values for both initial values $w(0)$ and frequencies, simulation studies show hysteresis shapes similar to those in Figure 7.1. This confirms again that the dynamic model (7.2) can be used to describe the backlash-like hysteresis. It also shows that the required shape of backlash hysteresis is dependent solely on the selection of a suitable parameter set $\{\alpha, c, B_1\}$.

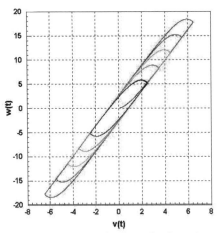

Fig. 7.1. Hysteresis curves given by (3.2) or (7.2) with $\alpha = 1$, $c = 3.1635$, and $B_1 = 0.345$ for $v(t) = kitsin(2.3t)$ with $k = 2.5, 3.5, 4.5, 5.5, 6.5$.

To construct the function approximator, the following membership functions $\mu_K(x) = \exp(-\sigma(x-k)^2)$ are adopted, where $k = -3, -2, -1, 0, 1, 2, 3$.

In this example, the initial values of $\hat{W}(0)$, $\hat{\xi}(0)$, $\hat{\sigma}(0)$, and $\hat{\theta}_f(0)$ are chosen as $\hat{W}(0) = \frac{1}{3.16}[-0.8, -0.6, -0.4, 0, 0.4, 0.6, 0.8]^T$, $\hat{\xi}(0) = [-3, -1, -1, 0, 1, 2, 3]^T$, $\hat{\sigma}(0) = [2, 2, 2, 2, 2, 2, 2]^T$, and $\hat{\theta}_f(0) = [0.1, 0.01, 0.01, 0.01]$.

In the simulations, the robust adaptive control law (6.2) – (6.9) was used, taking $k_d = 20$. Since the backlash distance is around 2.5, we can choose the upper bound ρ in (4.2) as $\rho = 4$ and we also choose $c_{min} = 3$, which results in $k^* = 4/3$. In the adaptation laws, we choose $\Gamma_i|_{i=1,2,3,4} = \{0.2, 0.2, 0.3, 0.01\}$ and $\eta = 0.5$ and the initial parameter $\phi = 0.8/3$. The initial state is chosen as $x(0) = 1.05$ and sample time is 0.005. In the simulation the initial value, $v(0)$, is required, which is selected as $v(0) = 0$.

Choosing the desired trajectory $x_d(t) = 12.5 sin(2.3t)$, simulation results are shown in Figures. 7.2 and 7.3. Figure 7.2 shows the tracking error for the desired trajectory and Figure 7.3 shows the input control signal $v(t)$. We see from Figure 7.2 that the proposed robust controller clearly demonstrates excellent tracking performance. We should mention that it is desirable to compare the control performance with and without considering the effects of hysteresis. Unfortunately, this comparison is not possible in this case as the control law (6.2) – (6.9) is designed for the entire cascade system.

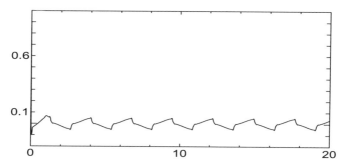

Fig. 7.2. Tracking error of the state with backlash hysteresis.

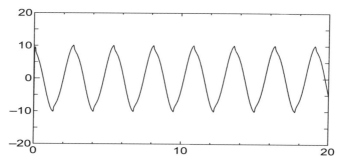

Fig. 7.3. Control signal $v(t)$ acting as the input of backlash hysteresis.

8 Conclusion

In this paper, a stable adaptive control architecture is proposed for a class of continuous-time nonlinear dynamic systems preceded by an unknown backlash-like hysteresis, where the backlash-like hysteresis is modeled by a dynamic equation. By showing the properties of the hysteresis model and by combining a universal function approximator with adaptive control techniques, an adaptive control scheme is developed without constructing the hysteresis inverse. The proposed adaptive control law ensures global stability of the adaptive system and achieves the desired tracking. Simulations performed on a nonlinear system illustrate and clarify the approach.

Appendix A: Proof of Lemma 2

Proof: Denoting $\hat{G} = G\left(\mathbf{x}, \hat{\xi}, \hat{\sigma}\right)$ and noticing $v_f(\mathbf{x}(t)) = v_f^*(\mathbf{x}) + \varepsilon_v$ with $|\varepsilon_v| \leq \varepsilon_h$, one has

$$\begin{aligned}
\tilde{v}_f(t) &= v_f(\mathbf{x}(t)) - \hat{v}_f(\mathbf{x}(t)) \\
&= v_f^*(\mathbf{x}) - \hat{W}^T \hat{G} + \varepsilon_v \\
&= W^{*T} G^* - W^{*T} \hat{G} + W^{*T} \hat{G} - \hat{W}^T \hat{G} + \varepsilon_v \\
&= W^{*T} \tilde{G} + \tilde{W}^T \hat{G} + \varepsilon_v \\
&= W^{*T} \tilde{G} - \hat{W}^T \tilde{G} + \hat{W}^T \tilde{G} + \tilde{W}^T \hat{G} + \varepsilon_v \\
&= \tilde{W}^T \tilde{G} + \hat{W}^T \tilde{G} + \tilde{W}^T \hat{G} + \varepsilon_v
\end{aligned} \quad (\text{A.1})$$

where $\tilde{G} = G^* - \hat{G} = G(\mathbf{x}, \xi^*, \sigma^*) - G\left(\mathbf{x}, \hat{\xi}, \hat{\sigma}\right)$. In order to deal with $\tilde{G}$, the Taylor's series expansion of G^* is taken about $\xi^* = \hat{\xi}$ and $\sigma^* = \hat{\sigma}$. This produces

$$G(\mathbf{x}, \xi^*, \sigma^*) = G\left(\mathbf{x}, \hat{\xi}, \hat{\sigma}\right) + G'_\xi \cdot (\xi^* - \hat{\xi}) + G'_\sigma \cdot (\sigma^* - \hat{\sigma}) + o(\mathbf{x}, \tilde{\xi}, \tilde{\sigma}) \quad (\text{A.2})$$

where $o(\mathbf{x}, \tilde{\xi}, \tilde{\sigma})$ denotes the sum of high-order arguments in a Taylor's serious expansion, and G'_ξ and G'_σ are derivatives of $G(\mathbf{x}, \xi^*, \sigma^*)$ with respect to ξ^* and σ^* at $(\hat{\xi}, \hat{\sigma})$. They are expressed as

$$\begin{aligned}
G'_\xi &= \frac{\partial G(\mathbf{x}, \xi^*, \sigma^*)}{\partial \xi^*}\bigg|_{\xi^* = \hat{\xi},\ \sigma^* = \hat{\sigma}} \\
G'_\sigma &= \frac{\partial G(\mathbf{x}, \xi^*, \sigma^*)}{\partial \sigma^*}\bigg|_{\xi^* = \hat{\xi},\ \sigma^* = \hat{\sigma}}
\end{aligned} \quad (\text{A.3})$$

Equation (A.2) can then be written as

$$\tilde{G} = G'_\xi \tilde{\xi} + G'_\sigma \tilde{\sigma} + o(\mathbf{x}, \tilde{\xi}, \tilde{\sigma}) \quad (\text{A.4})$$

Using equation (A.4), $\tilde{v}_f(t)$ in (A.1) can be expressed as

$$\begin{aligned}
\tilde{v}_f(t) &= \tilde{W}^T(G'_\xi \tilde{\xi} + G'_\sigma \tilde{\sigma} + o(\mathbf{x}, \tilde{\xi}, \tilde{\sigma})) + \hat{W}^T(G'_\xi \tilde{\xi} + G'_\sigma \tilde{\sigma} \\
&\quad + o(\mathbf{x}, \tilde{\xi}, \tilde{\sigma})) + \tilde{W}^T \hat{G} + \varepsilon_v \\
&= \tilde{W}^T(\hat{G} - G'_\xi \hat{\xi} - G'_\sigma \hat{\sigma}) + \hat{W}^T(G'_\xi \tilde{\xi} + G'_\sigma \tilde{\sigma}) + d_f \quad \text{(A.5)}
\end{aligned}$$

where

$$d_f = \tilde{W}^T(G'_\xi \xi^* + G'_\sigma \sigma^*) + W^{*T} o(\mathbf{x}, \tilde{\xi}, \tilde{\sigma})) + \varepsilon_v$$

Now let us examine d_f. First, using (A.4), the high-order term $o(\mathbf{x}, \tilde{\xi}, \tilde{\sigma})$ is bounded by

$$\begin{aligned}
\|o(\mathbf{x}, \tilde{\xi}, \tilde{\sigma})\| &= \|\tilde{G} - G'_\xi \tilde{\xi} - G'_\sigma \tilde{\sigma}\| \\
&\leq \|\tilde{G}\| + \|G'_\xi \tilde{\xi}\| + \|G'_\sigma \tilde{\sigma}\| \\
&\leq c_1 + c_2 \|\tilde{\xi}\| + c_3 \|\tilde{\sigma}\| \quad \text{(A.6)}
\end{aligned}$$

where c_1, c_2, and c_3 are some bounded constants due to the fact that a Gaussian function and its derivative are always bounded by constants (the proof is omitted here to save space). Secondly, it is obvious that there should exist constants $\bar{W}$, $\bar{\xi}$, and $\bar{\sigma}$ satisfying $\|W^*\| \leq \bar{W}$, $\|\xi^*\| \leq \bar{\xi}$, and $\|\sigma^*\| \leq \bar{\sigma}$. Finally, based on the facts:

$$\|\tilde{W}\| \leq \|W^*\| + \|\hat{W}\| \leq \|\bar{W}\| + \|\hat{W}\|$$

$$\|\tilde{\xi}\| \leq \|\xi^*\| + \|\hat{\xi}\| \leq \|\bar{\xi}\| + \|\hat{\xi}\|$$

$$\|\tilde{\sigma}\| \leq \|\sigma^*\| + \|\hat{\sigma}\| \leq \|\bar{\sigma}\| + \|\hat{\sigma}\|$$

The term d_f can be bounded as

$$\begin{aligned}
|d_f| &= \|\tilde{W}^T(G'_\xi \xi^* + G'_\sigma \sigma^*) + W^{*T} o(\mathbf{x}, \tilde{\xi}, \tilde{\sigma})) + \varepsilon_v\| \\
&\leq \|\tilde{W}\|\|G'_\xi\|\|\xi^*\| + \|\tilde{W}\|\|G'_\sigma\|\|\sigma^*\| + \|W^*\|(c_1 + c_2\|\tilde{\xi}\| + c_3\|\tilde{\sigma}\|) + \varepsilon_h \\
&\leq (\bar{W} + \|\hat{W}\|)c_2\bar{\xi} + (\bar{W} + \|\hat{W}\|)c_3\bar{\sigma} + \bar{W}c_1 + \bar{W}c_2(\bar{\xi} + \|\hat{\xi}\|) \\
&\quad + \bar{W}c_3(\bar{\sigma} + \|\hat{\sigma}\|) + \varepsilon_h \\
&= 2c_2\bar{W}\bar{\xi} + 2c_3\bar{W}\bar{\sigma} + c_1\bar{W} + \varepsilon_h + (c_2\bar{\xi} + c_3\bar{\sigma})\|\hat{W}\| + c_2\bar{W}\|\hat{\xi}\| + c_3\bar{W}\|\hat{\sigma}\| \\
&= [\theta^*_{f1}, \theta^*_{f2}, \theta^*_{f3}, \theta^*_{f4}] \cdot [1, \|\hat{W}\|, \|\hat{\xi}\|, \|\hat{\sigma}\|]^T \\
&= \theta^{*T}_f Y_f \quad \text{(A.7)}
\end{aligned}$$

where $\theta^*_{f1} = 2c_2\bar{W}\bar{\xi} + 2c_3\bar{W}\bar{\sigma} + c_1\bar{W} + \varepsilon_h$, $\theta^*_{f2} = c_2\bar{\xi} + c_3\bar{\sigma}$, $\theta^*_{f3} = c_2\bar{W}$, and $\theta^*_{f4} = c_3\bar{W}$.

References

1. N. J. Ahmad and F. Khorrami, Adaptive control of systems with backlash hysteresis at the input. Proc. of the American Control Conf., pp. 3018-3022, 1999
2. C. Chiu and T. P. Leung, Modelling and microcomputer control of a nonlinear pneumatic servomechanism. Trans Inst Measurement and Control, 10:71-78, 1988
3. S. S. Ge, C. C. Hang, and T. Zhang, Adaptive neural network control of nonlinear systems by state and output feedback. IEEE Trans. on Systems, Man, and Cybernetics-Part B: Cybernetics, 29:818-827, 1999
4. C. Han and Y. X. Zhong, Robust adaptive control of time-varying systems with unknown backlash nonlinearity. Proc. of the 13th IFAC Triennial World Congress, pp. 439-444, 1996
5. A. Isidori, Nonlinear Control Systems: An Introduction, 2nd edn. Springer, Berlin, 1989
6. J. W. Macki, P. Nistri, and P. Zecca, Mathematical models for hysteresis. SIAM Review, 35:94-123, 1993
7. M. Mata-Jimenez, B. Brogliato, and A. Goswami, On the control of mechanical systems with dynamic backlash. Proc. of the 36th Conference on Decision and Control, pp. 1749-1756, 1997
8. T. E. Pare and J. P. How, Robust stability and performance analysis of systems with hysteresis nonlinearities. Proc. of the American Control Conf., pp. 1904-1908, 1998
9. V. Popov, Hyperstability of Control System. Springer, Berlin, 1973
10. R. M. Sanner and J. J. E. Slotine. Gaussian networks for direct adaptive control. IEEE Trans. on Neural Networks, 3:837-863, 1992
11. J. J. E. Slotine and W. Li, Applied Nonlinear Control. Prentice-Hall, Englewood Cliff, NJ, 1991
12. Y. Stepanenko and C.-Y. Su, Intelligent control of pizoelectric actuators. Proc. of the 37th IEEE Conference on Decision and Control, pp. 4234-4239, 1998
13. C.-Y. Su and Y. Stepanenko, Adaptive control of a class of nonlinear systems with fuzzy logic. IEEE Trans. on Fuzzy Systems, 2:258-294, 1994
14. X. Sun, W. Zhang, and Y. Jin, Stable adaptive control of backlash nonlinear systems with bounded disturbance. Proc. of the 31th Conf. on Decision and Control, pp. 274-275, 1992
15. G. Tao and P. V. Kokotovic, Adaptive control of plants with unknown hystereses. IEEE Trans. on Automatic Control, 40:200-212, 1995
16. G. Tao and P. V. Kokotovic, Continuous-time Adaptive control of systems with unknown backlash. IEEE Trans. on Automatic Control, 40:1083-1087, 1995

Adaptive Control of a Class of Time-delay Systems in the Presence of Saturation

A.M. Annaswamy[1], S. Evesque[2], S.-I. Niculescu[3], and A.P. Dowling[2]

[1] Adaptive Control Laboratory, Department of Mechanical Engineering, MIT, Cambridge, MA 02139, USA
[2] Department of Engineering, University of Cambridge, Cambridge, UK
[3] HEUDIASYC, University of Compiegne, Compiegene, France

Abstract

The control of a class of systems with unknown parameters in the presence of time-delay and magnitude constraints on the input is considered. An example of such systems can be found in continuous combustion processes where the active control input corresponds to fuel injection, and the goal of the controller is to suppress the pressure oscillations that occur at several operating conditions. Stringent emission specifications and the danger of flame extinction impose severe constraints on the magnitude of the fuel. Time-delays due to convection, mixing, atomization, and vaporization of the fuel are also present in these systems. Simultaneously present are changes in the operating conditions due to demands on the load, performance, and emissions, and variations in the fuel composition. All of the above characteristics imply that an adaptive control approach that accommodates saturation constraints in the presence of a time-delay are required. In this chapter, control algorithms that accommodate these characteristics are presented. The behavior of these controllers is validated in the context of a combustion system subject to time-delays and input saturation.

1 Introduction

Typical control goals in dynamic systems are to ensure stabilization, command following, and disturbance rejection in a robust manner in the face of operating uncertainties. The field of adaptive control has focused on the realization of these control goals in problems where these uncertainties can be represented as unknown parameters in the dynamic model. Several kinds of dynamic systems, both linear and nonlinear, single- and multiple-input, continuous and discrete, and deterministic and stochastic, have been considered in the literature. One of the specific examples of nonlinearities is due to saturation. An ubiquitous constraint that is often present is on the allowable magnitude of the control input. This nonlinearity in turn mandates that any control design must acccommodate this constraint without sacrificing performance. Therefore the adaptive control design must address not only the effects of parametric uncertainty but also the effect of the control system.

Delay systems represent a class of infinite-dimensional systems where mechanisms related to transport, propagation, or other effects related to a significant time-lag are present. Time-delays are present in almost all physical systems, simply due to the fact that there is always a delay between the application of the control input and the response of the key variables, with the actual value dependent on the underlying physics. Because of their effect on the stability properties of the resulting closed-loop system, time-delay can be viewed as the nemesis of a controls engineer. Small delays may destabilize some systems, while large delays may stabilize others. Depending on the complexity of the dynamics of the underlying system, as the delay increases, the property of the closed-loop system may switch from stability to instability and back several times.

In this chapter, we address a class of systems with unknown parameters in the presence of saturation constraints and time-delay and present an adaptive controller that accommodates both the effects in its design. This class arises in the context of continuous combustion processes where large pressure oscillations arise due to resonant feedback interactions between acoustics and heat-release. In applications such as gas turbines, ramjets, and afterburners, under certain operating conditions, heat release stemming from combustion in an enclosed chamber excites the acoustic modes which in turn introduce perturbations in the heat-release thereby generating a resonant feedback loop. While large pressure oscillations have obvious implications for the stability of the combustion system, operating at these points is desirable since they often meet other performance objectives such as maximized thrust and minimized emissions. Active control has been shown to be an attractive technology for realizing the combined optimization of reduced emissions/increased thrust and reduced pressure oscillations. By modulating air-flow and/or fuel-flow in these systems on-line, it has been shown in many applications that pressure suppression is possible. Recent efforts in this field have focused on model-based control [1–3] where controllers that accommodate time-delays and parametric uncertainties have been developed. In this chapter, we consider the same class of systems with an additional constraint due to input saturation. Motivated by the problem of unsteady pressure suppression, we focus in this chapter primarily on stabilization, rather than command following. The same controllers are, however, applicable in the latter case as well.

2 Statement of the Problem

The problem is to control plants of the form

$$y = W_p(s)u(t-\tau) \tag{1}$$

where u is constrained in amplitude and is given by

$$u = \begin{cases} v & \text{if } |u| \leq u_0 \\ sgn(v)u_0 & \text{if } |u| \geq u_0 \end{cases} \qquad (2)$$

when $W_p(s)$ is unknown. It is however assumed that τ is known, and that u_0 is known. It is also assumed that the order is known, that the zeros of $W_p(s)$ are stable. In the context of the combustion problem, u represents the mass flow rate from a fuel injector that can be modulated using active control, and y denotes the pressure perturbation at a given point in the combustor. τ represents the transport delay from the point of fuel-injection to the burning plane. $W_p(s)$ is a transfer function that represents the combined effect of acoustics and heat-release, which can be viewed as a marginally stable system that produces sustained oscillations in the absence of control. Whether $W_p(s)$ is stable or marginally stable, it is desired to stabilize the system, and reduce y to zero. The saturation constraint is usually imposed on the fuel-injector since it is desirable to limit the amount of secondary injection so as to avoid flame extinction and minimize emission levels.

It should be noted that plants of the form of (1) have been addressed at length by a number of researchers when $\tau = 0$ [4–6]. In this paper, an extension to these results for a nonzero τ is addressed. By using a modified Smith Controller, and by adjusting its parameters appropriately, it is shown that the all signals that start within a compact set remain bounded for $\tau \leq \tau^*$, a finite value, and that the tracking error $e_1 = y_p$ tends to zero asymptotically.

To illustrate the results, the problem in the absence of delay is first presented and solved in section 3. The controller is then modified to address the effect of delay in section 4. To illustrate the idea behind the algorithm, we consider the controllers for the relative degree unity case. The same approach is then extended to plants with higher relative degree using an augmented error [7].

3 The Controller in the Presence of Saturation

We begin with the problem of controlling plants with unknown parameters subject to magnitude constraints on the control input. The controller structure and the tools used set the stage for the necessary solution in the presence of the time-delay.

The main difficulty due to the imposition of a saturation constraint is the introduction of a nonlinearity which needs to be accommodated in the controller design. In the context of the adaptive controller, the underlying closed-loop system is nonlinear to begin with, even without the presence of the saturation, since the control parameters are adjusted as nonlinear functions of the system variables. Therefore, it is important to ensure that even with the nonlinear gain introduced by the saturation, closed-loop stability is guaranteed. Below, we show that this can indeed be achieved, with the stability being global when the plant is open-loop stable and local with a domain of attraction when the plant poles lie in the closed right-half of the complex plane.

To illustrate the main idea, we begin with a simple case of a first-order plant, and proceed to the higher-order case of control with output feedback only.

3.1 A First-order Plant

Statement of the problem. A plant with an input-output pair $\{u(\cdot), x_p(\cdot)\}$ is described by the scalar equation

$$\dot{x}_p(t) = a_p x_p(t) + b_p u(t)$$

where a_p and b_p are unknown, but the sign of b_p is assumed to be known. The input $u(t)$ is additionally subject to the magnitude constraint

$$|u(t)| \leq u_0 \qquad (3)$$

where u_0 is a known constant. The aim is to determine a bounded control input $u(t)$ so that all signals in the system remain bounded, for a given set of initial conditions, and x_p tends to zero.

An adaptive controller is chosen as

$$v(t) = \theta(t) x_p(t)$$
$$u(t) = \begin{cases} v(t) & if \ |v(t)| \leq u_0 \\ u_0 \mathrm{sgn}(v(t)) & if \ |v(t)| > u_0 \end{cases} \qquad (4)$$

where $\theta(t)$ is time-varying. Then the closed-loop system can be described by

$$\dot{x}_p(t) = (a_p + b_p \theta(t)) x_p(t) + b_p \Delta u(t) \qquad (5)$$

where $\Delta u(t) = u(t) - v(t)$. Defining $e = x_p$, $\tilde{\theta}(t) = \theta(t) - \theta^*$, and choosing a negative gain $a_m < 0$ which represents the desired closed-loop pole, we obtain the error equation as

$$\dot{e}(t) = a_m e(t) + b_p \tilde{\theta}(t) x_p(t) + b_p \Delta u(t).$$

In order to remove the effect of Δu, which can be considered as a disturbance, we generate a signal $e_\Delta(t)$ as the output of a differential equation

$$\dot{e}_\Delta(t) = a_m e_\Delta(t) + k_\Delta(t) \Delta u(t) \qquad e_\Delta(t_0) = 0.$$

If $e_u(t) = e(t) - e_\Delta(t)$, we obtain

$$\dot{e}_u(t) = a_m e_u(t) + b_p \tilde{\theta}(t) x_p(t) + \kappa(t) \Delta u(t) \qquad (6)$$

where $\kappa(t) = b_p - k_\Delta(t)$. Equation (6) is in a standard error model form for which we can use the adaptive laws

$$\dot{\tilde{\theta}}(t) = -\gamma_1 \operatorname{sgn}(b_p) e_u x_p, \qquad \dot{\kappa}(t) = \gamma_2 e_u \Delta u \qquad (7)$$

where $\gamma_i > 0$ $i = 1, 2$. These adaptive laws can then be used to show that (i) e_u and the parameters θ and κ are bounded for all $t \geq t_0$. The boundedness of e_1 can then be established using the state equation in (5) for all initial conditions in a bounded set Ω. This set coincides with the entire state space if the plant is open-loop stable [5].

3.2 Output Feedback with $n^* = 1$

We consider in this section the adaptive control of an nth order plant using output feedback. The plant we consider is a single-input single-output system described by

$$y_p(t) = W_p(s)u(t) = \left[k_p \frac{Z_p(s)}{R_p(s)}\right] u(t)$$

where the input is subjected to a magnitude constraint as in equation (3). It is assumed the order and k_p are known, that the relative degree is one, and that $Z_p(s)$ is Hurwitz. The assumptions on relative degree and k_p is for ease of exposition. The same idea can be extended to higher relative degree as well [8]. Our objective is to let the plant output $y_p(t)$ follow a reference trajectory $y_m(t)$ as closely as possible, where the latter is described as

$$y_m(t) = W_m(s)r(t) = \left[k_p \frac{Z_m(s)}{R_m(s)}\right] r(t)$$

where $W_m(s)$ is strictly positive real. As in the previous section, we focus primarily on asymptotic stabilization of the plant. Extensions to tracking can be carried out as in [5].

A differentiator-free controller is chosen as in [7], and can be described by the following differential equation

$$\begin{aligned}
\dot{\omega}_1(t) &= \Lambda \omega_1(t) + lu(t) \\
\dot{\omega}_2(t) &= \Lambda \omega_2(t) + ly_p(t) \\
\omega(t) &= \left[\omega_1(t)^T, \omega_2(t)^T\right]^T \\
\theta(t) &= \left[\theta_1(t)^T, \theta_2(t)^T\right]^T \\
v(t) &= \theta(t)^T \omega(t) \\
u &= \begin{cases} v(t) & \text{if } |v(t)| \leq u_0 \\ u_0 \operatorname{sgn}(v(t)) & \text{if } |v(t)| > u_0 \end{cases}
\end{aligned} \qquad (8)$$

where $\theta_1, \omega_1 : \mathbb{R}^+ \to \mathbb{R}^n$, $\theta_2, \omega_2 : \mathbb{R}^+ \to \mathbb{R}^n$, Λ is an asymptotically stable matrix and $\det(sI - \Lambda) = \lambda(s)$. It is well known that a parameter θ^* exists such that if $\theta(t) = \theta^*$, the transfer function of the plant together with the controller matches that of the model. With the controller in (8), and the parameter errors

$$\tilde{\theta}_1(t) = \theta_1(t) - \theta_1^*, \qquad \tilde{\theta}_2(t) = \theta_2(t) - \theta_2^*, \qquad \tilde{\theta}(t) = \left[\tilde{\theta}_1^T(t), \tilde{\theta}_2^T(t)\right],$$

the overall system becomes

$$\dot{x}(t) = A_{mn}x(t) + b_{mn}\left[\tilde{\theta}^T(t)\omega(t) + \Delta u(t)\right] \qquad (9)$$
$$y(t) = h_{mn}^T x(t)$$

where $x = [x_p^T, \omega_1^T, \omega_2^T]^T$, $\Delta u(t) \triangleq u(t) - v(t)$, and $W_m(s) = h_{mn}^T(sI - A_{mn})^{-1}b_{mn}$ [5]. If $e_1 = y_p - y_m$, the error model is given by

$$e_1(t) = W_m(s)\left[\tilde{\theta}^T(t)\omega(t) + \Delta u(t)\right]. \qquad (10)$$

Boundedness of parameter errors. The adaptive law for adjusting $\tilde{\theta}(t)$ is first determined. First, we compensate for the $\Delta u(t)$ term by subtracting from the output error the signal

$$e_\Delta(t) = W_m(s)\Delta u(t)$$

where $k_\Delta(t)$ is a time-varying gain, adapted according to an adaptive law that is to be determined. Define $e_{u1} = e_1 - e_\Delta$, so that

$$e_{u1}(t) = W_m(s)\left[\tilde{\theta}^T(t)\omega(t)\right]. \qquad (11)$$

The adaptive law for adjusting $\tilde{\theta}$ is therefore chosen as

$$\dot{\tilde{\theta}}(t) = -e_{u1}(t)\omega(t). \qquad (12)$$

Choosing

$$V(e, \tilde{\theta}) = e^T P e + \tilde{\theta}^T \tilde{\theta}, \qquad (13)$$

we obtain, due to the Kalman-Yakubovich Lemma, that $\dot{V} = -e^T Q e \leq 0$ where Q satisfies the equations

$$A_m^T P + P A_{mn} = -Q \quad P b_{mn} = h_{mn}. \qquad (14)$$

Hence, $e(t)$ and $\theta(t)$ are bounded. In particular, $\|\tilde{\theta}(t)\| \leq \theta_m \qquad \forall t \geq t_0$ for a finite θ_m.

Boundedness of the state We now show that the state x is bounded. Define ρ as the ratio between the maximum and the minimum eigenvalues of P defined in (14) so that $\rho = \sqrt{\frac{\lambda_{\max}(P)}{\lambda_{\min}(P)}}$. Let $q_0 = \lambda_{\min}(Q)$, $p_b = \|b_{mn}^T P\|$, and

$$C_{4\times 5} = \begin{bmatrix} 0_{4\times 1} & I_{4\times 4} \end{bmatrix}.$$

Then, we have $\bar{\omega} = Cx$. We now prove the boundedness of x.
Theorem 1
The system in equation (9) has bounded solutions if

$$\text{(i)} \ \|x(t_0)\| < \frac{u_0}{\|\theta^{*T}C\|}$$
$$\text{(ii)} \ \sqrt{V(t_0)} < \frac{q_0}{2p_b\|C\|}. \tag{15}$$

Further

$$x^T(t)Px(t) \leq \lambda_{\min}(P)\left[\frac{u_0}{\|\theta^{*T}C\|}\right]^2 \quad \forall\, t \geq t_0$$

and the output error e_1 is given by

$$|e_1(t)| = O[\sup_{\tau \leq t} \Delta u(\tau)].$$

Proof: We consider the following two cases.
(a) Let $|v| \leq u_0$. Here $\Delta u = 0$, and hence $u(t) = v(t) = \theta^T(t)w(t)$. As a result,

$$\dot{x}(t) = A_{mn}x(t) + b_{mn}\tilde{\theta}^T w.$$

Define $W(x) = x^T P x$, whose time-derivative is given by

$$\dot{W}(x) = x^T\left[A_{mn}^T P + P A_{mn}\right]x + 2(b_{mn}^T Px)[\tilde{\theta}^T Cx]$$
$$\leq -q_0\|x\|^2 + 2p_b\|x\|^2\theta_m\|C\|$$
$$< 0$$

since $\|\tilde{\theta}(t)\| < \sqrt{V(t)}$ and from (ii) in (15).
(b) Let $|v| > u_0$. Hence,

$$u = u_0\text{sgn}(v) \text{ and } |v| > u_0$$

and (9) becomes

$$\dot{x}(t) = A_{mn}x(t) + b_{mn}u_0\text{sgn}(v(t)) - b_{mn}\theta^{*T}w(t).$$

Correspondingly,

$$\dot{W} = x^T\left[A_{mn}^T P + P A_{mn}\right]x - 2(b_{mn}^T Px)(\theta^{*T}Cx) + 2(b_{mn}^T Px)u_0\text{sgn}(v)$$
$$= -x^T Q x + 2\left|b_{mn}^T Px\right| u_0\text{sgn}(v)\text{sgn}(b_{mn}^T Px) - 2(b_{mn}^T Px)(\theta^{*T}Cx). \tag{16}$$

Two subcases follow:
Case I. $sgn(v) = -sgn(b_{mn}^T Px)$. Equation (16) becomes

$$\dot{W}(x) \leq -x^T Q x - 2|b_{mn}^T Px|\left(u_0 - \|x\|\|\theta^{*T}C\|\right)$$
$$< 0$$

due to (i) in (15).
Case II. $sgn(v) = sgn(b_{mn}^T Px)$. Here,

$$\dot{W}(x) \leq -q_0\|x\|^2 + 2|x^T Pb|u_0 - 2x^T Pbv + 2x^T Pb\tilde{\theta}^T Cx$$
$$\leq -\|x\|^2 (q_0 - 2p_b\theta_m \|C\|) + 2|b_{mn}^T Px|(u_0 - |v|). \quad (17)$$

The first term in (17) is negative due to (i) in (15), while the second is negative since $|v| > u_0$ in case (b).

That is, $\dot{W}(x) < 0$ on all points on the level surface and Theorem 2 follows.

Since x, the overall state of the adaptive system, is bounded, y_p, and hence e_1 is bounded. Similar arguments to the standard adaptive control problem yield that $\lim_{t\to\infty} \epsilon_{1u}(t) = \lim_{t\to\infty} e_2(t) = 0$. Hence,

$$\lim_{t\to\infty} |e_{u1}(t)| = \lim_{t\to\infty} |e_1(t) - e_\Delta(t)| = 0, \quad \text{or } |e_1(t)| = O[\sup_{\tau \leq t} \Delta u(\tau)].$$

For very small initial conditions, and small $r(t)$, it can be shown that $e_1(t) \to 0$ as $t \to 0$.

3.3 Output Feedback with $n^* = 2$

An extension of the above approach is to the case when the relative degree of the plant is two. Using the same controller as in (8), error equations (10) and (11) can be derived. We note that the adaptive law (12), however, cannot guarantee stability since the relative degree is two and hence, $W_m(s)$ is not strictly positive real. The stability questions that arise in this context are outlined below.

In addition to the term $\theta^T \omega$ used in the controller output, an additional signal $\dot{\theta}^T \omega_a/a$ is added, where

$$\omega_a = \frac{a}{s+a}\omega$$

and a is chosen such that

$$W_m'(s) \triangleq W^{-1}(s)W_m(s), \qquad W(s) = \frac{a}{s+a}$$

and W_m' is strictly positive real. With the resulting v given by

$$v = \theta^T \omega + \dot{\theta}^T \frac{\omega_a}{a} \quad (18)$$

we note that the plant output is of the form

$$y_p = W_p(s)\left(W^{-1}(s)\theta^T \omega_a + \Delta u\right). \quad (19)$$

Since the closed-loop system matches the model when $\theta = \theta^*$, the error equation becomes [1]

$$e_1 = W_m(s)\left(W^{-1}(s)\widetilde{\theta}^T \omega_a + \Delta u\right).$$

Choosing $e_\Delta = W_m(s)\Delta u$, and defining $e_u = e_1 - e_\Delta$, we obtain

$$e_u = W'_m(s)\widetilde{\theta}^T \omega_a. \tag{20}$$

Noting that (20) is similar to (11), the adaptive law

$$\dot{\theta}(t) = -e_{u1}(t)\omega_a(t) \tag{21}$$

and the same approach as in section 3.2 can be used to show the boundedness of all the signals. While the reader is referred to [8] for further details, we briefly state the key steps below.

The boundedness of e_u and $\widetilde{\theta}$ is established by using the error equation in (20) and the adaptive law in equation (21). To show the boundedness of e_1, the state equations can be shown to be of the form

$$\dot{X}' = A_{mn}X' + b_{mn}(\widetilde{\theta}^T \omega_a + \bar{\nu}) \tag{22}$$

where

$$\bar{\nu} = \frac{a}{s+a}\Delta u \tag{23}$$

and $\omega_a = CX'$, $X' = [x_p^T, \omega_1^T, \omega_2^T, \omega_a]^T$, and C is a constant matrix. To establish the proof of boundedness of X', the following cases must be considered. (a) $|v| \leq u_0$, (b-i) $|v| > u_0$, $sgn(X')^T P_o b_{mn} = -sgn(\bar{u})$ and (b-ii) $|v| > u_0$, $sgn(X')^T P_o b_{mn} = sgn(\bar{u})$. It is worth noting that when the signal v hits the saturation limit, its effect on $\bar{\nu}$ is delayed as can be seen in equation (23). Despite this delayed effect, it can be shown to lead to boundedness by evaluating W as in section 3.2 in the three cases above.

3.4 A Low-order Adaptive Controller

The controllers discussed in Sections 3.2 and 3.3 are given by (8), (12), (18), and (21), and are applicable for the cases when the relative degree is one or two. These equations show that the controllers are of order $2n$, and have about $2n$ adjustable parameters in θ. In many systems, it becomes more expedient to design a low-order controller if n is large. For example, in the combustion control problem, a dynamic model that is based on the wave approach [9] often results in a model as in equation (1) where $W_p(s)$ includes time-delays associated with the propagation of the acoustic waves. As a result,

---
[1] Since $r = 0$, without loss of generality, we assume that $e_1 \equiv y_p$.

the resulting rational transfer function is of a very high order. In these cases, an alternative approach can be adopted where a controller of the order of relative degree n^*, rather than the plant order n, can be derived [2]. In this section, this controller is briefly described. When $n^* = 2$,

$$\dot{\omega}_1(t) = -\lambda \omega_1(t) + u(t)$$
$$\omega(t) = [\omega_1(t), y_p(t)]^T$$
$$\theta(t) = [\theta_1(t), \theta_2(t)]^T \qquad (24)$$
$$v(t) = \theta(t)^T \omega(t) + \dot{\theta}^T \frac{\omega_a}{a}$$

where $\lambda > 0$ is a scalar constant. Using the same procedure as in sections 3.2 and 3.3, it can be shown that the underlying error model is of the form of equation (20). Therefore, the same adaptive law as in (12) leads to a closed-loop system that has bounded solutions as in Theorem 1. The details of the proof of stability can be found in [10].

We note that the controller in (24) is of the order of $n^* - 1$ rather than $2n$. For plants with higher relative degree, it can be shown that an extension of this controller is still applicable. The adaptive law, however, requires a more complex form than (21) since in this case, a parameter θ whose $n^* - 1$ derivatives are well defined is needed. The results corresponding to these controllers are presented in [16] and are summarized in an upcoming report [8].

4 The Adaptive Controller in the Presence of a Delay

4.1 The Controller Without Saturation Constraints

The problem is the control of a plant given by the input-output description

$$y_p(t) = W_p(s)[u(t-\tau)], \qquad W_p(s) = \frac{k_p Z_p(s)}{R_p(s)} \qquad (25)$$

where $W_p(s)$ is the transfer function of a finite-dimensional system whose order n is known, relative degree n^* is one, zeros are in $\mathbb{C}^-$, and its high frequency gain is known. The time-delay τ is assumed to be known as well. The plant poles and zeros are unknown and it is assumed that all poles have multiplicity one. The control goal, as in the above sections, is asymptotic stabilization of the plant. For ease of exposition, in what follows, we assume that the high frequency gain is known with $k_p = k_m = 1$.

The fixed controller. A unique approach for controlling systems With known time-delay was originated by Otto Smith in the 1950s [11] by compensating for the delayed output using input values stored over a time window of $[t-\tau, t]$ and estimating the plant output using a model of the plant. In [12], this idea was extended to include unstable plants as well using finite-time integrals of the delayed input values thereby avoiding unstable pole-zero cancellations that may occur in Smith's controller. The controller that we propose in this section is an adaptive version of the Smith controller and utilizes finite-time integrals as in [13,14]. To facilitate the derivation of the adaptive controller, we describe the fixed controller structure in this section.

Since only the plant input and output are accessible for measurement, a standard pole-placement controller is required. The presence of the time-delay motivates the use of an additional signal (denoted u_1 below) which attempts to anticipate the future outputs using a model of the plant. The resulting controller structure can be described as follows [13]:

$$u(t) = \frac{c(s)}{\Lambda(s)}u(t-\tau) + \frac{d(s)}{\Lambda(s)}y(t) + u_1(t) + \frac{k_m}{k_p}r(t)$$

$$u_1(t) = \frac{n_1(s)}{R_p(s)}u(t) + \frac{n_2(s)}{R_p(s)}u(t-\tau) \qquad (26)$$

where $\Lambda(s)$ is a hurwitz polynomial of degree n, c, d, and n_1, are polynomials of degree $n-1$, respectively, which satisfy the relations

$$c(s)R_p(s) + d(s)Z_p(s) = -n_2(s)\Lambda(s), \qquad (27)$$
$$n_1(s) = R_p(s) - R_m(s)Z_p(s). \qquad (28)$$

$R_m(s)$ is a monic Hurwitz polynomial of degree n^* and represents the desired closed-loop poles of the plant, and $n_2(s)$ is an $n-1$ degree polynomial. From the Bezout identity, it can be easily shown that c, d, and n_1 exist that satisfy (27) and (28). The controller structure in (26) can be shown to result in a closed-loop system with the transfer function $W_m(s)e^{-s\tau}$, where $W_m(s)$ represents the desired closed-loop transfer function.

A point to note above is the case when the plant is unstable. Since the controller in (26) introduces zeros at $R_p(s)$, it leads to unstable pole-zero cancellations for the case when the plant is open-loop unstable. As a result, a modification as suggested by [12] was introduced in [13]: generate $u_1(t)$ as

$$u_1(t) = \sum_{i=1}^{n} \alpha_i \left(\int_{-\tau}^{0} e^{-\lambda_i \sigma} u(t+\sigma) d\sigma \right) \triangleq \int_{-\tau}^{0} \lambda(\sigma) u(t+\sigma) d\sigma \qquad (29)$$

where

$$R_p(s) = \Pi_{i=1}^{n}(s-\lambda_i).$$

By taking Laplace Transform on both sides of (29), it can be shown that

$$\frac{n_1(s)}{R_p(s)} = \sum_{i=1}^{n} \frac{\alpha_i}{s-\lambda_i}, \quad \frac{n_2(s)}{R_p(s)} = \sum_{i=1}^{n} \frac{\alpha_i e^{\lambda_i \tau}}{s-\lambda_i}$$

We note that the realization of $u_1(t)$ using finite-time integrals requires that the plant poles have multiplicity one. For implementation purposes, we approximate $u_1(t)$ by by a summation so that

$$\int_{-\tau}^{0} (\lambda(\sigma) u(t+\sigma)) d\sigma \approx \sum_{j=1}^{p} \overline{\lambda}_i^* \overline{u}_i(t) \triangleq \overline{\lambda}^{*T} \overline{u}(t) \tag{30}$$

where $\overline{u}_i(t)$ represents the ith sample of $u(t)$ over the interval $[t-\tau, t]$, for $i = 1, \ldots, p$.

The above discussions show that the time-delay controller of the form

$$\begin{aligned}
\dot{w}_1 &= \Lambda_0 w_1 + \ell u(t-\tau) \\
\dot{w}_2 &= \Lambda_0 w_2 + \ell y(t) \\
u &= \theta_1^{*T} w_1 + \theta_2^{*T} w_2 + \overline{\lambda}^{*T} \overline{u} + r
\end{aligned} \tag{31}$$

stabilizes the plant in (1) if all plant parameters including the time-delay are known. In the next section, we show how an adaptive controller can be developed in the presence of a time-delay and parametric uncertainties.

The adaptive controller We now consider the control of the plant (1) when the transfer function $W_p(s)$ has unknown coefficients and the time-delay τ is known.

The form of the controller in (31) can be utilized to develop the adaptive controller, as was done in [14]:

$$\begin{aligned}
\dot{w}_1 &= \Lambda_0 w_1 + \ell u(t-\tau) \\
\dot{w}_2 &= \Lambda_0 w_2 + \ell y(t) \\
u &= \theta_1^T(t) w_1 + \theta_2^T(t) w_2 + r(t) + \overline{\lambda}^T(t) \overline{u}(t)
\end{aligned} \tag{32}$$

where $\overline{u}_i$ is the ith sample of $u(t)$ in the interval $[t-\tau, t)$, $i = 1, \ldots, p$. Expressing $\tilde{\theta} = \theta - \theta^*$, $\omega = [w_1^T, w_2^T, \overline{u}^T]^T$, $\theta = [\theta_1^T, \theta_2^T, \overline{\lambda}^T]^T$, $\theta^* = [\theta_1^{*T}, \theta_2^{*T}, \overline{\lambda}^{*T}]^T$, the closed-loop system equations that describe the plant and the controller are given by

$$\dot{X} = AX(t) + b\left[r(t-\tau) + \tilde{\theta}^T(t-\tau)\omega(t-\tau)\right], \qquad y = c^T X.$$

Defining X_m as the model state corresponding X when the parameter errors are zero, the state error e as $e = X - X_m$, the output error e_1 as $e_1 = y_p - y_m$, we obtain that the underlying error model is of the form

$$\dot{e} = Ae(t) + b\tilde{\theta}^T(t-\tau)\omega(t-\tau), \qquad e_1 = c^T e. \tag{33}$$

Equation (33) can be expressed also in the form

$$e_1(t) = W_m(s) e^{-s\tau} \left[\tilde{\theta}^T(t)\omega(t)\right]. \tag{34}$$

It is easy to see that when $\tau = 0$, equation (33) collapses to the standard error equations in adaptive control.

Suppose the *adaptive law* is chosen as in the delay-free case

$$\dot{\tilde{\theta}}(t) = -e_1(t)\omega(t-\tau). \tag{35}$$

The underlying error equations can be expressed from (34) and (35) as

$$\begin{cases} \dot{e}(t) = Ae(t) + b\tilde{\theta}^T(t)\omega(t-\tau) + b\omega(t-\tau)^T \left(\int_{-\tau}^{0} \dot{\tilde{\theta}}(t+\nu)d\nu\right) \\ \dot{\tilde{\theta}}(t) = -c^T e(t)\omega(t-\tau) \end{cases}. \tag{36}$$

It is quite evident that in the delay-free case ($\tau \equiv 0$), if the corresponding system is SPR, we may find a Lyapunov function as in (13). Based on it, we introduce the following Lyapunov-Krasovskii functional:

$$V(e,\tilde{\theta},\dot{\tilde{\theta}}) = e^T P e + \tilde{\theta}^T \tilde{\theta} + V_2(\dot{\tilde{\theta}}) \tag{37}$$

where

$$V_2(\dot{\tilde{\theta}}) = \int_{-\tau}^{0} \left(\int_{t+\nu}^{t} \dot{\tilde{\theta}}(\xi)^T \dot{\tilde{\theta}}(\xi) d\xi\right) d\theta. \tag{38}$$

It is clear that $V(e,\tilde{\theta},\dot{\tilde{\theta}})$ is positive definite and has an infinitesimal upper bound defined appropriately by the corresponding 'sup' norm in the space $\mathbf{R}^n \times \mathcal{L}^2([-\tau,0],\mathbf{R}^m)$. We shall now compute the derivative of V along the solutions of the model transformation (36):

$$\dot{V} = -e^T Q e + 2e_1(t)\omega^T(t-\tau)\left(\int_{-\tau}^{0} e_1(t+\nu)\omega(t+\nu-\tau)d\nu\right)$$

$$+ \int_{-\tau}^{0} \left[\|e_1(t)\omega(t-\tau)\|^2 - \|e_1(t+\nu)\omega(t+\nu-\tau)\|^2\right] d\nu \tag{39}$$

Denoting

$$\bar{a} = e_1(t)\omega(t-\tau), \qquad \bar{b} = e_1(t+\nu)\omega(t+\nu-\tau)$$

Equation (39) can be rewritten as

$$\dot{V} = -e^T Q e + \int_{-\tau}^{0} \left[2\bar{a}^T \bar{b} + \bar{a}^T \bar{a} - \bar{b}^T \bar{b}\right] d\nu$$

$$\leq -e(t)^T \left[Q - 2\tau \|\omega(t-\tau)\|^2 cc^T\right] e(t) \leq 0 \tag{40}$$

for bounded signals ω satisfying, at time t, the matrix inequality:

$$Q > 2\tau \|\omega(t-\tau)\|^2 cc^T. \tag{41}$$

Since ω is a dependent variable, it needs to be verified as to when condition (41) is satisfied. By using the facts that $e = X - X_m$, and $\|\omega(t)\| \leq C\|X(t)\|$, where C is a positive constant, it can be shown that for $\tau \leq \tau^*$, if

$$\|\omega(\zeta)\| \leq \gamma \qquad \zeta \in [t_0 - \tau, t_0] \tag{42}$$

for some γ, then $\dot{V} \leq 0$. In [15], this was accomplished using the form of (41), and using a step-by-step construction procedure that extended the negativity of the derivative $\dot{V}$ from one delay interval to the other. The boundedness of closed-loop signals can then be used to guarantee that $e_1(t) \to 0$ as $t \to \infty$. This stability result is summarized in the following theorem:

Theorem 2

Given γ for which equation (42) holds, a positive constant τ^* exists such that for the plant in (25), the controller given by (32) and (35) leads to global bounded solutions and $\lim_{t\to\infty} e_1(t) = 0$ for all $\tau \leq \tau^*$.

We note that for a stabilization problem, one can simply choose $r(t) = 0$ in which case $e_1(t) = y_p(t)$. Hence, the plant output tends to zero if r is set to zero in the controller in (32) and (35).

A low-order adaptive controller. Similar to the case when only saturation was present, a controller of order $n^* - 1$ can be derived in the presence of delay as well [2,16]. This is given by

$$\dot{\omega}_1 = \lambda \omega_1 + u(t - \tau) \qquad \lambda > 0$$
$$u = \theta_1 \omega_1 + \theta_2 y_p + \overline{\lambda}^T(t)\overline{u} + r \tag{43}$$

where $\lambda > 0$ is a scalar constant. For a small τ, it can be shown that the closed-loop system is stable for all signals that start within a compact set. The main point that has to be established with such a low-order controller is that the identities (27) and (28) are valid even with the limited degrees of freedom in the controller. The idea here is to choose $n_1(s)$ so that (28) holds, exploit the relation between n_1 and n_2 for small τ, and show that with the resulting n_2, equation (28) holds for small τ. The reader is referred to [16] for further details.

4.2 The Controller in the Presence of Saturation and Time-delay

The problem that we shall now consider is the control of plant in the presence of a time-delay as in (1) subject to input saturation constraints as in (2). We focus our attention on the case when the relative degree is one. Extensions to higher relative degree can be found in [8]. We propose the controller to be a combination of those in (8) and (31), and is of the form

$$\dot{\omega}_1 = \Lambda_0 \omega_1 + \ell u(t - \tau)$$
$$\dot{\omega}_2 = \Lambda_0 \omega_2 + \ell y(t)$$
$$v = \theta_1^T(t)\omega_1 + \theta_2^T(t)\omega_2 + r(t) + \overline{\lambda}^T(t)\overline{u}(t) \tag{44}$$

Since equation (44) can be represented as

$$u = \theta_1^T(t)\omega_1 + \theta_2^T(t)\omega_2 + r(t) + \overline{\lambda}^T(t)\overline{u}(t) + \Delta u(t) \tag{45}$$

where $\Delta u = u - v$, noting the similarity of (45) and (32), following the same lines as in Section 4.1, we obtain that

$$\dot{X} = AX(t) + b\left[r(t-\tau) + \widetilde{\theta}^T(t-\tau)\omega(t-\tau) + \Delta u(t-\tau)\right], \quad y = c^T X. \tag{46}$$

As in the saturation-free case [15], it can be shown that

$$\|\omega(t)\| \leq C\|X(t)\| \tag{47}$$

and that the error $e = X - X_m$ satisfies the equation

$$e_1(t) = W_m(s)\left[\widetilde{\theta}^T(t-\tau)\omega(t-\tau) + \Delta u(t-\tau)\right], \quad e_1 = c^T e \tag{48}$$

where $W_m(s)$ is strictly positive real. As in section 4.1, we generate a signal e_Δ as $e_\Delta(t) = W_m(s)\Delta u(t-\tau)$ which results in the error equation

$$e_u(t) = W_m(s)e^{-s\tau}\left[\widetilde{\theta}^T(t)\omega(t)\right] \tag{49}$$

where $e_u = e_1 - e_\Delta$. While the form of the error equation shows that the same adaptive law as in (35), the same procedure as in section 4.1 does not suffice in establishing an inequality as in (40) that depends only on the initial conditions of $\omega(t)$. The proof of boundedness of the closed-loop system can still be shown by proceeding as follows:

Defining e' as the state of the error equation (49), and

$$V = e'^T P e' + \widetilde{\theta}^T \widetilde{\theta} + V_2(\dot{\theta})$$

where V_2 is defined as in (38), we obtain that

$$\dot{V} \leq -e'^T(t)\left[Q - 2\tau\|\omega(t-\tau)\|^2 cc^T\right]e'(t) \leq 0$$

if

$$Q > 2\tau\|\omega(t-\tau)\|^2 cc^T.$$

Suppose

$$\sup_{\xi \in [t_0 - \tau, t_0]} \|\omega(\xi)\|^2 \leq \gamma \tag{50}$$

From (40), it follows that

$$\|\widetilde{\theta}(\xi)\| \leq \sqrt{V(\xi)} \leq \sqrt{V(t_0)} \quad \forall \xi \in [t_0, t_0 + \tau].$$

In order to establish the boundedness of the state of the system, we note that the governing equation satisfies (46) where w satisfies (47). We proceed as in the delay-free case and consider the cases (a) $|v(t-\tau)| \leq u_0$, (b-i) $|v(t-\tau)| > u_0$, $sgn(X(t)^T Pb) = -sgn(u(t-\tau))$, and (b-ii) $|v(t-\tau)| > u_0$, $sgn(X(t)^T Pb) = sgn(u(t-\tau))$.

Case (a) $|v(t-\tau)| \leq u_0$: Defining W as in the proof of Theorem 1, we obtain

$$\dot{W}(X) = X^T \left[A_{mn}^T P + P A_{mn}^T \right] X + 2(b_{mn}^T PX)[\tilde{\theta}^T(t-\tau)w(t-\tau)]$$
$$\leq -q_0 \|X\|^2 + 2p_b \|X\| \theta_m C \|w(t-\tau)\|.$$

From (50), it follows that if $\sqrt{V(t)} < q_0/(2p_b C)$, then $\dot{W}(t) < 0$ for all $\|X(t)\| > \gamma/C$. This in turn allows the extension of the inequality in (50) to the interval $[t_0, t_0 + \Delta]$ for a finite but nonzero Δ. Since t_0 is arbitrary, this implies that if case (a) holds, $X(t)$ is bounded for all time t.

(b) Let $|v(t-\tau)| > u_0$. Hence,

$$u(t-\tau) = u_0 sgn(v(t-\tau)) \text{ and } |v(t-\tau)| > u_0.$$

As before, we can show that

$$\dot{W} = -X^T QX + 2 \left| b^T PX(t) \right| u_0 sgn(v(t-\tau)) sgn(b^T PX(t)) \\ - 2(b^T PX(t))(\theta^{*T} w(t-\tau)). \quad (51)$$

Case I $sgn(v(t-\tau)) = -sgn(b^T PX(t))$: Equation (51) becomes

$$\dot{W}(\xi) \leq -q_0 \|X\|^2 - 2|b^T PX(t)| (u_0 - \|w(t-\tau)\| \|\theta^*\|) \\ < 0 \quad \forall \xi \in [t_0 - \tau, t_0]$$

if

$$\sup_{\xi \in [t_0-\tau, t_0)} \|w(\xi)\|^2 \leq \frac{u_0}{\|\theta^*\|}. \quad (52)$$

As before, this extends the boundedness of $X(t)$ and therefore $w(t)$ to the interval $[t_0, t_0 + \Delta]$.

Case II $sgn(v(t-\tau)) = sgn(b^T PX(t-\tau))$: Here,

$$\dot{W}(t) \leq -q_0\|X\|^2 + 2|X^T Pb|u_0 - 2X^T Pb v(t-\tau) + 2p_b \theta_m \|X\| \|w(t-\tau)\| \\ \leq 0$$

if $\|X\| \leq \gamma/C$. The above arguments show that if $\gamma < u_0/\|\theta^*\|$, all sufficient conditions in all the three cases are satisfied for all initial values in a compact domain with (i) w as in (50) and (ii) $\sqrt{V(t_0)} < q_0/(2p_b C)$. Therefore the boundedness of the closed-loop system signals follows for all initial conditions within this compact domain. Using Barbalat's Lemma, once again, it can be shown that $|e_1(t)| = O[\sup_{\tau \le t} \Delta u(\tau)]$. These results are summarized in the following theorem:

Theorem 3
The plant in (25), together with the controller in (32) and (35), has bounded solutions if

$$\begin{aligned}&\text{(i) } \sup_{\xi \in [t_0-\tau,t_0)} \|\omega(\xi)\|^2 < u_0/\|\theta^*\| \\ &\text{(ii) } \sqrt{V(t_0)} < \tfrac{q_0}{2p_bC}.\end{aligned} \quad (53)$$

Further, the output error e_1 is given by

$$|e_1(t)| = O[\sup_{\tau \le t} \Delta u(\tau)].$$

5 Simulation results

A wide class of combustion systems can be modeled as an enclosed chamber where the incoming flow is a mixture of fuel and air which burns over a certain region, with the hot gases exiting downstream to a turbine (see Figure 1). The underlying dynamics in such a system can be represented as a feedback

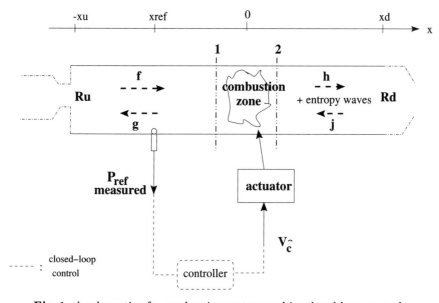

Fig. 1. A schematic of a combustion system and its closed-loop control.

system with the forward loop representing the acoustics and the feedback loop representing the flame dynamics (see Figure 2). We briefly describe this system and its control in the presence of saturation constraints below. The details of this feedback system together with the parameters can be found in [16] and the references therein.

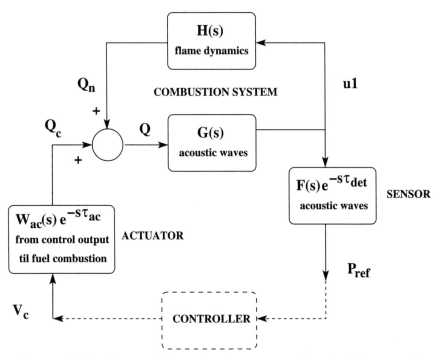

Fig. 2. A block diagram of the combustion system and its closed-loop control.

The transfer function $G(s)$ is the relation between the unsteady heat release $Q(t)$ and the unsteady velocity $u_1(t)$ at the flame, and represents the dynamics, while the transfer function $H(s)$ represents the flame dynamics. A fuel-injector that delivers a controlled heat-release $Q_c(t)$ is chosen as an actuator, and is driven by a voltage V_c. A pressure transducer is chosen as the sensor which measures the unsteady pressure P_{ref} at a particular location (see figure 1). τ_{ac} is the mixing and transport delay between injection and burning, while τ_{det} is the detection time-delay due to the pressure measurement location. $W_{ac}(s)$ represents the dynamics of the fuel-injector. The underlying transfer function of the system to be controlled is therefore between the input-output pair $\{V_c, P_{\text{ref}}\}$, and is of the form

$$P_{\text{ref}} = W_p(s)V_c(t-\tau), \qquad \tau = \tau_{ac} + \tau_{det}$$

where

$$W_p(s) = \frac{F(s)P_{du}(s)G(s)W_{ac}(s)}{1 - G(s)H(s)}$$

and $P_{du}(s)$ is the transfer function between the unsteady velocity u_1 and the pressure P_{ref}. From the properties of the acoustics and the flame dynamics, it can be shown that the transfer function $W_p(s)$ has stable zeros, a positive high frequency gain, and a relative degree equal to that of $W_{ac}(s)$ which is usually one or two. Due to the representation of the acoustics using a wave approach, $W_p(s)$, which can be viewed as a rational approximation of dynamics that includes time-delays, is of a high order. The uncontrolled system was simulated at an operating condition of Mach number= 0.08, equivalence ratio=0.7 and mean temperature of 287.6°K, which corresponds to a 250 kW combustor considered in [17]. The resulting pressure response is shown in Figure 3 which indicates the marginal stability of the system to be controlled. $W_{ac}(s)$ was chosen to be a first-order transfer function. A saturation

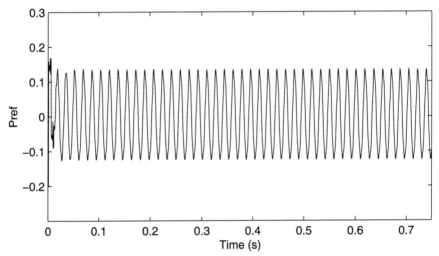

Fig. 3. The pressure response of the uncontrolled combustion system at a Mach number of 0.08 and equivalence ratio of 0.7.

constraint was imposed on V_c that corresponds to the use of a maximum of 15% of the main fuel.

A low-order controller was implemented as described in section 3.4. The response of the controlled system is presented in Figures 4 and 5. In Figure 4 the improvement due to the adaptive controller over the uncontrolled case is illustrated. A background noise that is 10% of the mean pressure was added to test the robustness properties of the controller. No time-delay or saturation were included. Next, a time-delay of 8 ms and a saturation constraint corresponding to a magnitude limit of 0.1 on V_c, equivalent to the use of a maximum of 15% of the main fuel were included in the plant. The evolution of the control parameter k_1 is indicated in the two cases, one where the saturation effect is included in the controller and the other where it is not,

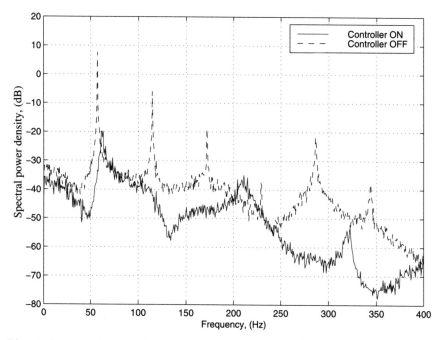

Fig. 4. A comparison of the pressure power spectra of the controlled and the uncontrolled combustion system, with background noise and no time-delay.

in Figure 5. The figure clearly illustrates the improvement in the pressure response when the saturation effect is included.

6 Summary

In this chapter, adaptive controllers that lead to stabilization in the presence of a time-delay and saturation were presented. The cases when either saturation only or time-delay only is present were considered first before addressing the general problem. It was shown that in each of these cases, for marginally stable plants, for all initial conditions starting within a compact set, the adaptive controller guarantees that the closed-loop system remains bounded, and that the output becomes small. The problem of saturation was treated by converting the nonlinearity into an input disturbance which was suitably accommodated by the adaptive law. The problem of time-delay was addressed by modifying the Lyapunov function to a Lyapunov-Krasovskii functional that accommodated the evolution of the adaptive parameter over an interval $[t_0 - \tau, t_0]$ rather than just at an instant. An extension to low-order controllers whose order depends on the plant relative degree rather than the plant order was presented. The latter was applied to a combustion control

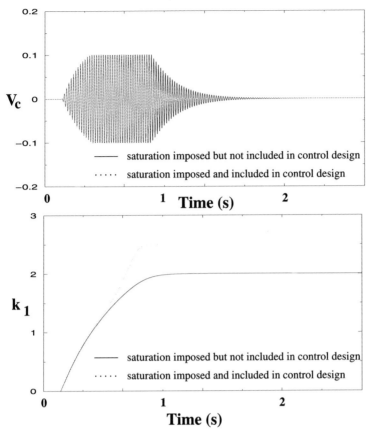

Fig. 5. The performance of the system with the adaptive controller that accommodates saturation in the control design, at a Mach number of 0.08, equivalence ratio of 0.7, and time-delay of 8 ms. The control input had a magnitude constraint of 0.1.

problem where the goal was to bring pressure oscillations down to zero in the presence of a delay and a saturation constraint on the fuel input injected.

Acknowledgements

The research reported here is a result of an international collaboration between the authors at the three institutions MIT, University of Cambridge, and University of Compiegne, carried out during numerous visits of the authors to the collaborating institutions. This work was also supported in part by Trinity College, University of Cambridge, UK, and a joint NSF/CNRS project under grant No. INT-9603271.

References

1. A.M. Annaswamy, O. El-Rifai, M. Fleifil, and A.F. Ghoniem. A model-based self-tuning controller for thermoacoustic instability. *Combustion, Science, and Technology*, 135:213–240, August 1998.
2. S. Evesque, A.P. Dowling, and A.M. Annaswamy. Adaptive algorithms for control of combustion. In *NATO RTO/AVT Symposium on Active Control Technology for Enhanced Performance in Land, Air, and Sea Vehicles*, Braunschweig, Germany, May 2000.
3. J. P. Hathout, A. M. Annaswamy, and A. F. Ghoniem. Control of combustion instability using fuel injection. In *NATO RTO/AVT Symposium on Active Control Technology for Enhanced Performance in Land, Air, and Sea Vehicles*, Braunschweig, Germany, May 2000.
4. S.P. Kárason. Adaptive control in the presence of input constraints. Master's thesis, MIT, Cambridge, MA, USA, 1993.
5. S.P. Kárason and A.M. Annaswamy. Adaptive control in the presence of input constraints. *IEEE Transactions on Automatic Control*, 39:2325–2330, 1994.
6. A.M. Annaswamy, M. Fleifil, Z. Ghoniem, and A.F. Ghoniem. A feedback model of thermoacoustic instability in combustion processes. Technical Report 9502, Adaptive Control Laboratory, MIT, Cambridge, MA, 1995.
7. K. S. Narendra and A. M. Annaswamy. *Stable Adaptive Systems*. Prentice-Hall, Inc., Englewood Cliffs, NJ, 1989.
8. S. Evesque, A.M. Annaswamy, S.-I. Niculescu, and A.P. Dowling. A stable low-order adaptive controller for systems with time-delays. Technical Report Tech. Report No. 2001, Adaptive Control Laboratory, MIT, Cambridge, MA, January 2001.
9. A.P. Dowling. The calculation of thermoacoustic oscillations. *Journal of Sound and Vibration*, 180(4):557–581, 1995.
10. S. Evesque, A.P. Dowling, and A.M. Annaswamy. Adaptive combustion control with saturation. In *37th AIAA/ASME/SAE/ASEE Joint Propulsion Conference*, page to appear, Salt Lake City, Utah, July 2001.
11. O.J. Smith. A controller to overcome dead time. *ISA Journal*, 6, 1959.
12. A.Z. Manitius and A.W. Olbrot. Finite spectrum assignment problem for systems with delays. *IEEE Transactions on Automatic Control*, AC-24 no. 4, 1979.
13. K. Ichikawa. Frequency-domain pole assignment and exact model-matching for delay systems. *Int. J. Control*, 41:1015–1024, 1985.
14. R. Ortega and R. Lozano. Globally stable adaptive controller for systems with delay. *International Journal of Control*, 47(No. 1):17–23, 1988.
15. S. I. Niculescu and A. M. Annaswamy. A simple adaptive controller for positive-real systems with time-delay. In *Proceedings of the IFAC Conference on Adaptive Systems*, Prague, Chekoslovakia, 2000.
16. S. Evesque. *Adaptive control of Combustion Oscillations*. PhD thesis, University of Cambridge, Cambridge, UK, November 2000.
17. P.J. Langhorne, A.P. Dowling, and N. Hooper. Practical active control system for combustion oscillations. *Journal of Propulsion and Power*, 6(3):324–333, 1990.

Adaptive Control for Systems with Input Constraints - A Survey

J.-W. John Cheng and Yi-Ming Wang

Department of Mechanical Engineering
National Chung Cheng University
Chia Yi, 621, Taiwan, R.O.C.

Abstract

Input constraints represent a class of most encountered nonlinearities and have been primary concerns in control design research since the 1950s. The accumulated outcomes since then have formed an established discipline, 'constrained control theory'. However, because most results have appeared in the form of technical papers distributed in journals and various conference proceedings, it is a formidable task for control practitioners to learn and effectively apply the theory to their daily work. The objective of this survey is to compile relevant information in one source in order to facilitate the dissemination of this important control theory. The survey was prepared around two main themes, a historical overview of the constrained control theory and a technical summary on the state of the art of adaptive constrained control designs. The great size of the subject prevents us from writing a comprehensive document on all aspects of the theory. Thus, in this survey we focus on constrained control designs for systems with input constraints. Furthermore, the presentation is confined to the theory for linear time invariant systems due to its maturity and wide range of application.

1 Introduction

Input constraints constitute a class of most encountered nonlinearities in control design. The control variables of all real-world control systems are constrained or limited due to the physical nature of the actuator (e.g., finite input voltage of electrical motors, finite capacity of a pump, etc.). Ignoring their existence may lead to severe performance deterioration and even instability in some cases. Their frequent occurrences in practice have made them a focus of attention in the control community since the 1950s. The design approaches of the 1950s were largely *ad hoc* in nature. The primary effort in the 1960s and 1970s emphasized saturating control using the optimal control theory. The book written by Ryan in 1982 [36] summarized such efforts. Since then, several constrained control approaches, including some adaptive ones, have emerged and gradually formed an established discipline, constrained control theory. So far most results are distributed in various journals and conference proceedings. Such widespread information presents an enormous barrier for practicing control engineers to learn, let alone apply, the theory in their daily work. Therefore, in this chapter, we set out to disseminate this important control theory by compiling relevant, important results in one single source. With the large amount of information available, instead of preparing a comprehensive document on all aspects of the theory, the presentation will focus on the constrained control theory for linear time-invariant systems with input constraints due to its maturity and wide range of application.

In Section 2, we first give a historical account of the development of several important topics in constrained control theory such as the existence of various effective control designs, especially non-adaptive ones. The input magnitude constraint will be emphasized. Section 3 constitutes the major part of the chapter, and contains a technical survey of adaptive constrained control designs. The discussion addresses not only the literatures on adaptive designs with saturating input but also those with high-order input constraints, e.g. rate and acceleration constraints. The chapter concludes in Section 4.

2 History of the Constrained Control Theory

The research on constrained control has a long history that can be traced back to the 1950s when several *ad hoc* designs were proposed. A list of references to the work in that decade can be found in Ryan 1982 [36]. From the 1960s, the research focus was shifted to investigation of fundamental issues such as controllability [27] and stabilizability with bounded input and to the development of more systematic design approaches to constrained control. Such a long history of development renders a chronological account impractical and of little use. Instead, a topic-oriented presentation is adopted in this brief overview. Important topics considered include the existence of stabilizing constrained control and various designs of effective and yet practical constrained control.

2.1 Existence of Stabilizing Magnitude Constrained Control

Conditions for the existence of stabilizing magnitude constrained control, or in other words, conditions for the null controllability with bounded input, were the primary research concern in the 1960s and 1970s. By the early 1980s, satisfactory answers to the existence issue had come to exist for linear systems. Notable results appeared in, for example, Schmitendorf and Barmish 1980 [39] for continuous linear time-varying systems and Sontag 1984 [40] where an algebraic analysis was used to investigate the bounded null controllability for linear time-invariant systems. Sontag's analysis applies equally well to both continuous-time and discrete-time cases. In the following, we compile Sontag's results on the null controllability for linear time invariant systems.

Linear time-invariant system, Σ

(Continuous-time system) $\quad\quad \dot{x}(t) = Ax(t) + Bu(t), \quad x(0) = x_0 \quad (2.1)$

(Discrete-time system) $\quad\quad x(t+1) = Ax(t) + bu(t), \quad x(0) = x_0 \quad (2.2)$

$$\text{where } A \in R^{n \times n}, B \in R^{n \times m}.$$

Input constraint set, $\Omega \subseteq R^m$

Ω being a bounded set in R^m which contains 0 in its interior.

Definitions of null controllability (NC) and asymptotically null controllability (ANC)

- Σ is Ω-ANC (asymptotically null controllable using controls in Ω) if and only if, for each initial state x_0, there is an input $u(\cdot)$ with $u(t) \in \Omega$ for all t such that $x(t) \to 0$ as $t \to \infty$.
- Σ is Ω-NC (null controllable using controls in Ω) if and only if, for each initial state x_0, there are a finite time $T < \infty$, and input $u(\cdot)$ with $u(t) \in \Omega$ for all $t \leq T$, such that $x(T) = 0$

Null controllability in terms of properties of matrices A and B

- When $\Omega = R^m$, Σ is R^m-NC (i.e. the null controllability in the usual sense when there are no contraints) if and only if $[B \ AB \ \cdots \ A^{n-1}B]$ has full rank n.
- When $\Omega \neq R^m$
  - Σ is Ω-NC if and only if Σ is R^m-NC and all eigenvalues λ of A have Re $\lambda \leq 0$ for continuous-time systems and $|\lambda| \leq 1$ for discrete-time systems.
  - Σ is Ω-ANC if and only if Σ is R^m-ANC (i.e. the asymptotic null controllability in the usual sense when there are no constraints) and all eigenvalues λ of A have Re $\lambda \leq 0$ for continuous-time systems and $|\lambda| \leq 1$ for discrete-time systems.

2.2 Designs of Effective Magnitude Constrained Control

2.2.1 Saturated linear controllers. Perhaps when confronted with magnitude constraints, we might wonder if saturated linear controls, for example, linear state feedback: $u(t) = sat(-kx(t))$, can do the job? Unfortunately, the answer is negative. Although feasible for systems with only stable poles and simple imaginary poles, or systems with stable poles and double integrators, global stabilizability with saturated linear control is impossible for systems with multiple integrators greater than 2 [22, 43].

Taking a different point of view, in a series of papers in the early 1990s, Lin and Saberi initiated an investigation into the possibility of semi-global stabilization using saturated linear controllers

Definition (Semi-global stabilizability)
A linear time-invariant system Σ is semi-globally exponentially stabilizable by saturated linear static state feedback if, for any *a priori* given bounded set of initial values $W \subset R^n$, there exists a saturated linear state feedback law such that x = 0 of the closed-loop system is locally exponentially stable and W is contained in the domain of attraction of the equilibrium x = 0.

It was shown that an Ω-ANC system could be semi-globally stabilized using saturated linear state feedback with appropriately designed low gains [28, 29]. Furthermore, a low-and-high gain approach was proposed later to remedy the inherent slow-convergence drawback of the low-gain design [30, 38].

2.2.2 Optimal control approach. In the search for globally stabilizing saturated control, the optimal control approach received the major research effort in the 1960s and 1970s. A comprehensive presentation on developments in constrained optimal control before 1980 can be found in [36]. Important optimal control techniques included constrained time optimal control, time-fuel optimal control, and quadratic optimal control. The relationship between bounded null controllability (Ω-NC) and time optimal control can be explained in the following theorem (see Theorem 2.5.3 in [36] for proof).

Theorem 2.1 (Existence of Ω-NC time optimal control)
If $x_0 \in \Omega$-NC set, then there exists a time optimal control which transfers the system from x_0 to the desired final state x_f in minimum time t_f^*.

Despite being a feasible stabilizing saturated control, practical use of time optimal control is limited because of the inherent deficiencies: (a) requiring solving TPBVP (two point boundary value problem) with unacceptably increasing complexity for high-order systems, and (b) resulting discontinuous, specifically, bang-bang type controls with possibly infinite switching.

Quadratic optimal control was another approach that also attracted great attention. However, it suffers from the same TPBVP difficulty as time optimal control. Frankena and Sivan 1979 [21] proposed a modification to the quadratic cost function by including an additional non-quadratic term. The additional non-quadratic term offers an extra design degree-of-freedom, allowing a reformulation of the TPBVP problem into an initial value problem. Bernstein 1995 [8] exploited the extra design degree-of-freedom further. Besides attaining reformulation of the TPBVP, he was successful in addressing another serious problem associated with quadratic optimal control, namely, the issue of singular control. The non-quadratic cost functional employed by Frankena and Sivan [22] as well as Bernstein [8] assumed the following cost functional formulation:

$$J = \int_0^\infty [x^T R x + h(x, u)] dt \qquad (2.3)$$

where R is a non-negative definite matrix and $h(.,.)$ a craftily selected non-quadratic term whose existence allows an extra degree of freedom so that reformulating the TPBVP into an initial problem and removing singular control become possible.

2.2.3 Nonlinear control designs with composition or combination of saturated linear functions.
The discontinuity properties in optimal controls as discussed above prompted intensive research efforts in the 1980s searching for globally stabilizing smooth saturated control. The first successfully constructed smooth saturated control was reported in Sontag and Sussmann 1990 [41] via an inductive procedure. However, its use is far from practical. A practically useful smooth nonlinear control design in the form of composition of saturated linear functions was conceived in Teel 1992 [44] with application to multiple integrator systems. The idea was later generalized to all Ω-ANC systems in Yang, Sussmann and Sontag 1992 [53] and Sussmann, Sontag and Yang 1994 [42] where the design was further extended to include combinations of saturated linear functions. The composition and combination types of saturated controls take the following forms:

Composition type:

$$u(t) = \sigma_n(f_n(x) + c_{n-1}\sigma_{n-1}(f_{n-1}(x) + c_{n-2}\sigma_{n-2}(f_{n-2}(x) \cdots c_1\sigma_1(f_1(x)) \cdots))) \qquad (2.4)$$

where $\sigma_i(.)$ are saturation functions, $f_i(x)$ linear functions, and c_i constants.

Combination type:

$$u(t) = a_1\sigma_1(f_1(x)) + a_2\sigma_2(f_2(x)) + \cdots + a_n\sigma_n(f_n(x)) \qquad (2.5)$$

where $\sigma_i(.)$ are saturation functions, $f_i(x)$ linear functions, and a_i non-negative constants with $a_1 + ... + a_n \leq 1$. Note that the combination type control can be treated as a neural network control with a single hidden layer.

2.2.4 Model predictive control. Model Predictive Control (MPC) represents a highly flexible and versatile approach to constrained control design. The rationale of MPC is to select a proper control based on its effect on the forecast behavior of the controlled plant. The predictive feature of MPC provides a convenient mechanism to include constraints such as input saturation, rate and higher order constraints and even state constraints. However, the moving horizon feature that suitable control is determined at each sampling time by optimally evaluating its effect on the future behavior of the plant makes MPC a more complex approach than traditional optimal control where the performance horizon is fixed in time.

Being inexperienced users of the MPC method, we refer interested readers to a very recent tutorial Rawlings 2000 [37] and the references therein, rather than presenting an incomplete survey on this important approach here.

2.2.5 Anti-windup control. Starting in the 1970s while the optimal control approach was being intensively investigated, other major research searching for globally stabilizing saturated control was focused on anti-windup control designs, aimed at maintaining consistency in the internal states of linear controllers by introducing proper compensation. Sifting the wealth of anti-windup related literature presents another considerable task if a reasonable survey is sought. Due to space and time limitations, the survey of anti-windup control designs is omitted here and interested readers are referred to several well written papers and the references therein, for example, [7, 48, 50, 26].

3 Adaptive Constrained Control

Besides the above mentioned activities on non-adaptive designs of constrained control, a number of studies have been devoted to designing adaptive constrained controls. The earliest literature addressing adaptive constrained control, to the authors' knowledge, seems to be Monopoli 1974 [31] in which the concept of augmented error was exploited to represent the effect of excessive control signal, i.e., the amount greater than the magnitude constraint. However, the stability proof was not complete. It was not until the 1980s that literature on adaptive magnitude constrained control with complete stability proofs started appearing.

A precursory remark. Two most commonly used adaptive control design methodologies are the self-tuning regulator (STR) and Model Reference Adaptive Control (MRAC). Although the direct STR approach can be interpreted as an MRAC in the unconstrained case [6], the validity of such equivalence in the constrained case needs further investigation. As a result, in the following presentation we classify the results into two categories, namely STR based and MRAC based. The results in each category are then surveyed based on their underlying control design approach. The presentation is conducted in two parts, one for designs with input magnitude constraint and the other for designs with input magnitude, rate and higher order constraints.

3.1 Adaptive Control Designs with Input Magnitude Constraint

3.1.1 A summary of past work

STR-based approaches

One-step-ahead design (minimum-variance control if considered in the stochastic context). The ease of the design rendered the one-step-ahead approach the first candidate for adaptive constrained control investigation. Available results include Evans, Goodwin and Betz 1980 [17], Payne 1986 [34], and Zhang and Evans 1987 [55]. The first two dealt with stable minimum-phase plants while the last one extend the results to type-one stable discrete-time systems with an even number of non-minimum-phase zeros. The inherent requirement that the controlled plant be of minimum phase limits the applicability of the approach.

Pole placement control with anti-windup compensation. In a series of papers (Abramovitch, Kosut and Franklin 1986 [1], Abramovitch and Franklin 1987 [2] and 1990 [3]) indirect adaptive control employing pole placement control with anti-windup compensation and least-squares identification was proposed. Peng and Chen 1993 [35] considered the same type of indirect adaptive control with a different anti-windup scheme. Walgama and Sternby 1993 [49] extended Abramovitch and Franklin's approach to a more general anti-windup scheme called the general conditioning technique [48, 50], which also includes the anti-windup scheme used in [35] as a special case. The integrated design of pole placement and anti-windup compensation represents a very useful methodology, but has the drawback that the analyses proposed were applicable only to open-loop stable plants.

Generic pole placement control. While Abramovitch and Franklin were developing indirect adaptive constrained control with pole placement and anti-windup, Zhang and Evans 1987 [56], Zhang 1993 [58], Feng, Zhang and Palaniswami 1994 [19] addressed indirect adaptive constrained control with pole placement design directly without including anti-windup compensation. The most general result was contained in [19] where adaptive constrained pole placement control was shown applicable to type-m stable systems. The stability results were in BIBO sense. Since 1995, Chaoui *et al.* [10, 11, 12] have presented a series of papers investigating the conditions under which the controller output will stop saturating and tracking performance will be maintained using an indirect adaptive control with a specially selected pole placement control. In [10], adaptive constrained control for type-1 plants was considered while [11] considered a saturating control with integral action for stable plants. All studies above mentioned adopted the indirect STR principle. There were a few others employing the direct adaptive control approach.

For example, direct adaptive constrained pole placement control for type-1 stable systems was investigated in [59] and [12] in continuous-time and discrete-time respectively.

MRAC-based approaches

As mentioned at the beginning of the section, MRAC was the approach under investigation in an early paper by Monopoli 1974 [31]. However, the amount of work on MRAC is very limited in contrast to that for STR-based results. Ohkawa and Yonezawa 1982 [33] considered a one-step-ahead MRAC design for stable plants. A set of convergence results was obtained which was later rediscovered in Payne 1986 [34] using the STR point of view. Two modified MRAC designs were mentioned in Wang and Sun 1992 [51]. Instead of seeking for global stability performance, Karason and Annaswamy 1994 [25] and Annaswamy and Karason 1995 [4] established local stability for constrained MRAC control without assumptions on the open-loop stability of the plant.

Other approaches

Different from STR and MRAC approaches discussed above, Yeh and Kokotovic 1996 [54] studied adaptive constrained control via a backstepping design. Application to plants with a chain of integrators in the observable canonical form and the parametric-strict-feedback form was considered.

From the above anatomical inquiry into the research activities on adaptive constrained control, one may quickly notice the abundance of literature on adaptive constrained pole placement control. Such prevalence becomes even more significant if we recognize the fact that one-step-ahead control can be viewed as a special case of pole placement control [6]. In the following we present the adaptive constrained pole placement design in more detail, because the approach has reached a level of maturity ready for practical applications. Specifically, indirect adaptive designs with and without anti-windup compensation will be reviewed.

3.1.2 Generic pole placement design without anti-windup consideration. The block diagram of a constrained indirect adaptive pole placement controlled system is depicted in Figure 3.1.

Plant model

Consider a linear time-invariant plant expressed in the following deterministic input-output ARX form:

$$A(z^{-1})y(t) = B(z^{-1})u(t) \quad (3.1)$$

where u(.), y(.) denote the input and output sequences, respectively, $A(z^{-1})$ and $B(z^{-1})$ are polynomials of the unit delay operator z^{-1}, explicitly,

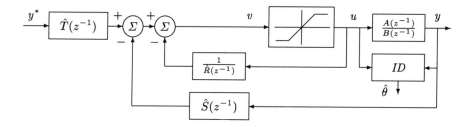

Fig. 3.1. Block diagram of indirect adaptive pole placement design.

$$A(z^{-1}) = 1 + a_1 z^{-1} + a_2 z^{-2} + \cdots + a_n z^{-n} \quad (3.2)$$

$$B(z^{-1}) = b_1 z^{-1} + b_2 z^{-2} + \cdots + b_m z^{-m} \quad (3.3)$$

For plants with d steps of dead time, $b_i = 0$ for $1 \leq i \leq (d-1)$. Furthermore, $u(t)$ is a magnitude-constrained signal, i.e.,

$$u(t) = \sigma_M[v(t)] \quad (3.4)$$

where $\sigma_M(.)$ is a saturation function with

$$\sigma_M(c) = sat(c, M) = \begin{cases} c & \text{if } |c| \leq M \\ M \cdot sgn(c) & \text{if } |c| \geq M \end{cases} \quad (3.5)$$

In the following, the subscript M in σ_M will sometimes be omitted when there is no fear of confusion. Also, the notation $\sigma_M(.)$ and $sat(\cdot, M)$ will be used interchangeably.

System identification

For the system identification purpose, the plant model is reformulated in the following regression form:

$$y(t) = \phi^T(t-1)\theta \quad (3.6)$$

where

$$\phi^T(t-1) = [-y(t-1), \ldots, -y(t-r), u(t-1), u(t-2), \ldots, u(t-r)] \quad (3.7)$$

$$\theta^T = [a_1, \ldots, a_r, b_1, \ldots, b_r] \quad (3.8)$$

and $r = max(m, n)$, $a_i = 0$ for $i > n$ and $b_i = 0$ for $i > m$.

Most commonly used identification algorithms are of least-squares type, for example, the recursive least squares and projection algorithms [6]. In this presentation, the projection algorithm is adopted:

$$\hat{\theta}(t) = P[\hat{\theta}(t-1) + \rho \frac{\phi(t-d)}{1+\|\phi(t-d)\|^2}(y(t) - \hat{\theta}(t-1)^T \phi(t-d))], t \geq 1 \quad (3.9)$$

where $P(\cdot)$ is a projection operator to a convex domain C in the parameter space [1] with $\hat{\theta}(0) = P(\hat{\theta}_0)$ for some bounded $\hat{\theta}_0$, $\|\cdot\|$ is the Euclidean norm on R^{2r}, and $0 < \rho < 2$.

The least squares estimation enjoys a set of very useful convergence properties as stated in the following well-known lemma. In fact, these convergence properties of the least squares estimation make possible all the indirect adaptive constrained control designs available so far.

Lemma 3.1 (Convergence lemma of least-squares type estimation)
Consider the estimate $\hat{\theta}(t)$ defined in (3.9). The following properties hold

(1) $\hat{\theta}(t) \in C$.
(2) For each fixed $k \geq 1$, $\Sigma_{t=k}^{\infty} \|\hat{\theta}(t) - \hat{\theta}(t-k)\|^2 < \infty$.
(3) $\lim_{t \to \infty} \frac{e^2(t)}{1+\|\phi(t-d)\|^2} = 0$, where $e(t) = y(t) - \hat{\theta}(t-1)^T \phi(t-d)$.

Proof: See, for example, [6].

Pole placement controller design via polynomial approach

Pole placement control via the polynomial approach [7] can be expressed as follows:
$$u(t) = \sigma[\hat{S}(z^{-1})[y^*(t) - y(t)] - \hat{R}'(z^{-1})u(t-1)] \quad (3.10)$$
with $y^*(t)$ being the reference input, $\sigma(\cdot)$ the saturation function, and
$$\hat{R}'(z^{-1}) = \hat{r}_1 + \hat{r}_2 z^{-1} + \cdots + \hat{r}_{r-1} z^{-(r-2)} \equiv z[\hat{R}(z^{-1}) - 1] \quad (3.11)$$
The polynomials $\hat{R}$ and $\hat{S}$ are obtained by solving the Diophantine equation:
$$\hat{A}(z^{-1})\hat{R}(z^{-1}) + \hat{B}(z^{-1})\hat{S}(z^{-1}) = A_{cl}(z^{-1}) \quad (3.12)$$
where $\hat{A}$ and $\hat{B}$ are estimates of plant polynomials A and B, and $A_{cl}(z^{-1})$ represents the desired stable closed-loop characteristic polynomial with degree $(2r-1)$:
$$A_{cl}(z^{-1}) = 1 + a_{cl,1}z^{-1} + a_{cl,2}z^{-2} + \cdots + a_{cl,2r-1}z^{-(2r-1)} \quad (3.13)$$

Stability and convergence properties

Theorem 3.1 (BIBO stability)
The input constrained indirect adaptive pole placement control system as described in (3.9) and (3.10) is BIBO stable subject to the following assumptions:

---

[1] C is to be defined later in which the estimated plant polynomials $\hat{A}(z^{-1}, t)$ and $\hat{B}(z^{-1}, t)$ are coprime.

(A1) $B(z^{-1})/A(z^{-1})$ is a stable type-m system, i.e., a system with m repeated poles at $z = 1$ plus the remaining poles lying inside the unit circle,
(A2) $A(z^{-1})$ and $B(z^{-1})$ are co-prime, and
(A3) there exists a known convex region C such that: $\theta^* \in C$ and for all $\hat{\theta} \in C$, the corresponding $\hat{A}(z^{-1})$ and $\hat{B}(z^{-1})$ are co-prime.

Proof: See [19].

Theorem 3.2 (Asymptotic closed-loop performance)
Consider the input constrained indirect adaptive pole placement control system subject to the assumptions A1-A3 as stated in Theorem 3.1. If the control output becomes unsaturated eventually, i.e., existing time t_1 such that for $t \geq t_1$,

$$u(t) = \hat{S}(z^{-1})[y^*(t) - y(t)] - \hat{R}'(z^{-1})u(t-1) \qquad (3.14)$$

then

$$\lim_{t \to \infty} \{A_{cl}(z^{-1})y(t) - \hat{B}(z^{-1}, t)\hat{S}(z^{-1}, t)y^*(t)\} = 0 \qquad (3.15)$$

Proof: See [19].

Remark 1. Theorem 3.2 states that if the adaptive controller comes out of saturation eventually, the desired closed-loop performance is obtained asymptotically. Theorems 3.1 and 3.2 are also valid for systems under certain disturbances and modeling uncertainties [19]. Stronger conclusions are possible pertaining to particular application situations. For example, Chaoui et al. 1996 [10] and 1998 [11] presented a pole placement design addressing constant regulation problems.

3.1.3 Anti-windup pole placement design via conditioning technique.
When the control signal becomes saturated, the controller's internal states are subject to the windup problem which may result in performance degradation. Anti-windup control (AWC) approaches were aimed at restoring the consistency of the internal states of the controller. A large class of different AWCs, including the Generalized AWC [7], can be treated as special cases of the Generalized Conditioning Technique (GCT) [48]. The rationale behind the conditioning technique can be interpreted as follows [48].

'Whenever the computed controller output $v(t)$ results in an actuator output $u(t) \neq v(t)$, the controller inputs $y^*(t)$ (the setpoint) and/or $y(t)$ (the process output) are modified such that if these modified inputs were used in the controller, the controller output $v(t)$ would be equal to the actuator output $u(t)$. By doing so, the effect of actuator nonlinearity such as the saturation is removed from the feedback loop and introduced into the conditioned controller inputs $y^*(t)$ and $y(t)$.'

The conditioning anti-windup technique is applied equally well in adaptive control case. Several studies on this front have been reported, first by

Abramovitch et al. 1986, 1987, 1990 [1, 2, 3] and later extended to the generalized conditioning technique by Walgama and Sternby 1993 [49]. A separate study employing a special form of the general conditioning technique was reported in Peng and Chen 1993 [35]. For the purpose of conveying the idea clearly, the following presentation adopts the original conditioning technique as in Abramovitch et al. [1, 2, 3]; and we refer the readers to [49] for the approach using generalized conditioning technique.

Simple conditioning technique

Consider the general linear controller, also known as a two degrees of freedom controller, without saturation:

$$R(z^{-1})v(t) = T(z^{-1})y^*(t) - S(z^{-1})y(t) \qquad (3.16)$$

Definitions of polynomials, $R(z^{-1})$, $S(z^{-1})$, and $T(z^{-1})$ are the same as those depicted in Figure 3.1 in the previous subsection. The functioning of the anti-windup conditioning technique can be explained as follows.

For the case when the controller output is within the actuator saturation limit, no modification to the setpoint is necessary. The actuator output $u(t)$ equals the controller output $v(t)$. In this case, the present value of linear controller $v(t)$ should be calculated using:

$$v(t) = [1 - R(z^{-1})]u(t) + t_0 y^*(t) + [T(z^{-1}) - t_0]y_m^*(t) - S(z^{-1})y(t) \qquad (3.17)$$

with the past control values replaced with actual actuator output and past setpoint values being the modified ones $y_m^*(t)$. When saturation occurs, the current setpoint is then modified to ensure consistency in the internal state. Specifically, the setpoint is modified such that the following equation is satisfied:

$$u(t) = sat(v(t)) = [1 - R(z^{-1})]u(t) + t_0 y_m^*(t) + [T(z^{-1}) - t_0]y*_m(t) - S(z^{-1})y(t) \qquad (3.18)$$

Note that in the above equation, $v(t)$ is calculated using (3.17). Subtracting the above two equations, we obtain the setpoint modification expression:

$$y_m^*(t) = y^*(t) + [u(t) - v(t)]/t_0 \qquad (3.19)$$

With $u(t) = sat(v(t))$ in mind, the equations (3.17), (3.18) and (3.19) provide an explicit formulation of a linear controller with setpoint conditioning. Furthermore, putting equation (3.19), the expression for $y_m^*(t)$, into equation (3.18), we obtain the following implicit formula for the linear controller with setpoint conditioning without reference to the modified setpoint $y_m^*(\cdot)$ [48].

$$(T(z^{-1})/t_0)v(t) = (T(z^{-1})/t_0 - R(z^{-1}))u(t) + T(z^{-1})y^*(t) - S(z^{-1})y(t) \qquad (3.20)$$

System identification and pole placement control

Both system identification and pole placement control are the same as those in the previous subsection.

Convergence properties

Theorem 3.3 (Boundedness of modified setpoint $y_m^*(.)$)
The modified setpoint $y_m^*(.)$ is bounded if the following conditions are satisfied:

(B1) the process is stable,
(B2) the polynomial $\hat{T}(z^{-1}, t)$ has zeros inside the unit circle at all times,
(B3) the estimate polynomials $\hat{A}$ and $\hat{B}$ are coprime at all times.

Theorem 3.4 (Asymptotic closed-loop performance)
Considering the process (3.1), assume $r = max(m, n)$ is known and $A(z^{-1})$ and $B(z^{-1})$ are coprime. Provided that the conditions (B1-B3) of Theorem 3.3 hold, the adaptive control system described by (3.16)-(3.19) or (3.20) has the following closed-loop performance when the controller is out of saturation,

$$\lim_{t \to \infty}[A_{cl}(z^{-1})y(t) - \hat{B}(z^{-1})\hat{T}(z^{-1})y_m^*(t)] = 0 \quad (3.21)$$

Proofs of Theorems 3.3 and 3.4 are available in [3].

3.2 Adaptive Control Design with Input Magnitude and Rate Constraints

One recent comment [9] on bridging the gap between control theory and practice considered input rate constraint the second most important nonlinearity in practice, next to the input magnitude constraint. Despite its importance, the input rate constraint has not received due attention. Among the five non-adaptive control design methodologies mentioned previously in this chapter, only MPC (model predictive control) sees a sizable collection of results addressing the rate constraint issue. However, the analysis of adaptive MPC with multiple constraints remains a challenge. In particular, when it comes to considering adaptive constrained control, we are left with a very limited number of papers [57, 18, 20, 32, 13, 14, 15]. Zhang and Evans 1988 [57], Feng, Palaniswami and Zhu 1992 [18], and Feng 1994 [20] considered adaptive input rate constrained pole placement control, among which Zhang and Evans [57] adopted discrete-time direct STR design while Feng et al. [18] and Feng [20] looked into discrete-time and continuous-time indirect STR approach. These studies were for input rate constraint only. No magnitude constraint was considered. The design issue with both magnitude and rate constraints were addressed in Moussas and Bitsoris 1993 [32] where an indirect STR with generalized one-step-ahead control was proposed. However no stability analysis was given. In recent years, Cheng and Wang conducted a series of studies [13, 14, 15, 52, 16] on adaptive control with multiple input constraints. Subjects of their investigation included the indirect STR with one-step-ahead control and direct STR with pole placement control. The proposed constrained STR with one-step-ahead control was applicable to both minimum phase plants [13, 52] and non-minimum phase plants [15] considering input magnitude and rate constraints. An extension to input magnitude, rate, and acceleration constraints under a size condition between the rate and acceleration constraints was considered in [15, 52]. Their most recent work addressed direct STR with pole placement control under input magnitude and rate constraints [16].

3.2.1 Adaptive constrained one-step-ahead control with input magnitude and rate constraints for minimum and non-minimum phase systems.

A key component in Cheng and Wang's study is a governing equation relating the prediction error to the 'input discrepancy' between adaptive constrained control and the corresponding non-adaptive control, independent of how the parameter estimates are obtained. Together with the prediction error convergence property of the least-squares type estimation algorithm described in previous Lemma 3.1, the governing equation leads to a successful analysis on the convergence and tracking performance of the adaptive constrained controller. In this subsection, we summarize the results of Cheng and Wang on adaptive one-step-ahead control for systems with input magnitude and rate constraints. Adopting the same analysis approach, similar convergence results could be obtained for constrained direct STR with pole placement design [16].

Plant model

Consider the following deterministic ARX model:

$$A(z^{-1})y(t) = z^{-d}B(z^{-1})u(t) \tag{3.22}$$

where $u(\cdot)$, $y(\cdot)$ denote the input and output sequences, respectively, $A(z^{-1})$ and $B(z^{-1})$ are polynomials of the unit delay operator z^{-1}, specifically,

$$A(z^{-1}) = 1 + a_1 z^{-1} + \cdots + a_n z^{-n} \tag{3.23}$$

$$B(z^{-1}) = b_0 + b_1 z^{-1} + \cdots + b_m z^{-m}, \ b_0 \neq 0 \tag{3.24}$$

and d represents a pure time delay. $u(t)$ is under the magnitude and rate constraints:

$$|u(t)| \leq M, \ M > 0 \quad \text{(magnitude constraint)} \tag{3.25}$$

$$|u(t) - u(t-1)| \leq r, \ 0 < r < 2M \quad \text{(rate constraint)} \tag{3.26}$$

Equivalently, the ARX process can be reformulated into the d-step ahead predictor form:

$$y(t+d) = \alpha(z^{-1})y(t) + \beta(z^{-1})u(t) \tag{3.27}$$

with

$$\alpha(z^{-1}) = \alpha_0 + \alpha_1 z^{-1} + \cdots + \alpha_{n-1} z^{-(n-1)} \tag{3.28}$$

$$\beta(z^{-1}) = \beta_0 + \beta_1 z^{-1} + \cdots + \beta_{m+d-1} z^{-(m+d-1)}, \ \beta_0 = b_0 \neq 0 \tag{3.29}$$

where α_i and β_i are known functions of unknown parameters a_i and b_i. For system identification purpose, the prediction formulation is conveniently expressed in the regression vector form:

$$y(t+d) = \theta^T \phi(t) \tag{3.30}$$

where $\theta, \phi \in R^{n+m+d}$ are defined by

$$\theta = (\alpha_0, \alpha_1, \ldots, \alpha_{n-1}, \beta_0, \ldots, \beta_{m+d-1})^T \tag{3.31}$$

$$\phi(t) = (y(t), y(t-1), \ldots, y(t-n+1), u(t), \ldots, u(t-m-d+1))^T \tag{3.32}$$

System identification

Any least-squares type parameter estimation algorithm is equally applicable in the ensuing analysis. For illustration, in this presentation we assume $\hat{\theta}(t)$ is computed recursively via the projection algorithm as described in the previous pole-placement section:

$$\hat{\theta} = P[\hat{\theta}(t-1) + \rho \frac{\phi(t-d)}{1 + \|\phi(t-d)\|^2}(y(t) - \hat{\theta}(t-1)^T \phi(t-d))], \ t \geq 1 \tag{3.33}$$

Besides the function described in the previous pole placement section, the projection operator $P(\cdot)$ has an additional property described below to ensure a non-zero estimate of β_0 and guaranteeing that $\hat{\beta}_0(t)$ and β_0 have the same sign:

$$P_{n+1}(v) = sgn(\beta_0) max\{b, v_{n+1} sgn(\beta_0)\} \tag{3.34}$$

where b is a known positive lower bound on the magnitude of β_0 and the sign of β_0 is also assumed known.

Adaptive one-step-aead control with magnitude and rate constraints

The design objective of constrained one-step-ahead control is to cause the actual output sequence $y(t)$ to follow $y^*(t)$ as closely as possible under the input constraints. Explicitly, the problem is formulated as a constrained optimization problem at each sampling time t as follows.

$$u(t) = Arg\ min_u\{[y^*(t+d) - \hat{y}(t+d)]^2 + (\hat{\beta}_0(t)\tilde{\rho})[u]^2 :$$
$$|u| \leq M,\ |u - u(t+1)| \leq r\} \quad (3.35)$$

where $(y^*(\cdot))$ is a desired bounded output sequence and $\hat{y}(t+d)$ is an adaptive d-step ahead predictor:

$$\hat{y}(t+d) = \hat{\theta}^T(t)\phi(t) \quad (3.36)$$

It can easily be verified that the adaptive one-step-ahead control $u(t)$ is

$$u(t) = sat\left\{\left(u(t-1) + sat\left\{\left(\frac{\hat{X}}{\hat{\beta}_0(t) + \tilde{\rho}} - u(t-1)\right),\ r\right\}\right),\ M\right\} \quad (3.37)$$

where
$$\hat{X} = y^*(t+d) - \hat{\alpha}(t, z^{-1})y(t) - \hat{\beta}'(t, z^{-1})u(t) \quad (3.38)$$
$$\hat{\beta}'(t, z^{-1}) = \hat{\beta}(t, z^{-1}) - \hat{\beta}_0(t) \quad (3.39)$$

and $sat(\cdot, \cdot)$ is the saturation function as defined before.

A prediction error governing equation

In the following, we state the governing equation relating the prediction error to the 'input discrepancy' between the adaptive constrained control and the corresponding non-adaptive one.

Lemma 3.2 (Prediction error governing equation)
With the adaptive constrained one-step-ahead control $u(t)$ in (3.37), consider the corresponding non-adaptive control $u^*(t)$:

$$u^*(t) = Arg\ min_u\{[y^*(t+d) - y(t+d)]^2 + (\beta_0\tilde{\rho})[u]^2 :$$
$$|u| \leq M,\ |u - u(t-1)| \leq r\} \quad (3.40)$$

There exists $\hat{\lambda}_1(t) \geq 0$, $\hat{\lambda}_2(t) \geq 0$, $\lambda_1^*(t) \geq 0$, and $\lambda_2^*(t) \geq 0$, and the following equation governing the prediction error and the 'input discrepancy' holds

$$e(t+d) = \gamma[u(t) - u^*(t)] + \frac{\hat{\lambda}_2(t)}{\hat{\gamma}(t)}\frac{1}{\hat{Z}(t)}u(t) - \frac{\lambda_2^*(t)}{\gamma}\frac{1}{Z(t)}u^*(t)$$
$$+ \frac{\hat{\lambda}_1(t)}{\hat{\gamma}(t)}\Delta u(t) - \frac{\lambda_1^*}{\gamma}\Delta u^*(t) \quad (3.41)$$

where
$$e(t+d) = y(t+d) - y^*(t+d) \quad (3.42)$$
$$\Delta u(t) = u(t) - u(t-1) \quad \text{and} \quad \Delta u^* = u^*(t) - u(t-1) \quad (3.43)$$
$$Z(t) \equiv \frac{\gamma^2}{\lambda_1^*(t) + \gamma^2} \quad \text{and} \quad \hat{Z}(t) \equiv \frac{\hat{\gamma}^2}{\hat{\lambda}_1(t) + \hat{\gamma}(t)^2} \quad (3.44)$$

with
$$\gamma = \beta_0 + \tilde{\rho} \quad \text{and} \quad \hat{\gamma}(t) = \hat{\beta}_0(t) + \tilde{\rho} \tag{3.45}$$

Furthermore,

$$\hat{\lambda}_1(t) > 0, \hat{\lambda}_2(t) > 0 \overset{imply}{\to} |u(t)| = M \tag{3.46}$$
$$\hat{\lambda}_1(t) = 0, \hat{\lambda}_2(t) > 0 \to |u(t)| = M \tag{3.47}$$
$$\hat{\lambda}_1(t) > 0, \hat{\lambda}_2(t) = 0 \to u(t) = u(t-1) + \Delta u(t); |\Delta u(t)| = r \tag{3.48}$$
$$\hat{\lambda}_1(t) = 0, \hat{\lambda}_2(t) = 0 \to u(t) = \hat{X}/\hat{\gamma}(t) \tag{3.49}$$

and

$$\lambda_1^*(t) > 0, \lambda_2^*(t) > 0 \to |u^*(t)| = M \tag{3.50}$$
$$\lambda_1^*(t) = 0, \lambda_2^*(t) > 0 \to |u^*(t)| = M \tag{3.51}$$
$$\lambda_1^*(t) > 0, \lambda_2^*(t) = 0 \to u^*(t) = u^*(t-1) + \Delta u^*(t); |\Delta u^*(t)| = r \tag{3.52}$$
$$\lambda_1^*(t) = 0, \lambda_2^*(t) = 0 \to u^*(t) = X/\gamma \tag{3.53}$$

where

$$X = y^*(t+d) - \alpha(z^{-1})y(t) - \beta'(z^{-1})u(t) \tag{3.54}$$
$$\hat{X} = y^*(t+d) - \hat{\alpha}(z^{-1}, t)y(t) - \hat{\beta}'(z^{-1}, t)u(t) \tag{3.55}$$

Proof: See [15].

Remark 2. Since the constrained one-step-ahead solution can be expressed as the composition of two saturation functions, $\hat{\lambda}_1, \hat{\lambda}_2, \hat{\lambda}_1^*,$ and $\hat{\lambda}_2^*$ can be identified as the Lagrange multipliers associated with two appropriately defined constrained optimization problems.

Convergence and tracking properties

To proceed, two assumptions about the process are in order:

(C1) the process is stable, and
(C2) the process parameters a_i and b_i are unknown but the orders n and m, the time delay d and $sgn(b_0)$ are assumed known. We also assumed there is a known constant b with $0 < b \leq |b_0|$.

Then we have the following theorems regarding the convergence properties of the adaptive constrained one-step-ahead control. The proofs can be seen in [15].

Theorem 3.5 (Input matching property)
Consider the stable, discrete-time, single-input and single-output process (3.22) with Assumptions C1 and C2 and the adaptive constrained one-step-ahead control defined in (3.33) and (3.35). The following 'input matching'

property between the adaptive control $u(t)$ (3.35) and its corresponding non-adaptive one $u^*(t)$ (3.40) holds:

$$\lim_{t\to\infty}[u(t)-u^*(t)]=0 \tag{3.56}$$

Theorem 3.6 (Asymptotic closed-loop performance)
Consider the same process and the adaptive control as those defined in Theorem 3.5. Let $y(t)$ denote the output of the adaptively controlled processed. Define $u^0(t)$ as the solution of the following unconstrained optimization:

$$u^0(t) = Arg\ min_u\left\{[y^*(t+d)-y(t+d)]^2\right\} \tag{3.57}$$

We have

$$\lim_{t\to\infty}(y^*(t)-y(t)-\tilde{\rho}u(t))=0 \ \ iff\ \ \lim_{t\to\infty}(u^0(t)-u^*(t))=0 \tag{3.58}$$

Remark 3. The conclusion of the above theorem states that the tracking performance of the adaptive constrained one-step-ahead control asymptotically achieves that of the corresponding non-adaptive control if the non-adaptive control is eventually out of constraints.

Remark 4. For minimum phase plants, an ordinary one-step-ahead control formulation without penalty on input effort, i.e., $\tilde{\rho}=0$, can be applied and perfect tracking performance is achieved asymptotically.

4 Conclusions

Input constraints represent the most frequently encountered nonlinearities in real-world control problems. However, because most results have appeared in the form of technical papers distributed in journals and various conference proceedings, it has been a very time-consuming and difficult job for control practitioners to learn and effectively apply the theory to their daily work. In order to facilitate the dissemination of the information, a survey on the state of the art of constrained control theory was conducted. The presentation focused on results for linear time-invariant systems.

In a topic-oriented overview of the constrained control theory, the fundamental issue regarding the existence of bounded stabilizing control was reviewed first, followed by a series of reviews of existing constrained control design methodologies. Specifically, saturated linear control designs, optimal control, and recent nonlinear designs involving compositions or combination of saturated linear controls were considered. References to tutorial and survey papers on other important design approaches such as model predictive control and anti-windup designs were provided.

Besides the overview of the theory, a technical summary on adaptive constrained control was provided. A historical account of the development of adaptive constrained control was presented first. A great majority of available literatures adopted the indirect STR adaptive control configuration with least squares type identification algorithms. Reflecting the situation and its value to practitioners, the technical summary was prepared with the aim of introducing the basic ideas and stability and convergence behaviors of the existing indirect STR constrained controls. The discussion was focused on the designs with both input magnitude and rate constraints.

Acknowledgment

The work is supported by the National Science Council in Taiwan, R.O.C. through the grant NSC 89-2622-E-194-009.

References

1. D. Y. Abramovitch, R. L. Kosut and G. F. Franklin. Adaptive control with saturating inputs, *IEEE 25th Conf. on Decision and Control*, Athens, 848-852, Dec 1986.
2. D. Y. Abramovitch and G. F. Franklin. On the stability of adaptive pole-placement controllers with a saturating actuator, *IEEE Conf. on Decision and Control*, Los Angeles, CA, 825-829, 1987.
3. D. Y. Abramovitch and G. F. Franklin. On the stability of adaptive pole-placement controllers with a saturating actuator, *IEEE Trans. Automatic Control*, 35, 303-306, 1990.
4. A. M. Annaswamy and S. P. Karason. Discrete-time adaptive control in the presence of input constraints, *Automatica*, 31, 1421-1431, 1995.
5. S. Abu el Ata-Doss, P. Fiani and J. Richalet. Handling input and state constraints in predictive functional control, *IEEE 30th Conf. on Decision and Control*, Brighton, 985-990, Dec 1991.
6. K. J. Astrom and B. Wittenmark. *Adaptive Control*, 2nd ed, Addison-Wesley, 1995
7. K. J. Astrom and B. Wittenmark. *Computer Controlled Systems - Theory and Design*, Prentice-Hall, New Jersey, 1984.
8. D. S. Bernstein. Optimal nonlinear, but continuous, feedback control of systems with saturating actuators, *Int. J. Control*, 62, no. 5, 1209-1216, 1995.
9. D. S. Bernstein. On bridging the theory/practice gap, *IEEE Control Systems Magazine*, Special Section on the Theory/Practice Gap in Aerospace Control, 64-70, Dec 1999.
10. F. Z. Chaoui, M. M'Saad, F. Giri, J. M. Dion and L. Dugard. Indirect adaptive control in presence of input saturation constraint, *IEEE 35th Conf. on Decision and Control*, Kobe, Japan, 1940-1944, 1996.
11. F. Z. Chaoui, M. M'Saad, F. Giri, J. M. Dion and L. Dugard. Adaptive tracking with saturating input and controller integral action, *IEEE Trans. Automatic Control*, 43, no. 11, 1638-1643, 1998.

12. F. Z. Chaoui, F. Giri, J. M. Dion, M. M-Saad and L. Dugard. Direct adaptive control subject to input amplitude constraint, *IEEE Trans. Automatic Control*, 45, no. 3, 485-490, 2000
13. J.-W. J. Cheng and Y.-M. Wang. Adaptive one-step-ahead control with input amplitude and rate constraints, *15th National Conf. on Mechanical Engineering of the Chinese Society of Mechanical Engineers*, Tainan, Taiwan, Nov 27-28, 1998.
14. J.-W. J. Cheng and Y.-M. Wang. Adaptive one-step-ahead control with input amplitude, rate, and acceleration constraints, *IEEE Int. Conf. on Control Applications*, Hawaii, USA, 22-26 Aug, 1999.
15. J.-W. J. Cheng and Y.-M. Wang. *Adaptive One-Step-Ahead Control for Minimum and Non-minimum Phase Systems with Input Magnitude and Rate Constraints*, under preparation.
16. J.-W. J. Cheng and Y.-M. Wang. *Direct Adaptive Pole Placement Control with Input Magnitude and Rate Constraints*, under preparation.
17. R. J. Evans, G. C. Goodwin and R. E. Betz. Discrete time adaptive control for classes of nonlinear systems, in *Analysis and Optimization of Systems*, A. Bensoussan and J. Lions (eds.), Springer-Verlag, Berlin, pp. 213-228, 1980.
18. G. Feng, M. Palaniswami and Y. Zhu. Stability of rate constrained robust pole placement adaptive control systems, *Systems & Control Letters*, 18, 99-107, 1992.
19. G. Feng, C. Zhang and M. Palaniswami. Stability of input amplitude constrained adaptive pole placement control systems, *Automatica*, 30, no 6, 1065-1070, 1994.
20. G. Feng. Input rate constrained, continuous-time, indirect, adaptive control, *Int. J. Systems Sci.*, 25, no 11, 1977-1985, 1994.
21. J. F. Frankena and R. Sivan. A nonlinear optimal control law for linear systems, *Int. J. Control*, 30, no. 1, 159-178, 1979.
22. A. T. Fuller. In-the-large stability of relay and saturating control systems with linear controllers, *Int. J. Control*, 10, no. 4, 457-480, 1969.
23. G. C. Goodwin and K. S. Sin. *Adaptive Filtering, Prediction and Control*, Prentice-Hall, New Jersey, 1984.
24. K. Hui and C. W. Chan. A stability analysis of controllers subject to amplitude and rate constraints, *IEEE Int. Conf. on Control Applications*, Hartford, CT, 1997, 259-264.
25. S. P. Karason and A. M. Annaswamy. Adaptive control in the presence of input constraints, *IEEE Trans. Automatic Control*, 39, no. 11, 2325-2330, 1994.
26. V. K. Mayuresh, J. C. Peter, M. Morari and C. N. Nett. A unified framework for the study of anti-windup designs, *Automatica*, 30, no. 12, 1869-1883, 1994.
27. E. Lee and L. Markus. *Foundations of Optimal Control Theory*, John Wiley, 1967.
28. Z. Lin and A. Saberi. Semi-global exponential stabilization of linear systems subject to 'Input saturation' via linear feedbacks, *Systems & Control Letters*, 21, 225-239, 1993.
29. Z. Lin and A. Saberi. Semi-global exponential stabilization of linear discrete-time systems subject to input saturation via linear feedbacks, *Systems & Control Letters*, 24, 125-132, 1995.

30. Z. Lin and A. Saberi. A semi-global low-and-high gain design technique for linear systems with input saturation - Stabilization and disturbance rejection, *Int. J. Robust and Nonlinear Control*, 5, 381-398, 1995.
31. R. V. Monopoli. Adaptive control for systems with hard saturation, *IEEE Conf. Decision and Control*, 841-843, 1975.
32. C. Moussas and G. Bitsoris. Adaptive constrained control subject to both input amplitude and input-velocity constraints, *Int. Conf. on Systems Man and Cybernetics*, Le Touquet, 601-606, Oct 17-20, 1993.
33. F. Ohkawa and Y. Yonezawa. A discrete model reference adaptive control system for a plant with input amplitude constraints, *Int. J. Control*, 36, no. 5, 747-753, 1982.
34. A. N. Payne. Adaptive one-step-ahead control subject to an input amplitude constraint, *Int. J. Control*, 43, no. 4, 1257-1269, 1986.
35. H.-J. Peng and B.-S. Chen. Robust discrete-time adaptive control subject to a control input amplitude constraint, *Int. J. Control*, 58, no 5, 1183-1199, 1993.
36. E. P. Ryan. *Optimal Relay and Saturating Control System Synthesis*, The Institution of Electrical Engineers, London, 1982.
37. J. B. Rawlings. Tutorial overview of model predictive control, *IEEE Control Systems Magazine*, Special Section on Industrial Process Control, pp. 38-52, June 2000.
38. A. Saberi, Z. Lin and A. R. Teel. Control of linear systems with saturating actuators, *IEEE Trans. Automatic Control*, 41, no. 3, 368-378, 1996.
39. W. E. Schmitendorf and B. R. Barmish. Null controllability of linear systems with constrained control, *SIAM J. Control and Optimization*, 18, no. 4, 327-345, 1980
40. E. D. Sontag. An algebraic approach to bounded controllability of linear systems, *Int. J. Control*, 39, no. 1, 181-188, 1984
41. E. D. Sontag and H. J. Sussmann. Nonlinear output feedback design for linear systems with saturating controls, *IEEE Conf. on Decision and Control*, Hawaii, USA, 3414-3416, Dec 1990
42. H. J. Sussmann, E. D. Sontag and Y. Yang. A general result on the stabilization of linear systems using bounded controls, *IEEE Trans. Automatic Control*, 39, no. 12, 2411-2425, 1994.
43. H. J. Sussmann and Y. Yang. On the stabilizability of multiple integrators by means of bounded feedback controls, *IEEE Conf. on Decision and Control*, Brighton, 70-72, Dec. 1991.
44. A. R. Teel. Global stabilization and restricted tracking for multiple integrators with bounded controls, *Systems & Control Letters*, 18, 165-171, 1992
45. A. R. Teel. *Feedback Stabilization: Nonlinear Solutions to Inherently Nonlinear Problems*, Ph.D. dissertation, University of California, Berkeley, CA, 1992.
46. A. R. Teel. Semi-global stabilizability of linear null controllable systems with input nonlinearities, *IEEE Trans. Automatic Control*, 40, no. 1, 96-100, 1995.
47. R. P. van-Til and W. E. Schmitendorf. Constrained controllability of discrete-time systems, *Int. J. Control*, 43, 943-956, 1986.
48. K. S.Walgama, S. Ronnback and J. Sternby. Generalisation of conditioning technique for anti-windup compensators, *IEE Proceedings-D*, 139, no 2, 109-118, 1992.

49. K. S. Walgama and J. Sternby. On the convergence properties of adaptive pole-placement controllers with antiwindup compensators, *IEEE Trans. Automatic Control*, 38, no 1, 128-132, 1993.
50. K. S. Walgama and J. Sternby. Conditioning technique for multiinput and multioutput processes with input saturation, *IEE Proceedings-D, Control Theory and Applications*, 140, no. 4, 231-241, 1993.
51. H. Wang and J. Sun. Modified model reference adaptive control with saturated inputs, *IEEE 31st Conf. on Decision and Control*, Arizona, 1992, 3255-3256.
52. Y.-M. Wang. *Adaptive One-Step-Control for Systems with Input Constraints*, M.S. thesis, National Chung Cheng University, 1999.
53. Y. Yang, H. J. Sussmann and E. D. Sontag. Stabilization of linear systems with bounded controls, *IFAC Nonlinear Control Systems Design Symposium*, Bordeaux, 51-56, 1992
54. P.-C. Yeh and P. Kokotovic. Adaptive tracking design for input-constrained linear system using backstepping, *American Control Conference*, 1995.
55. C. Zhang and R. J. Evans. Amplitude constrained adaptive control, *Int. J. Control*, 46, no. 1, 53-64, 1987.
56. C. Zhang and R. J. Evans. Adaptive pole-assignment subject to saturation constraints, *Int. J. Control*, 46, no. 4, 1391-1398, 1987.
57. C. Zhang and R. J. Evans. Rate constrained adaptive control, *Int. J. Control*, 48, no 6, 2179-2187, 1988.
58. C. Zhang. Discrete-time saturation constrained adaptive pole assignment control, *IEEE Trans. Automatic Control*, 38, no 8, 1250-1254, 1993.
59. C. Zhang and R. J. Evans. Continuous direct adaptive control with saturation input constraint, *IEEE Trans. Automatic Control*, 39, no 8, 1718-1722, 1994.

Robust Adaptive Control of Input Rate Constrained Discrete Time Systems

Gang Feng
Dept. of Manufacturing Engineering & Engineering Management
City University of Hong Kong
Hong Kong
Email: **megfeng@cityu.edu.hk**

Abstract

This chapter presents a design method and stability analysis for discrete time pole placement adaptive control systems with input rate saturation constraints. It is shown that with appropriate design, the robust adaptive control system is stable for a class of uncertain stable minimum phase plants in the sense that all the signals in the loop remain bounded in the presence of modeling uncertainties and disturbances. If the plant is free from modeling uncertainties and disturbances and the control input rate remains unsaturated for a period of time, then the closed loop adaptive control system will be asymptotically characterized by the desired closed loop poles.

1 Introduction

In practice, control systems frequently have to face some sort of nonsmooth nonlinearity of actuators and/or sensors, such as input amplitude saturation constraints, input rate saturation constraints, dead zones, backlash, hysteres, and so on. Many people have contributed a great deal of time to analysis and development of controllers for such systems. Lots of results have been established during the last few years [1-3, 7-8, 10-13, 16, 18-26] for both nonadaptive and adaptive control systems. See a survey chapter in this book for more references.

In the case of adaptive control, one of the stability and performance results was established in [26] for systems with input amplitude saturation constraints. The basic idea is to show that if the nonadaptive controller subject to input saturation constraint satisfies a finite gain stability condition, then the adaptive version of the same controller is also stable. A number of approaches have also been proposed in [20-23] for adaptive control of systems subject to dead zone, backlash, or hysteresis. The basic idea is to construct an adaptive inverse model for the nonlinearities and use it to cancel the nonlinearities in the system loop. However, for systems with input rate saturation constraints, it seems to the author that very few results have been obtained (even for nonadaptive control systems). The authors in [27] presented a stability result for a direct adaptive control scheme. They used the idea for the amplitude constrained adaptive control systems in [26] to show that if the nonadaptive controller subject to rate constraints satisfies a finite gain stability condition, then the adaptive version of the same controller is also stable for a class of minimum phase plants. Unfortunately, as in the case of amplitude constrained adaptive control, the technique used to obtain their stability results for direct adaptive control cannot easily be extended to the indirect adaptive control, and moreover the stability of the corresponding nonadaptive control systems is still pending. Recently we have shown in [4-6, 15] that the input rate constrained pole placement adaptive control system is stable for a class of systems under some specific conditions.

Here we consider the pole placement adaptive control systems with an input rate saturation constraint for a class of uncertain stable minimum phase discrete time plants, namely, the plants with all the poles and zeros strictly inside the unit circle of the q-plane. This is an improved result of the author's work in [6]. A new design method has been proposed which can guarantee the stability of the above mentioned adaptive control system. Furthermore, the knowledge of the upper bound on the uncertainty is not required *a priori* when using a new relative dead zone parameter estimation algorithm. It is shown that the closed loop adaptive control system will asymptotically be characterized by the desired closed loop poles if there is neither modeling uncertainties nor disturbances and if the control input rate remains unsaturated for a period of time. Moreover, it can be seen that the results are readily applied to the corresponding nonadaptive control systems.

This chapter is organized as follows. Section 2 presents new input rate constrained pole placement adaptive control scheme, section 3 establishes the main stability result for the described system. Simulation examples are provided to demonstrate the system performance in section 4, which is followed by concluding remarks in section 5.

2 Rate Constrained Pole Placement Adaptive Control

2.1 Plant Model

Consider a discrete-time single-input single-output plant

$$y(t) = \frac{B(q^{-1})}{A(q^{-1})}u(t) + H(q^{-1})u(t) + v(t) \tag{1}$$

where $y(t)$ and $u(t)$ are plant output and input respectively, $v(t)$ represents plant disturbances, $H(q^{-1})$ represents modeling uncertainties of the plant. $A(q^{-1})$ and $B(q^{-1})$ are polynomials in q^{-1}, written as

$$A(q^{-1}) = 1 + a_1 q^{-1} + \cdots + a_n q^{-n}$$
$$B(q^{-1}) = b_1 q^{-1} + \cdots + b_n q^{-n}$$

We shall consider pole placement control for a class of stable minimum phase plants subject to an input rate saturation constraint. We make the following assumptions.

A1: $B(q^{-1})/A(q^{-1})$ is a stable minimum phase model, namely, it has all the poles and zeros strictly inside the unit circle of the q-plane.

A2: The control input is symmetrically constrained in rate by a saturation function, i.e.

$$u(t) - u(t-1) = sat\{u'(t) - u(t-1)\}$$

where

$$sat\{x\} = \begin{cases} u_r & \text{if } x > u_r \\ x & \text{if } |x| \le u_r, \quad 0 < u_r < \infty \\ -u_r & \text{if } x < -u_r \end{cases}$$

A3: $A(q^{-1})$ and $B(q^{-1})$ are relatively prime.

A4: The modeling uncertainty $H(q^{-1})$ is BIBO stable and moreover satisfies
$$|H(1)| \le B(1)/A(1).$$

A5: The disturbance $v(t)$ is bounded.

Remark 1: Since the control input is no longer bounded contrary to the input amplitude constrained adaptive control case, a more restrictive assumption A1, namely, the plant is stably invertible or called minimum phase, is imposed here.

Remark 2: Assumption is standard for pole placement control systems [9].

Remark 3: Assumption A4 implies that the steady state response of the plant should be dominated by the modeled nominal part, i.e. $B(1)/A(1)$, not the modeling uncertainties $H(1)$. This assumption is reasonable because the modeling uncertainties are always supposed to be of less significance in plant dynamics.

Noting that the increment of control input is bounded, it follows directly from the plant model and assumptions that the plant has the following useful properties.

P1: $(1-q^{-1})y(t) := \Delta y(t)$ is bounded.

P2: $A(1) > 0$, for $A(q^{-1})$ is a stable polynomial.

P3: $B(1) > 0$, for $B(q^{-1})$ is a stable polynomial.

2.2 Pole Placement Control

This subsection presents a standard pole placement control for the nominal plant $B(q^{-1})/A(q^{-1})$ when its parameters are known *a priori* [9]. The control law without rate constraint is given as follows,

$$L(q^{-1})u(t) = P(q^{-1})[y^*(t) - y(t)] \qquad (2)$$

where $y^*(t)$ is a prespecified reference input, and the control polynomials $L(q^{-1})$ and $P(q^{-1})$ satisfy the following Diophantine equation,

$$A(q^{-1},t)L(q^{-1},t) + B(q^{-1},t)P(q^{-1},t) = A^*(q^{-1}) \qquad (3)$$

with the controller polynomials written as

$$L(q^{-1}) = 1 + l_1 q^{-1} + \cdots + l_{n-1} q^{-n+1}$$
$$P(q^{-1}) = p_0 + p_1 q^{-1} + \cdots + p_{n-1} q^{-n+1}$$

and $A^*(q^{-1})$ as a prespecified $(2n-1)$th order stable polynomial in q^{-1} which determines the desired closed loop poles. The block diagram of the pole placement closed loop control system can be shown below.

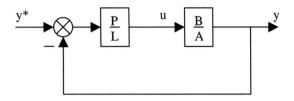

Fig. 1. Pole placement control.

It is well known that if the plant polynomials $A(q^{-1})$ and $B(q^{-1})$ are coprime, then the closed loop pole placement control system will be stable and the closed loop poles will be determined by the polynomial $A^*(q^{-1})$. With the input rate constraint, the control law is given as,

$$L(q^{-1})z(t) = P(q^{-1})[y^*(t) - y(t)] \tag{4}$$
$$u(t) - u(t-1) = sat\{z(t) - u(t-1)\} \tag{5}$$

The block diagram of the control system is as shown in Figure 2.

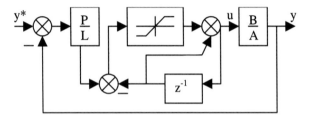

Fig. 2. Pole placement control with rate constraint.

2.3 Rate Constrained Pole Placement Adaptive Control

This subsection presents an adaptive version of the pole placement control system discussed in the previous subsection. We use a new adaptation algorithm with a relative dead zone. The plant (1) can be written in a regression form as

$$y(t) = \phi(t-1)^T \theta^* + \eta(t) \tag{6}$$

where

$$\phi(t-1) = [y(t-1), \cdots, y(t-n), u(t-1), \cdots, u(t-n)]^T$$
$$\theta^* = [-a_1, \cdots, -a_n, b_1, \cdots, b_n]^T$$
$$\eta(t) = A(q^{-1})[H(q^{-1})u(t) + v(t)] \tag{7}$$

For subsequent use we make the following assumptions:

A6: There exists a known convex region Ω such that (i) $\theta^* \in \Omega$; (ii) for all $\hat{\theta} \in \Omega$, the corresponding $\hat{A}(q^{-1})R(q^{-1})$ and $\hat{B}(q^{-1})$ are relatively prime, where $\hat{A}(q^{-1})$ and $\hat{B}(q^{-1})$ are polynomials defined above with coefficients from the vector $\hat{\theta}$, and $R(q^{-1}) := 1 - q^{-1}$. The term $R(q^{-1})$ will be discussed in the subsequent controller design.

A7: There exists an upper bounded function $\gamma(t)$ [14] such that
$$|\eta(t)| \le \gamma(t)$$
where $\gamma(t)$ satisfies $\quad \gamma(t) \le \varepsilon_1 \sup_{0 \le \tau \le t} \|\phi(\tau-1)\| + \varepsilon_2$
for some unknown constants $\varepsilon_1 > 0$ and $\varepsilon_2 > 0$.

Remark 4: Assumption A6 is consistent with that required in the standard pole placement adaptive control [9].

Remark 5: Since the control signal $u(t)$ is no longer constrained in amplitude, a constant upper bound for the plant modeling error $\eta(t)$ in equation (3) cannot be obtained. Instead assumption A7 is imposed as in [14].

Remark 6: In ordinary robust adaptive control algorithms such as in [5-6,14], both the parameters ε_1 and ε_2 of the upper-bounding function $\gamma(t)$ are supposed known. A weaker assumption is imposed here in assumption A7, which, we believe will extend the applicability of the robust adaptive control algorithms.

Let $\hat{\theta}(t)$ denote the estimate of the unknown parameter θ^* for the plant model (6). Defining the estimation error as
$$e(t) = y(t) - \phi(t-1)^T \hat{\theta}(t-1) \tag{8}$$
then we can use the following modified gradient algorithm with relative dead zone for robust parameter estimation,
$$\hat{\theta}(t) = proj\{\hat{\theta}(t-1) + a(t) \frac{\phi(t-1)}{1 + \sup_{0 \le \tau \le t} \phi(\tau-1)^T \phi(\tau-1)} e(t)\} \tag{9}$$

$proj\{\cdot\}$ is a projection operator [9] to constrain $\hat{\theta}(t)$ in the known convex region Ω, and the term $a(t)$ is a dead zone, which is defined as follows,
$$a(t) = \begin{cases} 0 & \text{if } |e(t)| \le \xi \hat{\gamma}(t) \\ \alpha f(\xi \hat{\gamma}(t), e(t)) / e(t) & \text{otherwise} \end{cases} \tag{10}$$

with
$$f(g,e) = \begin{cases} e - g & \text{if } e > g \\ 0 & \text{if } |e| \le g \\ e + g & \text{if } e < -g \end{cases}$$

and $0 < \alpha < 1$, $\xi = \dfrac{\xi_0 + 1}{1 - \alpha}$, $\xi_0 > 0$.

$\hat{\gamma}(t)$ is calculated by

$$\hat{\gamma}(t) = \begin{bmatrix} \hat{\varepsilon}_1(t-1) & \hat{\varepsilon}_2(t-1) \end{bmatrix} \begin{bmatrix} \sup_{0 \le \tau \le t} \|\phi(\tau-1)\| \\ 1 \end{bmatrix} \tag{11}$$

$$\hat{\varepsilon}_1(t) = \hat{\varepsilon}_1(t-1) + \frac{a(t)\beta |e(t)|}{1 + \sup_{0 \le \tau \le t} \phi(\tau-1)^T \phi(\tau-1)} \sup_{0 \le \tau \le t} \|\phi(\tau-1)\| \tag{12}$$

$$\hat{\varepsilon}_2(t) = \hat{\varepsilon}_2(t-1) + \frac{a(t)\beta \mid e(t) \mid}{1 + \sup_{0 \le \tau \le t} \phi(\tau-1)^T \phi(\tau-1)} \tag{13}$$

with $\beta>0$, and $\hat{\varepsilon}_1(0) = \hat{\varepsilon}_2(0) = 0$. It should be noted that $\hat{\varepsilon}_1$ and $\hat{\varepsilon}_2$ will be always positive and non-decreasing.

Remark 7: In the modified gradient algorithm for parameter estimation, the supreme of the regressor instead of the regressor itself is used which is essential for the convergence analysis of the adaptive algorithm.

As shown in [9], the projection operation does not alter the convergence properties of the original parameter estimation algorithms. Therefore, in the following analysis, the projection operation will be neglected.

The properties of the above modified gradient parameter estimator with relative dead zone are summarized in the following lemma.

Lemma 1. The modified gradient algorithm (8)•(13) applied to any plant, subject to the assumptions A6•A7, has the following properties irrespective of the control law,

(i) $\hat{\theta}(t)$ is bounded, and $\hat{\theta}(t) \in \Omega$.

(ii) $\hat{\varepsilon}_1(t)$ and $\hat{\varepsilon}_2(t)$ are bounded and nondecreasing, and thus $\hat{\varepsilon}_1$ converges to a constant, say $\bar{\varepsilon}_1$, and also $\hat{\varepsilon}_2$ converges to a constant.

(iii) $\| \hat{\theta}(t) - \hat{\theta}(t-1) \| \in l_2$.

(iv) $\tilde{f}(t) := \dfrac{f(t)}{[1+ \sup_{0 \le \tau \le t} \phi(\tau-1)^T \phi(\tau-1)]^{1/2}} := \dfrac{f(\xi \hat{y}(t), e(t))}{[1+ \sup_{0 \le \tau \le t} \phi(\tau-1)^T \phi(\tau-1)]^{1/2}} \in l_2.$

Proof: Define a Lyapunov function candidate

$$V(t) = \tilde{\theta}(t-1)^T \tilde{\theta}(t-1) + \beta^{-1} \tilde{\varepsilon}_1(t-1)^2 + \beta^{-1} \tilde{\varepsilon}_2(t-1)^2 \tag{14}$$

where $\tilde{\theta}(t) = \hat{\theta}(t) - \theta^*$, $\tilde{\varepsilon}_1(t) = \hat{\varepsilon}_1(t) - \varepsilon_1$, and $\tilde{\varepsilon}_2(t) = \hat{\varepsilon}_2(t) - \varepsilon_2$. It is noted that

$$e(t) = y(t) - \phi(t-1)^T \hat{\theta}(t-1) = -\phi(t-1)^T \tilde{\theta}(t-1) + \eta(t) \tag{15}$$

Then, its difference becomes
$V(t+1) - V(t)$

$= \tilde{\theta}(t)^T \tilde{\theta}(t) - \tilde{\theta}(t-1)^T \tilde{\theta}(t-1) + \dfrac{1}{\beta}\tilde{\varepsilon}_1(t)^2 - \dfrac{1}{\beta}\tilde{\varepsilon}_1(t-1)^2 + \dfrac{1}{\beta}\tilde{\varepsilon}_2(t)^2 - \dfrac{1}{\beta}\tilde{\varepsilon}_2(t-1)^2$

$= [\tilde{\theta}(t-1) + \dfrac{a(t)\phi(t-1)e(t)}{1+ \sup_{0 \le \tau \le t} \phi(\tau-1)^T \phi(\tau-1)}]^T [\tilde{\theta}(t-1) + \dfrac{a(t)\phi(t-1)e(t)}{1+ \sup_{0 \le \tau \le t} \phi(\tau-1)^T \phi(\tau-1)}]$

$- \tilde{\theta}(t-1)^T \tilde{\theta}(t-1) + \dfrac{1}{\beta}\tilde{\varepsilon}_1(t)^2 - \dfrac{1}{\beta}\tilde{\varepsilon}_1(t-1)^2 + \dfrac{1}{\beta}\tilde{\varepsilon}_2(t)^2 - \dfrac{1}{\beta}\tilde{\varepsilon}_2(t-1)$

$= \dfrac{2a(t)\phi(t-1)^T \tilde{\theta}(t-1)e(t)}{1+ \sup_{0 \le \tau \le t} \phi(\tau-1)^T \phi(\tau-1)} + \dfrac{a(t)^2 \phi(t-1)^T \phi(t-1)e(t)^2}{[1+ \sup_{0 \le \tau \le t} \phi(\tau-1)^T \phi(\tau-1)]^2}$

$+ \dfrac{1}{\beta}\tilde{\varepsilon}_1(t)^2 - \dfrac{1}{\beta}\tilde{\varepsilon}_1(t-1)^2 + \dfrac{1}{\beta}\tilde{\varepsilon}_2(t)^2 - \dfrac{1}{\beta}\tilde{\varepsilon}_2(t-1)$

$$= \frac{a(t)}{1+ \sup_{0\leq\tau\leq t} \phi(\tau-1)^T \phi(\tau-1)} [2e(t)\eta(t) - 2e(t)^2 + \frac{\phi(t-1)^T \phi(t-1)}{1+ \sup_{0\leq\tau\leq t} \phi(\tau-1)^T \phi(\tau-1)} a(t)e(t)^2]$$

$$+ \frac{1}{\beta}\tilde{\varepsilon}_1(t)^2 - \frac{1}{\beta}\tilde{\varepsilon}_1(t-1)^2 + \frac{1}{\beta}\tilde{\varepsilon}_2(t)^2 - \frac{1}{\beta}\tilde{\varepsilon}_2(t-1)^2$$

$$= \frac{a(t)}{1+ \sup_{0\leq\tau\leq t} \phi(\tau-1)^T \phi(\tau-1)} [2e(t)\eta(t) - 2e(t)^2 + \frac{\phi(t-1)^T \phi(t-1)}{1+ \sup_{0\leq\tau\leq t} \phi(\tau-1)^T \phi(\tau-1)} a(t)e(t)^2]$$

$$+ 2\tilde{\varepsilon}_1(t-1) \frac{a(t)|e(t)| \sup_{0\leq\tau\leq t} \|\phi(\tau-1)\|}{1+ \sup_{0\leq\tau\leq t} \phi(\tau-1)^T \phi(\tau-1)} + \frac{a(t)^2 |e(t)|^2 [\sup_{0\leq\tau\leq t} \|\phi(\tau-1)\|]^2}{[1+ \sup_{0\leq\tau\leq t} \phi(\tau-1)^T \phi(\tau-1)]^2}$$

$$+ 2\tilde{\varepsilon}_2(t-1) \frac{a(t)|e(t)|}{1+ \sup_{0\leq\tau\leq t} \phi(\tau-1)^T \phi(\tau-1)} + \frac{a(t)^2 |e(t)|^2}{[1+ \sup_{0\leq\tau\leq t} \phi(\tau-1)^T \phi(\tau-1)]^2}$$

$$= \frac{a(t)}{1+ \sup_{0\leq\tau\leq t} \phi(\tau-1)^T \phi(\tau-1)} \{2e(t)\eta(t) - 2e(t)^2 + \frac{\phi(t-1)^T \phi(t-1)}{1+ \sup_{0\leq\tau\leq t} \phi(\tau-1)^T \phi(\tau-1)} a(t)e(t)^2$$

$$+ 2\hat{\gamma}(t)|e(t)| - 2\gamma(t)|e(t)| + \frac{a(t)|e(t)|^2}{1+ \sup_{0\leq\tau\leq t} \phi(\tau-1)^T \phi(\tau-1)} [\sup_{0\leq\tau\leq t} \|\phi(\tau-1)\|]^2$$

$$+ \frac{a(t)|e(t)|^2}{1+ \sup_{0\leq\tau\leq t} \phi(\tau-1)^T \phi(\tau-1)} \}$$

$$\leq \frac{a(t)}{1+ \sup_{0\leq\tau\leq t} \phi(\tau-1)^T \phi(\tau-1)} \{2\hat{\gamma}(t)|e(t)| - 2e(t)^2 + 2\alpha e(t)^2\}$$

$$= \frac{2a(t)}{1+ \sup_{0\leq\tau\leq t} \phi(\tau-1)^T \phi(\tau-1)} \{(1-\alpha)e(t)^2 - \hat{\gamma}(t)|e(t)|\}$$

$$\leq -\frac{2a(t)}{1+ \sup_{0\leq\tau\leq t} \phi(\tau-1)^T \phi(\tau-1)} \{(1-\alpha)e(t)^2 - \frac{|e(t)|}{\xi}|e(t)|\}$$

$$= -\frac{2[(1-\alpha)\xi - 1]}{\xi} \frac{a(t)e(t)^2}{1+ \sup_{0\leq\tau\leq t} \phi(\tau-1)^T \phi(\tau-1)}$$

$$= -\frac{2\xi_0}{\xi} \frac{f(t)^2}{1+ \sup_{0\leq\tau\leq t} \phi(\tau-1)^T \phi(\tau-1)} \qquad (16)$$

where the fact that $a(t)e(t)^2 = \alpha f(t)e(t) \geq \alpha f(t)^2$ has been used. Therefore, following the same arguments as in [9,14], the results in Lemma 1 are thus proved. ◊◊◊

Given the parameter estimate $\hat{\theta}(t)$, and the corresponding plant polynomials $\hat{A}(q^{-1},t)$ and $\hat{B}(q^{-1},t)$, we can use the certainty equivalence principle to design adaptive controller. Here we propose the following adaptive controller,

$$R(q^{-1})\hat{L}(q^{-1},t)u(t) = \hat{P}(q^{-1},t)[y^*(t) - y(t)] \tag{17}$$

i.e.

$$\Delta u(t) = \hat{P}(q^{-1},t)[y^*(t) - y(t)] + \hat{L}'(q^{-1},t)\Delta u(t) \tag{18}$$

where $y^*(t)$ is a reference input, $\Delta u(t) := u(t) - u(t-1)$, $\hat{L}'(q^{-1},t) := 1 - \hat{L}(q^{-1},t)$. The controller polynomials $\hat{L}(q^{-1},t)$ and $\hat{P}(q^{-1},t)$ by solving the following Diophantine equation

$$R(q^{-1})\hat{A}(q^{-1},t)\hat{L}(q^{-1},t) + \hat{B}(q^{-1},t)\hat{P}(q^{-1},t) = A^*(q^{-1}) \tag{19}$$

where the controller polynomials can be written as

$$\hat{L}(q^{-1},t) = 1 + \hat{l}_1(t)q^{-1} + \cdots + \hat{l}_{n+1}(t)q^{-n-1}$$

$$\hat{P}(q^{-1},t) = \hat{p}_0(t) + \hat{p}_1(t)q^{-1} + \cdots + \hat{p}_{n+1}(t)q^{-n-1}$$

and $A^*(q^{-1})$ is a prespecified $(2n+2)$th order stable polynomial in q^{-1} which determines the desired closed loop poles. The block diagram of the adaptive control system is as shown in Figure 3.

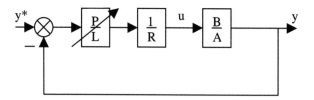

Fig. 3. Adaptive system without constraint.

If the system is subject to input rate constraint, the corresponding adaptive pole placement control law is:

$$\Delta u(t) = sat\{\hat{P}(q^{-1},t)[y^*(t) - y(t)] + \hat{L}'(q^{-1},t)\Delta u(t)\} \tag{20}$$

In this case, the adaptive control system block diagram can be depicted shown in Figure 4.

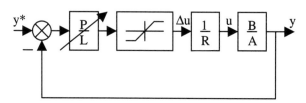

Fig. 4. Adaptive system with rate constraint.

It is noted that the adaptive control with rate constraint consists of the first three blocks in Figure 4.

Remark 8: The introduction of $R(q^{-1})$ in the controller is crucial in establishment of stability for the rate constrained adaptive control system, which will be seen in the next section.

3 Stability Analysis

The following lemma is useful for the establishment of our main stability result.

Lemma 2. For the input rate saturation constrained pole placement adaptive control system (1) and (20) with update laws (8)•(13), there exists a constant $\varepsilon>0$ such that
$$\hat{B}(1,t)B(1) \geq \varepsilon \qquad \forall t \geq 0 \tag{21}$$

Proof: Since the estimation property E1 ensures that $\hat{A}(q^{-1},t)R(q^{-1})$ and $\hat{B}(q^{-1},t)$ are relatively prime for all $t \geq 0$, we have $\hat{B}(1,t) \neq 0$. It then follows from the plant property P3 that there exists a constant $\varepsilon>0$ such that
$$|\hat{B}(1,t)B(1)| \geq \varepsilon \quad \forall t \geq 0 \tag{22}$$

In the $2n$-dimensional parameter space, the parameter estimate $\hat{\theta}(t)$, which is subject to equation (22) and the assumption A6, and the plant true parameter θ^* determine a straight line passing through them. Since Ω is a convex region, every point on this line between $\hat{\theta}(t)$ and θ^* should be in Ω for $\theta^* \in \Omega$ and $\hat{\theta}(t) \in \Omega$. We shall establish equation (21) by contradiction. For this it is assumed that the polynomial $\hat{B}(q^{-1},t)$ corresponding to $\hat{\theta}(t)$ satisfies, for some t,
$$\hat{B}(1,t)B(1) \leq -\varepsilon \tag{23}$$

It then follows from the above inequality and the plant property P3 that there exists a point $\theta' \in \Omega$ on the straight line between $\hat{\theta}(t)$ and θ^* such that the corresponding polynomial $B'(q^{-1})$ satisfies $B'(1) = 0$.

Since $R(1) = 0$, this means that $R(q^{-1})$ and $B'(q^{-1})$ are not relatively prime, which contradicts the fact $\theta' \in \Omega$. In other words, the assumption (23) does not hold. Then the result (21) is established in view of (22). ◊◊◊

Theorem 1. The input rate saturation constrained pole placement adaptive control system (1) and (20) with update laws (8)•(13) is stable in the sense that the plant input and output remain bounded.

Proof: It can be seen that boundedness of the plant output implies boundedness of the plant input since the plant is supposed to be stably invertible. Therefore we shall focus our attention on boundedness of the plant output. The proof will be completed by contradiction. For this we assume that $y(t)$ is unbounded. Then $\{y(t)\}$ must have a subsequence $\{y(t_k)\}$ such that
$$\lim_{t_k \to \infty} |y(t_k)| = \infty \tag{24}$$

$$|y(t_k)| > \max_{0 \leq \tau \leq t_k} |y(\tau)| \qquad (25)$$

Since $\Delta y(t)$ is bounded from plant property P1, without loss of generality, it is assumed that $y(t_k)$ is a positive monotone subsequence. Then equations (24) and (25) become

$$\lim_{t_k \to \infty} y(t_k) = \infty \qquad (26)$$

$$y(t_k) > \max_{0 \leq \tau \leq t_k} y(\tau) \qquad (27)$$

According to equation (20), the control law at t_k-i is

$$\Delta u(t_k - i) = sat\{\hat{P}(q^{-1}, t_k - i)[y^*(t_k - i) - y(t_k - i)] + \hat{L}'(q^{-1}, t_k - i)\Delta u(t_k - i)\}$$
$$= sat\{\bar{r}(t_k - i) - \hat{P}(1, t_k - i) y(t_k - i - 1)\} \qquad (28)$$

where

$$\bar{r}(t_k - i) = \hat{P}(q^{-1}, t_k - i) y^*(t_k - i) - \sum_{s=2}^{n} \hat{p}_s [y(t_k - i - s) - y(t_k - i - 1)]$$
$$+ \hat{L}'(q^{-1}, t_k - i)\Delta u(t_k - i)$$

Since $\Delta y(t_k)$ and $\Delta u(t_k\text{-}i)$ are bounded, and thus $\bar{r}(t_k - i)$ is bounded. Then it follows from equation (28) that $\Delta u(t_k\text{-}i)$ will be saturated as k approaches infinity. That is, when k approaches infinity,

$$\Delta u(t_k - i) = -\text{sgn}\{\hat{P}(1, t_k - i)\} u_r$$
$$= -\text{sgn}\{\hat{B}(1, t_k - i)\} u_r$$
$$= -\text{sgn}\{B(1)\} u_r \qquad (29)$$

where the Diophantine equation (19), $A^*(1) > 0$ as well as the result of Lemma 1 are used. Applying the control (29) to the plant (1) yields

$$\Delta y(t_k - i) = -\frac{B(1)}{A(1)} \text{sgn}\{B(1)\} u_r - H(1) u_r + v(t_k - i) - v(t_k - i - 1)$$
$$= -\frac{|B(1)|}{A(1)} u_r - H(1) u_r + v(t_k - i) - v(t_k - i - 1)$$

hence

$$y(t_k) - y(t_k - i) = -i[\frac{|B(1)|}{A(1)} + H(1)] u_r + v(t_k) - v(t_k - i - 1).$$

Since $A(1) > 0$, $v(t)$ is bounded, it follows from assumption A4 that there exists a sufficiently large, but still bounded i such that

$$i[\frac{|B(1)|}{A(1)} + H(1)] u_r \geq v(t_k) - v(t_k - i - 1)$$

then

$$y(t_k) < y(t_k\text{-}i). \qquad (30)$$

This contradicts equation (27). Thus the subsequence $\{y(t_k)\}$ satisfying (26) and (27) does not exist, and in consequence, $y(t)$ and thus $u(t)$ are bounded. ◊◊◊

It can be seen that the stability result established in the above theorem cannot guarantee that the system performance is characterized by the desired closed loop poles. However, as one would expect, if the system is free from the modeling uncertainties and disturbances, and the control input rate remains unsaturated for some period of time when time approaches infinity, then the closed loop system will be characterized by the desired poles within the same time period. The result is summarized in the following theorem.

Theorem 2. For the input rate saturation constrained pole placement adaptive control system (1) and (20) with update laws (8)•(13), if the plant model (1) satisfies $H = 0$, $v(t) = 0$, and thus the dead zone in the adaptation law equation (10) is removed, and furthermore at time t, the adaptive control $u(t-i)$ in (20) satisfies

$$\Delta u(t_k - i) = \hat{P}(q^{-1}, t_k - i)[y^*(t_k - i) - y(t_k - i)] + \hat{L}'(q^{-1}, t_k - i)\Delta u(t_k - i) \quad (31)$$

$$\text{for } i = 1, 2, \cdots, n.$$

Then

$$\lim_{t \to \infty} \{\hat{B}(q^{-1}, t)\hat{P}(q^{-1})y^*(t) - A^*(q^{-1})y(t)\} = 0. \quad (32)$$

Moreover, if the reference is a step signal, then the tracking error will approach zero as time tends to infinity, that is,

$$\lim_{t \to \infty} [y^*(t) - y(t)] = 0 \quad (33)$$

Proof: It can be seen that equation (31) can be rewritten as

$$q^{-i}\hat{L}(q^{-1}, t - i)R(q^{-1})u(t) = q^{-i}\hat{P}(q^{-1}, t - i)[y^*(t) - y(t)] \quad (34)$$

Multiplying equation (34) with $\hat{b}_i(t)$ and rearrange it leads to

$$\hat{b}_i(t)q^{-i}\hat{P}(q^{-1}, t)y^*(t) = \hat{b}_i(t)q^{-i}\hat{P}(q^{-1}, t)y(t) + \hat{b}_i(t)q^{-i}\hat{L}(q^{-1}, t)R(q^{-1})u(t)$$
$$+ \hat{b}_i(t)q^{-i}[\hat{P}(q^{-1}, t) - \hat{P}(q^{-1}, t - i)][y^*(t) - y(t)] \quad (35)$$
$$- \hat{b}_i(t)q^{-i}[\hat{L}(q^{-1}, t) - \hat{L}(q^{-1}, t - i)]R(q^{-1})u(t)$$

It is noted that equation (35) holds for $i=1, 2, \ldots, n$, then,

$$\hat{B}(q^{-1}, t)\hat{P}(q^{-1}, t)y^*(t) = \hat{B}(q^{-1}, t)\hat{P}(q^{-1}, t)y(t)$$
$$+ \hat{B}(q^{-1}, t)\hat{L}(q^{-1}, t)R(q^{-1})u(t) + \varepsilon_3(t) \quad (36)$$

where

$$\varepsilon_3(t) = \sum_{i=1}^{n} \hat{b}_i(t)q^{-i}\{[\hat{P}(q^{-1}, t) - \hat{P}(q^{-1}, t - i)][y^*(t) - y(t)]$$
$$- [\hat{L}(q^{-1}, t) - \hat{L}(q^{-1}, t - i)]R(q^{-1})u(t)\}$$

By using the Diophantine equation (19) and the error equation (8), we have

$$\hat{B}(q^{-1}, t)\hat{P}(q^{-1}, t)y^*(t) = \hat{B}(q^{-1}, t)\hat{P}(q^{-1}, t)y(t) + \hat{B}(q^{-1}, t)\hat{L}(q^{-1}, t)R(q^{-1})u(t)$$
$$+ \varepsilon_3(t) - R(q^{-1})\hat{L}(q^{-1}, t)e(t) + R(q^{-1})\hat{L}(q^{-1}, t)\hat{A}(q^{-1}, t - 1)y(t)$$
$$- R(q^{-1})\hat{L}(q^{-1}, t)\hat{B}(q^{-1}, t - 1)u(t)$$
$$= A^*(q^{-1})y(t) - \hat{L}(q^{-1}, t)R(q^{-1})e(t) + \varepsilon_3(t) + \varepsilon_4(t)$$

where

$$\varepsilon_4(t) = R(q^{-1})\hat{L}(q^{-1}, t)[\hat{B}(q^{-1}, t) - \hat{B}(q^{-1}, t - 1)]u(t)$$
$$- R(q^{-1})\hat{L}(q^{-1}, t)[\hat{A}(q^{-1}, t) - \hat{A}(q^{-1}, t - 1)]y(t)$$

It has been shown that the coefficients in polynomials $\hat{A}(q^{-1},t)$, $\hat{B}(q^{-1},t)$, $\hat{P}(q^{-1},t)$, and $\hat{L}(q^{-1},t)$ are bounded for all t, $\{y^*(t)\}$, $\{u(t)\}$ and $\{y(t)\}$ are also bounded. Then it follows that $\varepsilon_3(t)$ and $\varepsilon_4(t)$ tend to zero as time approach to infinity. It also follows from the results of Lemma 1 and Theorem 1 that $e(t)$ approaches zero as time tends to infinity for the case without using deadzone for the ideal situation. Therefore, we can conclude that

$$\lim_{t \to \infty} \{\hat{B}(q^{-1},t)\hat{P}(q^{-1},t)y^*(t) - A^*(q^{-1})y(t)\} = 0.$$

If the reference is a step signal, it follows that

$$\hat{A}(q^{-1},t)\hat{L}(q^{-1},t)R(q^{-1})y^*(t) = 0$$

then we have,

$$\lim_{t \to \infty} \{\hat{B}(q^{-1},t)\hat{P}(q^{-1},t)y^*(t) - A^*(q^{-1})y(t)\}$$

$$= \lim_{t \to \infty} \{\hat{B}(q^{-1},t)\hat{P}(q^{-1},t)y^*(t) + \hat{A}(q^{-1},t)\hat{L}(q^{-1},t)R(q^{-1})y^*(t) - A^*(q^{-1})y(t)\}$$

$$\lim_{t \to \infty} \{A^*(q^{-1})y^*(t) - A^*(q^{-1})y(t)\} = 0$$

This implies that

$$\lim_{t \to \infty} [y^*(t) - y(t)] = 0.$$

Thus the proof is completed. ◊◊◊

Remark 9: The stability and convergence results can be established for the corresponding input rate saturation constrained non-adaptive pole placement control systems using the technique developed in this section.

4 A Simulation Example

In this section, an example is presented to demonstrate the system performances. Consider a continuous time plant (Rohrs' example) [17] as follows:

$$G(s) = \frac{2}{s+1} \frac{229}{s^2 + 30s + 229}$$

where

$$\frac{B(s)}{A(s)} = \frac{2}{s+1}$$

is the nominal part and the other part is supposed to be unmodeled dynamics. With the sampling interval $T = 0.04$ sec., the discrete time expression of the plant can be obtained as:

$$G(q^{-1}) = \frac{0.0036q^{-1} + 0.0107q^{-2} + 0.0019q^{-3}}{1 - 2.0549q^{-1} + 1.3524q^{-2} - 0.2894q^{-3}}$$

and the discrete time equivalent of the nominal part is

$$\frac{0.0784q^{-1}}{1 - 0.9608q^{-1}}$$

With α=0.1, ξ=2, β=0.1, $u_r = 0.2$, the initial parameters $\hat{a}_1 = -0.5$, $\hat{b}_1 = 0.1$, the closed loop polynomial $A^*(q^{-1}) = (1-0.7q^{-1})(1-0.9q^{-1})$, then the simulation results are shown as in Figure 5.

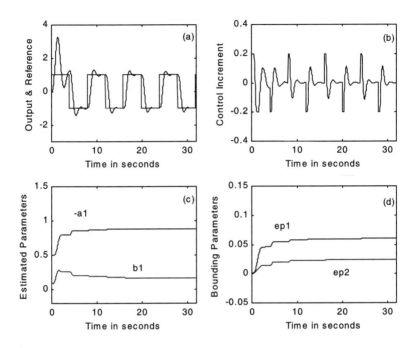

Fig. 5. Adaptive control with variable bounding parameters.

where (a) represents the reference input y^* and the plant output y, (b) the control increment Δu, (c) the estimated plant parameters $-\hat{a}_1$ and $\hat{b}_1$, and (d) the estimated bounding parameters $\hat{\varepsilon}_1$ and $\hat{\varepsilon}_2$. In the subsequent figures, (a)•(c) will represent the same variables as in Figure 5.

For comparison, the simulation has also been carried out using the ordinary robust estimation approaches with the fixed bounding parameters. In this case, $\varepsilon_1 = 0.05$ and $\varepsilon_2 = 0.1$ are chosen. With other conditions the same as in the first case, the results are shown in the Figure 6.

If the unmodeled dynamics is not present in the above example, with the same conditions as above but without using dead zone, the simulation results are shown in Figure 7.

A number of conclusions can be drawn from the above simulation results. The stability of the input rate saturation constrained pole placement adaptive control systems is maintained. The proposed approach with variable bounding parameters is better than the ordinary method with the fixed bounding parameters in the sense that the transient and tracking performance can be improved. This is mainly due to the conservatism of the method with the fixed bounding parameters for the sake of robustness. Furthermore, if the plant is free from unmodeled dynamics or disturbances and the control input is not saturated after some time instant, the output will track the reference input perfectly.

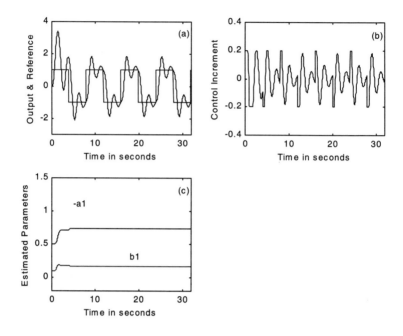

Fig. 6. Adaptive control with fixed bounding parameters.

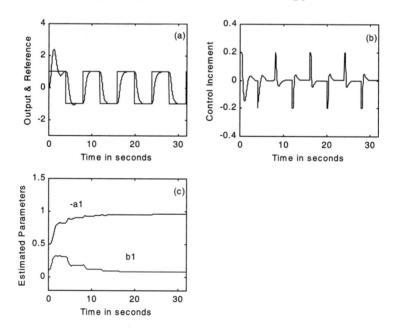

Fig. 7. Adaptive control in the absence of unmodeled dynamics.

5 Conclusions

This chapter presents stability analysis for the discrete time adaptive pole placement control systems with an input rate saturation constraint. It is shown that robust stability results can be established for stable minimum phase plants. It is also shown that when there is neither modeling uncertainties nor disturbances and the control remains unsaturated for a period of time, then the system performance is asymptotically characterized by the desired closed loop poles within this time period. Moreover the established stability and convergence results can be applied to the corresponding rate constrained nonadaptive pole placement control systems.

References

1. D.Y. Abramovitch and G.F. Franklin, On the stability of adaptive pole placement controllers with a saturating actuator, *IEEE Trans. on Automat. Control*, 35, 303-306, 1990.
2. B.S. Chen and S.S. Wang, The stability of feedback control with nonlinear saturating actuator: time domain approach, *IEEE Trans. Automat. Control*, 33, 483-487, 1988.
3. J.W. Cheng and Y.M. Wang, Adaptive one-step-ahead control with input amplitude, rate, and acceleration constraints, *Proc. IEEE Int. Conf. On Control Applications*, Hawaii, 1999.
4. G. Feng, C. Zhang and M. Palaniswami, Stability of input amplitude constrained adaptive pole placement control systems, *Automatica*, 30, 1065-1070, 1994.
5. G. Feng, Input rate constrained continuous time indirect adaptive control, *Int. J. Systems Science*, 25, 1977-1985, 1994.
6. G. Feng, M. Palaniswami and Y. Zhu, Stability of rate constrained indirect pole placement adaptive control, *Systems & Control Letters*, 18, 99-107, 1992.
7. G. Feng, C. Zhang and M. Palaniswami, Stability analysis of input constrained continuous time indirect adaptive control systems, *Systems & Control Letters*, 17, 209-215, 1991.
8. A.H. Glattfelder and W. Schaufelberger, Stability analysis of single loop control systems with saturation and antireset-windup circuits, *IEEE Trans. Automat. Control*, 28, 1074-1080, 1983.
9. G.C. Goodwin and K.S. Sin, *Adaptive Filtering, Prediction and Control*, Prentice Hall, Englewood Cliffs, NJ, 1984.
10. S.P. Karason and A.M. Annaswamy, Adaptive control in the presence of input constraints, *IEEE Trans. Automat. Control*, 39, 2325-2330, 1994.
11. N.J. Krikelis and S.K. Barkas, "Design of tracking systems subject to actuator saturation and integrator wind-up", *Int. J. Control*, 39, 667-682, 1984.
12. Z. Lin and A. Saberi, Semi-global exponential stabilization of linear systems subject to input saturation via linear feedback, *Systems & Control Letters*, 21, 225-239, 1993.
13. Z. Lin, A. A. Stoorvogel, and A. Saberi, Output regulation for linear systems subject to input saturation, *Automatica*, 32, 29-47, 1996.

14. R. Middleton, G.C. Goodwin, D. Hill, and D. Mayne, "Design issues in adaptive control", *IEEE Trans. Automat. Control,* 33, 50-58, 1988.
15. C. Moussas, and G. Bitsoris, Adaptive constrained control subject to both input amplitude and input-velocity constraints, *Proc. Int. Conf. On Systems, Man, and Cybernetics*, 601-606, Le Touquet, France, 1993.
16. A.N. Payne, Adaptive one step ahead control subject to an input amplitude constraint, *Int. J. Control,* 43, 199-207, 1986.
17. C.E. Rohrs, M. Athans, L. Valvani and G. Stein, Some design guidelines for discrete time adaptive control, *Automatica,* 20, 653-660, 1984.
18. A. Saberi, Z. Lin, and A.R. Teel, Control of linear systems subject to input saturation, *IEEE Trans. Automat. Control*, 41, 368-378, 1996.
19. H.J. Sussmann, E.D. Sontag, and Y. Yang, A general result on the stabilization of linear systems using bounded control, *IEEE Trans. Automat. Control*, 39, 2411-2425, 1994.
20. G. Tao and P.V. Kokotovic, *Adaptive Control of Systems with Actuator and Sensor Nonlinearities*, John Wiley & Sons, New York, 1996.
21. G. Tao and P.V. Kokotovic, Adaptive control of systems with backlash, *Automatica*, 29, 323-335, 1993.
22. G. Tao and P.V. Kokotovic, Adaptive control of plants with unknown dead-zones, *IEEE Trans. Automat. Control*, 39, 59-68, 1994.
23. G. Tao and P.V. Kokotovic, Adaptive control of plants with unknown hystereses, *IEEE Trans. Automat. Control*, 40, 200-212, 1995.
24. A.R. Teel, Global stabilization and restricted tracking for multiple integrators with bounded control, *Systems & Control Letters*, 18, 165-171, 1992.
25. C. Zhang and R.J. Evans, Robust amplitude constrained adaptive control, *Proc. IFAC Workshop on Robust Adaptive Control*, Newcastle, 1988.
26. C. Zhang and R. J. Evans, Adaptive pole-assignment subject to saturation constraints, *Int. J. Control,* 46, 1391-1398, 1987.
27. C. Zhang and R.J. Evans, Rate constrained adaptive control, *Int. J. Control,* 48, 2179-2187, 1988.

Adaptive Control of Linear Systems with Poles in the Closed LHP with Constrained Inputs

Dionisio A. Suarez-Cerda[1] and Rogelio Lozano[2]

[1] Instituto de Investigaciones Electricas, Av. Reforma No. 113,
Col. Palmira, 62490 Temixco, Morelos, MEXICO
[2] UTC, HEUDIASYC UMR-CNRS 6599, BP 20259,
60205 Compiegne, Cedex, FRANCE

Abstract

In this chapter we propose an indirect adaptive control strategy for possibly nonminimum phase systems but having poles in the closed LHP. The adaptive control is based on a control law for systems with amplitude-constrained inputs. We take a departure from standard adaptive control by using as starting point a control scheme for the known-parameters case. Since the state of the system is not supposed to be measurable, we use a state observer designed for the corresponding plant state space representation. Several simulations performed on different second-order systems show that the control scheme brings the state of the system to the origin using a control input with arbitrary bounds.

1 Introduction

One of the most common nonlinearities found in control engineering is probably the saturation function. Constraints on actuators are present in every application due to bounds on available power supply. Control of systems with saturated inputs has been a research topic for a long time but, recently, there has been renewed interest motivated by the work [9] and [8]. [9] has proposed a control strategy to stabilize an arbitrary number of integrators connected in cascade. The constraint on the control input is arbitrary and is obtained as a function of nested saturations. [8] extended this technique to cover any linear system with poles in the closed left-hand plane (LHP).

Several adaptive control schemes for systems with constrained inputs can be found in the literature. For simplicity of presentation, we distinguish three different types of plants:

1. stable plants;
2. systems with poles in the closed LHP; and
3. exponentially unstable plants, i.e. with at least one pole in the RHP.

In classical adaptive control theory the systems are rather distinguished by whether they are minimum phase or not regardless of the stability of the plant to be controlled. It is interesting to point out that by considering the control input limitations, stability of the system is brought forward to a more relevant system characteristic. Stable plants are clearly the most easy to deal with because one can readily obtain global asymptotic stability results. Adaptive controller for such systems has been proposed in [6].

Adaptive control of some classes of systems with poles in the closed LHP has been studied in [1, 2, 10]. In [1] an adaptive control scheme has been proposed for minimum phase, continuous-time, type-1 plants with constrained input. a priori knowledge on upper bounds for the plant parameters as well as on the disturbances allow them to prove boundedness of the plant output. An adaptive pole-placement controller for possibly nonminimum phase systems is proposed in [10]. They deal with type-m, discrete-time plants. It is assumed that the plant parameters belong to a known convex region in which the system is always completely controllable. This allows them to prove boundedness of the plant output but its behavior in the limit has not been analyzed.

Exponentially unstable plants (i.e. systems with at least one pole in the open RHP) can only be controlled locally when the input amplitude is constrained. This is the case even if the plant parameters are exactly known. In the adaptive control context, [3] presented a locally convergent algorithm for continuous-time minimum phase (possibly unstable) systems.

Adaptive controllers for systems with constrained input proposed to date, [6, 1, 2, 10, 3] can be viewed as extensions of standard adaptive control schemes. These classical adaptive schemes are based on control synthesis methods for the known-parameters case, which unfortunately do not consider any limitations in the control input.

In this chapter we will take a departure from standard adaptive control by using as starting point a control scheme for the known-parameters case, which is designed considering the input amplitude restriction as in [8]. We will therefore focus our attention to the class of plants considered in [8] which have all its poles in the closed LHP i.e. the plant should not have any poles with strictly positive real part. It is important to point out that standard adaptive control techniques recourse to the minimum phase assumption, which is inconsequential in the strategy proposed in [8]. The use of the minimum phase assumption in classical adaptive control simplifies matters considerably because we do not have to impose the condition that the plant estimated model be also minimum phase. The control input boundedness is secured from the assumption that the original system is minimum phase.

However, the computation of the control input in [8] requires the plant model to have all its poles in the closed LHP. There are two possible ways to guarantee that the estimated plant model belongs to the family of plants having all their poles in the LHP. The first approach is based on the introduction of persistence of excitation into the plant input and output so that the estimates converge to the plant parameter values. The second approach is to project the parameter estimates into the subspace Ω that correspond to the family of plants having all their poles on the LHP.

The first method is simpler to analyze but introduces extra disturbances into the system. In spite of its simplicity this approach presents some drawbacks. Given that the system input amplitude is bounded, the amount of excitation that can be induced in the system is limited. The lack of excitation will be accentuated if for some reason the output size is large compared to the maximum input size. This can produce poor estimates convergence rate which means that the estimates may lie outside the subspace Ω for a long time before entering Ω. The control strategy proposed in [8] could be used only after the estimates have already entered Ω which will produce poor transient behavior.

The second method, i.e. the one in which the estimates are projected into the subspace Ω, has the advantage over the former that a control input can be computed every time. On the other hand it is more complex to design and to implement because the subspace Ω is in general non-convex. This subspace is indeed convex only in the case of first-order systems.

In this chapter we will analyze the two approaches presented above. For clarity of presentation we will deal only with strictly proper second-order systems having poles on the LHP. The systems considered may be nonminimum phase. Another reason to restrict ourselves to second-order systems is that the projection into the subspace Ω mentioned above becomes more complex as the order of the system increases. Note however than for second-order systems, the subspace Ω is non-convex.

2 Control of Systems with Constrained Inputs

In this section we will describe control algorithms for second-order systems with saturated control inputs. For simplicity purposes, we will separately consider three different cases. We study first the case of stable plants with no pole at the origin. We then study the cases when the plant has one and two poles at the origin.

2.1 Plants with no Poles at the Origin

$$Hp(s) = \frac{k_1 s + k_2}{s^2 + 2\xi w_n s + w_n^2} \qquad (2.1)$$

where $\xi \geq 0$ and $w_n > 0$, and k_1, k_2 are such that there are no pole–zero cancelations in (2.1).

It is clear that this system is either asymptotically stable or is a harmonic oscillator when $\xi = 0$.

A convenient state space representation for (2.1) is

$$\dot{x} = \begin{bmatrix} -2\xi w_n & -w_n^2 \\ 1 & 0 \end{bmatrix} x + \begin{bmatrix} 1 \\ 0 \end{bmatrix} u \qquad (2.2)$$

$$y = \begin{bmatrix} k_1 & k_2 \end{bmatrix} x \qquad (2.3)$$

The following stabilizing constrained control law is proposed:

$$u = \sigma\{-x_1\} \qquad (2.4)$$

where the saturating function is

$$\sigma(\alpha) = \begin{cases} \alpha & \text{if } |\alpha| \leq \delta \\ \text{sign}(\alpha) \cdot \delta & \text{if } |\alpha| > \delta \end{cases} \qquad (2.5)$$

where $\delta > 0$.

The convergence analysis of this control system can be carried out by using the following positive definite function:

$$V = \frac{1}{2}x_1^2 + \frac{1}{2}w_n^2 x_2^2 \qquad (2.6)$$

whose time derivative is

$$\begin{aligned} \dot{V} &= x_1 \dot{x}_1 + w_n^2 x_2 \dot{x}_2 \\ &= -2\xi w_n x_1^2 + x_1 u \\ &= -2\xi w_n x_1^2 + x_1\{-x_1\} \qquad \text{(using (2.4))} \end{aligned} \qquad (2.7)$$

We will next use the La-Salle invariant set theorem to prove asymptotic stability (see [7], pp. 68–77). From (2.7), we can see that the largest invariant set such that $\dot{V} = 0$ is $x_1 = 0$ which implies that $u = 0$. From (2.2), it follows that $\dot{x}_2 = x_1 = 0$ which means that x_2 is constant. Therefore, by contradiction we conclude that $x_2 = 0$. So, applying the La-Salle theorem, we conclude that the equilibrium point at the origin is globally asymptotically stable. □

2.2 Double Integrator

Let us consider the double integrator

$$\ddot{y} = u \qquad (2.8)$$

A convenient state space representation for (2.8) is

$$\dot{x} = \begin{bmatrix} 0 & 0 \\ 1 & 0 \end{bmatrix} x + \begin{bmatrix} 1 \\ 1 \end{bmatrix} u \qquad (2.9)$$

$$y = \begin{bmatrix} -1 & 1 \end{bmatrix} \qquad (2.10)$$

$$\theta^{*T} = [\; b_1^* \;\; b_2^* \;\; -a_1^* \;\; -a_2^* \;] \tag{3.7}$$

$$\phi^T = [\; u_f^{(1)} \;\; u_f \;\; y_f^{(1)} \;\; y_f \;] \tag{3.8}$$

Consider the following least squares estimation algorithm:

$$e = y_f^{(n)} - \theta^T \phi \tag{3.9}$$
$$\dot{\theta} = P\phi e \tag{3.10}$$
$$\dot{P} = -P\phi\phi^T P, \qquad P(0) > 0 \tag{3.11}$$

where e is the scalar identification error and θ is the estimate of θ^*.
Let us define

$$\tilde{\theta} \triangleq \theta - \theta^* \tag{3.12}$$

The properties of this algorithm are presented in the following lemma.

Lemma 3.1. *The least squares estimation algorithm (3.9)–(3.11) together with the system parameterization (3.6) has the following properties*

1. $\tilde{\theta}^T \phi, e$ and $P\phi \in \mathcal{L}_2$,
2. P and θ converge,
3. $\tilde{\theta}^T P^{-1} \tilde{\theta} \in \mathcal{L}_\infty$.

Proof See [4].

3.2 Adaptive State Observation

The control algorithms proposed in the previous section depend on the state of the system. Since the state of the system is not supposed to be measurable, we will use a state observer designed for the corresponding plant state space representation given in the previous section. Furthermore, given that the plant parameters are unknown we will replace them by their estimates θ given above which are computed from the input–output representation of the plant. The observer structure is given next where $\{\hat{A}, \hat{B}, \hat{C}\}$ are the system matrices given in (2.2) or (2.18) using the estimates in (3.9)–(3.11).

$$\dot{\hat{x}} = \hat{A}\hat{x} + \hat{B}u + l[y - \hat{C}\hat{x}] \qquad \hat{x}(0) = \hat{x}_0 \tag{3.13}$$

where $\hat{x}_0$ is the initial state estimate vector and l is the *feedback gain vector*, which is chosen so that the observer dynamics is asymptotically stable.

The adaptive control scheme we will use is basically composed of the LS parameter estimation algorithm (3.9)–(3.11), the adaptive observer above and a state feedback control law as proposed in the previous section using $\hat{x}$ instead of x. Global convergence of such a scheme in the ideal case, i.e. without plant disturbances, can be ensured by introducing persistent excitation in the plant input and output to secure convergence of the plant estimates to their true values. Excitation could also be introduced only when the plant input and output do not approach zero, this technique has been called self-excitation in the literature. Alternatively we could obtain convergence results without introducing excitation into the system by redressing the plant estimates so that they correspond to admissible plant models as proposed in [4, 5]. The drawback of the latter technique is that the computations involved in the estimates modification become more complex as the order of the system increases.

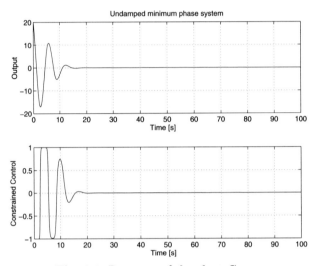

Fig. 3.1. Response of the plant G_1.

4 Simulations Results

We have studied the performance of the proposed adaptive control algorithm by carrying out a series of simulations on MATLAB.

In the simulations we have considered the following second-order plants:

1. Undamped minimum phase system

$$G_1(s) = \frac{s+3}{s^2+1} \tag{4.1}$$

2. Undamped nonminimum phase system

$$G_2(s) = \frac{s-3}{s^2+1} \tag{4.2}$$

3. Type-1 plant with a stable pole

$$G_3(s) = \frac{s-3}{s(s+1)} \quad (4.3)$$

The plant parameters have been estimated using a LS algorithm. The regressor vector has been composed with filtered input and output signals as in [5]. We have used a filter with tranfer function

$$G_f(s) = \frac{1}{s^2 + 9s + 20} \quad (4.4)$$

In all cases we have placed the observer poles at {-4, -3} using Ackerman's Rule.

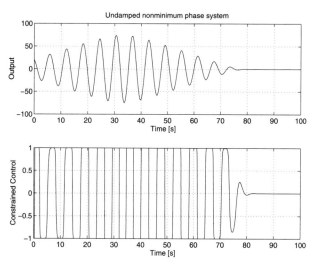

Fig. 4.1. Response of the plant G_2.

In the third case, i.e. for G_3, the least squares identification algorithm has been slightly simplified since we are not estimating the parameter a_2, which is known to be zero. The initial conditions used in the simulations were

$P(0) = 500 I_{4\times 4}$ Covariance matrix
$\theta(0) = [1\ 3\ 0\ 9]^T$ Parameter estimates
$\hat{x}(0) = [0\ 0]^T$ State estimate
$x_u(0) = [0\ 0]^T$ State of the control input filter
$x_y(0) = [0\ 0]^T$ State of the output filter
$x(0) = [5\ -5]^T$ State of the plant

In the singular case when $\{\hat{A}, \hat{C}\}$ happened to have a nonobservable mode, we used the last value of the observer gain vector l before the singularity appeared. We kept this value for l until the singularity disappeared

In the simulations we carried out we did not have to introduce excitation into the system. In all cases $\hat{x}$ and x both converged to the origin while the parameter estimates converged to constant values but not necessarily equal to their true values. In the case of plants G_1 and G_2, the estimate of the parameter a_1 converged to a small negative number. We have run simulations including a modification strategy such that the modified estimates of a_1 and a_2 were both strictly positive. However, the transient performance of the adaptive control was better without projecting the estimates to the admissible region. The results presented here were obtained without introducing excitation or parameter projection to the admissible region.

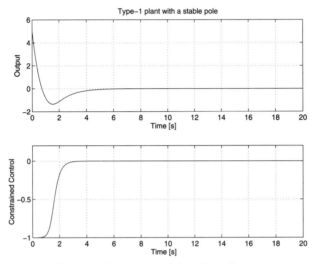

Fig. 4.2. Response of the plant G_3.

5 Conclusions

In this chapter we have proposed an indirect adaptive control strategy for possibly nonminimum phase systems but having poles in the closed LHP. The adaptive control is based on a control law for systems with amplitude-constrained inputs. Several simulations performed on different second-order systems showed that the control scheme brings the state of the system to the origin using a control input with arbitrary bounds. The simulation results presented did not require excitation of the system or projection of the parameter estimates into the admissible region, however this may not be the general case. Extension of the proposed strategy to higher order systems is feasible. The approach based on excitation presents no particular problem to cover higher order systems, nevertheless the approach based on projecting the estimates into the admissible region becomes more complex as the order of the system increases.

References

1. G. Feng and M. Palaniswami. Stability analysis of input constrained continuous time indirect adaptive control. *Systems and Control Letters*, 17:209–215, 1991.
2. G. Feng, C. Zhang, and M. Palaniswami. Stability of input amplitude constrained adaptive pole placement control systems. *Automatica*, 30:1065–1070, 1994.
3. S.P. Kárason and A.M. Annaswamy. Adaptive control in the presence of input constraints. *IEEE Transactions on Automatic Control*, 39:2325–2330, 1994.
4. R. Lozano and B. Brogliato. Adaptive control of a simple nonlinear system without a priori information on the parameters. *IEEE Transactions on Automatic Control*, 37:30–37, 1992.
5. R. Lozano and X.-H. Zhao. Adaptive pole placement without excitation probing signals *IEEE Transactions on Automatic Control*, 39:47–58, 1994.
6. A.N. Payne. Adaptive one step ahead control subject to an input amplitude constaint. *International Journal of Control*, 43, 1986.
7. J-J.E. Slotine, W. Li. Applied Nonlinear Control. Prentice–Hall International, Inc., 1991.
8. H. Sussmann, E.D. Sontag, and Y. Yang. A general result on the stabilization of linear systems using bounded controls. *IEEE Transactions on Automatic Control*, 39:2411–2425, 1994.
9. A.R. Teel. Global stabilization and restricted tracking for multiple integrators with bounded controls. *Systems and Control Letters*, 18:165–171, 1992.
10. C. Zhang and R.J. Evans. Continuous direct adaptive control with saturation input constraint. *IEEE Transactions on Automatic Control*, 39:1718–1722, 1994.

Adaptive Control with Input Saturation Constraints

Cishen Zhang[1]

Department of Electrical and Electronic Engineering
The University of Melbourne
VIC 3010, Australia

Abstract

All real dynamical systems are subject to hard limits on input, state and output variables. And most common hard limit constraints are on amplitude and changing rate of the system input. Recently, stability of dynamical systems with hard limit constraints attracted much attention from researchers. In this chapter, stability of discrete time adaptive control systems with hard limit constraints on input amplitude and changing rate is analyzed. It is shown that the adaptive control system is bounded-input bounded-output stable if the non-adaptive version of the control system satisfies a finite gain stability condition.

1 Introduction

It is known that all real dynamical systems are subject to hard limits on input, state and output variables. This is due to inherent physical constraints of the dynamical system and constraints in the controller actuators. Dynamical systems with hard limit constraints on the amplitude and changing rate of control inputs are the most common cases, where the hard limit constraint is modelled by a saturation nonlinearity.

In many situations, the saturation nonlinearity of dynamical systems dominates system performance. It is often the case that considerably more effort is involved in designing control loops to handle the saturation nonlinearity than is spent in designing the linear loop parts of the system. The study of dynamical systems with saturation nonlinearities is of practical and theoretical importance and this area has recently become very active. See, for example [2, 5, 7, 8, 10, 14, 15]

The problem of dynamical systems with saturation nonlinearities which attracts most concern is stability. Stability analysis of discrete time indirect adaptive control systems subject to saturation constraints can be found in [10], and that of continuous time direct adaptive control systems can be found in [15].

This chapter considers discrete time direct adaptive control of linear plants with saturation constraints on the amplitude and changing rate of control input and presents results on the stability of the adaptive control system. The main results of this chapter are based on those of [9, 11, 12] for amplitude constrained direct adaptive control systems and [13] for rate constrained direct adaptive control systems.

For adaptive control of linear systems with a saturation constraint on the control input, we consider that the plant is unknown and subject to a bounded disturbance. A constrained parameter estimator incorporating a deadzone is used to model the plant. Since the plant is subject to a saturation constrained control input, the bounded disturbance model is also able to deal with a wide class of unmodeled plant dynamics. For such a system, we show that the closed loop adaptive control system is BIBO (bounded-input/bounded-output) stable provided the closed loop system with the control input saturation and known plant parameters is finite gain stable. Moreover, if there are no disturbances and unmodeled dynamics then we show that the adaptive control converges to the corresponding non-adaptive control. This result relates stability of the adaptive control system to the corresponding non-adaptive control system and does not impose specific assumptions on the plant type. Thus it is quite general and is as strong as one can expect. The result can be applied to adaptive control of a class of stable and type-1 plants to obtain closed loop system stability.

For adaptive control of linear systems with saturation constraint on the changing rate of the control input, we extend the results on the adaptive control system with input saturation constraint to obtain closed loop system stability. If the corresponding non-adaptive control system with control rate constraint satisfies a finite gain stability condition, then the adaptive version of the controller yields a BIBO closed loop system and converges to the non-adaptive control.

This chapter is organized as follows. Section 2 formulates a discrete time direct adaptive control scheme subject to a saturation constraint on control input amplitude. Section 3 analyzes the stability of adaptive control systems with a saturation constraint on the amplitude of the control input. Section 4 applies the stability result of Section 3 to a class of type-1 plants. In Section 5, a direct adaptive control scheme subject to saturation constraint on control input changing rate is formulated. Then the stability result of Section 3 is extended to the rate constrained direct adaptive control system in Section 6. Section 7 applies the stability result of Section 6 to a class of stable plants.

2 Amplitude Constrained Direct Adaptive Control

This section briefly reviews a direct adaptive control scheme and uses it to introduce an amplitude constrained direct adaptive control system. Consider a linear discrete time plant

$$y(t) = \frac{b_0 z^{-d} B(z^{-1})}{A(z^{-1})} u(t) + v(t), \tag{2.1}$$

where $y(t)$ and $u(t)$ are the system output and control input, respectively, $v(t)$ represents an unknown but bounded disturbance signal, $b_0 \neq 0$ is a plant parameter, $B(z^{-1})$ and $A(z^{-1})$ are monic polynomials in the backward shift operator z^{-1} written as

$$A(z^{-1}) = 1 + a_1 z^{-1} + \cdots + a_n z^{-n},$$
$$B(z^{-1}) = 1 + b_1 z^{-1} + \cdots + b_{\tilde{m}} z^{-\tilde{m}}, \quad \tilde{m} < n.$$

Specify a reference model as

$$E(z^{-1})y(t) = z^{-d} H(z^{-1}) r(t), \tag{2.2}$$

where $E(z^{-1})$ is a strictly stable monic polynomial in z^{-1} written as

$$E(z^{-1}) = 1 + e_1 z^{-1} + \cdots + e_{\tilde{n}} z^{-\tilde{n}}, \quad \tilde{n} \leq n + d - 1.$$

Then there exist unique polynomials

$$F(z^{-1}) = 1 + f_1 z^{-1} + \cdots + f_{d-1} z^{-d+1},$$
$$G(z^{-1}) = g_0 + g_1 z^{-1} + \cdots + g_{n-1} z^{-n+1},$$

such that

$$E(z^{-1}) = F(z^{-1}) A(z^{-1}) + z^{-d} G(z^{-1}). \tag{2.3}$$

This equation can be used to write plant (2.1) as

$$\bar{y}(t+d) = \phi(t)^T \theta + w(t+d), \tag{2.4}$$

where

$$\bar{y}(t+d) = E(z^{-1}) y(t+d),$$
$$w(t+d) = F(z^{-1}) A(z^{-1}) v(t+d),$$

$$\phi(t) = [u(t), u(t-1) \cdots u(t-m+1), y(t) \cdots y(t-n+1)]^T$$
$$= [u(t), \phi'(t)^T]^T,$$

$$\theta = [\theta_1, \theta_2 \cdots \theta_{m+n}]^T = [\theta_1, \theta'^T]^T,$$
$$m = \tilde{m} + d - 1.$$

The following assumptions are made for the plant model (2.4).

Assumption A1. *There is a known constant θ_{1m} satisfying $|\theta_{1m}| \leq |\theta_1|$ and $\theta_{1m} \theta_1 > 0$.*

Assumption A2. *There is a known upper bound w^+ of the disturbance $w(t)$ such that $w^+ \geq |w(t)|$ for all t.*

Let $\hat{\theta}(t)$ denote the estimate of θ which is written as

$$\hat{\theta}(t) = [\hat{\theta}_1(t), \hat{\theta}_2(t) \cdots \hat{\theta}_{m+n}(t)]^T = [\hat{\theta}_1(t), \hat{\theta}'(t)^T]^T.$$

Define the estimation error as

$$e(t) = \bar{y}(t) - \phi(t-d)^T \hat{\theta}(t-1).$$

Also define a deadzone function

$$f(h, e) = \begin{cases} e - h, & if \quad e > h \\ 0, & if \quad |e| \leq h \\ e + h, & if \quad e < -h \end{cases}$$

$$0 < h < \infty.$$

A useful property of this deadzone function is

$$|f(h, e)| + h \geq |e|. \qquad (2.5)$$

Then, with $\hat{\theta}(0)$ given, the following least squares algorithm can be used to estimate θ.

$$\bar{\theta}(t) = \hat{\theta}(t-1) + \frac{P(t-d-1)\phi(t-d)}{1 + \phi(t-d)^T P(t-d-1)\phi(t-d)} f\left(\frac{w^+}{\sqrt{1-\alpha}}, e(t)\right),$$

$$\hat{\theta}(t) = \begin{cases} \bar{\theta}(t), & if \quad \bar{\theta}_1(t) sgn(\theta_{1m}) \geq |\theta_{1m}| \\ \bar{\theta}(t) + \frac{P_1(t-d)}{P_{11}(t-d)}(\theta_{1m} - \bar{\theta}_1(t)) & otherwise, \end{cases}$$

$$\alpha(t) = \begin{cases} \alpha, & if \quad f\left(\frac{w^+}{\sqrt{1-\alpha}}, e(t)\right) \neq 0 \\ 0, & otherwise, \end{cases}$$

$$\alpha \in (0, 1), \qquad P(0) = \sigma I, \qquad 0 < \sigma < \infty,$$

where $P_1(t)$ is the first column of $P(t)$ and $P_{11}(t)$ is the $(1,1)$th entry of $P(t)$. It is noted that $\hat{\theta}_1(t)$ is constrained by a projection operation.

Following from the parameter estimation results of [3] and [4], the estimate $\hat{\theta}(t)$ satisfies the following properties.

Property P1

$$\hat{\theta}(t) \text{ is bounded;}$$

Property P2

$$|\hat{\theta}_1(t)| \geq |\theta_{1m}| \quad \text{and} \quad \frac{1}{\hat{\theta}_1(t)} = \left|\frac{\theta_1}{\hat{\theta}_1(t)}\right| \frac{1}{\theta_1};$$

Property P3

$$\lim_{t\to\infty} \frac{f\left(\frac{w^+}{\sqrt{1-\alpha}}, \epsilon(t)\right)}{(1 + \phi(t-d)^T \phi(t-d))^{\frac{1}{2}}} = 0;$$

where

$$\epsilon(t) = \bar{y}(t) - \phi(t-d)^T \hat{\theta}(t-d). \tag{2.6}$$

Without the saturation constraint, the direct adaptive control law is written as

$$u(t) = \frac{r_f(t) - \phi'(t)^T \hat{\theta}'(t)}{\hat{\theta}_1(t)}, \tag{2.7}$$

where $r_f(t) = H(z^{-1})r(t)$ with $r(t)$ a reference input and $H(z^{-1})$ as specified in (2.2).

The adaptive control system is stable provided the plant is minimum phase [6, 4]. Moreover, if $w^+ = 0$ the closed loop system is characterized by the reference model (2.2) [3].

Now assume that the adaptive control (2.7) is subject to a saturation nonlinearity. The amplitude constrained direct adaptive control law can be written as

$$u(t) = sat\left\{\frac{r_f(t) - \phi'(t)^T \hat{\theta}'(t)}{\hat{\theta}_1(t)}\right\}, \tag{2.8}$$

where $sat\{\cdot\}$ is the saturation nonlinearity defined as

$$sat\{x\} = \begin{cases} u^+, & if \quad x > u^+ \\ x, & if \quad -u^+ \leq x \leq u^+ \\ -u^+, & if \quad x < -u^+, \end{cases}$$

$$0 < u^+ < \infty.$$

Remark 2.1. (i) The adaptive control law (2.8), when not saturated, is in the same form as (2.7). The saturation nonlinearity can be contained either in the plant or in the controller. In order to perform parameter estimation properly, the value of $u(t)$ is needed. To meet this requirement, the control saturation can be practically implemented inside the controller.

(ii) The disturbance term $w(t+d)$ in (2.4) now represents both system external disturbances and system unmodeled dynamics since $u(t)$ is subject to a saturation constraint. It is required that the system unmodeled dynamics be bounded-input bounded-output (BIBO) stable for $w(t)$ satisfying

Assumption A2. This allows the unmodeled dynamics to be nonlinear and time-varying.

(iii) An important property of the saturation nonlinearity is

$$sat\{k(t)x(t)\} = sat\{x(t) + e_k(t)\}, \qquad (2.9)$$

where $k(t)$ is a sector bounded time-varying gain satisfying $k_1 \leq k(t) \leq k_2$ with $0 < k_1 \leq k_2 < \infty$ and $e_k(t)$ is a bounded signal for all t. This property can be clearly seen from Figure. 2.1.

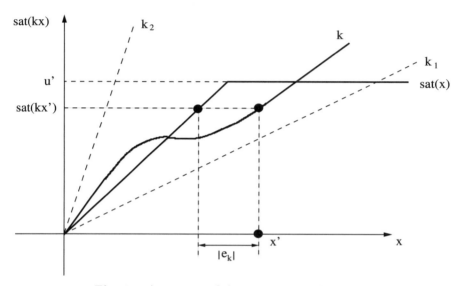

Fig. 2.1. A property of the saturation nonlinearity.

3 Stability of Amplitude Constrained Adaptive Control Systems

First consider an amplitude constrained control law employing the true system parameter given by

$$u^*(t) = sat\left\{\frac{r_f^*(t) - \phi'(t)^T \theta'}{\theta_1}\right\}, \qquad (3.1)$$

where $r_f^*(t)$ is a reference input. For the amplitude constrained adaptive control system (2.1) and (2.8) to be stable, it is reasonable to expect that the control $u^*(t)$ of (3.1), if known, would stabilize the plant (2.1). Thus, an assumption can be made as follows:

Assumption A3. *The closed loop feedback system (2.1) and (3.1) is finite gain stable in the sense that the output $y(t)$ satisfies*

$$|y(t+d)| \leq c_1 \sup_{0 \leq \tau \leq t} |r_f^*(\tau)| + c_2 \qquad (3.2)$$

for any filtered reference input $r_f^*(t) = H(z^{-1})r^*(t)$, $\forall t \geq 0$, and some $0 < c_1 < \infty$, $0 < c_2 < \infty$.

Then stability of the amplitude constrained adaptive control system can be obtained and the results are in the following two theorems.

Theorem 3.1. *The amplitude constrained adaptive control system (2.1) and (2.8) in conjunction with the estimation algorithm is stable in the sense that the output $y(t)$ is bounded, if Assumptions A1, A2 and A3 are satisfied and the reference input $r_f(t)$ is bounded.*

Proof. By Properties P1 and P2, the adaptive control law (2.8) can be written as

$$u(t) = \text{sat}\left\{ \left|\frac{\theta_1}{\hat{\theta}_1(t)}\right| \frac{r_f(t) - \phi'(t)^T \hat{\theta}'(t)}{\theta_1} \right\} \qquad (3.3)$$

and there exist $0 < k_1 \leq k_2 < \infty$ such that

$$k_1 \leq \left|\frac{\theta_1}{\hat{\theta}_1(t)}\right| \leq k_2$$

is satisfied. Following from property (2.9) of the saturation nonlinearity, we can write (3.3) into the form

$$u(t) = \text{sat}\left\{ \frac{r_f(t) + \theta_1 e_k(t) - \phi'(t)^T \hat{\theta}'(t)}{\theta_1} \right\} \qquad (3.4)$$

where $e_k(t)$ is bounded. Let

$$R(t+d) = r_f(t) + \theta_1 e_k(t) + \hat{\theta}_1(t)u(t) - \theta_1 u(t)w(t+d), \qquad (3.5)$$
$$r_f^*(t) = R(t+d) + \epsilon(t+d). \qquad (3.6)$$

$R(t)$ is bounded since every term in the right-hand side of (3.5) can be shown to be bounded. $r_f^*(t)$ in (3.6) is dependent on $r_f(t)$, $\hat{\theta}(t)$ and $\phi(t)$. Substituting (2.4), (2.6), (3.5) and (3.6) into (3.4), the adaptive control law (3.4) can be written into the form of the non-adaptive control law (3.1). This result allows the application of the finite gain stability condition (3.2) to the adaptive control system, yielding

$$|y(t+d)| \leq c_1 \sup_{0\leq \tau \leq t} |R(\tau+d) + \epsilon(\tau+d)| + c_2$$
$$\leq c_1 \sup_{0\leq \tau \leq t} |\epsilon(\tau+d)| + c_3, \qquad 0 < c_3 < \infty.$$

This inequality together with the deadzone property (2.5) further yields

$$|y(t+d)| \leq c_1 \sup_{0\leq \tau \leq t} \left| f\left(\frac{w^+}{\sqrt{1-\alpha}}, \epsilon(\tau+d)\right) \right| + c_1 \frac{w^+}{\sqrt{1-\alpha}} + c_3. \qquad (3.7)$$

Then it follows from the definition of $\phi(t)$, the boundedness of $u(t)$ and (3.7) that

$$\|\phi(t)\| \leq c_4 \sup_{0\leq \tau \leq t} \left| f\left(\frac{w^+}{\sqrt{1-\alpha}}, \epsilon(\tau+d)\right) \right| + c_5,$$
$$0 < c_4 < \infty, \qquad 0 < c_5 < \infty.$$

The boundedness of $y(t)$ then follows directly from Property P3 and the Key-Technical-Lemma of [3]. □

In the case that the system is free from the disturbance, i.e. $w(t) = 0$, a control tracking result can be obtained as follows.

Theorem 3.2. *If the amplitude constrained adaptive control system (2.1) and (2.8) satisfies the stability conditions in Theorem 3.1 and is free from the disturbance, i.e. $w(t) = 0$, then the adaptive control law (2.8) converges to the corresponding non-adaptive one as t approaches infinity, that is*

$$\lim_{t\to\infty} \left(u(t) - \text{sat}\left\{ \frac{r_f(t) - \phi'(t)^T \theta'}{\theta_1} \right\} \right) = 0. \qquad (3.8)$$

Proof. Two cases $|u(t)| < u^+$ and $|u(t)| = u^+$ are considered. For $|u(t)| < u^+$, the adaptive control law (2.8) implies

$$\phi(t)^T \hat{\theta}(t) = r_f(t).$$

It follows that

$$u(t) = \frac{r_f(t) + \epsilon(t+d) - \phi'(t)^T \theta'}{\theta_1}$$
$$= \text{sat}\left\{ \frac{r_f(t) + \epsilon(t+d) - \phi'(t)^T \theta'}{\theta_1} \right\}, \qquad (3.9)$$

where (2.4) and (2.6) are used. Since $\phi(t)$ is bounded and $w^+ = 0$, Property P3 can be applied to (3.9) to obtain (3.8).

For $|u(t)| = u^+$, it is considered that $u(t) = u^+$ without loss of generality. The adaptive control law (2.8) becomes

$$u(t) = sat\left\{\frac{r_f(t) + \epsilon(t+d) - \phi(t)^T\theta}{\hat{\theta}_1(t)} + u^+\right\}.$$

This implies

$$\frac{r_f(t) + \epsilon(t+d) - \phi(t)^T\theta}{\hat{\theta}_1(t)} \geq 0.$$

Applying Property P2 to this inequality yields

$$\frac{r_f(t) + \epsilon(t+d) - \phi(t)^T\theta}{\theta_1} \geq 0,$$

$$\frac{r_f(t) + \epsilon(t+d) - \phi'(t)^T\theta'}{\theta_1} \geq u^+,$$

$$sat\left\{\frac{r_f(t) + \epsilon(t+d) - \phi'(t)^T\theta'}{\theta_1}\right\} = u^+ = u(t). \quad (3.10)$$

Again, Property P3 can be applied to (3.10) to obtain (3.8). Finally, the results for the cases $|u(t)| < u^+$ and $|u(t)| = u^+$ together establish the theorem. □

Remark 3.1. (i) The output boundedness and control tracking results do not imply the output tracking property of conventional adaptive control systems, because of the existence of the the saturation nonlinearity. Moreover, these results do not guarantee desirable system performance.

(ii) By making use of the fact that the control is constrained in amplitude, the system unmodeled dynamics is dealt with by incorporating a fixed deadzone in the estimator to achieve a certain degree of robustness.

(iii) The system considered is highly nonlinear and its response is closely related to the reference input, the saturation constraint value u^+ and size of the deadzone $w^+/\sqrt{1-\alpha}$, the initial estimate value $\hat{\theta}(t)$ etc.

(iv) It turns out that the system stability is characterized by that of the corresponding non-adaptive control system. In the stability analysis, we did not assume the plant to be minimum phase which is normally required by conventional direct adaptive control algorithms. It will be shown in the next section that stable plants and stable type-1 plants with an even number (including 0) of non-minimum phase positive real zeros satisfy the system stability conditions.

4 Amplitude Constrained Adaptive Control of Type-1 Plants

This section applies the stability results of the amplitude constrained direct adaptive control system to a class of plants. It is straightforward to show that a stable plant satisfies stability conditions in Theorems 1 and 2. In this section, the attention is confined to a class of type-1 plants, i.e. plants with a single integrator. This class of plants is assumed to satisfy

Assumption A4. *The plant (2.1) has one pole at $z = 1$ and all the other poles lying inside the unit circle and has an even number (including 0) of zeros lying on $(1, \infty)$ of the real axis and no zero at $z = 1$ in the z-plane.*

The following two lemmas establish some useful properties of the type-1 plants satisfying Assumption A4.

Lemma 4.1. *Associated with the following polynomial*

$$Q(z^{-1}) = q_0 + q_1 z^{-1} + \cdots + q_{n-1} z^{-n+1} + q_n z^{-n}, \qquad q_0 \neq 0$$

there is a polynomial equation

$$q_0 z^n + q_1 z^{n-1} + \cdots + q_{n-1} z + q_n = 0. \qquad q_0 \neq 0. \tag{4.1}$$

Let N_z denote the number of real roots of (4.1) with value larger than 1. If (4.1) has no root at $z = 1$, then

$$sgn(q(1)) = sgn(q_0 + q_1 + \cdots + q_n) = sgn((-1)^{N_z} q_0).$$

Proof. Denote by z_i a real root of (4.1), $0 \leq i \leq r$ and by $(\sigma_l \pm j\omega_l)$ a pair of complex conjugate roots of (4.1), $0 \leq i \leq c$, where $r + 2c = n$. It is clear that $q_0 + q_1 + \cdots + q_n \neq 0$ and

$$sgn(q(1)) = sgn(q_0) \prod_{i=1}^{r} sgn(1 - z_i) \prod_{l=1}^{c} sgn((1 - \sigma_l - j\omega_l)(1 - \sigma_l - j\omega_l)).$$

Since

$$sgn(1 - z_i) = \begin{cases} 1, & z_i < 1 \\ -1, & z_i > 1 \end{cases}$$

and

$$sgn((1 - \sigma_l - j\omega_l)(1 - \sigma_l - j\omega_l)) = sgn((1 - \sigma_l)^2 + \omega_l^2) = 1,$$

it follows that

$$sgn(q(1)) = sgn((-1)^{N_z} q_0).$$

$\square$

Lemma 4.2. *If the plant (2.1) in conjunction with (2.4) satisfies Assumption A4, then it satisfies*

$$\sum_{i=1}^{n} \theta_{m+i} > 0 \qquad (4.2)$$

and

$$y(t+d) = y_1(t+d) + y_2(t+d), \qquad (4.3)$$

where

$$y_1(t+d) = y_1(t+d-1) + \theta_1 k u(t), \qquad 0 < k < \infty \qquad (4.4)$$

and $y_2(t+d)$ is bounded.

Proof. Assumption A4 implies

$$A(1) = A(z^{-1})|_{z^{-1}=1} = 0.$$

Substituting this into (2.3) yields

$$E(1) = G(1). \qquad (4.5)$$

It follows from (2.1), (2.4) and (2.3) that

$$G(z^{-1}) = \sum_{i=1}^{n} \theta_{m+i} z^{-i+1}.$$

Combining this with (4.5) yields

$$\sum_{i=1}^{n} \theta_{m+i} = E(1). \qquad (4.6)$$

Since $E(z^{-1})$ is specified to be a monic strictly stable polynomial, we can apply the result of Lemma 4.1 to (4.6) to obtain (4.2).

To establish (4.3) and (4.4), Assumption A4 can be used to write polynomial A as

$$A(z^{-1}) = (1 - z^{-1})A_1(z^{-1}), \qquad (4.7)$$

where $A_1(z^{-1})$ is a strictly stable polynomial. Then (4.7) can be used to write plant (2.1) in the form of (4.3) with

$$y_1(t) = \frac{b_0 B(1) z^{-d}}{A_1(1)(1-z^{-1})} u(t), \qquad (4.8)$$

$$y_2(t) = \frac{z^{-d} B_1(z^{-1})}{A_1(z^{-1})} u(t),$$

where $B_1(z^{-1})$ is a polynomial satisfying

$$b_0 \frac{B(1)}{A_1(1)} A_1(z^{-1}) + (1 - z^{-1})B_1(z^{-1}) = b_0 B(z^{-1}).$$

Since $A_1(z^{-1})$ is a stable polynomial and $u(t)$ is bounded, $y_2(t)$ is a bounded signal. Comparing (2.1) and (2.3) with (2.4) and noticing $F(z^{-1})$ is a monic polynomial, the following equation can be obtained

$$\theta_1 = b_0.$$

Let

$$k = \frac{B(1)}{A_1(1)}.$$

This can be used to write $y_1(t)$ in (4.8) in the form of (4.4). Since the plant (2.1) satisfies Assumption A4, $B'(z^{-1})$ and $A_1(z^{-1})$ are monic, and $A_1(z^{-1})$ is a stable polynomial, the result of Lemma 4.1 can be used to obtain $0 < k < \infty$. □

Then the following stability result can be established.

Theorem 4.1. *If the plant (2.1) in conjunction with (2.4) satisfies Assumption A4, then the amplitude constrained adaptive control system (2.1) and (2.8) is stable in the sense that*
  (i)   if $r_f(t)$ is bounded, $y(t)$ is bounded;
  (ii) if $w^+ = 0$, the control convergent result (3.8) is satisfied.

Proof. It is first shown that the control (3.1) employing the true parameter θ, when applied to the plant, satisfies the finite gain stability condition (3.2). Then the results of Theorems 3.1 and 3.2 are applied to establish this theorem.

Consider that the plant is subject to the control (3.1). Equation (4.4) can be used to obtain

$$\begin{aligned} y_1(t+d) - y_1(t+d-1) &= |b_0 k| \mathrm{sat}\left\{\frac{r_f^*(t) - \phi'(t)^T \theta'}{|\theta_1|}\right\} \\ &= |\theta_1 k| \mathrm{sat}\left\{\frac{\bar\theta}{|\theta_1|}\left(\frac{r_f^*(t) + \bar r(t)}{\bar\theta} - y_1(t)\right)\right\}, \end{aligned} \qquad (4.9)$$

where $\bar\theta = \sum_{i=1}^{n} \theta_{m+i} > 0$ by (4.2) and

$$\bar r(t) = \sum_{i=2}^{n} \theta_{m+i}(y_1(t) - y_1(t-i+1)) - \sum_{i=1}^{n} \theta_{m+i} y_2(t-i+1) + \sum_{i=2}^{m} \theta_1 u(t-i+1).$$

For any finite j, $(y_1(t+j) - y(t))$ is bounded by (4.4). Following form this and boundedness of $y_2(t)$, $\bar r(t)$ is bounded. It is also implied by (4.4) that there exists a constant c_6, $0 < c_6 < \infty$, such that

$$|y_1(t+j) - y_1(t)| \leq c_6, \quad 1 \leq j \leq d.$$

It follows form (4.9) that, if

$$|y_1(t)| > \sup_{0 \leq \tau \leq t} \left\{ \frac{r_f^*(\tau) + \bar{r}(\tau)}{\bar{\theta}}, c_6 \right\}$$

is satisfied, there must be

$$|y_1(t+d)| - |y_1(t+d-1)| = -\left|\theta_1 \ k \ \text{sat}\left\{\frac{\bar{\theta}}{|\theta_1|}\left(\frac{r_f^*(t) + \bar{r}(t)}{\bar{\theta}} - y_1(t)\right)\right\}\right| < 0.$$

Therefore, $y_1(t+d)$ eventually enters and remains inside the region described by

$$y_1(t+d) \leq \sup_{0 \leq \tau \leq t} \left|\frac{r_f^*(\tau) + \bar{r}(\tau)}{\bar{\theta}}\right| + c_6.$$

This inequality together with (4.3) and the boundedness of $y_2(t+d)$ and $\bar{r}(t)$ leads to the finite gain stability condition (3.2). Hence, the results of Theorems 3.1 and 3.2 can be applied to establish (i) and (ii). □

5 Rate Constrained Direct Adaptive Control

The stability analysis of rate constrained adaptive control follows the idea of the analysis of amplitude adaptive control. It is shown that if the non-adaptive controller subject to rate constraint satisfies a finite gain stability condition then the adaptive version of the same controller is also stable and the control signal in the adaptive case converges to the control signal in the non-adaptive case. Since the control signal subject to the rate constraint is no longer bounded, one is not able to treat the system modeling uncertainties as a bounded disturbance.

For rate constrained adaptive control, consider the plant (2.1) with the disturbance term $v(t) = 0$. If the system is not subject to the rate constraint, the parameter estimation and direct adaptive control algorithms basically follow from that in Section 2 with

$$w(t) = 0,$$

$$f(h, e(t)) = e(t).$$

The parameter estimation algorithm satisfies

Property P4
$$(\hat{\theta}(t) - \hat{\theta}(t-i)) \in l_2, \quad i < \infty.$$

Property P5

$$\frac{e(t+d)}{\sqrt{1+\phi(t)^T\phi(t)}} \in l_2.$$

It is assumed that the plant (2.1) satisfies Assumption A1 and

Assumption A5. *The plant is minimum phase in the sense that all the plant zeros are strictly inside the unit circle in the z-plane.*

The direct adaptive control law, without rate constraint, is as given by (2.7), which stabilizes the closed loop system. And the closed loop adaptive control system is characterized by the reference model (2.2).

For a rate constraint magnitude u_r, $0 < u_r < \infty$, assume that the adaptive control law is subject to a rate constraint such that

$$|\Delta u(t)| = |u(t) - u(t-1)| \leq u_r. \qquad (5.1)$$

This shows that the change of the control signal between sampling instants is limited by u_r. The direct adaptive control law (3.1), when subject to the rate constraint (5.1), becomes

$$u(t) = u(t-1) + sat\left\{\frac{r_f(t) - \phi'(t)\hat{\theta}'(t)}{\hat{\theta}_1(t)} - u(t-1)\right\}. \qquad (5.2)$$

The rate constrained adaptive control (5.2) is identical to (2.7) when $|\Delta u(t)| \leq u_r$. In general the adaptive control algorithm allows the rate constraint nonlinearity to be either in the plant or in the controller. It requires the value of $u(t)$ to be known to perform the parameter estimation properly.

6 Stability of Rate Constrained Direct Adaptive Control Systems

To analyze the stability of the rate constrained adaptive control system, a rate constrained control law employing the system true parameter θ is introduced as follows:

$$u(t) = u(t-1) + sat\left\{\frac{r_f^*(t) - \phi'(t)\theta'}{\theta_1} - u(t-1)\right\}. \qquad (6.1)$$

For the closed loop adaptive control system (2.1) and (5.2) to be stable, it is reasonable to expect that the control law (6.1), if known, would stabilize the system. Thus it is again assumed that the closed loop feedback control system (2.1) and (6.1) satisfies the finite gain stability condition A3. Then, stability of the closed loop direct adaptive control system is given in the following theorem.

Theorem 6.1. *If Assumptions A1, A3 and A4 are satisfied, the rate constrained adaptive control system (2.1) and (5.2), in conjunction with the proposed least squares estimation algorithm, is stable in the sense that*

(i) $y(t)$ and $u(t)$ are bounded;

(ii)
$$\lim_{t \to \infty} \left(u(t) - u(t-1) - \text{sat}\left\{ \frac{r_f(t) - \phi'(t)^T \theta'}{\theta_1} - u(t-1) \right\} \right) = 0;$$

(iii) if
$$\left| \frac{r_f(t) - \phi'(t)^T \hat{\theta}'(t)}{\hat{\theta}_1(t)} - u(t-1) \right| \leq u_r,$$

then
$$\lim_{t \to \infty} (\bar{y}(t+d) - r_f(t)) = 0.$$

Proof.

(i) Using
$$e(t+d) = \phi(t)^T \theta - \phi(t)^T \hat{\theta}(t+d-1),$$
the rate constrained adaptive control (5.2) can be written as

$$u(t) = \text{sat}\left\{ k(t) \frac{r_f(t) + e(t+d) + \phi(t)^T \Delta\hat{\theta}(t) + \hat{\theta}_1(t)\Delta u(t) - \phi(t)^T \theta}{\theta_1} \right\}$$
$$+ u(t-1) \qquad (6.2)$$

where
$$k(t) = \frac{\theta_1}{\hat{\theta}_1(t)},$$
$$\Delta\hat{\theta}(t) = \hat{\theta}(t+d-1) - \hat{\theta}(t).$$

In view of Properties P1 and P2, one can have $0 < k_1 \leq k_2 < \infty$ such that $k_1 \leq k(t) \leq k_2$ is satisfied. Then the saturation nonlinearity property (2.9) can be used to write (6.2) as

$$u(t) = \text{sat}\left\{ \frac{r_f(t) + e(t+d) + \phi(t)^T \Delta\hat{\theta}(t) + \hat{\theta}_1(t)\Delta u(t) + \theta_1 e_k(t) - \phi(t)^T \theta}{\theta_1} \right\}$$
$$+ u(t-1)$$
$$= \text{sat}\left\{ \frac{\hat{r}(t) - \phi'(t)^T \theta'}{\theta_1} - u(t-1) \right\} + u(t-1), \qquad (6.3)$$

where $e_k(t)$ is bounded and

$$\hat{r}(t) = r_f(t) + e(t+d) + \phi(t)^T \Delta\hat{\theta}(t) + \hat{\theta}_1(t)\Delta u(t) + \theta_1 e_k(t) - \theta_1 \Delta u(t). \quad (6.4)$$

Since $r(t)$ and $\Delta u(t)$ are bounded, there exists a constant γ_1, $0 < \gamma_1 < \infty$, such that

$$|\hat{r}(t)| \leq |e(t+d) + \phi(t)^T \Delta\hat{\theta}(t)| + \gamma_1. \quad (6.5)$$

Note that the rate constrained adaptive control (6.3) is in the same form as the non-adaptive control (6.1). It follows from the finite gain stability condition A3 that

$$\begin{aligned}
|y(t+d)| &\leq c_1 \sup_{0\leq\tau\leq t} |\hat{r}(\tau)| + c_2 \\
&\leq c_1 \sup_{0\leq\tau\leq t} |e(\tau+d) + \phi(\tau)^T \Delta\hat{\theta}(\tau)| + \gamma_2, \quad (6.6)\\
&\quad 0 < \gamma_2 < \infty.
\end{aligned}$$

Assumption A5 implies that the plant is stably invertible, i.e.

$$|u(t)| \leq \gamma_3 \sup_{0\leq\tau\leq t} |y(\tau+d)| + \gamma_4, \quad 0 < \gamma_3 < \infty, \quad 0 < \gamma_4 < \infty.$$

It follows that

$$\|\phi(t)\| \leq \gamma_5 \sup_{0\leq\tau\leq t} |y(\tau+d)| + \gamma_6, \quad 0 < \gamma_5 < \infty, \quad 0 < \gamma_6 < \infty. \quad (6.7)$$

Combining (6.7) with (6.6) yields

$$\|\phi(t)\| \leq \gamma_7 \sup_{0\leq\tau\leq t} |e(\tau+d) + \phi(\tau)^T \Delta\hat{\theta}(\tau)| + \gamma_8, \quad 0 < \gamma_7 < \infty, \quad 0 < \gamma_8 < \infty. \quad (6.8)$$

Squaring both sides of (6.8) and using the inequality $2(a^2 + b^2) \geq (a+b)^2$, the following inequalities can be obtained.

$$\begin{aligned}
\|\phi(t)\|^2 &\leq 2\gamma_7^2 \sup_{0\leq\tau\leq t} |e(\tau+d) + \phi(\tau)^T \Delta\hat{\theta}(\tau)|^2 + 2\gamma_8^2 \\
&\leq 4\gamma_7^2 \sum_{\tau=0}^{t} \left(e(\tau+d)^2 + \|\phi(\tau)\|^2 \|\Delta\hat{\theta}(\tau)\|^2\right) + 2\gamma_8^2 \\
&\leq 4\gamma_7^2 \sum_{\tau=0}^{t} \left(\|\phi(\tau)\|^2 \left(\frac{e(\tau+d)^2}{1+\phi(\tau)^T\phi(\tau)} + \|\Delta\hat{\theta}(\tau)\|^2\right)\right) + \gamma_9,
\end{aligned}$$
$$(6.9)$$

where the estimation Property P5 is used and $0 < \gamma_9 < \infty$. Further using Properties P4 and P5 and applying Gronwall's Lemma [1] to (6.9), the boundedness of $\phi(t)$ is obtained. Thus (i) is established.

(ii) First consider the case $|\Delta u(t)| < u_r$ for an infinitely large t. The adaptive control is

$$u(t) = \frac{r_f(t) - \phi'(t)^T \hat{\theta}'(t)}{\hat{\theta}_1(t)}$$

$$= \frac{r_f(t) - \phi(t)^T \theta + e(t+d) + \phi(t)^T \Delta\hat{\theta}(t) + \hat{\theta}_1(t) u(t)}{\hat{\theta}_1(t)}.$$

This equation implies

$$r_f(t) - \phi(t)^T \theta + e(t+d) + \phi(t)^T \Delta\hat{\theta}(t) = 0. \tag{6.10}$$

The boundedness of $\phi(t)$ together with Properties P4 and P5 yields

$$\lim_{t \to \infty} (e(t+d) + \phi(t)^T \Delta\hat{\theta}(t)) = 0. \tag{6.11}$$

It follows from (6.10) that for $|\Delta u(t)| < u_r$

$$u(t) = \frac{r_f(t) - \phi'(t)^T \theta'}{\theta_1}$$

$$= u(t-1) + \text{sat}\left\{ \frac{r_f(t) - \phi'(t)^T \theta'}{\theta_1} - u(t-1) \right\}. \tag{6.12}$$

Next, consider the case $|\Delta u(t)| = u_r$. Without loss of generality, consider $\Delta u(t) = u_r$, which yields

$$u(t) = u(t-1) + \text{sat}\left\{ \frac{r_f(t) - \phi(t)^T \theta + e(t+d) + \phi(t)^T \Delta\hat{\theta}(t)}{\hat{\theta}_1(t)} + u_r \right\}.$$

This implies

$$\frac{r_f(t) - \phi(t)^T \theta + e(t+d) + \phi(t)^T \Delta\hat{\theta}(t)}{\hat{\theta}_1(t)} \geq 0. \tag{6.13}$$

This inequality can be rewritten as

$$\frac{r_f(t) - \phi(t)^T \theta + e(t+d) + \phi(t)^T \Delta\hat{\theta}(t) - \theta_1 u(t-1)}{\hat{\theta}_1(t)} \geq \frac{\theta_1}{\hat{\theta}_1(t)} \Delta u(t). \tag{6.14}$$

Note that $\frac{\theta_1}{\hat{\theta}_1(t)} > 0$ and $\Delta u(t) = u_r$. Multiplying both sides of (6.14) by $\frac{\hat{\theta}_1(t)}{\theta_1}$ and using (6.11) yield

$$\frac{r_f(t) - \phi'(t)^T \theta'}{\theta_1} - u(t-1) \geq u_r.$$

It follows that

$$\text{sat}\left\{\frac{r_f(t) - \phi'(t)^T \theta'}{\theta_1} - u(t-1)\right\} = u_r,$$

$$\begin{aligned} u(t) &= u(t-1) + u_r \\ &= u(t-1) + \text{sat}\left\{\frac{r_f(t) - \phi'(t)^T \theta'}{\theta_1} - u(t-1)\right\}. \end{aligned}$$

Hence, (ii) is established by combining the results obtained for $|\Delta u(t)| < u_r$ and $|\Delta u(t)| = u_r$.

(iii) It follows from (ii) that for an infinitely large t and

$$\left|\frac{r_f(t) - \phi'(t)^T \hat{\theta}'(t)}{\hat{\theta}_1(t)} - u(t-1)\right| \leq u_r,$$

$$u(t) = \frac{r_f(t) - \phi'(t)^T \theta'}{\theta_1}$$

is satisfied. Hence

$$\bar{y}(t+d) = \phi(t)^T \theta = r_f(t).$$

□

Remark 6.1. (i) The input $r^*(t)$ to the non-adaptive control system (6.1) is assumed to be any arbitrary signal. This enables the feedback signal $e(t+d) + \phi(t)^T \Delta\hat{\theta}(t)$ to be considered as a part of the system input. Thus the system stability is derived using the finite gain stability condition without knowing *a priori* the boundedness of this signal. However, the input $r_f(t)$ to the rate constrained direct adaptive control (5.2) is assumed to be bounded in order to establish the boundedness of $y(t)$ and $u(t)$.

(ii) Theorem 6.1 shows that properties of the rate constrained adaptive control system, as time approaches infinity, are characterized by those of the non-adaptive control system employing the system true parameters in the controller. In general, the established stability results do not imply desirable system performance. The closed loop system performance depends upon the rate constraint size, the reference input etc. which deserves further study.

7 Rate Constrained Adaptive Control of Stable Plants

In this section, it is shown that a stable plant satisfying Assumptions A1 and A5, when subject to non-adaptive control in the form (6.1), will satisfy the finite gain stability condition A3. Then the rate constrained direct adaptive control of the stable plant is stable.

Consider the non-adaptive rate constrained control system (2.1) and (6.1) and assume

Assumption A6. *The plant (2.1) is stable, i.e. all its poles lie strictly inside the unit circle in the z-plane.*

Then the following theorem is obtained.

Theorem 7.1. *If the plant (2.1) with $v(t) = 0$ satisfies Assumptions A1, A5 and A6, then the rate constrained direct adaptive control system (2.1) and (5.2) is stable in the sense that the system satisfies the stability conditions of Theorem 6.1.*

Proof. Introduce $\tilde{y}(t)$ and $\tilde{u}(t)$ which satisfy

$$y(t) = z^{-d} B(z^{-1}) \tilde{y}(t), \tag{7.1}$$

$$(1 - z^{-1}) u(t) = \tilde{u}(t). \tag{7.2}$$

Associated with the plant (2.1) and the non-adaptive control law (6.1), the non-adaptive closed loop system can be written as

$$\tilde{y}(t) = \frac{1}{(1 - z^{-1}) A(z^{-1})} \tilde{u}(t), \tag{7.3}$$

$$\begin{aligned}
\tilde{u}(t) &= \operatorname{sat}\left\{ \frac{r_f^*(t) - \phi'(t)^T \theta'}{\theta_1} - u(t-1) \right\} \\
&= \operatorname{sat}\left\{ \frac{r_f^*(t) - \phi(t)^T \theta}{\theta_1} + \tilde{u}(t) \right\} \\
&= \operatorname{sat}\left\{ \frac{r_f^*(t) - E(z^{-1}) y(t+d)}{\theta_1} + (1 - z^{-1}) A(z^-) \tilde{y}(t) \right\} \\
&= \operatorname{sat}\left\{ \frac{r_f^*(t) - E(z^{-1}) B(z^{-1}) \tilde{y}(t)}{\theta_1} + (1 - z^{-1}) A(z^-) \tilde{y}(t) \right\} \\
&= \operatorname{sat}\left\{ \frac{r_f^*(t)}{\theta_1} - (E(z^{-1}) B'(z^{-1}) - (1 - z^{-1}) A(z^-)) \tilde{y}(t) \right\} \quad (7.4)
\end{aligned}$$

Now, the system (7.3) and (7.4) can be considered as a type-1 plant subject to a saturation constrained control law. Since $E(z^{-1})$ and $B'(z^{-1})$ are stable polynomials, there exists an $\epsilon > 0$ such that

$$\left(E(z^{-1})B'(z^{-1}) - (1-z^{-1})A(z^{-1})\right)_{z^{-1}=1} > \epsilon.$$

It then follows from Theorem 4.1 that there exist $0 < \gamma_9 < \infty$ and $0 < \gamma_{10} < \infty$ such that

$$\tilde{y}(t) \leq \gamma_9 \sup_{0 \leq \tau \leq t} \left|\frac{r^*(t)}{\theta_1}\right| + \gamma_{10}.$$

In view of (7.1) and Assumption A3, the output of the non-adaptive system satisfies the finite gain stability condition. Thus Theorem 6.1 can be readily applied to show that the rate constrained direct adaptive control of the stable plant is stable. □

8 Conclusion

This chapter has analyzed the stability of discrete time direct adaptive control systems subject to saturation constraints on the input amplitude and changing rate. It is shown that the saturation constrained adaptive control system is BIBO stable provided its non-adaptive version satisfies a finite gain stability condition. The result obtained is quite general without referring to specific plants, and is readily applicable to a class of type-1 plants for the amplitude constrained adaptive control case and to a class of stable plants for the rate constrained adaptive control case.

References

1. C.A. Desoer and M. Vidyasagar, *Feedback Systems: Input-output Properties*, Academic Press, New York, 1975.
2. G. Feng, C. Zhang, and M. Palaniswami, Stability of input constrained adaptive pole placement control systems, *Automatica*, 30, 1065-1070, 1994.
3. G.C. Goodwin, and K. Sin, *Adaptive Filtering, Prediction and Control*, Prentice Hall, Englewood Cliffs, N.J., 1984.
4. G.C. Goodwin, R.L. Leal, D.Q. Mayne and R.H. Middleton, Rapproachment between continuous and discrete model reference adaptive control, *Automatica*, 22, 199-207, 1986.
5. Saberi, A., Z. Lin and A.R. Teel, Control of linear systems with saturation actuators, *IEEE Trans. on Automatic Control*, 41, 368-378, 1996.
6. C. Samson, Stability analysis of adaptive control systems subject to bounded disturbances, *Automatica*, 19, 81-86, 1983.
7. H.J. Sussmann, E.D. Santag and Y. Yang, A general result on the stability of linear systems using bounded controls, *IEEE Trans. on Automatic Control*, 39, 2411-2425, 1994.
8. A.R. Teel, A. R., Global stabilization and restricted tracking for multiple integrators with bounded controls, *Systems & Control Letters*, 18, 165-171, 1992.
9. C. Zhang and R.J. Evans, Amplitude constrained adaptive control, *International Journal of Control*, 46, 53 - 64, 1987.

10. C. Zhang and R.J. Evans, Adaptive pole assignment subject to saturation constraints, *International Journal of Control*, 46, 1391-1398, 1987.
11. C. Zhang and R.J. Evans, Robust amplitude constrained adaptive control, *Proc. IFAC Workshop on Robust Adaptive Control*, Newcastle, Australia, pp.195-199, 1988.
12. C. Zhang and R.J. Evans, Amplitude constrained direct self tuning control, *Prepints, IFAC Symposium on System Identification and Parameter Estimation*, Beijing, China, 373 - 377, 1988.
13. C. Zhang and R.J. Evans, Rate constrained direct adaptive control, *International Journal of Control*, 48, 2179 - 2187, 1988.
14. C. Zhang, Discrete time saturation constrained adaptive pole assignment control, *IEEE Trans on Automatic Control*, 38, 1250-1254, 1993.
15. C. Zhang and R.J. Evans, Continuous direct adaptive control with saturation constraint, *IEEE Trans on Automatic Control*, 39, 1718-1722. 1994.

Adaptive Control of Linear Systems with Unknown Time Delay

Changyun Wen[1], Yeng Chai Soh[1], and Ying Zhang[2]

[1] School of Electrical and Electronic Engineering,
  Nanyang Technological University,
  Nanyang Ave. 639798, SINGAPORE
  e-mail: ecywen@ntu.edu.sg
[2] Division of Automation Technology,
  Gintic Institute of Manufacturing Technology,
  Singapore 638075

Abstract

This chapter discusses adaptive control of continuous time systems with unknown time delay. In the controller design, the delay is taken into consideration by using rational approximation. For the implementation of controllers, no a priori knowledge, which is related to modeling errors due to the approximation of time delay, plant unmodeled dynamics and bounded external disturbances, is required. It is shown that, the resulting adaptive controller can ensure global boundedness of the overall closed loop system even in the presence of modeling errors. A small mean tracking error can also be ensured.

1 Introduction

As we know, time delay exists in most practical systems and the presence of time delay has a significant influence on system performance. Thus, in the controller design for such systems, the delay should be explicitly considered to enable better performance. In the face of uncertainties in the delay and other parameters, the controller design is rather challenging and an adaptive scheme is a possible effective method to employ. However, this problem has not been adequately addressed so far and there are only a limited number of results available, among which most consider only systems having known time delay and unknown parameters in discrete time systems [1, 2]. Some results for continuous time systems with unknown time delay have been obtained [3]. The time delay was approximated by an all-pass rational transfer function. It was argued by the authors that this type of approximation is preferable if good closed loop performance is required and a first-order, or at most a second-order, transfer function for the approximation will be sufficient. By incorporating a relative deadzone originating from the robust adaptive control theory in [4] and [5], the authors developed an adaptive algorithm based on the approximate model. It was shown that a globally stable system can be ensured in spite of the presence of modeling error due to the time delay approximation and unmodeled dynamics. In order to implement the relative deadzone, an overbounding function of the modeling error must be obtained. Thus some parameters related to the modeling error, for example gain of the overbounding function, must be chosen by the designers. To do this, an assumption of prior knowledge of the modeling error is imposed. The stability condition, on the other hand, requires the gain of the overbounding function to be sufficiently small. Clearly, the choice of such parameters makes complicated to implement the algorithm.

Recently, it was shown in [6 - 8] and [19] that a pole placement adaptive algorithm without using any a priori knowledge of modeling errors is still robust against unmodeled dynamics and external bounded disturbances. This algorithm, with a slight modification, is applied to continuous time systems with unknown time delay in this chapter. The internal model principle, including the integral action as a special case, is employed to track/reject a general class of deterministic set-points/disturbances. It is shown that global stability can be ensured even without the use of a relative deadzone. A small mean tracking error property is also established. The removal of a relative deadzone in the adaptive scheme relaxes the requirement for prior knowledge of the modeling error.

Because of the all-pass rational approximation, the resulting nominal model for controller design is nonminimum phase. As such, all robust direct model reference adaptive control algorithms such as those developed in [9 - 12] are not applicable.

The remaining of the chapter is organized as follows. Section 2 presents the unknown plant and formulates the control problem. The adaptive controller is described in Section 3. System stability and tracking property are established in Section 4. Finally, Section 5 concludes the chapter.

2 System Models

The class of systems with unknown time delay to be considered is modeled as follows:

$$A_o(p)y = B_o(p)HDu + d + \omega \tag{1}$$

where A_o and B_o are polynomials of the following forms

$$A_o(p) = p^{n_o} + a_{n_o-1,o}p^{n_o-1} + ... + a_{0,o}$$
$$B_o(p) = b_{m_o,o}p^{m_o} + b_{m_o-1}p^{m_o-1} + ... + b_{0,o},$$

with $m_o < n_o$ and $p = \frac{d}{dt}$; D denotes a pure time delay operator e^{-sT} with time delay T seconds; H is the unmodeled dynamics; d is a deterministic disturbance and w is a bounded non-deterministic disturbance. The disturbance $d(t)$ satisfies

$$S(p)d(t) = 0$$

where $S(p)$ is a known polynomial having non-repeated roots on the stability boundary, i.e.

$$S(p) = p^{n_s} + s_{n_s-1}p^{n_s-1} + \ldots + s_0$$

Remark 2.1. If $d(t)$ is a constant offset, $S(p) = p$. When $d(t)$ is a sinewave of angular frequency w_0, then $S(p) = p^2 + w_0^2$.

In order to take the time delay into consideration, an all-pass transfer function

$$\frac{B_r}{A_r} = \left(\frac{1 - \frac{sT}{2r}}{1 + \frac{sT}{2r}}\right)^r \tag{2}$$

is used to approximate it. It was shown in [3] that it is sufficient to let order r be 1, or at most 2.

Equation (1) can now be rewritten as

$$Ay = Bu + \eta \tag{3}$$

where

$$A(p) = A_o A_r$$
$$= p^n + a_{n-1}p^{n-1} + \ldots + a_0 \tag{4}$$
$$B(p) = B_o B_r$$
$$b_m p^m + b_m p^m + \ldots + b_0 \tag{5}$$

with $n = n_o r, m = m_o r$ and

$$\eta = B_o[HA_r D - B_r]u + A_r d + A_r w \tag{6}$$

denoting the effect of modeling error due to the rational approximation, unmodeled dynamics and external disturbances.

To eliminate the effects of the deterministic disturbance, a filter $\frac{S}{S'}$ is used where $S'(p)$ is chosen to be stable, 'close to' $S(p)$ and have the same order as $S(p)$. For example when $S(p) = p$, $S'(p) = p + \epsilon'$ with ϵ' being a small positive constant so that the filter $\frac{S}{S'}$ operates as a differentiator only at low frequencies. To reduce the effects of modeling errors including the bounded noise and high frequency unmodeled dynamics, an extra low pass filter $\frac{1}{E}$ is introduced, where E is a monic Hurwitz polynomial given below

$$E(p) = p^\gamma + e_{\gamma-1}p^{\gamma-1} + \ldots + e_0$$

and the order γ is not less than n. With the use of the composite filter $\frac{S}{F}$ where $F = S'E$, the following filtered variables are defined:

$$y_f(t) = \frac{S}{F}y(t) \tag{7}$$

$$u_f(t) = \frac{S}{F}u(t) = \frac{1}{E}\bar{u} \quad \text{where } \bar{u} = \frac{S}{S'}u \tag{8}$$

$$\eta_f(t) = \frac{S}{F}\eta(t) = \frac{SB_o(HA_rD - B_r)u}{F} + \frac{SA_r\omega}{F} \tag{9}$$

Regarding this filter, the comments in *Remark 3.1* of [3] still hold here. Now operating this filter on equation (3), we get

$$Ay_f(t) = Bu_f(t) + \eta_f(t) \tag{10}$$

On the plant, we have

Assumption 2.1

- A1: n, i.e. n_o, is known.
- A2: The coefficients of the polynomials $A(p)$ and $B(p)$ are inside a known compact convex region.
- A3: The unmodeled dynamics H is stable and the disturbance $\omega(t)$ is bounded.
- A4: The plant when augumented by $1/S$ has a right coprime fractional representation in the ring of causal stable operators. In other words, there exist causal stable operators ξ and Ω such that

$$\xi \frac{A_o S}{J} + \Omega \frac{B_o HD}{J} = 1 \tag{11}$$

where J is a given Hurwitz polynomial with the same degree as $A_o S$.

Remarks 2.2

1. b_m can be zero and its sign is not required. This implies that the (unknown) true numerator degree of the nominal plant model may be less than m_o used in the design.
2. Except for stability condition, there is no requirement for prior knowledge of unmodeled dynamics

To derive a proper adaptive control algorithm, equation (10) is rewritten as

$$\begin{aligned} p^n y_f &= (p^n - A)y_f + Bu_f(t) + \eta_f(t) \\ &= \phi^T(t)\theta_* + \eta_f(t) \end{aligned} \tag{12}$$

where

$$\phi^T(t) = [y_f(t), ..., p^n y_f(t), u_f(t), ..., p^m u_f(t)] \tag{13}$$
$$\theta_*^T = [-a_0, ..., -a_{n-1}, b_0, ..., b_m] \tag{14}$$

Note that equation (12) is differentiator free since the order of E, i.e. γ, is not less than n. Also from Assumption 2.1, we have $\theta_* \in \mathcal{C}$ where $\mathcal{C}$ is a compact convex subspace of $\Re^{n+m}$ and

$$\|\theta_1 - \theta_2\| \le k_\theta \tag{15}$$

where k_θ is a constant depending on the size of $\mathcal{C}$ and $\|.\|$ denotes the Euclidean norm.

Concerning the modeling error η_f, we can obtain

Lemma 2.1. For all members of the class of systems satisfying Assumption 2.1, there exists a constant $\epsilon \ge 0$ such that for all t

$$|\eta_f(t)| \le \epsilon \sup_{0 \le \tau \le t} \|x(\tau)\| + d_0 \tag{16}$$

where

$$x^T(t) = [\frac{y}{F}, ..., p^{n+n_s-1}\frac{y}{F}, \frac{\bar{u}}{F}, ..., p^{\gamma+n_s-1}\frac{\bar{u}}{F}], \tag{17}$$

d_0 is a constant bounding $\frac{SA_r}{F}\omega(t)$ and an exponentially decaying term depending on initial conditions.

Proof: Suppose $V(p)$ is a stable polynomial of the form

$$V(p) = p^{\gamma+n_s-1} + v_{\gamma+n_s-2}p^{\gamma+n_s-2} + ... + v_0$$

From (9), we have

$$\eta_f(t) = \frac{S'B_o(HA_rD - B_r)}{V}V\frac{\bar{u}}{F} + \frac{SA_r\omega}{F}$$
$$= \frac{S'B_o(HA_rD - B_r)}{V}x_u + \frac{SA_r\omega}{F}$$

where

$$x_u = v_0\frac{\bar{u}}{F} + v_1p\frac{\bar{u}}{F} + ... + v_{\gamma+n_s-2}p^{\gamma+n_s-2}\frac{\bar{u}}{F} + p^{\gamma+n_s-1}\frac{\bar{u}}{F}$$
$$= v^T x \tag{18}$$

where

$$v^T = [0, 0, ..., v_0, v_1, ..., v_{\gamma+n_s-2}, 1]$$

Following a similar argument as in the proof of Lemma 6.1 in [3], we have

$$|\eta_f(t)| \leq \beta(t) \quad \text{for all } t \tag{19}$$

where $\beta(t)$ satisfies

$$\dot{\beta}(t) = -\alpha\beta(t) + \epsilon_1|x_u(t)| + d'_0; \quad \beta(0) = \beta_0 \tag{20}$$

for some $\alpha \geq 0, \epsilon_1 \geq 0, d'_0 \geq 0$ and $\beta_0 \geq 0$. (20) gives

$$\beta(t) \leq \epsilon_2 \sup_{0 \leq \tau \leq t} |x_u(\tau)| + d_0 \tag{21}$$

with $\epsilon_2 \geq 0$. From (18), it is easily seen that

$$|x_u(\tau)| \leq k_e \|x(\tau)\| \tag{22}$$

where k_e is a constant depending on the coefficients of $V(p)$.

Substituting (22) and (21) to (19), we get (16).
◊

Remarks 2.3

1. As discussed in [3], the constant ϵ in (16) can be made sufficiently small by proper choice of the bandwith and the order of the filter $\frac{1}{E}$.
2. While being aware of their existence, we do not assume any knowledge of the constants ϵ and d_0. In [3], such knowledge was needed for the implementation of their adaptive scheme. On the other hand, the stability of their adaptive algorithm requires a sufficiently small ϵ. Clearly the choice of such an ϵ is critical and also difficult. Thus the relaxation of such a requirement is important and significant.
3. Also note that vector $x(t)$ can be obtained from measurement. This vector will be used for the adaptive controller design in the next section.

Suppose y^* is a given reference set-point for output y. The control problem is to design a controller for the class of plants satisfying Assumption 2.1 so that the closed loop system is stable in the sense that all signals in the system are bounded for arbitrary bounded y^* and initial conditions, and the tracking error is small in some sense. In addition, when the modeling error disappears the adaptive controller should retain the properties of earlier unmodified conventional adaptive controllers without any additional requirement.

3 Adaptive Control Scheme

Note that from (3) the modeled part of the plant for controller design is nonminimum phase. Thus direct model reference adaptive control schemes cannot be applied. An indirect adaptive control scheme is proposed in this section. The adaptive controller consists of two independent modules: a parameter estimator and a linear controller designed based on the *Certainty Equivalence Principle* [1]. We now present these two modules separately.

3.1 Parameter Estimator:

The following estimation algorithm with projection is introduced to the estimator:

$$\dot{\hat{\theta}}(t) = \mathcal{P}\{\frac{\phi(t)e(t)}{1+x^T(t)x(t)}\} \tag{23}$$

where $\hat{\theta}$ is the estimate of θ, $x(t)$ is defined in (17), $e(t)$ is the prediction error defined as

$$e(t) = p^n y_f(t) - \phi^T(t)\hat{\theta}(t) \tag{24}$$

$\mathcal{P}\{.\}$ denotes a projection operation proposed in [13]. Such an operation can ensure the estimated parameter vector $\hat{\theta}(t) \in \mathcal{C}$ for all t if $\hat{\theta}(0) \in \mathcal{C}$.

Now some useful properties of the estimator in (23) and (24) can be stated as in the following lemma.

Lemma 3.1. Suppose M_0 is a positive constant s.t. $d_0/M_0 \leq \delta$ where δ is a sufficiently small positive constant. The estimator (23) and (24), applied to plants given in (1), has the following properties.

1. If $\|x(t)\| \geq M_0$ and $\sup_{0\leq\tau\leq t}\|x(\tau)\| = \|x(t)\|$ for all $t \geq t_0$, then
   (a)
   $$\begin{aligned}\tilde{e}(t) &= \frac{e(t)}{(1+x^T(t)x(t))^{1/2}} \\ &\leq k_1 + \epsilon + \delta \qquad \text{for } t \geq t_0\end{aligned} \tag{25}$$
   where k_1 is a constant depending on k_θ in (15) and
   (b)
   $$\int_{t_0}^{t} \tilde{e}^2(\tau)d\tau \leq k_2 + \alpha_1(t-t_0) + \alpha_2(t-t_0) \qquad \text{for } t \geq t_0 \tag{26}$$
   where
   $$k_2 = \frac{1}{2}k_\theta^2, \tag{27}$$
   $$\alpha_1 = (k_1 + 2\epsilon)\epsilon, \tag{28}$$
   $$\alpha_2 = (k_1 + 2\delta)\delta, \tag{29}$$

2.
   $$\|\dot{\hat{\theta}}(t)\| \leq k_\phi|\tilde{e}(t)| \tag{30}$$
   where k_ϕ is a constant.

Proof:

1. (a) From (24) and (12), we get
$$e(t) = -\tilde{\theta}^T(t)\phi(t) + \eta_f(t) \tag{31}$$
where
$$\tilde{\theta} = \hat{\theta} - \theta_*$$
Then (16) and (15) give
$$|e(t)| \leq k_\theta \|\phi(t)\| + \epsilon \sup_{0 \leq \tau \leq t} \|x(\tau)\| + d_0 \tag{32}$$
Now suppose
$$S'(p) = p^{n_s} + s'_{n_s-1}p^{n_s-1} + \ldots + s'_0$$
Then noting that $S(p) = p^{n_s} + s_{n_s-1}p^{n_s-1} + \ldots + s_0$, we have
$$\phi^T(t) = [\frac{Sy}{F}, \ldots, p^{n-1}\frac{Sy}{F}, \frac{S'\bar{u}}{F}, \ldots, p^m\frac{S'\bar{u}}{F}]$$
$$= [(p^{n_s} + s_{n_s-1}p^{n_s-1} + \ldots + s_0)\frac{y}{F}, \ldots,$$
$$(p^{n+n_s-1} + s_{n+n_s-1}p^{n+n_s-2} + \ldots + s_0 p^{n-1})\frac{y}{F},$$
$$(p^{n_s} + s'_{n_s-1}p^{n_s-1} + \ldots + s'_0)\frac{\bar{u}}{F}, \ldots,$$
$$(p^{m+n_s} + s'_{n_s-1}p^{m+n_s-1} + \ldots + s_0 p^m)\frac{\bar{u}}{F}]$$
Thus, there exists a constant k_ϕ such that
$$\|\phi(t)\| \leq k_\phi \|x\| \tag{33}$$
Under the assumption of the lemma, using (32) and (33), we get,
$$|\tilde{e}(t)| \leq \frac{k_\theta k_\phi \|x(t)\| + \epsilon \|x(t)\| + d_0}{(1 + x^T(t)x(t))^{1/2}}$$
$$\leq k_1 + \epsilon + \delta$$
for $t \geq t_0$

(b) We consider the function $v(t) = \frac{1}{2}\tilde{\theta}^T(t)\tilde{\theta}(t)$

Then (23), (31), (16) and (33) yield
$$\dot{v}(t) \leq -\tilde{e}^2(t) + \frac{|\eta_f(t)||e(t)|}{1 + x^T(t)x(t)}$$
$$\leq -\tilde{e}^2 + \frac{\epsilon \sup_{0 \leq \tau \leq t} \|x(\tau)\| + d_0}{(1 + x^T(t)x(t))^{1/2}}\tilde{e}(t)$$
where equality holds save for those times when the projection operator $\mathcal{P}$ is invoked from the property of the projection operator [13]. Under the assumption of the lemma, we have
$$\int_{t_0}^{t} \tilde{e}^2(\tau)d\tau \leq -\int_{t_0}^{t} \dot{v}d\tau + (\alpha_1 + \alpha_2)\int_{t_0}^{t} d\tau$$
$$\leq k_2 + \alpha_1(t - t_0) + \alpha_2(t - t_0)$$

2.

$$\|\dot{\hat{\theta}}(t)\| \leq \frac{\|\phi(t)\|\|e(t)\|}{1+\|x(t)\|^2}$$
$$\leq k_\phi |\tilde{e}(t)|$$

Again the first inequality is an equality save for when the projection is invoked.

◊

Remarks 3.2

1. α_1 in (28) can be made small by reducing ϵ and α_2 in (29) by making a sufficiently large number M_0. M_0 is used here for the purpose of stability analysis only. It is not a design parameter.
2. The least squares version is more commonly used in practical algorithms [1]. Similar properties for this estimator can be derived by defining a different Lyapunov type function $v(t)$ as in [14], but the analysis is more tedious.
3. The properties in Lemma 3.1 are independent of the control laws to be employed.
4. When the modeling error disappears, $\eta_f = 0$ and $d_0 = 0$. In this case, $\epsilon = 0$ and $\delta = 0$. Then the properties given in Lemma 3.1 are exactly the same as those of unmodified conventional estimators [1]. As for indirect adaptive control, an adaptive controller can be decomposed into 'modules' of parameter estimation and control law synthesis. Since our modification is only on the estimator part and thus the adaptive controller will retain the results of unmodified conventional adaptive controllers without any additional requirement in the absence of unmodeled dynamics and disturbances.

3.2 Control Law Synthesis

For the module of controller synthesis, there are many possible choices of schemes such as the classic three term control law, pole assignment, linear quadratic optimal control and so on [1]. Here we use a pole assignment strategy for analysis. The internal model principle is used to achieve better tracking performance. To this end, suppose the trajectory y^* satisfies

$$Sy^* = 0 \tag{34}$$

The control $u(t)$ is then given by

$$\hat{L}S(\frac{u}{F}) = -\hat{P}(\frac{y}{F}) + \hat{G}(\frac{y^*}{F}) \tag{35}$$

where $\hat{G}$ is a stable transfer function giving feedforward action and its choice was discussed in [3], $\hat{L}$ and $\hat{P}$ are polynomials of the form

$$\hat{L}(p) = p^{n_l} + \hat{l}_{n_l-1}p^{n_l-1} + \ldots + \hat{l}_0$$
$$\hat{P}(p) = \hat{p}_{n_p}p^{n_p} + \hat{p}_{n_p-1}p^{n_p-1} + \ldots + \hat{p}_0$$

and determined from the following Diophantine equation.

$$\hat{A}\hat{L}S + \hat{B}(t)\hat{P}(t) = A^* \tag{36}$$

where A^* is a monic polynomial of degree $n+\gamma+n_s$ and its zeros are chosen to be the required closed loop poles according to guidelines in [15], $\hat{A}, \hat{B}$ are the estimates of A, B. The degrees n_l and n_p are set to γ and $n + n_s - 1$, respectively.

The resulting controller can be implemented by transfering (35) to the following form

$$\begin{aligned} u &= \frac{S'}{S}[(F - \hat{L}S')(\frac{\bar{u}}{F}) - \hat{P}(\frac{y}{F}) + \hat{G}(\frac{y^*}{F})] \\ &= \frac{S'}{S}[\hat{\psi}^T x + \hat{G}\frac{y^*}{F}] \end{aligned} \tag{37}$$

where the fact that $S'u = S\bar{u}$ has been used. Vector x is defined in (17) and vector ψ contains the controller parameters, i.e.

$$\psi^T = [-\hat{p}_0, \ldots, -\hat{p}_{n+n_s-1}, \hat{k}_0, \ldots, \hat{k}_\gamma]$$

with $\hat{k}_i, i = 0, \ldots, \gamma$ being the coefficients of the polynomial $F - \hat{L}S'$.

As in all indirect adaptive pole assignment control, a technical point is that we must ensure the stabilizability of the estimated plant models. One approach, for example in [5] and [18], is to constrain the parameter estimates to a convex region in which all models in the region are stabilizable. In the present chapter, this approach is utilized and the following assumption is imposed.

Assumption 3.1

- A5: For all $\hat{\theta} \in \mathcal{C}$, $\hat{A}S$ and $\hat{B}$ are uniformly coprime.

Remark 3.3. Assumption 3.1 is needed only when a pole assignment control law is employed. It can be removed when other approaches ensuring stabilizability of the estimated models are employed [16,17].

From Assumption A5, equation (36) gives a bounded solution for $\hat{L}, \hat{P}, \forall t$.

4 Stability Analysis

In this section, the closed loop adaptive system is analyzed. As a starting point, we derive an equation to describe the closed loop system. This can be achieved by considering the estimator and the controller equations. From (24), we get

$$e = p^n y_f - [(p^n - \hat{A})y_f + \hat{B}u_f]$$

This gives

$$\hat{Q}(\frac{y}{F}) = \hat{R}(\frac{\bar{u}}{F}) + e \tag{38}$$

where

$$\hat{Q} = \hat{A}S$$
$$= p^{n+n_s} + \hat{q}_{n+n_s-1}p^{n+n_s-1} + ... + \hat{q}_0$$
$$\hat{R} = \hat{B}S'$$
$$= \hat{r}_{m+n_s}p^{m+n_s} + \hat{r}_{m+n_s-1}p^{m+n_s-1} + ... + \hat{r}_0$$

For the controller in (35), we have

$$\hat{T}(\frac{\bar{u}}{F}) = -\hat{P}(\frac{y}{F}) + \hat{G}(\frac{y^*}{F}) \tag{39}$$

where

$$\hat{T} = \hat{L}S'$$
$$= p^{\gamma+n_s} + \hat{t}_{\gamma+n_s-1}p^{\gamma+n_s-1} + ... + \hat{t}_0$$

Then from (38) and (39), the closed loop system can be described as

$$\dot{x}(t) = \bar{A}_c x(t) + b_1 e(t) + b_2 r(t) \tag{40}$$

where

$$b_1^T = [0, ...0, 1, ..., 0]$$
$$b_2^T = [0, ...0, 0, ..., 1]$$
$$r(t) = \hat{G}\frac{y^*}{F} \tag{41}$$

$$\bar{A}_c = \begin{bmatrix} 0 & 1 & 0 & ... & & 0 & & 0 & ... & & 0 \\ & 0 & \ddots & \vdots & & \vdots & & \vdots & \vdots & & \\ & & & & & 1 & & & & & \\ -\hat{q}_0 & ... & & & -\hat{q}_{n+n_s-1} & \hat{r}_0 & ... & \hat{r}_{m+n_s}, & 0 & ... & 0 \\ 0 & ... & & & & 0 & 1 & 0 & & ... & 0 \\ \vdots & & & & & & & \ddots & & & \vdots \\ & & & & & & & & & & 1 \\ -\hat{p}_0 & ... & & & -\hat{p}_{n+n_s-1} & -\hat{t}_0 & & & & & -\hat{t}_{\gamma+n_s-1} \end{bmatrix} \tag{42}$$

Since $\hat{G}(t)$ is chosen to be bounded, then $\|r(t)\| \leq c_g\|y^*(t)\|$ where c_g is a constant.

From Lemma 3.1, we obtain

Lemma 4.1. The matrix $\bar{A}_c(t)$ defined in (42) satisfies

1. $\bar{A}_c(t)$ is bounded $\forall t$.
2. If $\|x(t)\| \geq M_0$ and $\sup_{0 \leq \tau \leq t} \|x(\tau)\| = \|x(t)\|$ for all $t \geq t_0$ then

$$\int_{t_0}^t \|\dot{\bar{A}}_c(\tau)\|^2 d\tau \leq \bar{k}(k_2 + \alpha_1(t-t_0) + \alpha_2(t-t_0))$$

for $t \geq t_0$

where $\bar{k}$ is a constant.
3. The eigenvalues of $\bar{A}_c(t)$ are the zeros of $A^*S'\,\forall t$.

Proof:

1. This follows from Assumption A2.
2. From Assumption A2 and Lemma 3.1, we obtain the result.
3. This is easy to verify from (36) and (42).

$\diamond$

From Lemma F.2.2 in [15] and Lemma 4.1 above, we can readily show that $\exists c > 0, \sigma > 0$ such that the transition matrix of the homogeneous part of (40), denoted as $\Phi(t,\tau)$, satisfies

$$\|\Phi(t,\tau)\| \leq ce^{-\sigma(t-\tau)} \qquad \text{for } t \geq \tau \geq t_0 \tag{43}$$

if $\|x(t)\| \geq M_0$, $\sup_{0 \leq \tau \leq t} \|x(\tau)\| = \|x(t)\|$ $\forall t \geq t_0$, and for all $\epsilon \leq \bar{\epsilon}^*, \delta \leq \bar{\delta}^*$ where bounds $\bar{\epsilon}^*, \bar{\delta}^*$ are sufficiently small numbers to ensure $(\alpha_1 + \alpha_2) \leq k_0$. α_1, α_2 are given in (28) and (29), k_0 is a sufficiently small number as given in Lemma F.2.2 in [15].

Now notice that for any bounded initial conditions $x(0)$, set points y^* and disturbances $w(t)$, there always exists a number M_0 such that $\|x(0)\| \leq M_0, \|r(t)\|_\infty \leq M_0$ and $\frac{d_0}{M_0} \leq \delta$ for a sufficiently small δ given in Lemma 3.1, where $r(t)$ is given by (41). In our stability analysis, we will use such an intermediate sufficiently large constant M_0. This enables us to apply Lemma 3.1 and the exponential stability property of $\bar{A}_c(t)$ in the closed loop equation (40).

Clearly, the closed loop system is stable if $\|x(t)\| \leq M_0$ for $t > 0$. Thus the only situation which can cause instability is that $\|x(t)\| \geq M_0$ for some $t \in \Re^+$. In this case, there must exist a time instant t_0 such that $\|x(t_0)\| =$

M_0. To start the stability analysis in this situation, we now examine a special case that the trajectory $\|x(t)\|$ satisfies certain conditions.

Lemma 4.2. Suppose that for all $t \geq t_0$ $\|x(t)\| \geq M_0$ and $\sup_{0 \leq \tau \leq t} \|x(\tau)\| = \|x(t)\|$. Then consider the adaptive system consisting of continuous-time systems modeled by (1) and the adaptive controller designed in (23), (24), (35) to (37). Under Assumptions 2.1 and 3.1, there exists a constant ϵ_1^* such that for all $\epsilon \leq \epsilon_1^*$ the closed loop system ensures that

$$\|x(t)\| \leq M \qquad \forall t \geq 0 \tag{44}$$

for all finite initial states, any bounded $y*$ and arbitrarily bounded external disturbances, where $M = \sqrt{c_1 M_0^2 + c_2}$ and c_1, c_2 are constants.

Proof:
The general solution of (40) is

$$x(t) = \Phi(t, t_0)x(t_0) + \int_{t_0}^{t} \Phi(t, \tau)(b_1 e(\tau) + b_2 r(\tau))d\tau \tag{45}$$

Under the assumptions of the above lemma, (43) holds. Thus using (43) and the inequality

$$|e(\tau)| \leq (1 + \|x(\tau)\|)|\tilde{e}(\tau)|$$

we have

$$\|x(t)\| \leq ce^{-\sigma(t-t_0)} M_0$$
$$+ \int_{t_0}^{t} ce^{-\sigma(t-\tau)}[|\tilde{e}(\tau)|\|x(\tau)\| + |\tilde{e}(\tau)| + M_0]d\tau \tag{46}$$

After squaring both sides of (46) and following similar steps to [5], [7] and [8] which involve the use of Schwartz inequality, Grownwall Lemma and Lemma 3.1, we get for all $t \geq 0$,

$$\|x(t)\|^2 \leq (c_1 M_0^2 + c_2) \tag{47}$$

for all $\epsilon \leq \epsilon_1^*, \delta \leq \delta^*$, where c_1, c_2, ϵ_1^* and δ^* are nonnegative constants with $c_1 > 1, \epsilon_1^* \leq \bar{\epsilon}^*$ and $\delta^* \leq \bar{\delta}^*$. $\bar{\epsilon}^*$ and $\bar{\delta}^*$ are given constants to ensure (43) is satisfied. ◊

It is the estimator properties in Lemma 3.1 that give the results of Lemmas 4.1 and 4.2. However, these properties are not sufficient to establish the global bondedness of signals in a general case. Thus we need to further explore the parameter estimator and this gives Lemma 4.3 as follows.

Lemma 4.3. Suppose M_0 is a positive constant s.t. $d_0/M_0 \leq \delta$. The estimator (23) and (24), applied to plants given in (1), has the following properties:

1. If $\|x(t)\| \geq M_0$ for all $t \geq t_0$, $\|x(\tau_1)\| \leq \sqrt{c_1 M_0^2 + c_2}\forall \tau_1 \in [0, t_1]$, also for all $t > t_1$ if $\sup_{0 \leq \tau \leq t}\|x(\tau)\| = \|x(t)\|$, then

   (a)
   $$\tilde{e}(t) \leq k_1 + \epsilon(\sqrt{c_1} + \sqrt{c_2}) + \delta \qquad \text{for } t \geq t_0$$
   and
   (b)
   $$\int_{t_0}^{t} \tilde{e}^2(\tau) d\tau \leq k_2 + \bar{\alpha}_1(t - t_0) + \alpha_2(t - t_0) \qquad \text{for } t \geq t_0$$
   where
   $$\bar{\alpha}_1 = (k_1 + 2(\sqrt{c_1} + \sqrt{c_2})\epsilon)(\sqrt{c_1} + \sqrt{c_2})\epsilon$$

2.
$$\|\dot{\hat{\theta}}(t)\| \leq |\tilde{e}(t)| \tag{48}$$

Proof:

1. From the lemma assumptions, we have
$$\sup_{0 \leq \tau \leq t_1} \|x(\tau)\| \leq \sqrt{c_1 M_0^2 + c_2}$$

   (a) It follows from the proof of Lemma 3.1.1(a) and the assumptions of Lemma 4.3.
   (b) From the proof of Lemma 3.1.1(b) and assumptions of Lemma 4.3, we have
   $$\int_{t_0}^{t} \tilde{e}^2(\tau) \leq -\int_{t_0}^{t} \dot{v}_i d\tau + \bar{\alpha}_1 \int_{t_0}^{t_1} d\tau + \alpha_1 \int_{t_1}^{t} d\tau + \alpha_2 \int_{t_0}^{t} d\tau$$
   $$\leq k_2 + \bar{\alpha}_1(t - t_0) + \alpha_2(t - t_0)$$

2. It is the same as Lemma 3.1.2.

Note that the properties in the above lemma are quite similar to those in Lemma 3.1 except that the constants c_1 and c_2 appear here. After establishing the above preliminaries, we are now in a position to state our main results.

Theorem 4.1. Consider the adaptive system consisting of plant (1), estimator (23) and (24) and controller (35) to (37). Under Assumptions 2.1 and 3.1 there exists a positive constant ϵ^* such that for all $\epsilon \leq \epsilon^*$:

1. the closed loop system is globally stable in the sense that all signals remain bounded $\forall t$ for all finite initial states, any bounded y^* and arbitrarily bounded external disturbances;

2. if $\hat{G} = \hat{P}$ and $\omega(t) = 0$, then for all $t, T > 0$,

$$\lim_{T \to \infty} \frac{1}{T} \sup_{T > 0} \int_t^{t+T} |y(\tau) - y^*(\tau)| d\tau \leq \kappa \epsilon \tag{49}$$

where κ is a positive constant. Futhermore if $\eta(t) = 0 \forall t$, then

$$\lim_{t \to \infty} |y(t) - y^*(t)| = 0 \tag{50}$$

Proof:

1. From (40), $\|x(.)\|$ is continuous and thus we can divide the time interval $[0, \infty)$ into two subsequences $\Re_i^+ = [s_i, \tau_i]$ and $\Re_i^- = (\tau_i, s_{i+1})$ with $\tau_0 = 0$ such that

$$[0, \infty) = (\cup_{i=1}^{\infty} \Re_i^+) \cup (\cup_{i=0}^{\infty} \Re_i^-) \tag{51}$$

$$\|x(t)\| \geq M_0, t \in \Re_i^+; \quad \|x(t)\| < M_0, t \in \Re_i^- \tag{52}$$

In (52), M_0 is an intermediate number satisfying $\|x(0)\| \leq M_0$, $\|r(t)\|_\infty \leq M_0$ and $\frac{d}{M_0} \leq \delta$ with $r(t)$ given in (41) and δ being a sufficiently small positive number. Clearly, such an M_0 always exists.

$\|x(t)\|$ can ensured to be bounded if we can show that it is bounded in $\Re_i^+, \forall i \geq 1$, which can be done through induction. Using Lemmas 4.1 and 4.2, and noting that $\|x(s_1)\| = M_0$, we can show that $\exists \epsilon^* > 0$ such that $\forall \epsilon \leq \epsilon^*$,

$$\sup_{0 \leq \tau \leq t, t \in \Re_1^+} \|x(\tau)\| \leq M \tag{53}$$

where M is a constant which depends on M_0 only. Then assuming (53) holds $\forall t \in \Re_k^+$, it can be shown that (53) is also true $\forall t \in \Re_{k+1}^+$ from the fact that $\|x(s_{k+1})\| = M_0$ and Lemma 4.3.

After establishing the boundedness of $x(t)$, we can have $\eta_f(t), \phi(t), e(t), \hat{\theta}$ and $\bar{u}(t)$ bounded from (16), (33), (30) and (37). From (40), $\dot{x}$ is bounded and thus $p^{\gamma+n_s} \frac{\bar{u}}{F}$ is bounded. Then $\bar{u} = F \frac{\bar{u}}{F}$ is bounded.

We now establish the boundedness of y and u. From (3) and (6), we get

$$\frac{AS}{F} y(t) = \frac{B}{F} S' \bar{u}(t) + \frac{\epsilon_1 BH_1 + \epsilon_2 AH_2}{F} S' \bar{u}(t) + \frac{AS}{F} \omega(t) \tag{54}$$

Since $\bar{u}$ is bounded, then $p^i(\frac{AS}{F} y)$, i.e. $p^{i+n+n_s}(\frac{y}{F}) + ... + a_0 s'_0 p^i(\frac{y}{F})$, is bounded for $i = 0, 1, ..., \gamma - n$. From this fact and the boundedness of $p^k(\frac{y}{F}), k = 0, 1, ..., n + n_s$, we can sucessively show that $p^i \frac{y}{F}$ is bounded for $i = n + n_s + 1, ..., \gamma + n_s$. Thus $y = F\frac{y}{F}$ is bounded.

Also from (1), we have $\frac{B(1+\epsilon_1 H_1)+\epsilon_2 AH_2}{J}u$ bounded for any Hurwitz polynomial J of degree $n + n_s$. Using (11), we obtain

$$u = (\xi\frac{AS}{J} + \Omega\frac{B(1+\epsilon_1 H_1)+\epsilon_2 AH_2}{J})u$$

for causal stable operators ξ and Ω. Therefore u is bounded.

2. Now we establish the tracking properties of the system when $w(t) = 0$. Having proven the boundedness of all states in the closed loop system and noting that d_0 at (9) is exponentially decaying, it can be shown from (26) that the prediction error $e(t)$ satisfies

$$\int_{t_0}^{t} e^2(t) \leq \beta_1 + \beta_2 \epsilon(t - t_0) \tag{55}$$

where β_1 and β_2 are constants.

Define $x_1(t)$ as

$$x_1^T(t) = [\frac{1}{F}(y - y^*), ..., p^{n+n_s-1}\frac{1}{F}(y - y^*), \frac{\bar{u}}{F}, ..., p^{\gamma+n_s-1}\frac{\bar{u}}{F}], \tag{56}$$

By noting that $Sy^* = 0$, the closed loop system equation (40) becomes

$$\dot{x}_1(t) = \bar{A}_c x_1(t) + b_1 e(t) \tag{57}$$

Then from the solution of (57) and the stability of $\bar{A}_c$, we have

$$\|x_1(t)\| \leq ce^{-\sigma t}\|x_1(0)\| + \int_0^t k_1 e^{-\sigma(t-\tau)}|e(\tau)|d\tau \tag{58}$$

Squaring both sides of (58) and using the Schwartz inequality, we get

$$\|x_1(t)\|^2 \leq k_2 e^{-\sigma t} + k_3 \int_0^t e^{-\sigma(t-\tau)}|e(\tau)|^2 d\tau \tag{59}$$

So

$$\int_t^{t+T} \|x_1(\tau)\|^2 d\tau \leq k_2 \int_t^{t+T} e^{-\sigma\tau} d\tau$$

$$+ k_3 \int_t^{t+T}\int_0^\tau e^{-\sigma(\tau-\tau_1)}|e(\tau_1)|^2 d\tau_1 d\tau \tag{60}$$

Now we consider the last term in (60). Define $f_1(t)$ and $f_2(t)$ as

$$f_1(t) = \int_t^{t+T}\int_0^\tau e^{-\sigma(\tau-\tau_1)}|e(\tau_1)|^2 d\tau_1 d\tau$$

$$f_2(t) = e^{-\sigma(t+T)}\int_0^{t+T} e^{\sigma\tau}\int_{\tau-T}^\tau |e(\tau_1)|^2 d\tau_1 d\tau$$

Clearly, $f_1(-T) = \dot{f}_1 = f_2(-T) = \dot{f}_2 = 0$ if $e(t) = 0 \forall t < 0$ and $T > 0$. Thus $f_1(t) = f_2(t) \forall t \geq -T$. This is because they satisfy the same differential equation with the same initial conditions, i.e.

$$\frac{d^2 f_1}{dt} = -\sigma \frac{df_1}{dt} + e^2(t+T) - e^2(t)$$

$$\frac{d^2 f_2}{dt} = -\sigma \frac{df_2}{dt} + e^2(t+T) - e^2(t)$$

Using (55), we get

$$f_1(t) = e^{-\sigma(t+T)} \int_0^{t+T} e^{\sigma\tau} \int_{\tau-T}^{\tau} |e(\tau_1)|^2 d\tau_1 d\tau$$

$$\leq \int_0^{t+T} e^{-\sigma(t+T-\tau)} (\beta_1 + \beta_2 \epsilon T) d\tau$$

$$\leq \beta_3 + \beta_4 \epsilon T \qquad (61)$$

Substituting (61) in (60) gives

$$\int_t^{t+T} \|x_1(\tau)\|^2 d\tau \leq k_4 + k_5 \epsilon T \qquad (62)$$

Using the Schwartz Inequality, we have

$$\int_t^{t+T} \|x_1(\tau)\| d\tau \leq (\int_t^{t+T} d\tau)^{1/2} (\int_t^{t+T} \|x_1(t)\|^2 d\tau)^{1/2}$$

$$\leq \sqrt{T}(\sqrt{k_4} + \sqrt{k_5 \epsilon}\sqrt{T}) \qquad (63)$$

Since the right side of (63) is a nondecreasing function of T,

$$\sup_{T>0} \int_t^{t+T} \|x_1(t)\| d\tau \leq \sqrt{k_4}\sqrt{T} + \sqrt{k_5 \epsilon T}$$

Therefore for $\epsilon > 0$, we have

$$\lim_{T \to \infty} \frac{1}{T} \sup_{T>0} \int_t^{t+T} \|x_1(t)\| d\tau \leq k_6 \epsilon \qquad (64)$$

Following the same arguments as when establishing the boundedness of y, u and the proof of (64), we can show that $\dot{x}_1$ and $\bar{u}$ satisfy a similar inequality to (64). By considering the equality

$$\frac{AS}{F}(y - y^*) = \frac{B(1 + \epsilon_1 H_1) + \epsilon_2 A H_2}{F} S' \bar{u}(t)$$

we can show that $p^i \frac{y-y^*}{F}$ for $i = 1, ..., \gamma + n_s$ also satisfy the same form of inequality as (64). Thus $y - y^* = F\frac{y-y^*}{F}$ satisfies (49).

When $\eta = 0$, then $\epsilon = 0$. In this case, $e(t) \in L_2^+$. It follows from the above analysis and also Barbalat's lemma that

$$\lim_{t \to \infty} x_1(t) = 0 \qquad (65)$$

By the same arguments as the proof of (49), we obtain (50) ◇

Remarks 4.1. Two of the key points which enable us to prove the stability result are the use of an intermediate number M_0 and the division of the time interval $\Re_+$. The idea is that, when $t \in \Re_i^-$ $\|x(t)\|$ is automatically bounded by $M_0 < M$, and when $t \in \Re_i^+$, it is shown that $\|x(t)\| \leq M$ for $\epsilon \leq \epsilon^*$. The division of $\Re_+$ allows $\|x(t)\|$ to have the same 'initial condition' for all the time intervals examined in $\Re_i^+$, i.e. $\|x(s_i)\| = M_0$ where s_i is the boundary between $\Re_i^-$ and $\Re_i^+$. Thus the bound M depends only on M_0. As for M_0, it is neither a design parameter nor affects system stability. Thus we do not need to know it while being aware of its existence and role as an auxiliary constant in proving our result.

5 Conclusion

In this chapter, the problem of controlling a continuous time system with unknown time delay is addressed. To achieve better performance, the time delay should be considered in the controller design. This is done by using an all-pass rational approximation. The approximation gives rise to a non-minimum phase model for the design and thus the direct model reference adaptive control strategy is not applicable. Here an indirect adaptive pole assignment scheme is proposed. It has been shown that the above adaptive control scheme can globally stabilize a system with modeling errors due to rational approximation on the time delay and unmodeled dynamics as well as bounded external disturbances. Small mean tracking error is ensured if the disturbances and set-points are purely deterministic. For implementation of the proposed control scheme, no prior knowledge is required from the modeling error.

References

1 G. C. Goodwin and K.S. Sin, *Adaptive Filtering Prediction and Control*, Prentice-Hall, New Jersey, 1984.

2 K.J. Astrom and B. Wittenmark, *Adaptive Control*, Addison-Wesley, 1989.

3 C.E. de Souza, G.C. Goodwin, D. Mayne and M. Palaniswami, An adaptive control algorithm for linear systems having unknown time delay, *Automatica*, 24, 327-341, 1988.

4 G. Kreisselmeier and B.D.O. Anderson, Robust model reference adaptive control, *IEEE Trans. Automatic Control*, 31, 127 - 133, 1986.

5 R.H. Middleton, G.C. Goodwin, D.J. Hill and D.Q. Mayne, Design issues in adaptive control, *IEEE Trans. Automatic Control*, 33, 50-58, 1988.
6 C. Wen, An indirect robust continuous-time adaptive controller with minimal modifications, *Automatica*, 31, 293-296, 1995.
7 C. Wen and D.J. Hill, Global boundedness of discrete-time adaptive control just using estimator projection, *Automatica*, 28, 1143-1157, 1992.
8 C. Wen and D.J. Hill, Robustness of adaptive control without deadzones, data normalization or persistence of excitation, *Automatica*, 25, No. 6, 943-947, 1989.
9 L. Praly, Robustness of model reference adaptive control, in *Proc. III Yale Workshop on Adaptive Systems*, 224-226, 1983.
10 P.A. Ioannou and K.S. Tsakalis, A robust direct adaptive controller, *IEEE Trans. Automatic Control*, 31, 1033 - 1043, 1986.
11 B.E. Ydstie, Stability of discrete MRAC - revisited, *Syst. Control Lett.*, 13, 429-438, 1989.
12 S.M. Naik, P.R. Kumar and B.E. Ydstie, Robust continuous-time adaptive control by parameter projection, *IEEE Trans. Automatic Control*, 37, 182-197, 1992.
13 J.B. Pomet and L. Praly, Adaptive nonlinear regulation: estimation from the Lyapunov Equation, *IEEE Trans. Automatic Control*, 37, 729-740, 1992.
14 Goodwin, G.C. and D.Q. Mayne, A parameter estimation perspective of continuous model reference adaptive control, *Automatica*, 23, 57-70, 1987.
15 Middleton, R.H. and G.C. Goodwin, *Digital Control and Estimation - A Unified Approach*, Prentice Hall, New Jersey, 1990.
16 de Larminat, P. and H.F. Raynaud, A robust solution to the admissibility problem in indirect adaptive control without persistency of excitation, *Int. J. of Adaptive Control and Signal Processing*, 2, No.2, 95-110, 1988.
17 Giri, F., M. M'Saad, L. Dugard and J.-M. Dion, Robust adaptive regulation with minimal prior knowledge, *IEEE Trans. Automatic Control*, 37, 305-315, 1992.
18 Kreisselmeier, G. An approach to stable indirect adaptive control, *Automatica*, 21, No.4, 425-433, 1985.
19 Wen, C. Robust adaptive tracking using the internal model principle, *Int. J. of Control*, 64, 127-140, 1996.

Index

Abrupt failures 86, 88
Ackerman's Rule 357
Activation functions 49, 54, 58
Actuator failures 86, 114, 124, 128
Actuator dynamics 88
Adaptive constrained control 316
– backstepping design 318
– model reference adaptive control 318
– one-step-ahead control with input magnitude constraint 317
– one-step-ahead control with input magnitude and rate constraints 324
– pole placement control with input magnitude constraint 317, 318, 321
– pole placement control with input rate constraints 323
– self-tuning regulators 317, 323
Adaptive control 31, 289, 354, 383, 388, 400
Adaptive control law 280
Adaptive controller 45, 222
Adaptive critic 71
Adaptive dead zone inverse 31, 32, 34, 40, 42, 45, 46
Adaptive friction compensation 211, 213, 214, 222
Adaptive inverse model 333
Adaptive laws 121, 127, 146, 150
Adaptive observer 86, 89
Admissible plant models 356

Aircraft 111, 114, 154
Amplitude constrained adaptive control 366-368, 370, 372-373, 380
Amplitude constrained direct adaptive control 362, 365, 370
Approximation error 92
Approximation of jump functions 55
ARMA 160, 162, 172
Asymptotic stability 352
Asymptotically stable 355
Augmented neural net 49, 59
Autoregressive moving average (see also ARMA) 160, 162

Backlash 1, 2, 191, 199
– angle 200
– control 205
– estimation of state of 207
Backlash gap 3
Backlash gap estimation 21
Backlash with rubber 11
Backlash-like hysteresis 273–275
Bayes' theorem 181, 182
Bi-modal system 162
Bochner Integration 256
Bounded-input bounded-output (BIBO) stability 362, 365, 380
Boundedness 294, 303, 342
Bounds 358
Brunovsky form 66

$C_0^\infty(0, T)$ 263, 270

Index

Caratheodory integral 249
Closed-loop error dynamics 67
Closed-loop poles 343
Collocated 96
Column stacking operator 163
Combustion control 309
Combustion systems 290, 305
Constrained control 352, 354
Constrained inputs 358
Constraints, state-variable 195
Control tracking 368-369
Convex region 337, 341
Corresponding adaptive pole placement control 341
Coulomb friction 216

Dahl model 220
Data association 164
Data matrix 161–162
Dead zone 31, 32, 33, 34, 40, 42, 45, 46, 92, 362, 364
Dead zone inverse 31, 32, 33, 34, 40, 42, 45, 46, 62
Dead zone model 8
Dead zone model, revised 11
Deadzone compensation 49, 52, 61, 64
Deadzone nonlinearity 49, 61, 73
Deadzone precompensator 64, 66
Describing function 16
Describing function analysis 23
Detection 164
Detection observer 86
Diagnostic observer 86
Diophantine equation 335, 340
Discriminant function 185
Distributed parameter system 86
Disturbance 362-363, 365, 368, 373
Domain of attraction 291
Drag friction 216
Dual-observer 214, 225, 226, 228

Elastic shaft 2, 28

E-mod 68, 70
Ergodic chain 178
Expectation maximization 179, 184
Exponential model 217

Failures 85
Fault accommodation 95, 103
Fault detection 186
Fault diagnosis 86
Fault diagnostic architecture 86
Filtered signals 354
Finite gain stability 362, 367, 372-374, 376, 378-380
Finite-time integrals 300
Forbidden region (in state space) 192, 196
Fourier Integrals 25
Friction 191, 196
Fuel-injector 306
Function approximators 278
Functional link neural network 55

Galerkin approximation 261
Garding inequality 254
Gelfand triple 253
Gradient algorithm 337
Gronwall inequality 263
Gronwall's Lemma 377
Guard mode 188

Hard limits, representation of 194
Hard nonlinearity 191
– modeling 193
Harmonic oscillator 351
Hidden Markov model 177–183
Hidden mode 177
Higher order systems 359
Higher relative degree 302
Hurwitz polynomial 354
Hybrid system 183
Hysteresis operators, dynamic 249–257
Hysteresis operators, static 249–255

Index   405

Identifiability canonical form   174
Identification   21
Implicit function   95
Incipient failures   86, 88
Input rate saturation   342
Input rate saturation constraints   333
Instability   31, 43
Internal model principle   384
Irreducible chain   178

Jump linear system   187

Kalman filter   193
Kalman-Yakubovich lemma   294
Kinetic energy   99
Kronecker product   164, 171

$L^\infty(0,T;\mathbb{R}^\kappa)$ 263
$L^1(\tau) \otimes C(S)$ 265, 270
$L^1(T;C(S))$ 256–265
$L^2(\tau)$ 267
$L^2(0,T;V)$ 270
$L^\infty_w(\tau, Z^*)$ 257–267
La-Salle theorem   352
Least squares   161, 364, 375
Limit stops   191, 197
Linear system   32, 383
Linear-in-the-parameters (LIP)   217
Linearly parametrized approximators   100
Low-order adaptive controller   297, 302
LS   161, 181
Lur'e   90, 102
LuGre model   213, 214, 215, 220
Lyapunov   31
Lyapunov function   68
Lyapunov-based control   276
Lyapunov-Krasovskii functional   301, 308

Magnitude limit   307
Marginal probability   180

Markov chain   178
Markov parameter   175
Mat   166
Matched filter   177
Matching conditions   116, 118, 125, 133, 143
Matching parameters   117, 119, 121, 141, 144
Measure-valued control   251
Mechanical systems   96
Minimum phase   365, 369, 374
Mode parameters   164, 167, 169, 172, 184
Model predictive control   158
Modeling errors   86, 91, 383, 384, 385
Modeling uncertainties   343
Motion systems   49, 51

Neural networks   49, 53
Neural net approximation   54, 55
Neural net basis functions   49, 58
Neurons   49, 53
NN controller   66
Non-adaptive constrained control   313
– anti-windup control   316
– model predictive control   315
– optimal control   314
– saturated linear control   313
– smooth nonlinear control - compositions/combinations of saturated linear functions   315
Nonlinear filtering   186
Nonlinear robot function   52
Nonlinear system   32, 45, 66
Nonlinearity, hard (see hard nonlinearity)
Nonminimum phase   384, 400
Non-smooth nonlinearity   333
Null controllability   313
Nullspace   165, 166

Observability canonical form   174

Observer 198
– reduced-order 208
On-line approximator 90, 97
Open-loop stable 291, 293
Open-loop unstable 299
Output feedback 115, 142, 151, 154
Output tracking 115, 145

Parameter estimation 337, 364-365, 373-374
Parameter space 90
Parameter-dependent Lyapunov function 87, 99
Parameterization problem 192
Periodic chain 178
Performance 343
Persistence of excitation 94
Persistency of excitation 165, 172, 178
Persistent excitation 31, 43, 355
Phase plane 4
Phase Plane Backlash Model 6
Phase transitions 250
Piecewise continuous functions 49, 55, 59
Piecewise linear 184
Plant-model matching 115, 116, 118, 124, 133, 144
Pole assignment 384, 391, 392, 400
Pole placement adaptive control 342
Pole placement control 334
Potential energy 99
Pressure suppression 290
Probability measures 257
Projection 90, 101, 389, 390
Projection operator 337
PZT bimorph 253

Quadratic data matrix 164, 169, 180, 183

Rate constrained adaptive control 373-376, 378-380

Rate constrained direct adaptive control 373-374, 378-380
Rational approximation 384, 385, 400
RBF neural networks 225
$rca(S)$ 257
Realization 176, 178
Reconfiguration 86
Recurrent state 178
Reference model 114, 115
Regressor 98, 338
Regressor vector 357
Relative dead zone 337
Relative degree 296
Relaxed controls 249–250
Residual 164
Riccati equation, state-dependent 192
Robot arm 72

Saturation 361-362, 365-367, 369, 375, 379-380
Saturation constraints 289, 307, 333
Saturating functions 353
Saturated control 351, 353
SDARE method 192
Second-order systems 96
Separatrix 185
Servo mechanism 222
SETARMA 159
Shaft torque 2
σ modification 101
Sigmoid jump approximation basis functions 57, 58
Signature 171
Simulations 26, 60, 71
Singular vector 182
Smith controller 299
Stability 31, 341
Stabilizing control 128
Stable 293
Stable minimum phase 334
Stable plants 351

Stably invertible 342
State feedback 116, 117, 119, 124, 126, 134, 145, 149, 154
State tracking 114, 119
State-dependent algebraic Riccati equation (SDARE) 192
State-variable constraints 195
Static friction 215
Stationary distribution 178
Stiction 216
Strictly positive real 87, 90, 98, 296
Switching interval 177
Symmetric parameter identification 167

The adaptive control system 340
The convergence properties 338
The Dead Zone Model 8
Threshold autoregressive model 184
Throughput error 65
Time profile 88
Time delay 289, 307, 383, 384, 385, 399
TLS 160

Total least squares (*see also* TLS) 160
Transfer function 42, 43
Transition mode 173
Transition period 173
Transition probability matrix 178, 183
Transport delay 306
Tuning law 68
Tustin model 217
Two-mass system 9
Type-1 plant 362, 369-370, 379-380

Unbiased estimates 172
Unmodeled dynamics 362, 365, 369, 383, 384

Vec 163
Viscous friction 216

$W^{1,2}(0, T; \mathbb{R}^n)$ 263
Weak* convergence 263–264, 267

Young measures 249–264